Lecture Notes in Civil Engineering

Volume 385

Lecture Notes in Civil Engineering (LNCE) publishes the latest developments in Civil Engineering—quickly, informally and in top quality. Though original research reported in proceedings and post-proceedings represents the core of LNCE, edited volumes of exceptionally high quality and interest may also be considered for publication. Volumes published in LNCE embrace all aspects and subfields of, as well as new challenges in, Civil Engineering. Topics in the series include:

- Construction and Structural Mechanics
- Building Materials
- Concrete, Steel and Timber Structures
- Geotechnical Engineering
- Earthquake Engineering
- Coastal Engineering
- Ocean and Offshore Engineering; Ships and Floating Structures
- Hydraulics, Hydrology and Water Resources Engineering
- Environmental Engineering and Sustainability
- Structural Health and Monitoring
- Surveying and Geographical Information Systems
- Indoor Environments
- Transportation and Traffic
- Risk Analysis
- Safety and Security

To submit a proposal or request further information, please contact the appropriate Springer Editor:

- Pierpaolo Riva at pierpaolo.riva@springer.com (Europe and Americas);
- Swati Meherishi at swati.meherishi@springer.com (Asia—except China, Australia, and New Zealand);
- Wayne Hu at wayne.hu@springer.com (China).

All books in the series now indexed by Scopus and EI Compendex database!

Nuridah Sabtu
Editor

Proceedings of AWAM International Conference on Civil Engineering 2022—Volume 2

AICCE, Sustainability and Resiliency

Springer

Editor
Nuridah Sabtu
School of Civil Engineering
Universiti Sains Malaysia
Nibong Tebal, Pulau Pinang, Malaysia

ISSN 2366-2557 ISSN 2366-2565 (electronic)
Lecture Notes in Civil Engineering
ISBN 978-981-99-6017-0 ISBN 978-981-99-6018-7 (eBook)
https://doi.org/10.1007/978-981-99-6018-7

This Springer imprint is published by the registered company Springer Nature Singapore Pte Ltd.
The registered company address is: 152 Beach Road, #21-01/04 Gateway East, Singapore 189721, Singapore

Paper in this product is recyclable.

Editorial

Chief Editor

Dr. Nuridah Sabtu

Editors

Dr. Herni Halim
Dr. Noorhazlinda Abd Rahman
Dr. Nik Azimatolakma Awang
Dr. Mohamad Fared Murshed
Dr. Nurul Hana Mokhtar Kamal

Committee Members

Patron

Prof. Dr. Mohd Suffian Yusoff

Advisory

Ts. Dr. Mastura Azmi
Assoc. Prof. Ir. Ts. Dr. Leong Lee Vien

Chairman

Dr. Mohamad Fared Murshed

Co-chairman/Technical

Dr. Zul Fahmi Mohamed Jaafar
Sr. Dr. Abdul Hakim Salleh
Mr. Mohd Fouzi Ali
Mr. Shamsul Ishak
Mr. Rasidi Razak
Mr. Mohammed Nizam Mohd. Kamal
Mr. Abdul Muluk Tajudin
Mr. Mohd Mazlan Kamis @ Mahmad
Mr. Dziauddin Zainol Abidin

Secretary

Dr. Nurul Hana Mokhtar Kamal
Ms. Ilya Ainaalis Mohd Noor

Finance

Dr. Rosnani Alkarimiah
Ms. Umi Syahida Hassan

Publication

Dr. Nuridah Sabtu
Dr. Noorhazlinda Abd Rahma
Dr. Herni Halim
Dr. Nik Azimatolakma Awang

Event

Wellbeing

International Advisory Board

Local Advisory Board

Collaborators

REDAC
RIVER ENGINEERING AND URBAN DRAINAGE RESEARCH CENTRE

United Nations Educational, Scientific and Cultural Organization

UNESCO Chair on Ecohydraulics for Sustainable Water Infrastructures for SDG 6 in the Asia and the Pacific Region
Universiti Sains Malaysia
Penang, Malaysia

Kyutech
Kyushu Institute of Technology

Keynote Speakers

Dr. Tetsushi Kanda

Title: "Sustainable Supply Chain in Construction Industry and Approach for Carbon Neutralization in Construction Materials"

Biography

Dr. Tetsushi Kanda is a Deputy Director at Kajima Technical Research Institute. Dr. Kanda received his Ph.D. in Civil Engineering from the University of Michigan, USA, 1998. KaTRI is well known to initiate industrializing emerging advanced structural materials in construction such as ultra-high strength concrete and high-performance fiber reinforced cementitious composites in its history, with which Dr. Kanda took a leading role in large scale application. Dr. Kanda's research interests are extended to shrinkage cracking controlling technologies and carbon neutral technologies for structures. In these research areas, Dr. Kanda maintains an approach in which advanced fundamental technologies are transferred to construction projects and optimized there.

Prof. Dr. Nor Azazi Zakaria

Title: "Integrated Water Resources for Sustainability Management of Stormwater Towards SDG 6"

Biography

Professor Dr. Nor Azazi Zakaria has served in Universiti Sains Malaysia since 1994 and received his MSc. in 1990 and Ph.D. in December 1994. He then established the River Engineering and Urban Drainage Research Centre (REDAC) in 2001 and has since remained as the Director. Under the leadership of Prof. Dr. Nor Azazi, REDAC acts as a Self-Sustaining Centre as expected by USM in this corporatization era. REDAC is the first research centre at the USM Engineering Campus and was awarded the 1st Higher Institution Centre of Excellence (HICoE) in Service Thrust in USM as well as in Malaysia in October 2014 and received a special grant from the Ministry of Education (MOE) with a niche area on Stormwater Management.

Professor Dr. Nor Azazi is currently the Vice Chairman of the Executive Committee for the Malaysian Stormwater Organisation (MSO), the Malaysian National on Irrigation and Drainage (MANCID), and also UNESCO chair holder on Ecohydraulics for Sustainable Water Infrastructures for SDG 6 in the Asia and the Pacific Region. Furthermore, he has been actively organizing short courses, seminars, briefings, exhibitions, acting as guest and invited speaker and actively delivering keynotes at International Conferences in the area related to Sustainable Urban Drainage Management and Green Technology.

Prof. Dr. Hemanta Hazarika

Title: "Sustainable Early Warning System for Mitigating Rainfall Induced Landslide Disasters"

Biography
Hemanta Hazarika is currently a professor in the Graduate School of Engineering and Department of Interdisciplinary Science and Innovation, Kyushu University, Fukuoka, Japan. He obtained his Bachelor of Technology (B.Tech.) degree in Civil Engineering from the Indian Institute of Technology (IIT), Madras, India in 1990. He obtained his master's degree in Geotechnical Engineering in 1993 from Nagoya University, Japan and Ph.D. from the same university in 1996. Currently he is also adjunct professor of IIT Madras, India.

Professor Hazarika's research activities include disaster prevention and mitigation, soil-structure interaction, stability of soil-structures during earthquakes and tsunami, ground improvement, application of recycled waste and lightweight materials in constructions, stability of cut slopes, and landslides and their protection.

He has more than 350 technical papers in various international journals, international conferences, workshops and symposia to date. He has also authored two books on Soil Mechanics: "Soil Mechanics Fundamentals and Applications" (Published by CRC Press, USA) and its Japanese version (Published by Kyoritsu Publisher, Tokyo). In addition, Prof. Hazarika also served as the editor of four books: (1) "Advances in Sustainable Construction and Resource Management", (2) "Geotechnical Hazards from Large Earthquakes and Heavy Rainfall", (3) "Earthquake Hazards and Mitigations" and (4) "Scrap tire derived geomaterials—opportunities and challenges".

Professor Hazarika has several years of teaching, research as well as practical and consulting experiences both in Japan and outside Japan. He is currently a foreign expert of the world's first research centre on liquefaction research called "The National Research Centre for Liquefaction", which was established in Palu, Indonesia aftermath of the devastating damage by liquefaction during the 2018 Palu

Earthquake. He also successfully guided a world bank project on "Improving the Resilience of Nepal's Strategic Roads Network" as the leader of the project.

He also successfully organized the first International Symposium on Construction Resources for Environmentally Sustainable Technologies (CREST 2020) in March this year. Currently, Prof. Hazarika is the chairman of Asian Technical Committee on "Geotechnical Mitigation and Adaptation to Climate Change-induced Geo-disasters in Asia-Pacific Regions" of International Society of Soil Mechanics and Geotechnical Engineering (ISSMGE). He is also the founding president of a general incorporated corporation called GLOSS (Global Society for Smart Geo-Sustainnovation) which was established in August 2021.

About the Conference

AWAM International Conference on Civil Engineering (AICCE) is the flagship conference of the School of Civil Engineering, Universiti Sains Malaysia (SoCE). SoCE was established in 1989, and since then, SoCE has organized numerous seminars, conferences and short courses. These activities have encouraged and contributed towards the creation of knowledge, innovativeness and exchange of ideas. To date, SoCE has successfully organized five national-level conferences (AWAM'99, AWAM'01, AWAM'04, AWAM'07 and AWAM'09) and three AICCE in collaboration with other collaborating partners which were Geohazard Information Zonation (GIZ) for the AICCE-GIZ'12, Construction Industry Development Board (CIDB) for eco-AICCE'15, Toyohashi University of Technology (TUT) for IGNITE-AICCE'17 and another collaboration with CIDB for AICCE'19.

In 2022, SoCE hosted the AICCE 2022 in Penang, Malaysia, during 15–17 February 2022. Following the success of the previous AICCE in 2019, this AICCE'22 provided another platform for academics and industries internationally for future collaboration opportunities. The scope of the conference was SUSTAINABILITY AND RESILIENCY: RE-ENGINEERING THE FUTURE.

Sustainability and Resiliency: Re-engineering the Future

Sustainability and resiliency are key concepts to new approaches in urban development. In recent years, the integrated use of sustainability and resiliency concepts in engineering are becoming more pronounced as a response to the increasing environmental threats due to climate changes and natural disasters. The application of sustainability and integration of resiliency are important to support and develop natural and built environments in a robust and enduring way.

Therefore, the AWAM International Conference in Civil Engineering (AICCE'22) served as the platform to bring together academics, scientists, engineers, research scholars, government officials, practitioners and learned individuals from multi-disciplines to present and discuss the most recent research, innovations, trends and concerns, practical challenges encountered as well as engineering solutions adopted in built environment. It aspires to advocate the integrated use of these key concepts in natural and built environment and endeavours towards re-engineering a better future.

Contents

Highway Engineering

Experimental Study of HMA Absorption According to Aggregate Types and Asphalt Grade

Eman A. Al-Ghalibi, Israa F. Al-Saadi, Safaa A. Mohamad, and Abbas F. Jasim

Abstract Different aggregate types tend to absorb asphalt differently, and these absorption and permeability properties can reduce costs and environmental pollution. At present, virgin aggregates (VA) with high-quality properties are not often used. Even though the quality of recycled asphalt pavement (RAP) and recycled concrete aggregates (RCA) is low in some cases, most studies tend to use them. There are many tests used to calculate asphalt absorption. Most studies have been concerned with only the total absorption causes and have not addressed their selective natures. Therefore, studying the asphalt absorption phenomenon regarding the aggregates and asphalt properties is important. Three types of aggregates with different absorption properties were used to achieve this research goal: VA, RAP, and RCA. In addition, three asphalt grades 40–50, 60–70, and 85–100 were used to determine the relationship between asphalt and water absorption. The results showed that aggregates with a high degree of water absorption could absorb at a higher rate, and this relationship affects aggregate behavior and the layers' future performance. The asphalt absorption rate increased with an increase in the viscosity of the asphalt at the mixing temperature, noting some increases were due to selective absorption. Furthermore, this study observed that the absorption ratio of both natural aggregates and recycled aggregates is related to the aggregate surface area. Based on these results, recycled aggregates are recommended to reduce the depletion of natural aggregates. The absorption ratio time also reduces waste, thus reducing economic and environmental costs.

Keywords Recycle aggregates · Bulk impregnated specific gravity · Asphalt absorption

---

E. A. Al-Ghalibi · S. A. Mohamad · A. F. Jasim (✉)
Highway and Transportation Engineering Department, Mustansiriyah University, Baghdad, Iraq
e-mail: abbas.jasim@uomustansiriyah.edu.iq

I. F. Al-Saadi
Department of Construction and Projects, University of Baghdad, Baghdad, Iraq

N. Sabtu (ed.), *Proceedings of AWAM International Conference on Civil Engineering 2022—Volume 2*, Lecture Notes in Civil Engineering 385,
https://doi.org/10.1007/978-981-99-6018-7_1

1 Introduction

When asphalt is absorbed into the absorbed aggregate pores, preferential composition changes occur due to the absorbed asphalt inside the pores having a different composition from the non-absorbent asphalt (effective asphalt) that connects the aggregate particles, this part of the content of the asphalt binder that appears as a coating on outside the total particles. This is asphalt content, which controls the performance of the mixture. Thus, there is selectivity in the absorption, with the ratio of absorbent and non-absorbent asphalt creating a different composition than the original asphalt. In other words, absorption differences may occur because the active asphalt film has different properties due to the asphalt added during the mixing.

Furthermore, aggregates that absorb asphalt during the construction and early life of an asphalt pavement have been observed. This can lead to premature raveling or cracking or both if the hot mix asphalt (HMA) combinations that appear rich during production become leaner [1]. Asphalt absorption must be investigated in relation to the mineral aggregate and asphalt cement binder qualities. The amount and extent of asphalt absorption depend on such asphalt properties as viscosity composition, as measured by separation methods (yielding asphaltenes, resins, saturates, etc.), molecular size distribution (GPC), and wetting properties, such as contact angle, surface tension, and time [6]. Several of these parameters are interdependent and are dependent on other variables, such as the temperature. Aggregate properties thought to be important in the absorption process are porosity or void fraction, pore size distribution (internal and/or external), surface areas, surface roughness, presence of fines, aggregate size distribution (gradation), and surface chemical composition of the aggregate. Other factors, such as the presence or absence of water, also influence the degree of absorption. The aggregate can be classified as coarse, which is defined by the Iraqi Standard Specification for Roads and Bridges [8] as the portion of the combined aggregate retained on a 4.75 mm no. 4 sieve, and fine, which is defined as passing through a 4.75 mm no. 4 sieve [3].

Two methods may be used to find absorbed asphalt by aggregates; the first method applies effective specific gravity (ESG; resulting from the Rice specific gravity method), and the second uses bulk impregnated specific gravity (BISG). ESG and BISG are similar in concept but vary in some properties [3].

According to ASTM C-127 [1], mass, apparent specific gravity, and water absorption are all measured for coarse aggregate [1]. This can be done by using Rice's specific gravity method to calculate ESG of aggregate, which is used to calculate asphalt absorption in the loose HMA mixture [4]. Aggregates can be considered individually or in combination when discussing bulk and apparent specific gravity. ESG relates exclusively to the total combined aggregate structure in a mix of HMA [7]. ESG and asphalt absorption by the total aggregate weight can be calculated from Eqs. 1 and 2, respectively.

$$G_{\text{eff.}} = \frac{100 - P_b}{\frac{100}{G_{mm}} - \frac{P_b}{G_b}} \tag{1}$$

$$Abs\% = \left\{ \frac{1}{Gb_{agg.}} - \frac{1}{G_{eff.}} \right\} * G_b * 100 \quad (2)$$

where

$G_{eff.}$ Effective specific gravity
Abs% Absorption of bitumen
G_{mm} Maximum theoretical specific gravity
P_b Bitumen percentage
G_b Specific gravity of bitumen
$Gb_{agg.}$ Bulk specific gravity of aggregate

In order to design and regulate bituminous mixtures, the U.S. Army Corps of Engineers developed and used the BISG method. With different aggregates, the asphalt-to-water absorption ratio varies greatly but follows an established pattern. Based on the BISG of the mixture, aggregate bulk specific gravity, and asphalt specific gravity, the asphalt absorption of the HMA mixture may be determined [5].

2 Research Objectives and Methodology

The main objective of this work is to determine the effect of asphalt and aggregates' properties on absorption ratio. Furthermore, the study aims to determine the methods and rates of absorption that can effectively reduce the negative effects of absorption on asphalt mixtures' performance. Figure 1 shows the methodology followed to fulfill this study's research objectives.

3 Materials Used

3.1 Virgin Asphalt Cement

This study used three asphalt cement grades. The grades 40–50 and 85–100 came from the Daura refinery, while the 60–70 grade came from the Nasiriyah refinery. Table 1 lists the physical properties of the asphalt cement.

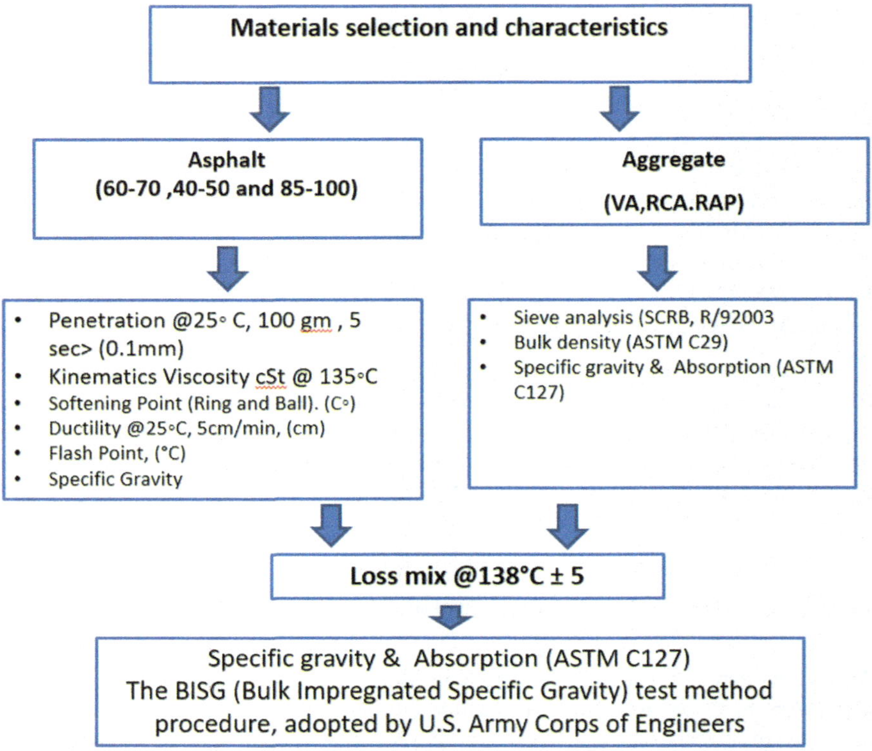

Fig. 1 Research methodology

Table 1 Physical properties of asphalt cement

Property	Virgin asphalt cement			RAP	RCA
	(40–50)	(60–70)	(85–100)		
Binder content after extraction (%)	–	–	–	3.8	–
Bulk specific gravity (gm/cm^3)	–	–	–	2.25	2.28
Apparent specific gravity (gm/cm^3)	–	–	–	2.37	2.48
Percent water absorption (%)	–	–	–	5.06	8.17
Penetration @25°C, 100 gm, 5 s > (0.1 mm)	46	63	91	–	–
Kinematics viscosity cSt @ 135 °C	410	370	280	–	–
Softening point (ring and ball) (°C)	51	46	41	–	–
Ductility @25 °C, 5 cm/min, (cm)	130	130	128	–	–
Flash point, (°C)	272	251	243	–	–
Specific gravity	1.04	1.03	1.03	–	–

3.2 Recycled Asphalt Pavement (RAP)

The RAP material was collected from the Al-Adel neighborhood in the west of Baghdad city. The optimum asphalt content was 4.6% using a penetration grade of 40–50. The recycled aggregates obtained from old pavement or crushed concrete are different from natural aggregates in that they are composed of adhered mortar and original aggregates.

3.3 Recycled Concrete Aggregate (RCA)

RCA is obtained from large parts of broken concrete block materials. The RCA in this study was collected from the waste materials of the College of Engineering Al-Mustansiriyah University laboratory. The size of the aggregate is mentioned in the next paragraph.

3.4 Aggregate Gradation

The VA is from Taji, a town north of Baghdad. According to SCRB/R9 [8], this aggregate combination was used to form a 12.5 mm surface layer mixture as indicated in Fig. 2. Nominal maximum sieve size for coarse materials in this research ranged between 3/4 inch (12.5 mm) and no. 4 (the finest) (4.75 mm). As a result, the stipulated requirements for surface layer type IIIA (SCRB R/9) were met when aggregate gradation was selected [8].

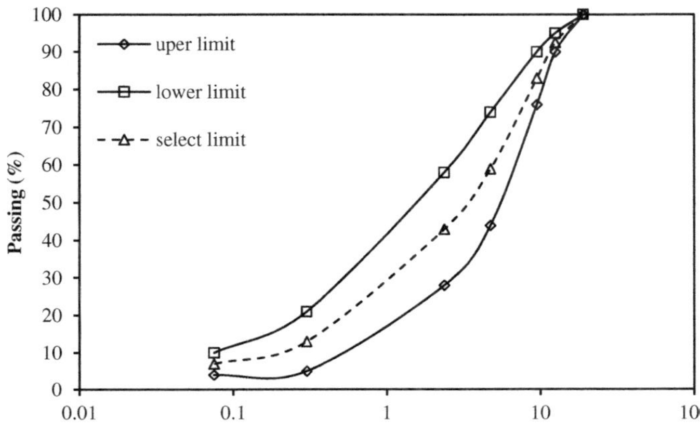

Fig. 2 Sieve analysis of used aggregate

4 Experimental Work

The experimental program conducted in this work was designed through laboratory tests to study the effects of aggregate types and asphalt grade on asphalt absorption in HMA. About 63 samples were prepared in the laboratory to achieve the current work's objectives. This section, which is divided into two parts, presents the selected materials' main properties. The first part contains virgin materials, including asphalt cement and coarse aggregate, while the second part contains recycled materials, including RAP and RCA. The standard test method of ASTM C-127 [1] was used to determine water absorption, followed by the bulk and apparent specific gravity for each coarse aggregate sample. A method for determining bulk specific gravity and water absorption was developed during absorption-time studies based on the following equations:

$$\%\,\text{Abs} = \frac{B - A}{A} * 100 \tag{3}$$

$$Gsb = \frac{A}{B - C} * 100 \tag{4}$$

$$Gsa = \frac{A}{A - C} * 100 \tag{5}$$

where

A Oven dry weight
B SSD weight
C Weight in water

To compute the realistic maximum asphalt absorption using individual coarse aggregate fractions [3], the BISG test method procedure, adopted by the U.S. Army Corps of Engineers [2] and the Oklahoma DOT [7] and authorized by the Austroads Guide to Pavement Technology [2], was used. The BISG (Gimp) and asphalt absorption (Abs.) can be computed from Eqs. 6 and 7.

$$G_{\text{imp}} = \frac{A}{(D - E) - (B - C)} \tag{6}$$

$$\text{Abs\%} = \left\{ \frac{1}{Gb_{\text{agg.}}} - \frac{1}{G_{\text{imp}}} \right\} * G_b * 100 \tag{7}$$

where

D Weight of aggregate and asphalt in air at room temperature
E Weight of aggregate and asphalt in water at 25 °C
G_{imp} Bulk impregnated specific gravity

G_b Bitumen specific gravity
$Gb_{agg.}$ Bulk specific gravity of aggregate

5 Results and Discussion

5.1 *Evaluation of Water Absorption for Mixes with Different Types of Aggregates*

The evaluation of water absorption by hydrostatic weighing averaged three tests of each aggregate sample. After 24 h of aggregate immersion, the aggregate's water absorption increased, and the variation was different between the aggregate types. Table 1 represents the results of specific gravity and absorption aggregate, laboratory separated coarse fractions, and 1/2 in.

A series of time-absorption studies was performed on laboratory aggregate 3/8 in. to no. 4 sieve fractions. In these tests, oven-dried samples (100–120 g) showed different attachments in air and water (at 25 °C) after immersion times of 1/4, 24, 48, 72, 96, and 120 h. At the end of the last weighing in water, samples were surface dried and weighed in air. Apparent and bulk specific gravities and water absorption were calculated and given in Table 2 for all aggregate types.

The results indicate that water absorption depends on the nature and size of the aggregates. In general, RCA absorbed more water than natural aggregate due to its rough surface texture and having a higher surface area. The nature of aggregate and the voids causes water absorption. Generally, the natural aggregate composites exhibit large water absorption due to the presence of microvoids at aggregate, and

Table 2 Bulk specific gravity and water absorption of aggregate

Aggregate types	Bulk specific gravity	Apparent specific gravity	Absorption (%Abs)	Porosity (%)
VA	2.60	2.69	1.38	3.60
	2.46	2.96	1.27	3.15
	2.28	2.40	2.21	5.06
Average	2.45	2.69	1.62	3.98
RCA	2.25	2.45	3.66	8.25
	2.307	2.49	3.29	7.60
	2.27	2.48	3.80	8.64
Average	2.28	2.48	3.59	8.17
RAP	2.27	2.38	1.94	4.41
	2.21	2.33	2.38	5.26
	2.27	2.40	2.42	5.51
Average	2.25	2.37	2.25	5.06

RAP water absorption is lower than that of RCA, as shown in Table 2. Both RAP and VA had the same trend of water absorption. The difference in porosity can explain this. RCA porosity is 8.17%; however, RAP porosity is 5%. The water absorption of natural aggregates decreases with the increase in aggregate size. This can be explained by aggregates' crushing effect, increasing pore connectivity, creating cracks, and generating a water absorption increase coefficient. This is due to the amount of adhered mortar, which is higher for small fractions than too coarse ones [8].

5.2 Evaluation of Asphalt Absorption

The results presented in Fig. 3 and Table 3 show that the 100–85 asphalt gives higher absorption values, as expected. This means that the greater the viscosity of the asphalt used, the higher the absorption ratio of the aggregate; as represented using the specific gravity method, the penetration grade bitumen 85–100 has a thermoplastic property which is similar to a synthetic resin material which gets softer with heating and hardens when cooled. This grade of bitumen 85/100 has very strong viscosity compared to other grades [8]. The bulk impregnation represents the realistic maximum value of asphalt absorption, and the overall absorption amount was found to be lower with the increased viscosity of asphalt. The required mixing temperature and absorption temperature are critical variables.

Asphalt absorption is also plotted against water absorption, as shown in Fig. 4. There is a specific general relationship between water absorption values and asphalt. RCA has a higher water absorption than RAP; however, its asphalt absorption ratio is almost the same, and this can be explained by the pore size of the concrete aggregate being a greater pore, while RAP has smaller pores. However, there are some exceptions to the general trend. VA and RCA have almost the same water absorption;

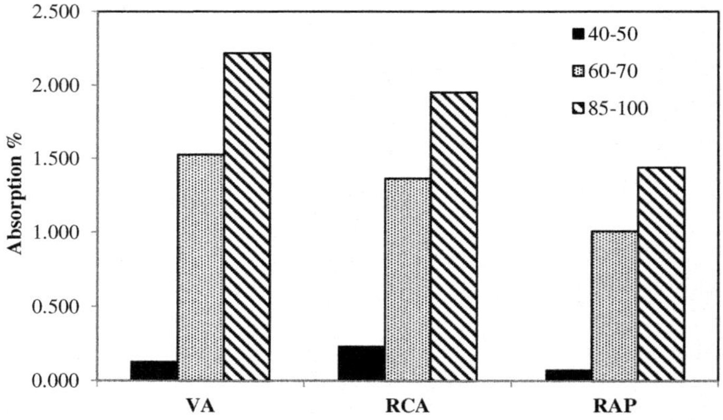

Fig. 3 Asphalt absorption—BISG method

Table 3 Correlation between water and asphalt absorption

Asphalt grade	VA (Wabs% = 3.42)	RCA (Wabs% = 3.80)	RAP (Wabs% = 2.25)	Equations	R^2
40–50	0.130	0.233	0.073	$y = 0.3338x + 0.364$	0.8222
60–70	1.527	1.364	1.007	$y = 0.0954 - 0.1483$	0.7148
85–100	2.217	1.952	1.439	$y = 0.4894x + 0.364$	0.7974

[a]The average of three tests of each sample of aggregate was conducted

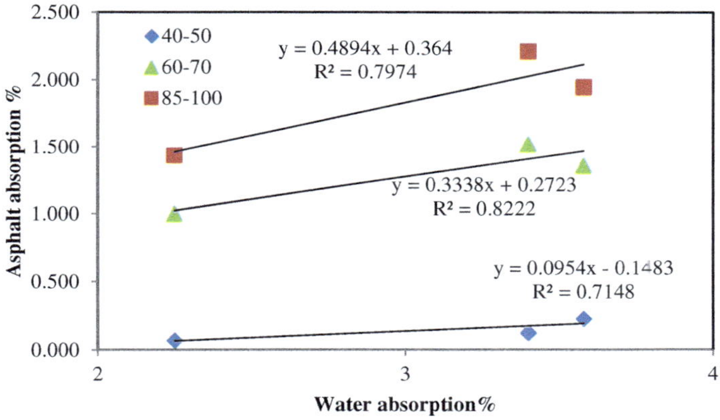

Fig. 4 Relationship between water and asphalt absorption for different grades of asphalt

however, asphalt cement absorbs VA much less than RCA. This can be explained by looking at the pore size distribution for the two aggregates, as VA contains a large amount of pores.

An analysis of variance (ANOVA) was conducted for the data presented in Table 3, and the results are reported in Table 4. Both asphalt cement and aggregates significantly affect asphalt absorption at ex = 0.05. The interaction (asphalt cement × aggregates) was also found to be significant at ex = 0.05. Figures 4 and 5 show the variation of asphalt absorption values with the viscosity (of the original asphalt cement) at 138 °C.

The above table shows the ANOVA of the laboratory results obtained, and there were no statistically significant differences in the absorption ratio for the types of aggregates, as the probability value (0.056) is greater than 0.05 due to the convergence of absorption rates.

Table 5 shows that the difference in absorption values is due to the difference in the asphalt gradient, where the probability of the 40–50 asphalt gradient is 0.001 and the probabilities for the 60–70 and 85–100 gradients are less than 0.05, with the probability of the 60–70 gradient indicating the convergence of absorption rates.

Table 4 ANOVA for absorption data from experiment

Asphalt absorption	Mean square	F	Significance	Statistical significance
Asphalt	29.425	51.143	0.005	Statistically
Aggregate	5.320	9.247	0.056	Not statistically
Aggregate and asphalt	3.895	6.770	0.077	Not statistically

Table 5 Chevy dimensional comparisons test

Asphalt grade		Mean difference	Significance	Statistical significance
4050	60–70	11.315	0.001	Significant
	85–100	12.687	0.001	Significant
60–70	40–50	11.315	0.001	Significant
	85–100	1.373	0.340	Not significant
85–100	40–50	12.687	0.001	Significant
	60–70	1.373	0.340	Not significant

5.3 *Effective, Apparent, and Bulk Specific Gravities*

The E-value criterion of the ESG for each type of aggregate cannot be found unless the maximum specific gravity of HMA is first determined using the Rice specific gravity method.

This paper adopted BISG (Gimp) instead of ESG for two reasons. First, the value of air space, calculated in the laboratory during the functional mix formula preparation, was restricted by 4% and between ISSRB specifications. Because of the overall absorption process, the air vacuum changes from time to time, which results in reduced effective asphalt content. The main changes started with mixing at the factory and continued until the road was opened to traffic. For this, it is best to use the most realistic standard for asphalt absorption, which can be handled with BISG. When applying BISG, the value of the air vacuum can be maintained approximately constant. Second, the BISG definition is similar to the ESG definition.

Equation (8) calclates $G_{eff.}$

$$G_{\text{eff.}} = \frac{100 - P_s}{\frac{100}{G_{mm}} - \frac{P_s}{Gb_{agg.}}} \quad (8)$$

where

G_{eff} Effective specific gravity

P_s Aggregate, percent by total dry weight of the mixture G_{mm} = Maximum theoretical specific gravity of mix

Table 6 Descriptive statistics for specific gravity

N		Minimum	Maximum	Mean	Std. deviation	Variance
G_{imp}	9	2.29	5.20	3.2837	0.90092	0.812
G_{eff}	9	1.49	1.80	1.5861	0.10383	0.011
G_b	9	2.25	2.44	2.3200	0.09042	0.008

Different G_{eff} values were found from the above equation. All G_{imp} values were obtained from Eq. 6 or observed. Those values found through laboratory work were used as input data in the SPSS program. Table 6 shows the comparison matrix and correlations between the observed and calculated G_{imp} values. Significant correlations between observed and calculated values were apparent due to the value of p, which was found to be 0.01.

6 Conclusions

This study was conducted to determine the effect of the material properties in asphalt mixtures and the extent to which these properties impact absorption ratio. Regarding the asphalt used, the asphalt absorption rate increased with an increase in the viscosity of the asphalt at the mixing temperature, noting some increases were due to selective absorption. The asphalt absorption ratios for natural aggregates and recycled aggregates showed that the absorption ratio is related to the aggregate surface area. This is a good indicator of the amount of absorption. Mainly, the aggregation of asphalt mixture performance provides an index for using recycled aggregate to reduce the depletion of natural aggregates, and knowing the optimal ratio reduces waste and decreases economic and environmental costs.

Also, this research examined the relationship between density, specific weight, and water absorption and compared different mixture types. The results showed that aggregates with a high degree of water absorption could absorb at a higher rate, and this relationship affects aggregate behavior and the layers' future performance. Finally, during the statistical analysis of the laboratory results, good mastery was observed between the observed and expected G_{imp} values, and a strong correlation between the two models was achieved.

Acknowledgements The authors would like to acknowledge the partial support provided by Mustansiriyah University through Highway and Transportation Engineering Department.

Author Contributions Eman and Israa conducted the data analysis and wrote the paper draft; Abbas and Safaa revised the paper draft and provided comments to data analysis and paper draft.

References

1. ASTM, C.: Standard test method for density, relative density (specific gravity), and absorption of coarse aggregate C-127. Annual Book of ASTM Standards (2004)
2. Austroads: Absorption of bituminous binder into aggregate (Report No. AGPT-T052-05) (2005)
3. Fadhil, T.H., Mohammed, H.A., Ahmed, T.Y.: An empirical relationship between asphalt and water absorption of coarse aggregates in HMA. J. Eng. Sustain. Dev. **19**(1), 132–146 (2015)
4. INTITUTE, A.: The asphalt handbook. MS-4 Seventh Edition. USA. 788p (2017)
5. Kandhal, P.S., Khatri, M.A.: Relating asphalt absorption to properties of asphalt cement and aggregate. Transp. Res. Record 76–76 (1992)
6. Lee, D.-Y., Guinn, J.A., Khandhal, P.S., Dunning, R.L.: Absorption of asphalt into porous aggregates (1990)
7. Oklahoma, D.: Method of test for bulk impregnated specific gravity of aggregate (Report No. OHD L-7). D. Oklahoma (2003)
8. SCRB, T.S.C.f.R.a.B.: Iraqi Standard Specifications for Roads and Bridges (ISSRB) R9, Revised Edition (2003)

Evaluation on the Physical Properties of Asphaltene-Modified Bitumen

Nurul Aliah Amirah Mazalan, Mohd Khairul Idham Mohd Satar, Azman Mohamed, Haryati Yaacob, Mohd Zul Hanif, and Ashraf Abdalla M. Radwan

Abstract Bitumen is composed of four primary fractions, which are saturate, asphaltene, resin, and aromatic. All fractions are important in determining the linear viscoelastic behavior of bitumen including asphaltene. Changes in asphaltene content affected the stiffness and viscosity of bitumen. This study assesses the influenced of asphaltene on the physical properties of bitumen penetration grade 60/70 bitumen. Various asphaltene contents with 1%, 2%, 3%, 4%, and 5% were added and tested for the laboratory test. The test includes a penetration test, a softening point test, and a rotational viscosity test. Asphaltene-modified bitumen was compared to control bitumen to determine the optimum asphaltene content required. Conventional bitumen penetration grade 60/70 and PG 76 bitumen were used as a control bitumen in this study. The results show that increased asphaltene contents elevate the stiffness and viscosity of bitumen. Thus, the recommended amount of asphaltene content to be added into the bitumen is 5% of the total weight of bitumen. The findings show that asphaltene is an efficient and affordable additive to enhance the properties of bitumen.

Keywords Asphaltene · Physical properties · Asphaltene-modified bitumen · Stiffness · Viscosity

---

N. A. A. Mazalan · M. K. I. Mohd Satar (✉) · A. Mohamed · H. Yaacob · M. Z. Hanif · A. A. M. Radwan
School of Civil Engineering, Faculty of Engineering, Universiti Teknologi Malaysia, 81310 Skudai, Malaysia
e-mail: khairulidham@utm.my

A. A. M. Radwan
Department of Civil Engineering, Faculty of Engineering, University of Tripoli, Tripoli, Libya

N. Sabtu (ed.), *Proceedings of AWAM International Conference on Civil Engineering 2022—Volume 2*, Lecture Notes in Civil Engineering 385,
https://doi.org/10.1007/978-981-99-6018-7_2

1 Introduction

Recently, researchers are paying attention to the utilization of waste materials to boost bitumen properties because of the growing public worries about sustainability and given the high cost of certain modifiers [1, 2]. Asphaltene is one of the waste materials in oil sand bitumen production obtained through the de-asphaltation process. Despite its high rate of production, asphaltene is typically regarded as a waste with few significant industrial uses [3, 4]. However, there have been some recent attempts to utilize asphaltene, such as the development of a gasification technology to convert asphaltene into gas fuel [5]. Unfortunately, gasification is a costly method that releases a large number of emissions into the environment [3].

The literature defines two broad strategies for altering the asphaltene quality of bitumen. First, utilize a mechanical agitator to mix bitumen and asphaltene-rich substances [6, 7]. Second, reintroduce the asphaltene removed from the bitumen using n-heptane extraction into the binder [8, 9]. Adjustable asphaltene content is a benefit of the second approach, which has led to its increased popularity and the development of the concept of modified bitumen.

SARA fractions, which consist of saturate (S), aromatic (A), resin (R), and asphalt (A), may be split into SARA fractions, providing an efficient tool for examining the traditional bitumen structure. The 12-component bitumen is used to consider the physical–mechanical and rheological properties of the bitumen. The study made use of Li and Greenfield's [10] SARA-based molecular model of SHRP AAA-1 asphalt. Table 1 presents the molecular properties of the components of asphalt, and Fig. 1 shows the molecular structures.

Table 1 Molecular compositions and characteristics of the assessed base asphalt [10]

SARA fractions	Molecules	Molecules number	Molecular formula	Molar mass (g/mol)	Mass fraction (%)
Saturates (S)	Squalane	4	$C_{30}H_{62}$	422.9	5.2
	Hopane	4	$C_{35}H_{62}$	482.8	5.8
Aromatics (A)	PHPN	11	$C_{35}H_{44}$	464.8	15.7
	DOCHN	13	$C_{30}H_{46}$	406.8	16.2
Resins (R)	Pyridinohopane	4	$C_{36}H_{57}N$	530.9	6.2
	Thio-isorenieratane	4	$C_{40}H_{60}S$	572.9	7.0
	Trimethylbenzene-oxane	5	$C_{29}H_{50}O$	414.7	6.4
	Quinolinohopane	4	$C_{40}H_{59}N$	554.0	6.8
	Benzobisbenzothiophene	15	$C_{18}H_{10}S_2$	290.4	13.4
Asphaltene (A)	Phenol	3	$C_{42}H_{54}O$	575	5.3
	Pyrrole	2	$C_{66}H_{81}N$	888.5	5.5
	Thiophene	3	$C_{51}H_{62}S$	707.2	6.5

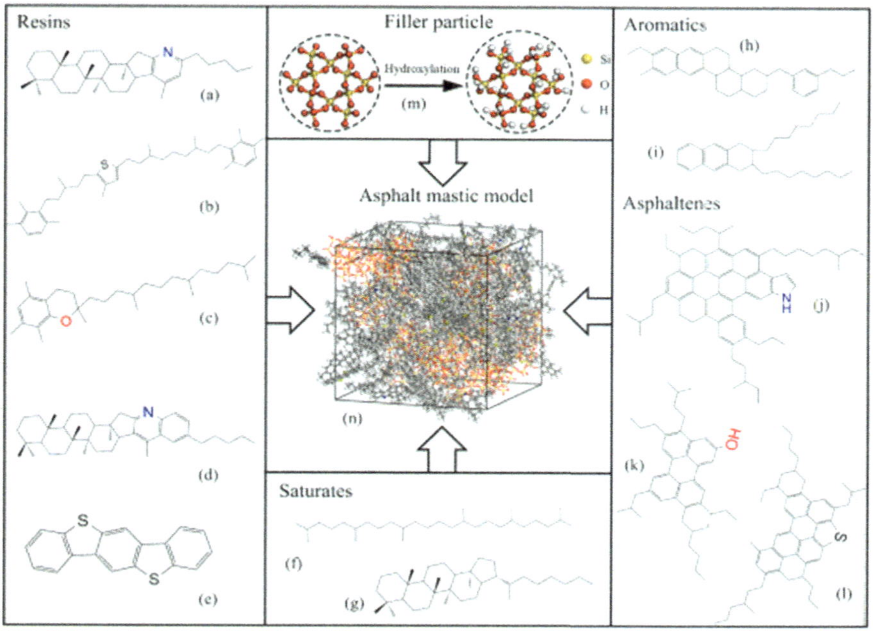

Fig. 1 Asphalt mastic molecular structures

1.1 Components of Bitumen

Bitumen is a mixture of hydrocarbon molecules that can be divided into distinct chemical groups. Bituminous elements are made up of asphaltene and maltenes (petrolenes). Maltenes are aromatic chemicals, polar aromatics (resins), and saturated aromatics (paraffins). There is some cross-pollination among these four parties. To test the chemical characteristics of bitumen, it must first be extracted. There are numerous methods for quantifying and identifying the organic bitumen structure. The most common method is chromatographic panel separation. Several chromatographic techniques have been used to separate the primary components of bitumen. These processes generate fractions with similar names, such as saturates and aromatics, but no chemical relationship [11]. Because bitumen materials vary in solubility, this separation can be achieved using appropriate solvents and chromatographic columns. The total of these components' factors has distinct physical and chemical properties, as determined by their quantitative ratios.

Saturated materials (paraffins)

Saturated compounds are aliphatic hydrocarbons with alkyl naphthenes and aromatic alkyls. These are usually non-polar, colorless viscose oils. They have a similar molecular weight to aromatics. In saturated formulations, waxy and non-waxy chemicals are employed. It comprises 5–20% of saturated formulations [12].

Oil aromatics

Aromatic oils are asphaltene anticoagulants with aromatic or petroleum lateral chains. The category is a viscous dark brown liquid with a molecular weight of 300–2000. Aromatics have polar carbon chains.

Polar aromatics (resins)

These molecules are usually composed of carbon and hydrogen, with a dash of oxygen, nitrogen, and sulfur. They are solid or semisolid and dark brown in color. The resin-to-asphaltene ratio can be as high as in the sol1 or gel2 modes. Their molecular weight ranges from 500 to 50,000. Bitumen's resins provide its adhesive properties [12].

Asphaltene

Asphaltene is brittle solids that contain nitrogen, sulfur, and oxygen in addition to carbon and hydrogen. They usually have polar compounds with high molecular weight aromatic complexes. Asphaltene is a mixture of compact aromatic rings and aliphatic lateral chains, with aromatic rings ranging from 6 to 20 (units), and often more.

1.2 Relation Between Asphaltene and Bitumen Physical Properties

It is widely acknowledged that asphaltene is significant as a road building material for the performance of a binder. No simple connection has been found between asphaltene and any performance parameters, neither for asphaltene quantity nor for any asphaltene chemical properties. Asphaltene is also well known to be crucial for the viscosity and stiffness of the binder. This is mostly caused by interactions between molecules of asphaltene and a contribution from other molecules [13]. So, it is not possible to find links between only the bitumen fraction and the physical properties; instead, all the bitumen molecules must be taken into account.

It is more difficult to determine solubility characteristics for complex mixes, such as bitumen, than for pure substances. On such materials, it is not feasible to measure the HSP directly. However, it is possible to indirectly infer the solubility characteristics of complicated substances by monitoring their interaction with liquids with known solubility values. Utilizing the well-known concept that a solvent with a solubility parameter equal to the solubility parameter of the substance is a good solvent, while a solvent with a different solubility parameter is a bad solvent, the method identifies the optimal solvent. The standard operating procedure is to select a group of liquids with a known and realistic range of solubility parameters, then test their solubility with each of them [14]. It is apparent, according to the Hansen Solubility Parameter, that the polar and hydrogen bonding interactions between bitumen molecules are weak yet significant. The existence of interactions with hydrogen bonding has been verified by methods using selective hydrogen bonding feature blocking in asphaltene [15]. From the dielectric constant, the total polarizability may be calculated. The non-polar dispersive (electronic) polarizability of London may be measured by measuring the refractive index. The main causes of polarization in materials with a higher permittivity at zero frequency will be Keesom and Debye attraction energies.

Overall, bitumen has a low level of total polarizability. Bitumen includes a minor proportion of insoluble n-heptane molecules with a slightly greater overall polarizability, which may contribute to Debye and Keesom interactions. The asphaltene percentage is more London dispersive (electronic) polarizable than bitumen as a whole. The degree of non-polar London dispersion polarizability rises as molecule size and aromaticity increase [16]. However, a kind of interaction resulting from variation in electron density called the London dispersive forces is the strongest interactions in bitumen, especially between bitumen molecules. They are weaker but more dominant in non-polar hydrocarbons than polar and hydrogen bonding. The most prevalent characteristic of the asphaltene fraction is its strong interaction with London dispersion. However, it is crucial not to associate chemical bonds with these interactions. The interactions should not be thought of as everlasting ties but rather as a commonplace "sticky" attraction between the molecules since they are constantly disrupted and rebuilt.

Van der Waals interactions have a substantial impact on the physical properties of bitumen since they are responsible for the substance's coherence and stickiness. At temperatures below about 70 °C, bitumen's non-Newtonian behavior may be attributed to the existence of Van der Waals interactions.

An example of how bitumen molecular interactions have been associated with bitumen elastic properties has previously been published [17]. In this situation, the entire bitumen and not just the asphaltene fraction was used.

2 Materials and Test Method

2.1 *Materials*

At a specific weight, bitumen penetration grades 60/70, PG 76, and modified bitumen with varying asphaltene content were prepared. The modified bitumen with 1%, 2%, 3%, 4%, and 5% asphaltene content was designated as 60/70-A, 60/70-B, 60/70-C, 60/70-D, and 60/70-E. Next, the bitumen penetration grade 60/70, PG 76, and modified bitumen samples with different asphaltene percentages were analyzed by implementing a physical properties test.

Sample preparation

Samples were prepared by mixing bitumen with various percentages of asphaltene content. As per standard technique, 500 g of bitumen was heated to 150 °C in a container fitted with a mechanical stirrer. Then, the desired amounts of asphaltene were added in different percentages. Heating and stirring were continued for 15 to 20 min until a homogeneous mixture was achieved.

2.2 *Test Method*

Figure 2 represents the framework of this research.

Penetration test

The penetration test is when a typical needle weighing 100 g will vertically penetrate a sample at 25 °C for five seconds at a distance or depth of tenths of a millimeter. The hardness of the bitumen samples was evaluated using a penetration test, in compliance with ASTM D5/D5M [18]. The average of three readings was then recorded for each sample.

Softening point test

The softening point of bitumen shows the bitumen's thermal sensitivity. This test shows how bitumen behaves when the temperature rises. It was done in accordance with ASTM D36/D36M [19]. The temperature at which the bitumen becomes sufficiently soft to enable 3.5 g of steel balls to fall from a height of 25 mm to strike the base plate was determined as the softening point.

Viscosity test

The ASTM D4402/D4402M standard [20] was used to calculate viscosity, which represents the flow resistance and internal friction of bitumen. Viscosity was measured with a Brookfield Thermosel viscometer at various temperatures, including 135 °C and 165 °C. At a fixed temperature and rotational speed of 20 rpm, a cylindrical spindle (size number 27) was placed into the bitumen. The rotational viscosity was calculated using spindle torque measurements. The low friction between the bitumen sample and the spindle indicates the low viscosity of the bitumen, and vice versa.

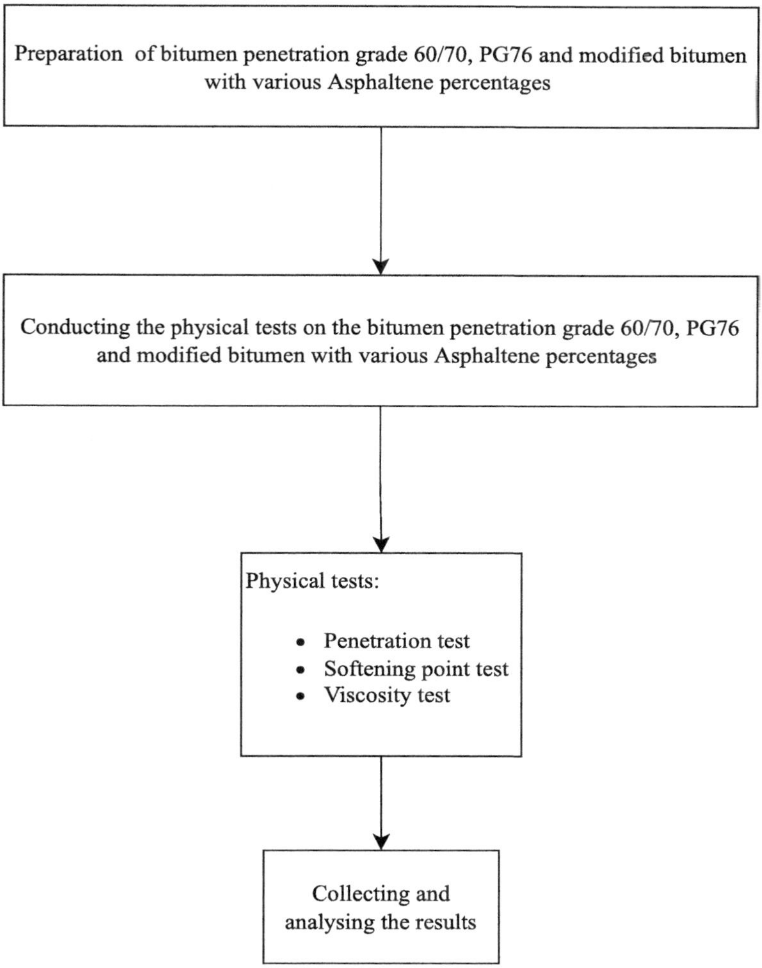

Fig. 2 Research framework

3 Results and Analysis

3.1 Effect of Asphaltene on Penetration

The amount of asphaltene influenced the depth to which the needle penetrated the sample. Modified bitumen with the amount of asphaltene increased by 1.0%, 2.0%, 3.0%, 4.0%, and 5.0% led to a decrease in the average reading of penetration of 46dmm, 45dmm, 44dmm, 43dmm, and 29dmm compared to the control samples, which are bitumen penetration grade 60/70 and PG76 with 58dmm and 50.2dmm

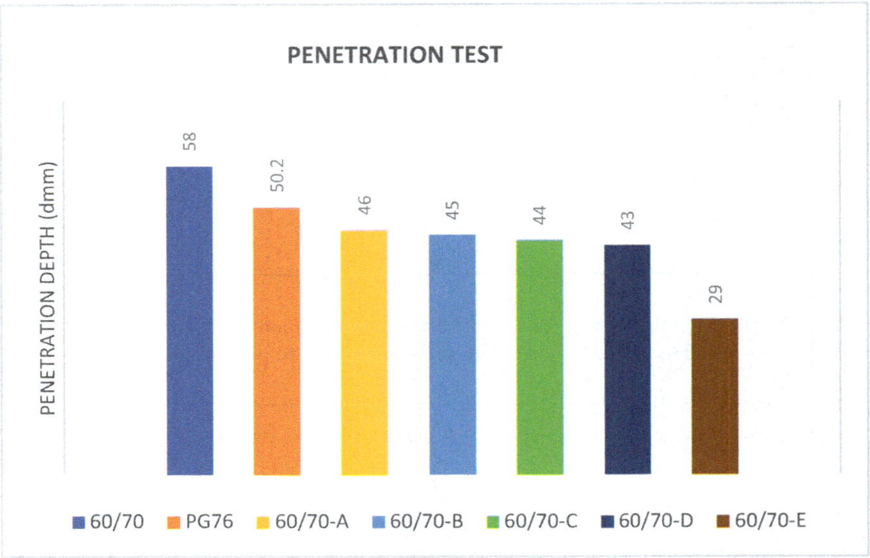

Fig. 3 Effect of asphaltene on penetration

penetration readings as shown in Fig. 3. In other words, penetration is an indicator of the bitumen viscosity. Bitumen's viscosity reduces as penetration rises, resulting from a decrease in asphaltene content. When penetration reduces, it indicates that the bitumen got stiffer, exhibited less elongation and flow, and increased in viscosity, as indicated by the softening point.

3.2 Effect of Asphaltene on Softening Point

The association between the asphaltene percentage and the softening point value revealed that an increase in asphaltene content by 1%, 2%, 3%, 4%, and 5% led to enhancing the softening point by 2.96% or 51.5 °C, 1.49% or 50.75 °C, 9.43% or 54.95 °C, and 11.32% or 56 °C, respectively, when compared with bitumen penetration grade 60/70 as shown in Fig. 4. Due to the positive relationship between bitumen viscosity and softening point, an increase in asphaltene concentration causes a rise in modified bitumen viscosity because the bitumen becomes more rigid and has less elongation and flow as represented by penetration. However, when compared to PG76, the softening point test results with various asphaltene percentages showed lower values. Although, the asphaltene content increased by 1.0%, 2.0%, 3.0%, 4.0%, and 5.0%, the softening point decreased by 22.10% or 51.5 °C, 23.56% or 50.75 °C, 15.68% or 54.95 °C, and 13.8% or 56 °C, respectively, when compared with PG76.

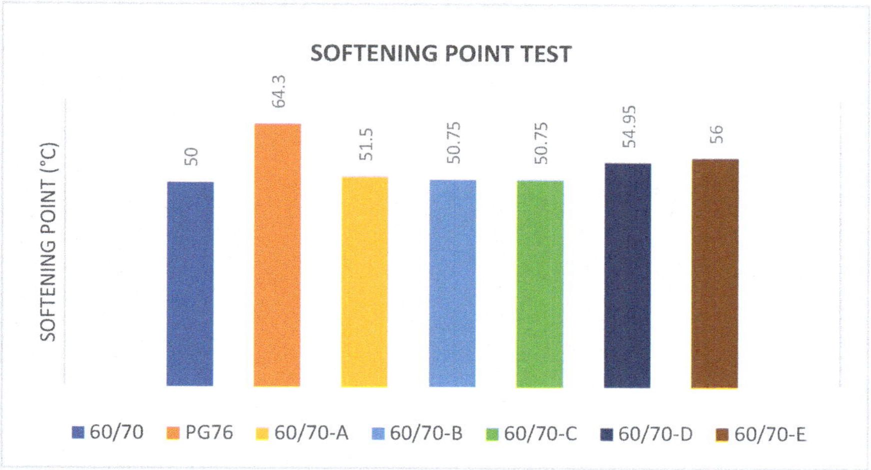

Fig. 4 Effect of asphaltene on softening point

This shows that PG76 is more viscous compared to modified bitumen with various amounts of asphaltene.

3.3 *Effect of Asphaltene on Viscosity*

Viscosity is a distinguishing property of bitumen. It measures the internal friction of a fluid when one layer is pushed to move relative to another. Viscosity is a value that is extremely temperature-dependent. Rotational viscosity (RV) was assessed at two different temperatures which are 135 °C and 165 °C. The RV at 165 °C is less than the RV at 135 °C, which is reduced by 100 MPa S or 66.67% for bitumen penetration grade 60/70 and 200 MPa S or 40% for PG76. As mentioned earlier, the connection between temperature and viscosity is inverse. The link between asphaltene concentration and rotation viscosity is also taken into account compared with bitumen penetration grade 60/70 and PG76. While asphaltene content increased by 5%, RV at 135 °C increased by 85.71% or 500 MPa S, asphaltene content increased by 4%, RV at 135 °C increased by 66.67% or 400 MPa S and increased by 40% or 300 MPa S for asphaltene content 1% to 3%, whereas RV at 165 °C increased by 100% and 120% or 300 MPa S and 400 MPa S, respectively, for 4% and 5% asphaltene content added. For asphaltene content added 1%, 2%, and 3%, the RV at 165 °C increased by 66.67% or 200 MPa S compared to bitumen penetration grade 60/70. Nevertheless, PG76 shows higher RV values for both 135 °C and 165 °C temperatures, which are 600 MPa S and 400 MPa S, respectively, when compared to modified bitumen.

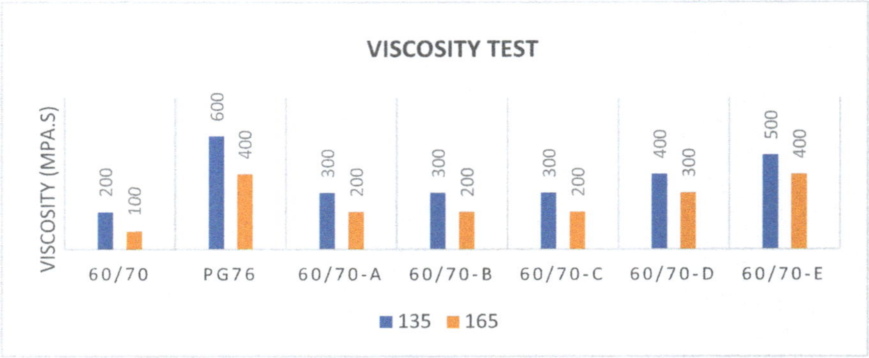

Fig. 5 Effect of asphaltene on viscosity

Thus, as shown in Fig. 5, it was noted that the addition of asphaltene raises the viscosity of the bitumen. Also, it was found that the effect of asphaltene on viscosity was more noticeable when the amount of asphaltene was higher and the temperature was lower. Additionally, when the temperature rose, it was discovered that the viscosities of various bitumen exhibited a converging tendency, as seen in Fig. 5. This demonstrates that the influence of asphaltene percentage on the viscosity of the binder was observed to decrease as the temperature rose. Based on the three tests, modified bitumen with 5% asphaltene content showed the highest increment in the physical properties of bitumen compared to others. Besides, 5% asphaltene content is the best amount of asphaltene to modify bitumen to replace PG76 since it is cheaper and easier during storage and application.

4 Conclusion

In this research, different asphaltene contents have been demonstrated to affect the physical properties. It is recommended to improve the asphaltene content for further research in the future. The bitumen property (which represents the physical properties) increases due to the increment of the amount of asphaltene content. It has been noticed that a rise in the percentage of asphaltene in modified bitumen alters its physical properties, as discussed below:

- The increasing viscosity of modified bitumen is described by the rising value of the softening point and the rotational viscosity.
- The increasing stiffness of modified bitumen is described by the decreasing value of penetration readings.
- The recommended asphaltene content is bitumen with 5% asphaltene content added.
- Asphaltene is a low-cost additive that improves the properties of bitumen.

Acknowledgements The authors would like to acknowledge the financial support from Ministry of Higher Education Malaysia under Fundamental Research Grant Scheme (FRGS) (FRGS/1/2023/TK01/UTM/02/1) and Universiti Teknologi Malaysia for the funding under UTM Fundamental Research (UTMFR) (Q.J130000.3822.22H49).

References

1. Kalantar, Z.N., Karim, M.R., Mahrez, A.: A review of using waste and virgin polymer in pavement. Constr. Build. Mater. **33**, 55–62 (2012)
2. Choudhary, J., Kumar, B., Gupta, A.: Utilization of solid waste materials as alternative fillers in asphalt mixes: a review. Constr. Build. Mater. **234**, 117271 (2020)
3. Ashtari, M.: New pathways for asphaltene upgrading via oxy-cracking in liquid phase. In: Chemical Engineering; University of Calgary: Calcar, ab, Canada (2016)
4. Meisen, A.: Bitumen beyond combustion (bbc) project phase 1 report; bbc project-phase 1 report 0416x.docx; alberta innovates corp.: edmonton, ab, Canada (2017)
5. Alipour, M., Kurian, V., Dhir, S., Gupta, R.: Analysis of syngas cooler fouling from asphaltene gasification. Fuel Process. Technol. **152**, 7–14 (2016)
6. Varanda, C.P.I., Ribeiro, J., Silva, A., Silva, C.M.: Influence of. Poly-phosphoric acid on the consistency and composition of formulated bitumen: standard characterization and nmr insights. J. Anal. Methods chem. (2016)
7. Mangiafico, S., Di Benedetto, H., Sauzéat, C., Olard, F., Pouget, S., Planque, L.: Effect of colloidal structure of bituminous binder blends on linear viscoelastic behavior of mixtures containing reclaimed asphalt pavement. Mater. Des. **111**, 126–139 (2016)
8. Hofko, B., Eberhardsteiner, L., Füssl, J., Grothe, H., Handle, F., Hospodka, M., Grossegger, D., Nahar, S.N., Schmets, A.J.M., Scarpas, A.: Impact of maltene and asphaltene fraction on mechanical behavior and microstructure of bitumen. Mater. Struct. **49**, 829–841 (2016)
9. Xu, Y., Zhang, E., Shan, L.: Effect of sara on rheological properties of bitumens. J. Mater. Civ. Eng. **31**, 04019086 (2019)
10. Li, D.D., Greenfield, M.L.: Chemical compositions of improved model asphalt systems for molecular simulations. Fuel **115**, 347–356 (2014)
11. Dean, W., Deborah, H., Kathy, P., Debbie, Z.: Evaluation of some open column chromatographic methods for separation of bitumen components. **66**(1), 44–50 (1987).https://doi.org/10.1016/0016-2361(87)90210-9
12. Didier, L.: The colloidal structure of bitumen: Consequences on the rheology and on the mechanisms of bitumen modification. **145**(1-2), 42–82 (2009).https://doi.org/10.1016/j.cis.2008.08.011
13. Kim, S.S., Branthaver, J.F.: Polarities of SHRP asphalts calculated from ion exchange chromatography separation data. Fuel Sci. Technol. Int. **14**(3), 265–393 (1996)
14. Redelius, P.: Bitumen solubility model using Hansen solubility parameter. Energy Fuels **18**(4), 1087–1092 (2004). https://doi.org/10.1021/ef0400058
15. Juyal, P., Merino-Garcia, D., Andersen, S.I.: Effect on molecular interactions of chemical alteration of petroleum asphaltene. Energy Fuels **19**, 1272–1282 (2005)
16. Asa, L.L.: A fundamental adhesion model for asphalt. Doctoral Thesis, Division of Highway and Railway Engineering, Stockholm, Sweden (2014)
17. Redelius, P.G., Soenen, H.: Correlation between bitumen polarity and rheology. Road Mater. Pavement Des. **6**(3), 385–405 (2005)
18. ASTM D5/D5M.: Standard Test Method for Penetration of Bituminous Materials, Annual book of ASTM standards. ASTM International, West Conshohocken, USA (2013)

19. ASTM D36/D36M.: Standard Test Method for Softening Point of Bitumen (Ring-and-Ball Apparatus). Annual Book of ASTM Standards. ASTM International, West Conshohocken, USA (2014)
20. ASTM D4402/D4402M.: Standard Test Method for Viscosity Determination of Asphalt at Elevated Temperatures Using a Rotational Viscometer. Annual Book of ASTM Standards. ASTM International, West Conshohocken, USA (2015)

Effects of Beta-Carotene Palm Oil on Asphalt Binder Properties and Coatability Performance

Tai Han Mok, Tai Guo, Tracy Leh Xin Wong, Mohd Rosli Mohd Hasan, Mohd Asyraf Kassim, and Abdulnaser Al-Sabaeei

Abstract Bio-asphalt has a high potential to be implemented as a petroleum-based asphalt replacement since it is a renewable and environmentally friendly resource that has identical rheological properties to the conventional binder. Excessive crude oil consumption and fluctuating prices urged researchers to adopt alternatives such as bio-asphalts to replace conventional binders. In this study, the beta-carotene bio-oil (CBO) partially replaced the conventional asphalt binder at 5, 10 and 15%. Penetration test, rotational viscosity test, specific gravity test, and mass loss after rolling thin film oven (RTFO) test were conducted to determine the physical and rheological properties of bio-asphalts. FTIR test is used to identify the chemical compositions and functional groups of the CBO bio-asphalt. The binder coatability was assessed by static water immersion and water boiling tests. The addition of CBO has lowered the viscosity and specific gravity while increasing the penetration and the mass loss after RTFO. The FTIR analysis shows that the functional groups of CBO bio-asphalt were mainly phenol, alcohols, ketones, aldehydes, and carboxylic acids. The CBO bio-asphalt enhanced the service characteristic in terms of bonding and coatability behavior. The CBO has a high potential for use as a rejuvenator in the production of asphalt mixtures incorporating reclaimed asphalt pavement (RAP), as it improves the mixture's workability and adhesion characteristics.

Keywords Bio-asphalt · Beta-carotene bio-oil · Coatability · Rheological properties · Low-carbon material · Alternative material source

---

T. H. Mok · T. Guo · T. L. X. Wong · M. R. Mohd Hasan (✉)
School of Civil Engineering, Universiti Sains Malaysia (Engineering Campus), 14300, Nibong Tebal, Penang, Malaysia
e-mail: cerosli@usm.my

M. A. Kassim
School of Industrial Technology, Universiti Sains Malaysia, 11850 Minden, Penang, Malaysia
e-mail: asyrafkassim@usm.my

A. Al-Sabaeei
Department of Civil and Environmental Engineering, Universiti Teknologi PETRONAS, UTP, 32610 Seri Iskandar, Perak, Malaysia

N. Sabtu (ed.), *Proceedings of AWAM International Conference on Civil Engineering 2022—Volume 2*, Lecture Notes in Civil Engineering 385,
https://doi.org/10.1007/978-981-99-6018-7_3

1 Introduction

Asphalt binder or bitumen can be defined as a viscoelastic material that is black in color and with high viscosity. It has been widely used since the nineteenth century for many applications, including highways, airport runways, sideways, carriageways, and even sports areas. The demand for bitumen has escalated due to expeditious development, urbanization, and industrialization. This raw material is highly beneficial for road construction and maintenance as it has a superb ability to withstand stress and provides riding comfort. Generally, the production of asphalt mixtures involves a small quantity of asphalt binder with roughly 5% by weight of aggregates. However, the huge stretch of road networks globally requires an enormous amount of asphalt binder in road construction work [1]. As a result, this caused a rise in demand for substitute bituminous materials in the asphalt pavement industry.

Bitumen is a by-product of crude oil refining and is considered a non-renewable resource. The increase in crude oil consumption, the fluctuations in crude oil prices, and the limited utilization of conventional petroleum urged the industry and researchers to search for alternative solutions. Green resource has been widely considered by researchers as it is in line with the United Nations Sustainable Development Goals. Over the years, the asphalt pavement industry has invented several approaches, known as hot or cold recycling techniques, as well as warm or cold asphalt mixtures, to deal with limited resources. An alternative approach is implemented in road material modification by using biomass materials for partial or full replacement in bio-asphalt production to reduce the dependency on bitumen [2–4].

Researchers have adopted biomass sources in asphalt modifications due to their accessibility and environmental friendliness [5]. Biomass material is a viable alternative because it can produce renewable and environmentally friendly bio-oil with similar rheological properties to conventional asphalt. Bio-oil may be yielded globally from a variety of biomass sources, including algae [6], switch grass [7], corn stover [8], swine manure [9–11], waste wood resources [1, 3, 12, 13], and waste cooking oil [2, 4, 14]. In Malaysia, when palm oil is extracted, a large amount of biomass is generated, including the mesocarp fruit fibers (MF), palm kernel shells (PKS), palm oil mill effluent (POME), and the empty fruit bunches (EFB). These palm oil wastes could be biomass sources for producing bio-oil, especially the EFB [15–20].

It has been observed that considerable study has been conducted over the years in quest of a modifier or substitute for the asphalt binder. However, research on bio-asphalt utilizing palm oil-based products is currently limited. Thus, in this experimental work, beta-carotene bio-oil (CBO) extracted from palm oil was utilized to partially substitute bitumen at 5%, 10%, and 15% by weight of asphalt binder, respectively. The physical and rheological properties of the asphalt binder were examined.

Table 1 Physical properties of the conventional binder

Properties	Units	Specification range	Results
Penetration (at 25 °C, 100g, 5s)	0.1 mm	60–70	64
Softening point	°C	48–56	49.5
Ductility (at 25 °C)	cm	> 100	> 100
Density	g/cm^3	1.01–1.06	1.03

Fig. 1 Physical appearance of CBO at room temperature

2 Materials and Methods

2.1 Materials

The beta-carotene bio-oil (CBO) was selected to prepare the bio-asphalt. The CBO is high-quality palm oil, which is only made available for the Malaysian market. Besides, the preparation of hot-mix asphalt (HMA) also involved a conventional 60/70 PEN binder and granite aggregates. Table 1 summarizes the physical properties of the conventional binder; meanwhile, the physical appearance of CBO is shown in Fig. 1.

2.2 Preparation of Asphalt Binder Samples

In this study, four asphalt binder types were prepared, including an unmodified base asphalt with penetration grade 60/70 and three bio-binders with the incorporation of CBO at 5%, 10%, and 15% by weight. For the preparation of bio-asphalt, virgin asphalt binder and palm bio-oil were mixed using a propeller mixer at 145 °C, with a speed of 1000 revolutions per minute (rpm) for 30 min. Table 2 shows the binder blends composition for the tested asphalt binder.

Table 2 Binder blends composition

Designation	60/70	CBO		
		5%	10%	15%
CS	✓			
5% CBO	✓	✓		
10% CBO	✓		✓	
15% CBO	✓			✓

2.3 Binder's Properties Test

Specific Gravity Test. This test is used to determine the change in densities of the bio-asphalts compared to control asphalt. The specific gravity of the bio-asphalt was determined according to ASTM D70 [21] by using a pycnometer bottle of 24 to 30 ml. First, the pycnometer was cleaned, dried, and weighed along with the stopper to obtain the initial weight. Then, it was filled with distilled water, closed with a stopper, and kept in a water bath for half an hour. The bio-asphalt was weighed, heated, and poured into the empty pycnometer bottle until half full. The weight was recorded after half an hour of allowing the pycnometer with stopper and bio-asphalt to cool to room temperature. Then, the remaining space was filled with distilled water and submerged in water for another half an hour. Finally, the pycnometer filled with bio-asphalt and distilled water was weighed. The specific gravity of bitumen can be calculated from the weights obtained using Eq. 1.

$$\text{Specific Gravity} = \frac{(C - A)}{[(B - A) - (D - C)]} \tag{1}$$

where A is the mass of the pycnometer and stopper, B is the mass of the pycnometer filled with water, C is the mass of the pycnometer partially filled with asphalt, and D is the mass of the pycnometer filled with asphalt and water.

Penetration Test. This test measured the consistency of the asphalt binders with a penetrometer by determining the binder's grade with its penetration value. Penetration value is the vertical distance penetrated by the needle with 100 g penetrated sample perpendicularly into the bituminous material at 25 °C for five seconds. The procedures of penetration tests should conform with ASTM D5 [22].

Rotational Viscosity Test. A rotational viscometer is used to assess the viscosity of asphalt binders at high temperatures. The torque required to keep the spindle in the asphalt binder revolving at a consistent speed, as measured by the rotational viscometer, is directly connected to the binder's viscosity. Thus, through the RV test, the data acquired by the rotational viscometer was used to determine the mixing and compaction temperatures. The control and bio-asphalt binders were tested at four different temperatures, including 120 °C, 140 °C, 160 °C, and 180 °C. A #27 spindle was used with a constant rotating speed of 20 rpm. For each control and bio-asphalt binder, samples were tested for three repetitions to obtain an average viscosity reading. The standard procedure referred to is in accordance with ASTM D4402 [23].

RTFO Mass Change. The RTFO test is performed in accordance with ASTM D2872 [24] to mimic the short-term aging of the asphalt binder by studying the impact of heat and air on a rotating film of the asphalt binder. Asphalt binder experiences short-term aging, especially during storage, transportation, mixing, and handling. The RTFO quantifies the volatiles lost (mass change) during the aging process. Generally a negative percentage represents mass loss; meanwhile, a positive percentage represents mass gain. First, the RTFO bottles were weighed and labeled. Then, the RTFO bottles and asphalt binder were preheated inside the oven for 4 h at 163 °C, respectively. The binder was subsequently poured into the bottles at a weight of roughly 35 g. The bottles were rotated to achieve a uniform coating of asphalt binder on the surface of the bottle. The coated bottles were weighed after cooling to ambient temperature. The bottles were then spun for 85 min at 15 rpm in the RTFO oven. A rate of 4000 ± 200 mL per minute of airflow was maintained during the test. After that, the RTFO bottles were set aside for cooling and weighed. The mass change of the asphalt binder can be determined by using Eq. 2.

$$\text{Mass Change} = \frac{m_2 - m_1}{m_1 - m_b} \times 100 \tag{2}$$

where m_1 is the mass of the RTFO bottle filled with asphalt sample before aging, m_2 is the mass of the RTFO bottle plus the asphalt sample after aging, and MB is the mass of the empty RTFO bottle.

2.4 *Physicochemical Analysis*

The Fourier transform infrared spectroscopy (FTIR) test was carried out to identify the functional groups present in control and CBO bio-asphalt binders. Generally, the FTIR test produced a spectrum with peaks that illustrated transmittance against wavenumber. The presence of any functional group in the bio-asphalt can be defined from the sharp peak observed at a specified wavenumber.

2.5 *Binder Coatability Characteristics*

Static Water Immersion Test. The percentage of asphalt binder that remained coated on aggregates after water immersion was measured in a static water immersion test. This test was conducted by mixing 100 g of oven-dried aggregates with 5.5 g of asphalt binder. The coated aggregates were oven-cured for 2 h. After curing, the loose mix sample was transferred to a glass container for water immersion at 25 °C for 16–18 h. Water may infiltrate through the thin asphalt binder film due to weak adhesion between the binder layer and aggregates, resulting in the stripping or removal of coated asphalt binder on aggregates. A visual inspection of the covered region determines the asphalt binder's adhesion strength. This test was conducted in accordance with AASHTO T182-84 [25].

Water Boiling Test. Based on ASTM D3625 [26], the adhesion loss in a loose mixture of binder-coated aggregate is evaluated using the influence of boiling water. A 250 g loose asphalt mixture was prepared and subjected to a boiling effect in a beaker filled with water for approximately 10 min. Before adding the mixture to the boiling water, the temperature of the mixture was below the water's boiling point but not less than 85 °C. After the test, the excessive asphalt binder floating on the water surface was skimmed off to avoid recoating. After that, the sample was placed on a white paper towel to cool to room temperature. The level of stripping of the sample after cooling was conducted by visual observation.

3 Results and Discussion

3.1 Binder Properties

Specific Gravity Result. The specific gravity values of asphalts are recorded in Fig. 2. The results showed that the specific gravity values for CBO bio-asphalts were lower than the specific gravity value of CS. The increment of CBO content led to a decrease in specific gravity values, which might be attributed to the higher light component content in CBO compared to bitumen. Overall, the CS has the highest specific gravity, while the lowest was achieved by 15% CBO bio-asphalt with specific gravity values of 1.034 and 1.013, respectively. All samples remain within the required specification of specific gravity for binders to be used in asphalt pavement applications.

Penetration Test Results. The results presented in Fig. 3 indicate that the penetration values increase for each bio-asphalt with increasing CBO content. The penetration values for CS, 5% CBO bio-asphalt, 10% CBO bio-asphalt, and 15% CBO bio-asphalt were 64 dmm, 111 dmm, 148 dmm, and 160 dmm, respectively. The CBO bio-asphalt reported higher penetration values than CS, thus showing that the CBO-modified binder had lower stiffness. Furthermore, the increased amount of bio-asphalt significantly affected the penetration value. That could be attributed to the increased light hydrocarbon components in CBO that softened the binders.

In conclusion, the highest penetration value was obtained at 15% CBO bio-asphalt, and the lowest penetration value has remained as the CS. The penetration index varies from − 2 for high-temperature susceptible bitumen to + 2 for highly blown low-temperature susceptible bitumen, as given in Table 3. The softening point and penetration value affect the penetration index of an asphalt binder. Table 4 shows

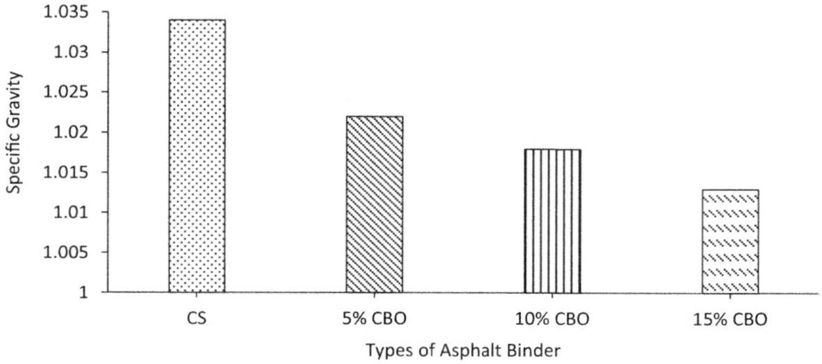

Fig. 2 Specific gravity values for the control and bio-asphalt binders

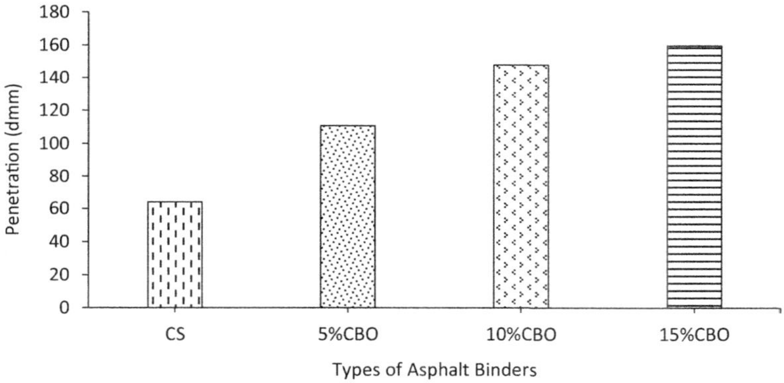

Fig. 3 Penetration test results

that the 5% CBO bio-asphalt is acceptable for the tropical climate based on the penetration index obtained. In contrast, for 10% CBO bio-asphalt and 15% CBO bio-asphalt, it could best possibly be used as temperature susceptible asphalt binder. The CBO could be used as a softening agent or asphalt rejuvenator in the production of asphalt mixtures incorporating reclaimed asphalt pavement (RAP).

Rotational Viscosity Test Results. The RV test results are demonstrated in Fig. 4. In general, the viscosity of all samples drops dramatically as the temperature rises,

Table 3 Typical values of penetration index [27]

Asphalt binder type	Penetration index
Blown asphalt binder	> + 2
Conventional paving asphalt binder	− 2 to + 2
Temperature susceptible asphalt binder (tars)	< − 2

Table 4 Control and bio-asphalt penetration index

Type of binder	Pen grade	Penetration index	Best possible use
CS	60/70	– 0.7439	Conventional paving asphalt binder
5% CBO	100/120	– 0.8051	Conventional paving asphalt binder
10% CBO	140/160	– 2.6121	Temperature susceptible asphalt binder (tars)
15% CBO	150/170	– 3.8804	Temperature susceptible asphalt binder (tars)

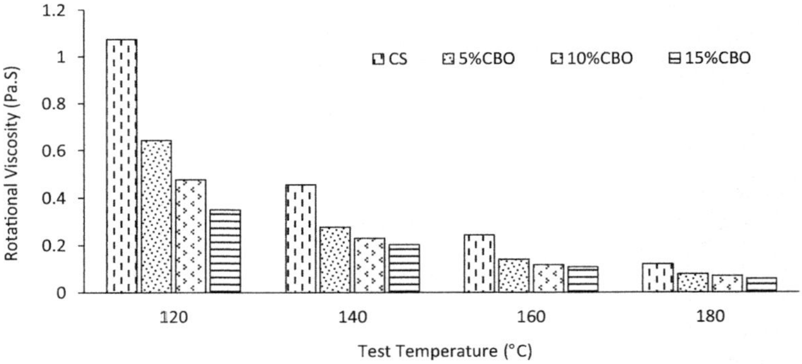

Fig. 4 Rotational viscosity test results

regardless of their dosages and ranges of testing temperature. It was observed that the CBO-modified asphalt attained lower viscosity as compared to CS. This can be due to the decreasing stiffness modulus of binders, especially at higher temperatures and with higher CBO content, which results in lower shear resistance. Thus, it can be deduced that the asphalt binder is more workable with the incorporation of CBO due to its lower viscosity and better fluidity than CS. The improved fluidity also resulted in less energy consumption throughout road pavement construction. All the samples achieved the JKR requirement with viscosity values lower than 3 Pa·s at 135 °C.

On the other hand, the Asphalt Institute recommends that the viscosity-temperature line should cross the viscosity ranges of 0.17 ± 0.02 Pa·s (mixing temperature range) and 0.28 ± 0.03 Pa·s (compaction temperature range) when determining the laboratory mixing and compaction temperatures of an asphalt binder. The optimum mixing and compaction temperatures for all the bio-asphalt binders are given in Table 5. Based on the outcomes, it is obvious that adding CBO will greatly reduce the energy demand for asphalt mixing and compaction which also leads to lower emissions and lower production costs for asphalt mixtures. The CBO application as an asphalt rejuvenator in the production of asphalt mixtures incorporating highly aged milled asphalt pavement would improve the workability of mixtures and rejuvenate their adhesion characteristics, as shown in the water boiling test (Fig. 9).

Mass Loss after RTFO. Figure 5 presents the mass loss test results for both control and bio-asphalt binders. The findings displayed that the asphalt binders met the JKR

Table 5 Mixing and compaction temperatures of asphalt mixture depending on optimal viscosity

Binder	Temperature (°C)	
	Mixing	Compaction
CS	169–175	153–159
5% CBO	153–158	138–144
10% CBO	147–154	133–138
15% CBO	143–151	125–134

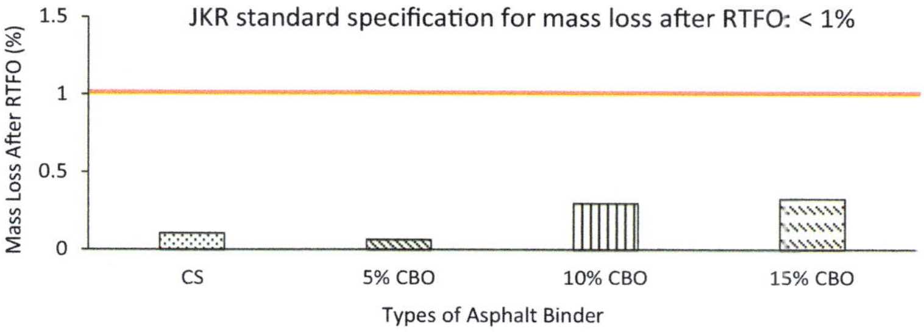

Fig. 5 RTFO mass loss test results

mass loss specification of less than 1%. Besides, the mass loss increased with the increasing content of CBO. The 5% CBO has shown a comparable mass loss with the CS. The slight changes in mass loss could be due to the polymerization of CBO at high temperatures due to the presence of palmitic acid in CBO, which functions as a bio-polymer. In conclusion, the lowest mass loss obtained was 5% CBO bio-asphalt and the highest mass loss obtained after RTFO was 15% CBO bio-asphalt. All the CBO bio-asphalts have fulfilled the JKR standard specification.

3.2 *Comparison Between Control Asphalt and CBO Bio-Asphalt Based on FTIR Analysis*

Figure 6 depicts the absorbance spectra of control asphalt and bio-asphalt samples. The shape of the spectra was similar between control asphalt and CBO bio-asphalts, which could indicate the compatibility of CBO with petroleum-based asphalt and lead to similar chemical behavior. The finding is in line with the previous studies, whereby, due to the low concentration of bio-oil incorporation in the asphalt, similar results were achieved between control asphalt, poppy capsule pulp bio-asphalt, and molasses bio-asphalt [1, 28]. Furthermore, the similar spectra indicated no substantial differences between functional groups, even when the dehydration process occurred during the mixing of bio-oil with the conventional binder [28].

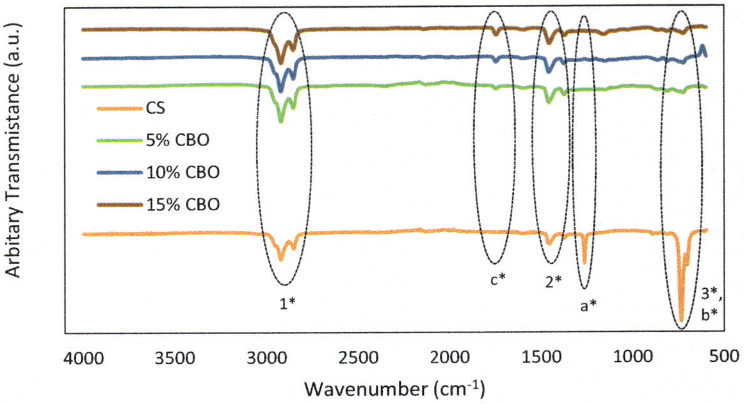

1* C-H stretch of the alkanes group (3066 cm^{-1} to 2800 cm^{-1})
2* CH_2 and CH_3 bend of the aliphatic group (1500 cm^{-1} to 1370 cm^{-1})
3* C-H plane blending of aromatic compound (900 cm^{-1} to 700 cm^{-1})
a* Peaks disappeared in CBO bio-asphalts (1300 cm^{-1} to 1000 cm^{-1})
b* Reduction of the peak in CBO bio-asphalts (900 cm^{-1} to 700 cm^{-1})
c* Peak increased in 10% and 15% CBO bio-asphalt Increased aromatic compound, ketone, aldehyde and carboxylic acid (2000 cm^{-1} to 1500 cm^{-1})

Fig. 6 Comparative FTIR spectra for control asphalt and CBO bio-asphalt

The spectra of control asphalt and bio-asphalts of CBO showed peaks at 3066 cm^{-1} to 2800 cm^{-1}, 1500 cm^{-1} to 1370 cm^{-1}, and 900 cm^{-1} to 700 cm^{-1}, which correspond respectively to the C–H stretch of the alkanes group, the CH_2 and CH_3 bend of the aliphatic group, and C-H plane blending of aromatic compound. Interestingly, the peak between 1300 cm^{-1} and 1000 cm^{-1} in CBO bio-asphalt mixture disappears significantly, suggesting the reduction of ester or phenol content in the bio-asphalt. Similarly, the absorbance for wavenumber range 900 cm^{-1} to 700 cm^{-1} was reduced in bio-asphalts. The study discovered that the addition of CBO in asphalt binder significantly increased the FTIR spectra at a peak between 2000 cm^{-1} and 1500 cm^{-1} in 10% and 15% CBO as compared to 5% and CS. An increase of the spectra at this range suggests increasing and improving aromatic compounds, ketone, aldehyde, and carboxylic acid in the bio-asphalt mixture. These findings are consistent with the experimental findings reported by Al-Sabaeei et al. [29].

3.3 Service Characteristics of Asphalt Mixtures

Static Water Immersion Test. The coatability index of the binder is identified by this test after being immersed in water at a constant temperature. The degree of coating is assessed by observing the coated area on the aggregates and with graphical help, as shown in Fig. 7. After conditioning, a good coatability should have more than 95% of

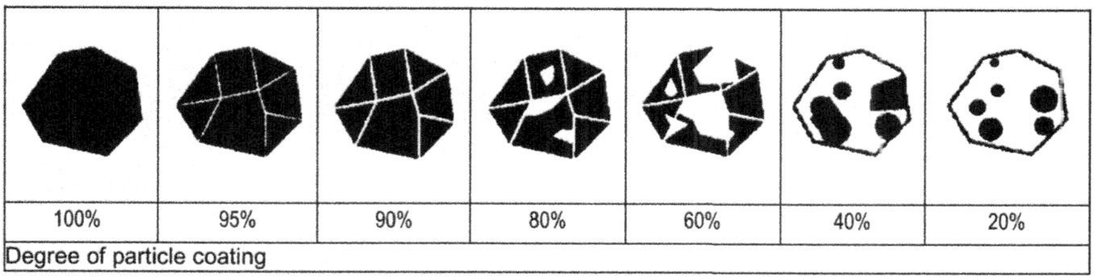

Fig. 7 Graphical help for assessment of the degree of coating [30]

the coated region remaining. Figure 8 depicts the appearances of the remaining coated region on aggregates after 16 h immersed in water at 25 °C, and Table 6 summarizes the results. From observation, all the control and bio-asphalt binders showed more than 95% of coated aggregates after the static immersion test. This indicates that the bio-asphalt maintained adequate adhesion properties between the asphalt film and aggregate in the presence of moisture, reflecting the sufficient moisture resistance of bio-asphalt binders.

Water Boiling Test. Figure 9 summarizes the percentage of the remained coated area of aggregates after being subjected to the boiling effect. From visual observation, the implementation of bio-oil improved the coatability of the loose asphalt mixture as compared to CS. Besides, the higher the bio-oil content, the aggregate area, covered with asphalt binder increases. The appearances of samples after the water boiling effect are illustrated in Fig. 10. These results implied that CBO bio-asphalt could strengthen the bonding between asphalt binder and aggregates.

4 Conclusions

Bio-asphalt with three dosages of CBO is investigated in this study. Laboratory testing and data analysis were performed to determine the binders' physical, rheological, and physicochemical properties. Additionally, asphalt-aggregate coatability and bonding characteristics were examined using water immersion and boiling tests.

Interpretation of the laboratory results concluded that the utilization of bio-oil significantly results in a lower viscosity while increasing the penetration value as compared to the conventional asphalt binder. This indicates the reduction of asphalt stiffness, especially at higher content of bio-oil. The specific gravity test results indicated that the addition of CBO decreases its specific gravity with increasing CBO content. Among the incorporations, 5% CBO has a comparable mass loss after conducting the RTFO test.

Besides, the FTIR test demonstrated the presence of C-H stretch of the alkanes group, C = O group, the CH_2 and CH_3 bend of the aliphatic group, and C-H plane blending of the aromatic compound according to their wavenumber absorbance. All bio-asphalts showed excellent bonding characteristics and coatability index

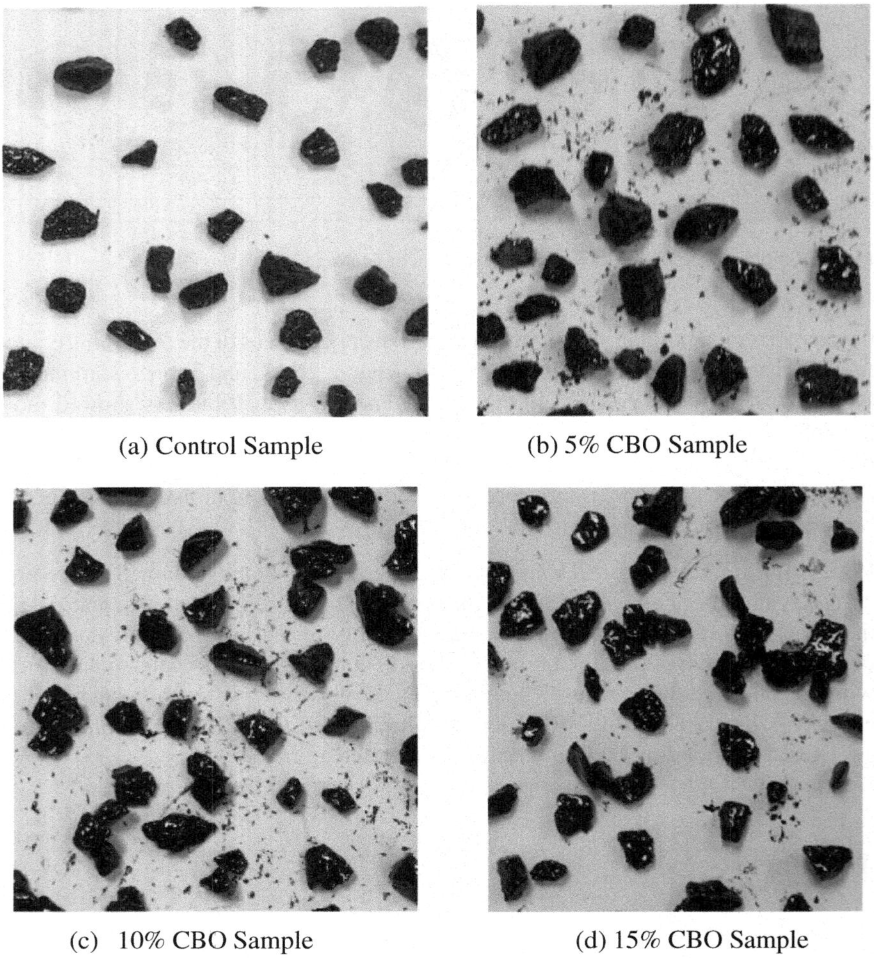

Fig. 8 Samples after static water immersion test

throughout the static water immersion and water boiling tests, indicating improved bonding strength, adhesion, and moisture damage resistance of bio-asphalt compared to the conventional binder.

The laboratory testing and data analysis found that the CBO is more suitable for being used as a rejuvenator due to its softening ability. The CBO would help to improve the mixture workability and adhesion characteristics. However, further study should be carried out to understand the influence of CBO on the durability and the long-term performance of the asphalt mixture.

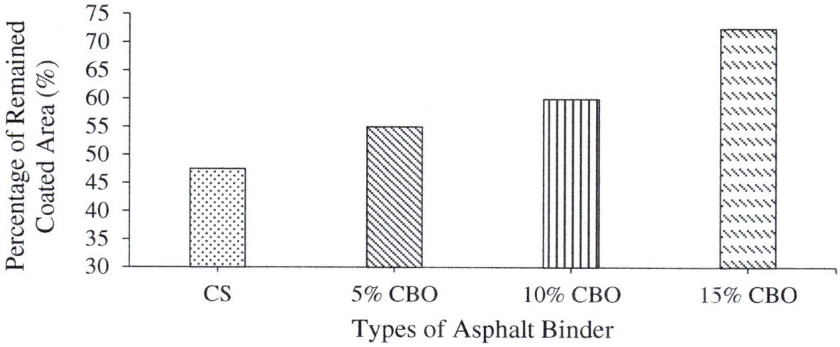

Fig. 9 Percentage of remained coated area of asphalt mixture after water boiling

Table 6 Static water immersion test results

Binders	Percentage of remained coated area (%)
CS	> 95
5% CBO	> 95
10% CBO	> 95
15% CBO	> 95

(a) Control Sample

(b) 5% CBO Sample

(c) 10% CBO Sample

(d) 15% CBO Sample

Fig. 10 Samples after water boiling test

Acknowledgements The author sincerely acknowledges Universiti Sains Malaysia for offering financial support via Research University Individual (RUI) Grant 1001.PAWAM.8014140 to conduct this research work. The authors also express their appreciation to all material suppliers and technicians of the Highway Engineering Laboratory at the School of Civil Engineering, Universiti Sains Malaysia, for their continuous assistance.

References

1. Gürer, C., Elmacı, A., Alagöz, O., Yılmaz, N.: Rheological behavior of bituminous binders replaced by poppy capsule pulp-based bio-oil. Constr. Build. Mater. 264 (2020)
2. Wen, H., Bhusal, S., Wen, B.: Laboratory Evaluation of waste cooking oil-based bioasphalt as an alternative binder for hot mix asphalt. J. Mater. Civ. Eng. **25**(10), 1432–1437 (2013)
3. Yang, X., Yu, Z., Dai, Q.: Performance evaluation of asphalt binder modified by bio-oil generated from waste wood resources. Int. J. Pavement Res. Technol. **6**(4), 431–439 (2013)
4. Sun, Z., Yi, J., Huang, Y., Feng, D., Guo, C.: Properties of asphalt binder modified by bio-oil derived from waste cooking oil. Constr. Build. Mater. **102**, 496–504 (2016)
5. Wang, H., Ma, Z., Chen, X., Hasan, M.R.H.: Preparation process of bio-oil and bio-asphalt, their performance, and the application of bio-asphalt: a comprehensive review. J. Traffic Transp. Eng. **7**(2), 137–151 (2020)
6. Abdul Latif, N.I.S., Ong, M.Y., Nomanbhay, S.: Hydrothermal liquefaction of Malaysia's algal biomass for high-quality bio-oil production. Eng. Life Sci. **19**(4), 246–269 (2019)
7. Yue, Y., Lin, Q., Irfan, M., Chen, Q., Zhao, X., Li, G.: Characteristics and potential values of bio-products derived from switchgrass grown in a saline soil using a fixed-bed slow pyrolysis system. BioResources **12**(3), 6529–6544 (2017)
8. Yang, W., Fu, P., Yi, W.M.: Catalytic fast pyrolysis of corn stover in a fluidized bed heated by hot flue gas: physicochemical properties of bio-oil and its application. Int. J. Agric. Biol. Eng. **10**(5), 226–233 (2017)
9. Fini, E.H., Kalberer, E.W., Shahbazi, A., Basti, M., You, Z., Ozer, H., Aurangzeb, Q.: Chemical characterization of biobinder from swine manure: sustainable modifier for asphalt binder. J. Mater. Civ. Eng. **23**(11), 1506–1513 (2011)
10. Mills-Beale, J., You, Z., Fini, E., Zada, B., Lee, C.H., Yap, Y.K.: Aging influence on rheology properties of petroleum-based asphalt modified with biobinder. J. Mater. Civ. Eng. **26**(2), 358–366 (2014)
11. Wang, H., Wang, L., Zhang, J., Jin, Y., Cao, Y.: Effects of pyrolysis temperature and reaction time on the performance of swine-manure-derived bio-binder. Transp. Res. Part D: Transp. Environ. **89**, 102608 (2020).
12. Yang, X., You, Z., Dai, Q., Mills-Beale, J.: Mechanical performance of asphalt mixtures modified by bio-oils derived from waste wood resources. Constr. Build. Mater. **51**, 424–431 (2014)
13. Ingrassia, L.P., Lu, X., Ferrotti, G., Canestrari, F.: Chemical, morphological and rheological characterization of bitumen partially replaced with wood bio-oil: towards more sustainable materials in road pavements. J. Traffic Transp. Eng. **7**(2), 192–204 (2020)
14. Zahoor, M., Nizamuddin, S., Madapusi, S., Giustozzi, F.: Sustainable asphalt rejuvenation using waste cooking oil: a comprehensive review. J. Clean. Prod. **278**, 123304 (2021)
15. Abdullah, N., Sulaiman, F., Gerhauser, H.: Characterisation of oil palm empty fruit bunches for fuel application. J. Phys. Sci. **22**(1), 1–24 (2011)
16. Abdullah, N., Safana, A.A., Sulaiman, F., Abdullahi, I.I.: A comparative analysis of physical and chemical properties of biochars and bio-oils obtained from pyrolytic process of mesocarp fibre and empty fruit bunch. Solid State Phenom. **268 SSP**(November), 387–392 (2017)

17. Chang, S.H.: Bio-oil derived from palm empty fruit bunches: fast pyrolysis, liquefaction and future prospects. Biomass Bioenerg. **119**, 263–276 (2018)
18. Pogaku, R., Hardinge, B.S., Vuthaluru, H., Amir, H.A.: Production of bio-oil from oil palm empty fruit bunch by catalytic fast pyrolysis: a review. Biofuels **7**(6), 647–660 (2016)
19. Rosli, N.S., Harum, S., Jahim, J.M., Othaman, R.: Chemical and physical characterization of oil palm empty. Malays. J. Anal. Sci. **21**(1), 188–196 (2017)
20. Yiin, C.L., Yusup, S., Mohamed, M., Udomsap, P., Yoosuk, B., Sukkasi, S.: Stabilization of empty fruit bunch derived bio-oil using solvents. J. Sustain. Dev. Energy Water Environ. Syst. **4**(1), 38–47 (2016)
21. ASTM D70-09: Standard test method for density of semi-solid bituminous materials (pycnometer) (2009)
22. ASTM D5/D5M-13: Standard test method for penetration of bituminous materials (2013)
23. ASTM D4402/D4402M-15: Standard test method for viscosity determination of asphalt at elevated temperatures using a rotational viscometer (2015)
24. ASTM D2872-19: Standard test method for effect of heat and air on a moving film of asphalt (rolling thin-film oven test) (2019)
25. AASHTO T182: Standard method of test for coating and stripping of bitumen-aggregate mixtures (1993)
26. ASTM D3625/D3625M-20: Standard practice for effect of water on asphalt-coated aggregate using boiling water (2020)
27. Read, J., Whiteoak, D.: The Shell Bitumen Handbook. United Kingdom (2003)
28. Hareru, W., Ghebrab, T.: Rheological properties and application of molasses modified bitumen in hot mix asphalt (HMA). Appl. Sci. **10**(6), 1931 (2020)
29. Al-Sabaeei, A., Napiah, M., Sutanto, M., Alaloul, W., Yusoff, N.I.M., Khan, M.I., Saeed, S.M.: Physicochemical, rheological and microstructural properties of Nano-Silica modified bio-asphalt. Constr. Build. Mater. **297**, 123772 (2021)
30. Hugener, M., Beltzung, F., Angst, C.: Affinity between bitumen and aggregates: improvement of the static test method. 13–15 (2012)

Coatability and Bonding Characteristics of Asphalt Binders Modified with Alkylamines-Based and Polyalkylene Glycol-Based Bonding Promoters

Hanizah Osman, Muhamad Zulfatah Izham Muhamad Rodzey, Tracy Leh Xin Wong, Mohd Rosli Mohd Hasan, and Abdulnaser M. Al-Sabaeei

Abstract Conventional asphalt binder is widely used to produce asphaltic concrete mixtures. The rheological weakness of conventional asphalt binder has developed an interest among researchers in modifying conventional asphalt binder by using various kinds of additives. Bonding promoters are used as additives in asphalt binder that have shown a significant effect in enhancing the physical and rheological properties as well as its performance. This study aims to enhance the coatability of modified asphalt binder and bonding characteristics by using bonding promoters, namely polyalkylene glycol-based (PLG) and alkylamines-based (ALM). The laboratory study was divided into three phases. The first stage involved the determination of the physical properties and performance of modified asphalt binders. The coatability index was evaluated in the second phase, whereas the third stage evaluated the asphalt-substrate bonding performance. The physical properties showed that incorporation of bonding promoter reduces the penetration values, which indicates the increment of the ability to stiffen the binder. However, the penetration grade of all modified binders is still within 60/70. Meanwhile, the incorporation of bonding enhancers increased the modified binders' rutting performance via DSR test by over 40% for 0.5% ALM. In terms of coatability performance, the static water immersion test shows insignificant changes resultant from the addition of PLG and ALM. Moreover, under the action of boiling

H. Osman · M. Z. I. Muhamad Rodzey · T. L. X. Wong · M. R. Mohd Hasan (✉)
School of Civil Engineering, Universiti Sains Malaysia (Engineering Campus), 14300 Nibong Tebal, Penang, Malaysia
e-mail: cerosli@usm.my

H. Osman
Jabatan Pendidikan Politeknik dan Kolej Komuniti (JPPKK), Aras 4-7, Galeria PJH, Jalan P4W, Persiaran Perdana, 62100 Putrajaya, Malaysia

A. M. Al-Sabaeei
Department of Civil and Environmental Engineering, Universiti Teknologi PETRONAS, 32610 Bandar Seri Iskandar, Perak, Malaysia

N. Sabtu (ed.), *Proceedings of AWAM International Conference on Civil Engineering 2022—Volume 2*, Lecture Notes in Civil Engineering 385,
https://doi.org/10.1007/978-981-99-6018-7_4

sodium carbonate, the coatability index of all modified asphalt binder reduces as it is used to accelerate the moisture damage effect. Lastly, both bonding promoters can improve tensile strength and resist shear stress according to asphalt-substrate bonding performance.

Keywords Resilient infrastructure · Pavement distress · Physical and rheological properties · Coatability · Bonding promoter

1 Introduction

Asphalt mixtures consist of aggregates, asphalt binders, and mineral fillers. The cohesiveness of bitumen is a critical component that affects the bond between aggregates and bitumen, which affects the strength of asphalt mixtures. The asphalt binder-aggregate bond strength and workability are enhanced when the asphalt binder's viscosity and surface tension are reduced [1]. The asphalt binder's adhesion and cohesion bond are strong indicators of an asphalt mixture's efficiency [1]. The interaction between the particles at the interface of two substances is called adhesion [2]. On the other hand, cohesion is the intermolecular force which keeps the molecules bonded under liquid or solid phases [3]. Moisture exposure on an asphalt mixture leads to stiffness and strength degradation known as "moisture damage". Under mechanical loading, moisture damage can cause adhesion failure between the aggregate and the asphalt, which is expressed by stripping. Moisture damage reduces pavement quality and could lead to pavement distress such as rutting, fatigue cracking, and raveling [4].

Generally, moisture severely affects the adhesion and cohesion of the binder-aggregates and asphalt, respectively. This results in the formation of the stripping and raveling phenomena between the binder and the aggregates. The mechanism of moisture damage involves two major steps. First, the moisture transport occurs when the vapor or liquid moisture penetrates through the asphalt binder and mixture until the binder-aggregate interface. Secondly, changes of the material's structure will deplete load-carrying capacity [5]. Under the effect of the infiltrated moisture at fractures and joints, asphalt can deteriorate. In such situations, the anti-stripping additives can elevate the bitumen-aggregates physiochemical bonds as they can lower the bitumen surface tension and consequently enhance adhesion and water resistance. The thermodynamic phenomena for the physical–chemical adhesion between aggregate and asphalt binder are governed by the material surface free energy. The adhesion of the asphalt binder to the aggregates is enhanced by using hydrated lime with the asphalt mix [6].

When the mastic stiffness decreases due to its migration through asphalt layers and due to water diffusion, bond cohesion is more vulnerable to failure [7]. The moisture presence deteriorates the bond molecule of the asphalt film and, consequently, leads to cohesion failure [8]. Additionally, the emulsification process can also be included as one of the cohesion failure mechanisms. Water at the asphalt-aggregate level can reduce the asphalt-aggregate surface adhesion. When emulsifier effects approach the aggregate surface, therefore a cohesive failure mechanism can contribute to an adhesive failure [9]. On the other hand, the adhesion failure separates the asphalt coating from the aggregate when water occurs at the asphalt-aggregate interface. Loss in adhesion will consequently reduce adhesive strength between asphalt and aggregate surfaces [9]. However, for years, moisture damage has been a point of conflict among researchers [5]. Conventional asphalt road pavement can be less sturdy and more prone to inclement weather, especially rain, where damage induced by moisture is one of the most common reasons for bonding failure [10].

The aging of the pavement structures and materials in the short and long term is another concern that may reduce the bonding strength as it reduces the adhesion of the asphalt binder specimens. The short-term aging process begins when the asphalt is exposed to heat and air during production and paving work [11]. With the advancement in the design life, aged roads deteriorate and reduce their performance characteristic [12]. Long-term aging caused asphalt binders to stiffen and embrittle, which leads to a high potential of cracking [13]. In terms of chemicals, asphalt binder oxidation is also highly related to the aging process. Exposure of asphalt binder in pavement mixture toward oxygen may include oxide, its chemical composition [14]. Additionally, the bonding characterization can be used to predict the negative consequences associated with the aging process of the asphalt binder and aggregates [15]. The problem of bonding reduction between asphalt binder and aggregates due to aging could be potentially diminished by incorporating bonding enhancers.

A bonding promoter that can be utilized in either solid or liquid form enhances the adhesive bond between the binder-aggregates and the cohesive bond of the asphalt binder itself. Adhesion promoters are substances, primarily fatty acids or polyamides, that can improve the adhesion within the asphalt binder and the aggregates, extending the asphalt mixture's lifespan [2] along with a reduction in the contact energy between the bituminous polar liquid and the solid mineral phases through intermolecular interactions between residual water and the polar aggregate surface [16]. Additionally, the cohesion and adhesion bond between the aggregate and asphalt binder is likely related to aggregate-binder bonding performance. Previous researchers have also reviewed the modified binders incorporated with bonding promoters such as Paved Bond Lite, Morlife 5000 [10], and amine-based adhesion promoters [2]. Anti-stripping agents, surfactants, and adhesion promoters, on the other hand, could be used to enhance the long-term performance and durability of asphalt pavements [15, 16].

The ability of the binder to withstand moisture damage is a critical parameter in determining bond strength. Asphalt bond strength (ABS) is a simple test performed directly on the surface of asphalt aggregate that can assist in determining the moisture sensitivity of asphalt mixtures [4, 8]. The bitumen bond strength (BBS) or ABS test is simple, quick, repeatable, reliable, and practical [8, 9, 17]. This study was conducted to determine the coatability and bonding characteristics of the modified asphalt binders in the presence of alkylamines-based (ALM) and polyalkylene glycol-based (PLG) bonding promoters as they have high cohesion and adhesion characteristics. An experimental study was carried out to quantify both bonding characteristics under different moisture conditions.

2 Materials and Methods

This part is divided into five sections. Materials and sample preparation are presented in the first and second sections, respectively. The physical properties and performance test are illustrated in the third section, while the coatability test is presented in the fourth section. The last section presents asphalt-substrate bonding performance tests.

2.1 *Materials*

The material parts are divided into three sections. The asphalt binder is presented in the first section, and the aggregates are presented in the second section. The last section shows the bonding promoters.

Asphalt Binder. A conventional asphalt binder with a 60/70 penetration grade was used for the asphalt modification. The asphalt binder was obtained from a local company named Kamunting Premix Plant Sdn. Bhd. The physical properties of the asphalt binder are summarized in Table 1. Theoretically, bitumen is in a semi-solid form at room temperature that must be heated before utilized in the mixing and testing process. A common binder comprises 90–98% by weight of hydrocarbons. The hydrocarbons present in the form of saturated branched or unbranched alkanes and cyclic alkene, and unsaturated aromatics [18], are also related to a colloidal system that is made up of asphaltenes micelles which is covered with stabilizing phase of polar resins [19].

Aggregates. The physical properties of the granite aggregates used are summarized in Table 2. The granite aggregates were taken from stockpiles and washed. Before preparing a loose asphalt mixture, coarse granite aggregates were sieved through a 9.5 mm sieve and retained at a 6.3 mm sieve prior to being used for preparing a loose asphalt mixture. The aggregates were heated for 24 h at 100 °C before being used for the sample preparation. Prior to the coatability tests, the aggregates were coated with binders and placed in an oven for 2 h before the test to ensure the aggregates were fully coated with the binders.

Table 1 Physical properties of the conventional asphalt binder

Test	Standard	Standard requirement	Result
Penetration	ASTM D5	60–70	66
Softening point	ASTM D36	42–52	46
Penetration index (PI)	–	– 2 to 2	– 1.745
Ductility	ASTM D113	> 100	> 100
Elastic recovery test	ASTM 6084	> 10%	10
Performance grade (PG)	ASTM D6373	–	64

Table 2 Granite aggregates properties

Properties		Results	Test methods
Aggregate crushing value (%)		23.26	BS 812–110
Flakiness index (%)		22.81	BS 812–105
Elongation index (%)		31.19	ASTM 4791–10
Los Angeles abrasion value (%)		23.07	AASHTO T96
Coarse aggregate specific gravity	Oven-dried (OD)	2.593	AASHTO T85
	Saturated surface dry (SSD)	2.61	
	Apparent	2.637	
	Water absorption (%)	0.635	
Fine aggregate specific gravity	Oven-dried (OD)	2.533	AASHTO T 84
	Saturated surface dry (SSD)	2.579	
	Apparent	2.655	
	Water absorption (%)	1.825	

Bonding Promoters. Two liquids bonding promoters, namely polyalkylene glycol-based (PLG) and alkylamines-based (ALM), as illustrated in Fig. 1, were used in this study. Both bonding promoters were manufactured by Ingevity located in South Carolina, the USA. Two percentages of these bonding promoters were used, which are 0.5–1.0% by weight of asphalt binder. The dosages were selected based on the recommendation from the manufacturer, which is between 0.25 and 1.0%. Despite its low concentration, based on a previous study, 0.5% was able to improve binder resistance to fatigue and low-temperature failure [20]. The basic properties of the bonding promoters are shown in Table 3. Practically, the binder can be modified by either adding the additive directly into the mixture or exactly to the base binder [21]. The addition of both ALM- and PLG-modified asphalt binders was prepared at a dosage of 0.5–1.0% using a high shear mixer at 1000 rpm for 30 min at 160 °C [20].

2.2 *Sample Preparation*

The sample preparation was divided into two parts. In the first part, preparation of pull-out test specimens was illustrated. The preparation for hear compression test is shown in the second part.

Preparation of Pull-Out Test Specimens. The cube blocks were initially prepared before preparing specimens for the pull-out test. A plastic mold with a 50 mm × 50 mm × 50 mm was used to prepare the concrete cube. The concrete blocks were then cut into half by using an automatic saw cutter. Those concrete blocks were inserted with two Perspex plates and wrapped around using aluminum foil. Those blocks were tied with rubber bands to avoid any movement. The concrete blocks

(a) (b)

Fig. 1 Appearances of bonding promoters: **a** polyalkylene glycol-based bonding promoter (PLG), **b** alkylamines-based bonding promoter (ALM)

Table 3 Basic properties of bonding promoters

Properties	Polyalkylene glycol-based	Alkylamines-based
Ingredients	Polyalkylene glycol mixture and glycol ethers	Polyethylene polyamines
Physical form	Dark brown liquid	Dark black liquid
Odor	Ammoniacal	Fishy
Specific gravity	1.04	1.09
Boiling point	> 220 °C	255 °C
Flashpoint	155 °C	145 °C

were then heated for 30 min at 80 °C before the binder sample was poured into the space between the concrete blocks. Figure 2 shows prepared specimens for the pull-out test.

Preparation of Shear Compression Test Specimens. The concrete blocks were cast using the same method in the pull-out test. Concrete cubes with 50 mm × 50 mm × 50 mm dimensions were cut diagonally at 60° using an automatic saw cutter. The specimens were tied using a rubber band with two Perspex plates inserted between the blocks. Binder samples were then poured in between the specimens. The specimens containing sample materials were left to cool at room temperature before conducting the test. Figure 3 shows prepared concrete cubes for the shear compression test.

Fig. 2 Concrete cubes prepared for pull-out test

Fig. 3 Concrete cubes prepared for shear compression test

2.3 *Physical Properties and Performance Tests*

The physical properties of the modified asphalt binders were determined by a series of tests, namely penetration test, softening point test, ductility test, and elastic recovery test. A penetration test was conducted to measure the penetration grade of the asphalt binder in assessing the asphalt binder's consistency. Theoretically, a higher penetration grade indicates a softer asphalt binder. This test have been carried as per the ASTM D5 [22]. Practically, the sample was melted and cooled under controlled conditions. Besides, the heating process should not be more than 90 °C. The test was conducted at 25 °C where the load of 100 g was applied for five seconds.

The softening point test is conducted to identify asphalt binder stiffness with the use of ring-and-ball apparatus based on ASTM D36 [23]. The softening points are the temperatures at which the asphalt binder softens to allow an enveloped ball in the asphalt binder to sag down about 25 mm below the weight of the steel ball. Two horizontal asphalt binder disks were cast in a brass ring and warmed with a monitored rate of 5 °C/min in a liquid bath. As the temperature rises, the asphalt binder viscoelastic properties became softer and less viscous.

The bitumen was classified based on the penetration test results. The temperature sensitivity in terms of the penetration index (PI) was determined for the control and modified asphalt binders. A low PI reflects a higher temperature sensitivity which means that temperature changes, the consistency changes rapidly.

The penetration index assesses the asphalt binder's temperature susceptibility. Formulations were derived from the standard values of the softening point and penetration using Eq. (1) [17, 18].

$$PI = (1952-500\log(Pen_{25})-20 \times SP)/(50 \times \log(Pen_{25})-SP-120) \quad (1)$$

where

Pen_{25} penetration value at 25 °C.
SP softening point

Ductility test has been carried out according to ASTM D113 [24] in order to determine the bitumen cohesiveness and elasticity. Data were typically expressed as the distance in cm. The asphalt binder was heated until it had sufficient fluidity to pour into the mold. It was allowed to cool at room temperature before placing it in the ductile machine. The sample was then pulled horizontally at a uniform speed of 5 cm/minutes until it broke and temperature (25 ± 0.5 °C) while continuously immersed in water. The ductility value indicates that the distance of the sample has been pulled until it breaks.

The capability of the asphalt binder to recover its original length after elongating to a specific distance at a specified rate known as elastic recover test was carried out as per ASTM D6084 [25]. The prepared specimen was cooled down at room temperature for 30 min. After that, it is immersed in a water bath at 25 ± 0.5 °C for 85 min. The sample is elongated to 20 cm at a 5 cm/min rate and held in stretched positions for 5 min before cutting in half. When the one-half sample specimen is withdrawn until the two shattered ends contact, the specimen is kept undisturbed for an hour before obtaining the results. Subsequent estimation of elastic recovery percentage is determined from Eq. (2).

$$\%\text{Elastic recovery} = ((E-x)/x) \times 100 \quad (2)$$

where

E Original length of the specimen (cm)
x Elongation of the specimen with severed ends just touching (cm)

A HAAKE RheoStress 6000 dynamic shear rheometer (DSR) in accordance with ASTM D7175 [26] was used to measure the performance of asphalt binders. The binder performances were observed using the temperature sweep mode, changes in complex modulus (G^*), phase angle (δ), rutting resistance factor ($G^*/\sin\delta$), and performance grade temperature. All samples were tested at elevated temperatures ranging from 46 °C to 82 °C in 6 °C increments. During the test, all samples were tested at the standard frequency of 10 rad/s (1.59 Hz). The susceptibility of the modified binder to rutting and fatigue cracking was evaluated.

2.4 *Coatability Test*

Two coatability tests are presented in two sections. The static water immersion test is presented in the next section, and the boiling water test is presented in the second section.

Static Water Immersion Test. The static water immersion test evaluated the percentage of asphalt binders that remain coated after being immersed in water. This test was conducted following BS EN 12697-07 [27]. Practically, 5.5 g of asphalt binder was used to coat 100 g of aggregate, ranging from 6.3 to 9.5 mm. Then, the loose mixture was immersed in 400 ml distilled water for 16 h and kept at a constant temperature of 25 °C. The loose mixture was then dried, and the percentage of the total visible area of aggregate that remains coated with bitumen was observed visually. A percentage of coated asphalt binders higher than 95% is considered a good coatability performance.

Boiling Water Test. The boiling water test was conducted to determine the loss of adhesion of loose asphalt mixture caused by boiling water. According to ASTM D3625, this test was conducted using 250 g of loose asphalt mixture prepared at the mixing temperature of 135 °C [28]. The temperature of the mixture must not be less than 80 °C after cooling at room temperature. Then, the mixture was placed in boiling water for 10 min. The mixture was then dried and allowed to cool at room temperature. The average percentage of the mix stripped of asphalt was visually observed and evaluated. A coated percentage higher than 95% is considered good coatability and is less likely to strip.

2.5 *Asphalt-Substrate Bonding Performance Tests*

Pull-Out Test. A pull-out test was conducted according to ASTM D4541 [29] to determine the tensile strength of the asphalt binder and the asphalt-substrate performance through a pulling force. Repeated extensive loads were applied to the sample materials to extend the original length of the sample. The test was conducted in three conditions: dried condition, immersed in distilled water, and immersed in sodium carbonate solution. Figure 4a shows a steel clamp specially designed and fabricated

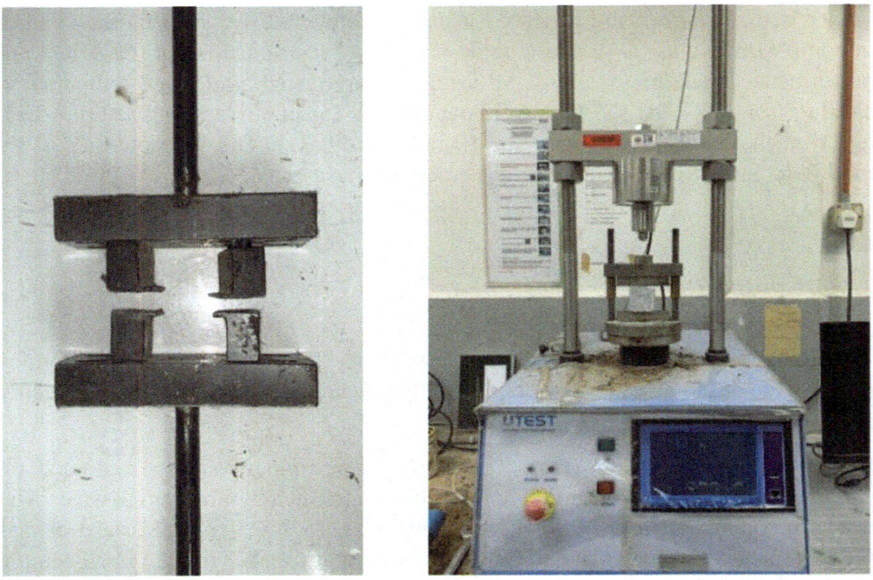

(a) (b)

Fig. 4 Apparatus for bonding performance tests **a** pull-out test jig, **b** Marshall machine used and indirect tensile strength jig for the shear compression test

for this test. The steel clamp was used to hold the specimen in position during the testing process.

Shear Compression Test. The shear compression test was conducted to evaluate the shearing phenomenon of asphalt binder under severe moisture conditions through asphalt-substrate performance. The specimens were immersed in sodium carbonate solution for 24 h before being tested before subject to moisture damage. The vertical force was applied to the specimens containing sample materials using the Marshall stability test, as shown in Fig. 4b until the specimens failed.

3 Results and Discussion

The results and discussion is divided into three parts. The first part discusses physical properties and performance of asphalt binder. The second part shows the coatability test results. The asphalt-substrate bonding performance test is presented in the last part.

3.1 Physical Properties and Performances of Asphalt Binder

The part is divided into four sections. The first section presents the physical properties of asphalt binder. The ductility test and elastic recovery tests are illustrated in the second section. The third section presents the viscoelastic properties of asphalt binder. The last section presents rutting resistance and failure temperature.

Physical Properties of Asphalt Binder. Figure 5 shows the penetration test results conducted for all types of binders. From the results obtained, the utilization of PLG and ALM bonding promoters had no significant effect in terms of penetration value. All the modified binders achieved the penetration values within the same range as the conventional binder grade. That indicates that the use of bonding promoters ALM and PLG does not affect the physical properties and consistency of the asphalt binder in terms of its penetration grade. Rossi et al. [16] specified that the primary function of bonding promoters is to modify the interfacial energy in such a way that the presence of water, even in small amounts, does not weaken the bitumen-aggregate bonding, which is not directly associated with the penetration value of the asphalt binder. However, at the current stage, the influence of the bonding promoter on the chemical properties has not been finalized yet in this study. Based on the test results, the lowest penetration value among all tested binders was shown at 0.5% ALM, with a reduction of 9% compared to the control binder. That reveals that 0.5% ALM-modified asphalt binder exhibited the highest hardness and best resistance at the intermediate temperatures among other tested binders. Overall, it can be stated that the penetration grade of all binders remains within 60/70 penetration grade.

Figure 6 shows softening point test results for all types of binders. All binders show softening points with minor differences from the plotted graph. About 0.5% of ALM-modified binder has the highest softening point, which is 48 °C, while 1.0% ALM has the lowest softening point, which is 45 °C. On the other hand, the control

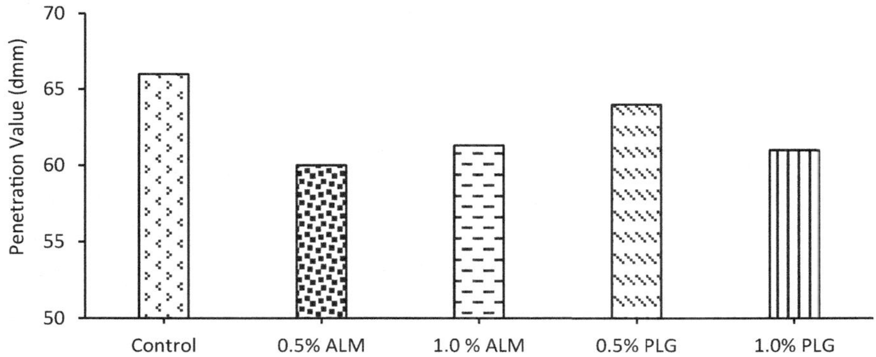

Fig. 5 Penetration test results

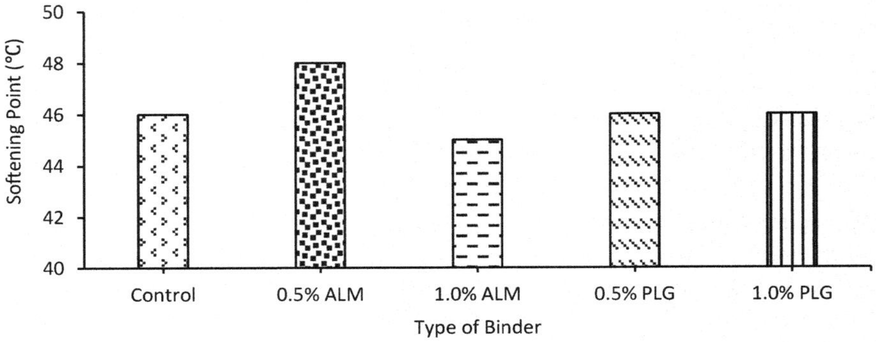

Fig. 6 Softening test results

binder, 0.5% PLG, and 1.0% PLG-modified binder have the same softening point value of 46 °C. The utilization of 0.5% PLG and 1.0% PLG shows no changes in their softening value which could be attributed to the asphaltene ratio of the binder and because the stiffness of the modified binders is equivalent to the control binder. Thus, the utilization of PLG as a bonding promoter did not affect the binder's physical properties in terms of softening point and binder stiffness at high temperatures. The 0.5% ALM-modified binder shows better properties in terms of stiffness, and thus, it is expected to sustain high temperatures, especially during hot weather.

The penetration index (PI) is determined based on the penetration and softening point test. Table 4 shows the penetration index of all modified binders. The result shows that the utilization of 0.5% ALM, 0.5% PLG, and 1.0% PLG meets the conventional paving asphalt binder requirement as shown in Table 5. That indicates that a lower temperature sensitivity is expected for these binders particularly at high-temperature applications. On the other hand, the utilization of 1.0% ALM is classified as temperature susceptible asphalt binder as referred to the typical asphalt binder penetration index by Hunter et al. [30]. Therefore, the asphalt binder with a high penetration value is suitable for cold climates as it has lower strength in sustaining high temperatures in hot countries such as Malaysia.

Table 4 Penetration index of modified binder [31]

Type of binder	Penetration (dmm)	Penetration index	Best possible use
Control	66.0	− 1.745	Conventional paving asphalt binder
0.5% ALM	60.0	− 1.437	Conventional paving asphalt binder
1.0% ALM	61.3	− 2.039	Temperature susceptible asphalt binder (tars)
0.5% PLG	64.0	− 1.703	Conventional paving asphalt binder
1.0% PLG	61.0	− 1.959	Conventional paving asphalt binder

Table 5 Typical asphalt binder penetration index [31]

Asphalt binder type	Penetration index
Blown asphalt binder	> 2
Conventional paving asphalt binder	-2 to $+2$
Temperature susceptible asphalt binder (tars)	< -2

This indicates that incorporating ALM and PLG bonding enhancers is appropriate for higher temperature sensitivity in hot climate countries. In cold climates, asphalt cement with a PI less than -2.0 is brittle at low temperatures and susceptible to transverse cracking. For PI greater than 1.0, permanent deformations via rutting have been established at elevated temperatures [32]. Because of its rapid brittleness and susceptibility to transverse cracking, the 1.0% ALM-based bonding enhancer is unsuitable for low-temperature conditions.

Ductility Test and Elastic Recovery Test. Figures 7a, b show the results of the ductility test and elastic recovery test, respectively. Based on the result, both ALM- and PLG-modified binders show higher ductility than the control binder. The utilization of the modified binders resulted in a ductility value of 150 cm, indicating a high resistance to breaking when stretched reaching the maximum limit. Besides, all the modified binders show a higher percentage of elastic recovery than the control binder. It indicates that the modified asphalt binders have higher tensile properties and retain their original shape. In conclusion, it was found that the addition of ALM and PLG bonding promoters became the ductility and elasticity of the asphalt binders increased, which is vital to achieving to improve the resistance of the asphalt binder toward fatigue cracking. The asphalt modification will eventually be extending the service life and durability of the pavement.

Viscoelastic Properties of Asphalt Binders. Based on Superpave specifications, the complex modulus (G^*) and phase angle (δ) values derived from DSR results are used as signifiers of permanent deformation. Asphalt binder is a viscoelastic material with a defined elastic to viscous response ratio [33]. Meanwhile, in asphalt mixture, the deformation is classified into two categories: elastic deformation and viscous deformation [34].

The binder G^* response to temperatures hints at its elastic storage and viscoelastic dissipation properties. According to Fig. 8a, all modified asphalt binders show G^* values higher than that for base asphalt along the test temperatures range of 46–82 °C. The ALM- and PLG-modified binders exhibit a higher G* value, providing them with sufficient rigidity to resist rutting. This increased stiffness is due to the binders' elastic behavior. Conversely, a lower G* value indicates that the asphalt is more prone to permanent deformation, which can be attributed to the binders' viscous characteristics. Furthermore, as shown in Fig. 8b, the δ observation for the control asphalt binder generally showed an increase continuously with increasing temperature, indicating dominant viscous behavior. From 58 °C to 82 °C, the modified binder with 1.0% ALM had the lowest δ, indicating an improvement in permanent deformation resistance compared to the base and other modified binders, while the binder with 1.0% PLG

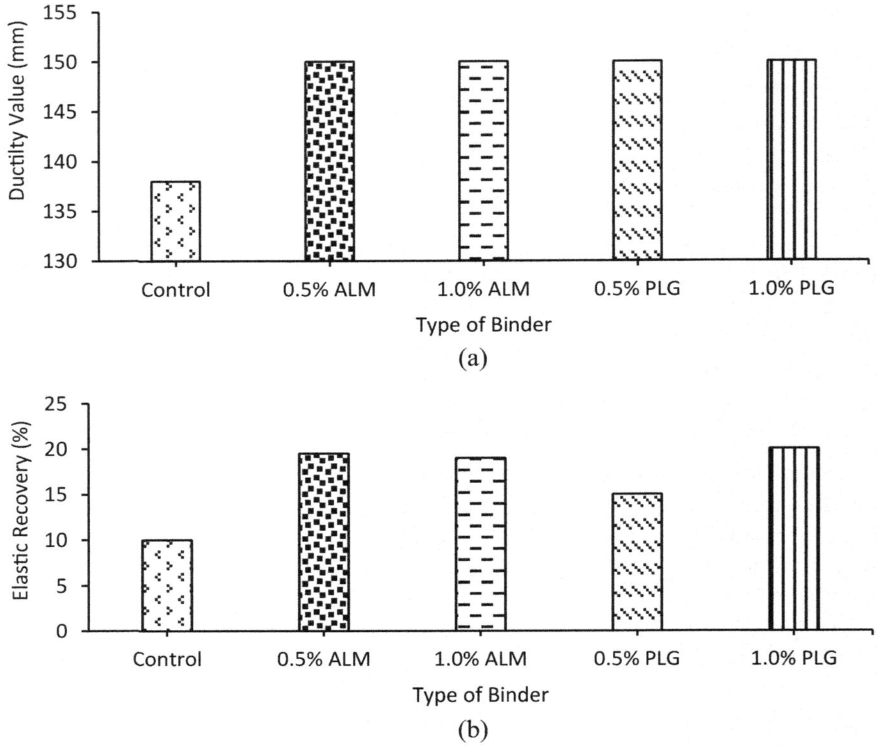

Fig. 7 Results of the **a** ductility test and **b** elastic recovery test

had the highest δ, indicating the lowest deformation resistance. At high temperatures, the phase angle of the polyalkylene glycol-based and alkylamines-based was significantly increased. Similarly, referring to the asphalt binder low strain–stress behavior, the purely plastic and elastic δ was 90° and 0°, respectively [35]. This occurrence is attributed to the interconnected polymer networks of the asphalt binder due to the increase in its viscous component which could be associated with the increased plasticity.

Rutting Resistance and Failure Temperature. Binder's exposure to high temperature and binder's low hardening under oxidization are the most common indicators of rutting potential. The combination of these two issues could be mitigated. However, the binder modifications by adding polymer and anti-strip additives (ASA) can increase its stiffness and reduce it vulnerability to high temperature and consequently, increasing its viscosity and elasticity [20]. Figure 9a shows that the incorporation of 0.5% ALM, 0.5% PLG, and 1.0% ALM to the binder enhanced its rutting resistance

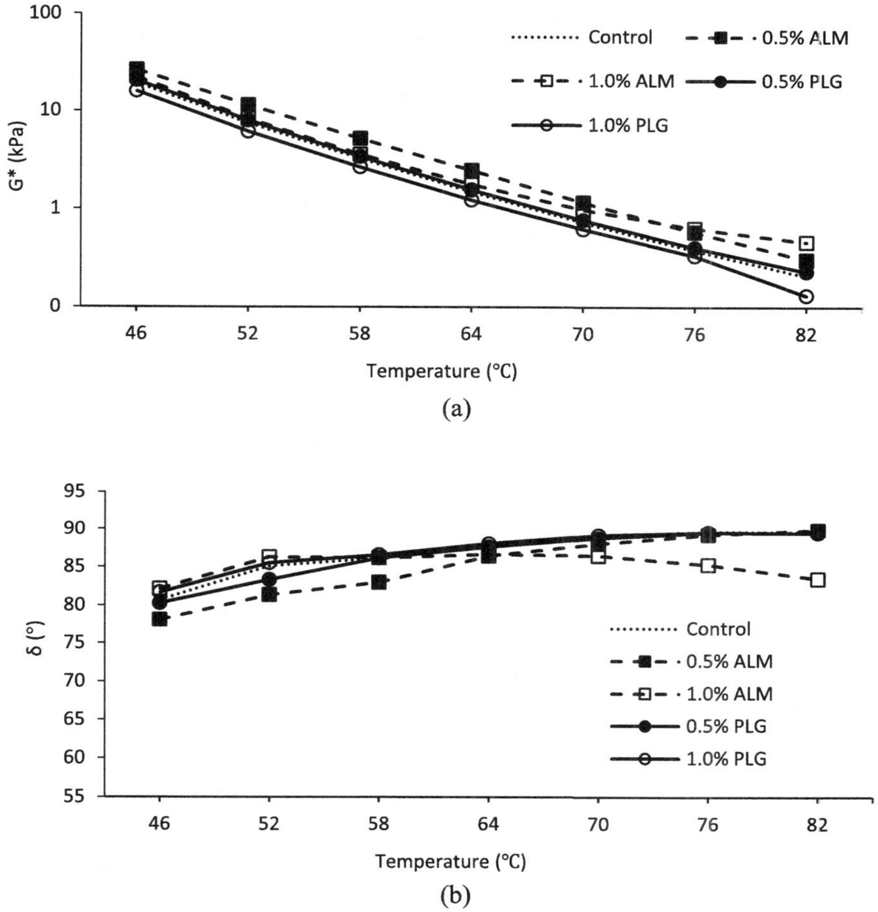

Fig. 8 Effect of ALM-based and PLG-based bonding enhancers on **a** G^* and **b** δ

$G^*/\sin\delta$ except for the case of using 1.0% PLG that showed a smaller rutting resistance. The findings suggest that polyalkylene glycol-based and alkylamines-based enhancers increased the rutting resistance.

The lone nucleophilic nitrogen in alkylamines increases the stiffness and bonding strength of the bitumen. This attributed to its reactivity where it can share its electron with the bitumen's negatively charged carboxylic groups (COO). It was found that the PLG's high viscous index increased the stiffness within bituminous phases. As a result, asphalt binders have become more rigid and have improved $G^*/\sin\delta$. Adding amine-based anti-stripping agent to Advera-modified binder can increase its fatigue and low-temperature resistance. For a higher portion of added ASA, binder's rutting factor is reduced [20]. This suggests that adding ASA to bituminous phases improves fatigue and rutting resistance. However, this enhancement depends on the amount of

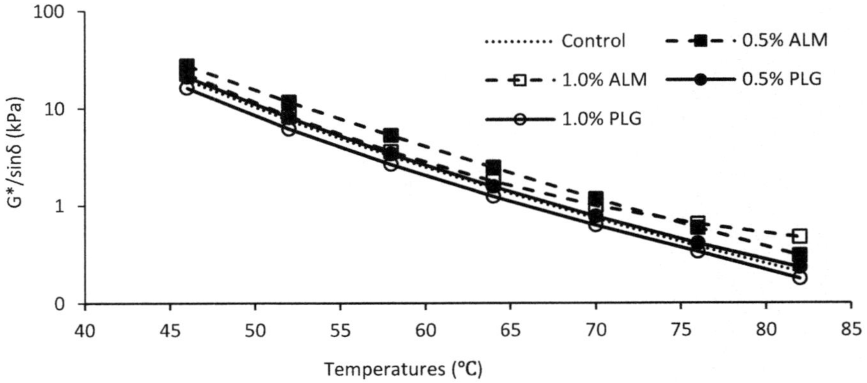

(a)

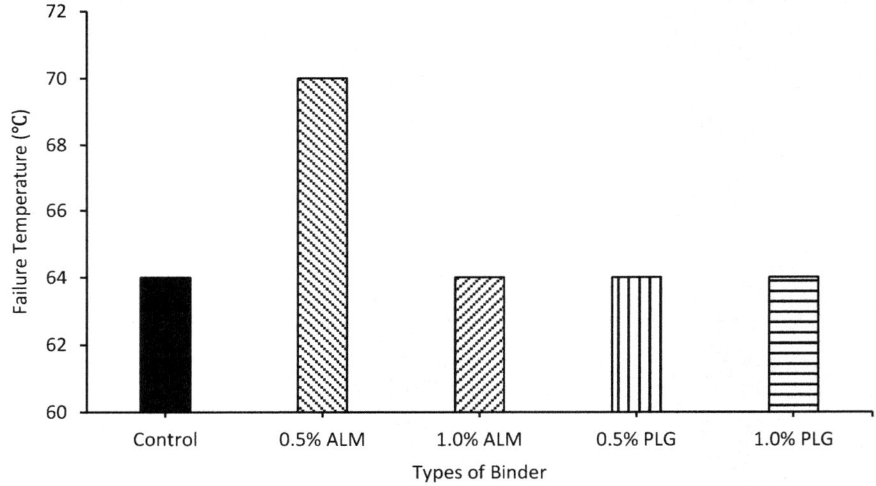

(b)

Fig. 9 **a** Rutting resistance and **b** failure temperature of ALM-based and PLG-based bonding enhancers

the added anti-strip additives and its concentration which agrees with the findings of the current research.

According to the Superpave specifications, the asphalt binder failing temperature is known as the limit where $G^*/\sin\delta$ is below 1.0 kPa. A higher failing asphalt binder temperature means a lesser deformation susceptibility under elevated pavement temperatures and vice versa. According to Fig. 9b, adding polyalkylene glycol-based and alkylamines-based to the asphalt binder did not affect the 64 °C failing temperature for 0.5% and 1.0% polyalkylene glycol-based. The same observation was noticed for 1.0% alkylamines-based additive. At 70 °C, the 0.5% alkylamines-based additive exhibited temperature failure. This indicates that the 0.5% ALM-modified binder shows an improvement of one high-temperature performance grade compared to other tested binders in this study. The findings show that binder rate failure was not affected by using bonding enhancer except in the case of 0.5% alkylamines-based additive. This is expected as the good improvement in terms of penetration, and softening was observed for the 0.5% alkylamines-based additive only. Similar results were obtained for the moisture and rheological susceptibility for the water-bearing warm asphalt mixtures containing liquid anti-stripping additives. Even though anti-stripping additives inclusion in warm-mix asphalts at high temperature can enhance the rutting resistance, there were no changes on the binder performance grade [20].

Further to that, amine-based liquid ASA at different ratios produced more noticeable changes in the performance grade for PG 70-28 and PG 76-28 polymer-modified binders compared to PG 64-22 [20]. The reported results show that using 0.5% of polyalkylene glycol-based and alkylamines-based enhanced rutting resistance. In this regard, the $G^*/\sin\delta$ of 0.5% polyalkylene glycol-based and alkylamines-based demonstrated adequate rutting resistance, outperforming the control binder and both 1.0% polyalkylene glycol-based and alkylamines-based.

3.2 *Coatability Tests*

Two coatability tests are presented. The static water immersion test is presented in the first subsection. The boiling water test is presented in the second subsection.

Static water immersion test. Significantly in this research, distilled water and sodium carbonate solutions were used to immerse binders. The utilization of sodium carbonate solvent accelerates the moisture damage condition to access the binder's coatability under severe moisture damage conditions. Table 6 shows the results of all the binder samples under both distilled water and sodium carbonate solutions. The results show that all binders immersed in distilled water have a good coatability index of more than 95%. The results of the previous researcher's study stated good affinity of all modified mixtures of PBL and M5000 with asphalt binder, in which 100% of the remaining aggregate is coated. Nevertheless, it is necessary to make a comparison of the influence of different adhesion promoters as well as at different doses, where no dehydration of the asphalt binder from the aggregates of the modified asphalt mixture appears to be significant [10]. The modified binders can have the same bonding strength as the control sample. They have an identical percentage of the remained coated area after the water immersion. On the other hand, when immersed in sodium carbonate solvent, the unmodified binder reduces bonding properties as the percentage of remained coated area is lower than 95%; meanwhile, the modified binders had coated area higher than 95%. This indicates that both ALM and PLG bonding promoters increase the bonding performance of the modified asphalt binder in terms of the coatability index.

Table 6 Results of static water immersion test

Type of binders	Conditioning medium or solutions	
	Distilled water	Sodium carbonate (%)
	Percentage of remained coated area (%)	
Control	> 95	< 95
0.5% ALM	> 95	> 95
1.0% ALM	> 95	> 95
0.5% PLG	> 95	> 95
1.0% PLG	> 95	> 95

Boiling water test. The boiling water test was conducted by using distilled water and sodium carbonate. Table 7 summarizes the results of the boiling water test for all types of binders, including the control binder. All the modified asphalt binders show uniform and excellent coatability performance under the action of boiling distilled water as more than 95% of the aggregates were remained coated. However, the control sample possesses a lower coatability index which is lower than 95%. These results signify that the utilization of both ALM and PLG bonding promoters enhanced the bonding performance in terms of adhesion bond between aggregates and binders. On the other hand, the modified asphalt binders showed a lower coatability index when subjected to the immersion of boiling sodium carbonate solvent as the most severe moisture damage condition. Such reduction in the coatability performance could be attributed to the chemical interactions among the sodium carbonate solvent, asphalt, and aggregate that result in losing the adhesion bonding. The 0.5% ALM achieved the highest percentage of the remained coated area of 40%; meanwhile, the control sample possesses the lowest coatability index of 10%. Hence, both ALM and PLG bonding promoters enhance the bonding performance compared to the control sample. The findings appear to be consistent with previous research by Sheng et al. [10], as the use of adhesion promoters resulted in all the modified asphalt mixtures, regardless of adhesion promoters, retaining more than 95% of the initial coating. The results show that the control sample retained 95% of the initial coating. Adhesion promoters improved the asphalt mixture's resistance to moisture damage. Therefore, even after being subjected to the action of boiling water, all the modified asphalt mixtures demonstrated superior adhesion.

3.3 Asphalt-Substrate Bonding Performance Test

The asphalt-substrate bonding performance test is investigated through two tests. The first test is the pull-out test. The second test is the shear compression test.

Pull-Out Test. Based on the obtained maximum tensile force, the value of pull-out strength was calculated and shown in Fig. 10. From the results obtained, the utilization of ALM and PLG bonding promoters showed an increase in pull-out

Table 7 Results of boiling water test

Type of binders	Percentage of remained coated area (%)/solutions	
	Distilled water (%)	Sodium carbonate (%)
Control	50	10
0.5% ALM	96	40
1.0% ALM	96	30
0.5% PLG	96	15
1.0% PLG	96	15

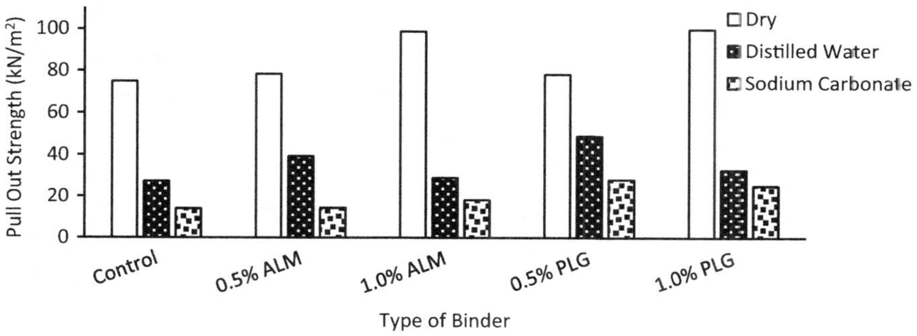

Fig. 10 Results of pull-out strength

strength compared to the control sample over all the testing conditions. Furthermore, both the 1.0% ALM and 1.0% PLG-modified binder yielded the highest pull-out strength among the asphalt binders. It can be concluded that the pull-out strength will increase with an increasing percentage of incorporated bonding promoters due to the higher ductility value and the elastic recovery percentage of the modified asphalt binders. The modified asphalt binders show no adhesive or cohesive failure upon the maximum elongation in bonding performance. Thus, the usage of bonding promoters effectively improves the bonding performance in terms of their cohesion and adhesion bond.

Furthermore, the rate of strength change was calculated before identifying the pull-out strength changes in verifying the significance of the results obtained. The effect of bonding promoter on the increment rate of pull-out strength was determined and summarized in Table 8. The ALM- and PLG-modified binder resulted in a higher increment rate of strength change than the conventional asphalt binder. Thus, it is concluded that the modified binder improved the bonding performance without changing the physical properties of the asphalt binder itself. While principally cohesive failures are witnessed, a small reduction in pull-off strength is evaluated compared to dry conditions, whereas modified failures result in larger reductions in strength. This could imply that when a modified failure occurs, it begins with a loss of adhesion at a considerably weaker point of the asphalt-aggregate interface. As a

Table 8 Effect of bonding promoter on the increment rate of pull-out strength

Condition	Type of binder/rate of strength change (%)			
	0.5% ALM	1.0% ALM	0.5% PLG	1.0% PLG
Dry	4.73	31.68	4.50	33.37
Distilled water	43.40	5.94	78.62	19.97
Sodium carbonate	1.84	30.13	101.77	80.57

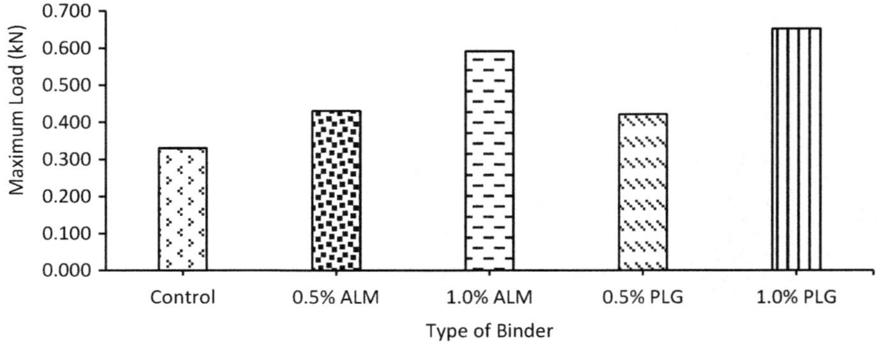

Fig. 11 Results of maximum shear load

result of the reduced bonding area, the asphalt film is subjected to increased tensile stress, resulting in cohesive failure [9].

Shear Compression Test. Figure 11 shows the results of the maximum shear load obtained for all types of binders. From the results obtained, both ALM and PLG bonding promoters showed significant results compared with the control sample. The usage of 1.0% ALM and 1.0% PLG shows a higher maximum load upon failure. This indicates that using a higher percentage of bonding promoter can subsequently increase the shearing resistance of the asphalt binders when subjected to severe moisture damage conditions. The displacement that occurred after undergoing compression is illustrated in Fig. 12. Aside from that, no adhesive or cohesion failure occurred for any modified asphalt binders, including the control binder; thus, the bonding performance of the asphalt binders remains in place after the shearing process occurs.

The shear strength was calculated for all types of binders from the maximum load values. Figure 13 shows the shear strength increased with the increasing content of each bonding promoter. The control sample had the lowest shear strength of 132 kN/m^2 among the asphalt binders. The utilization of ALM and PLG bonding promoters increased the bonding performance, hence resisting the shearing phenomenon under severe moisture damage conditions. Modified asphalt with 1.0% PLG exhibited the highest shear strength among all tested binders with an improvement of 97% compared to control binder. Both bonding promoters can withstand any shearing issue in the asphalt pavement by enhancing cohesion and adhesion bonding characteristics due to their sufficient ductility index and high elastic recovery.

(a) Sample Condition Before Testing

(b) Sample after Shear Compression Test (Displacement occurred)

Fig. 12 Displacement of sample after undergoing compression

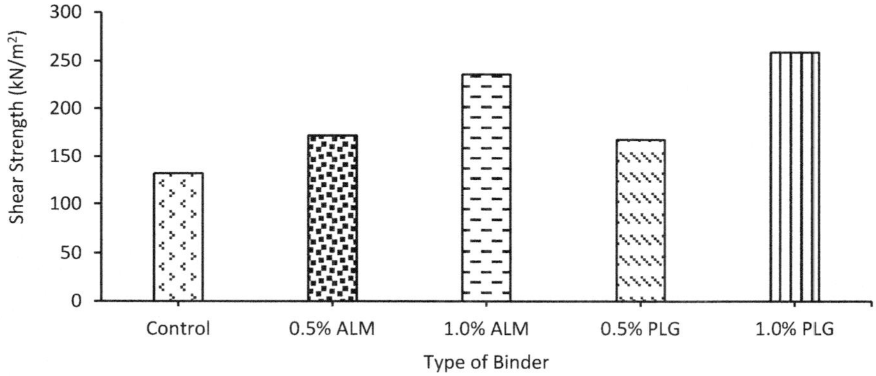

Fig. 13 Results of shear strength

4 Conclusions

The incorporation of ALM and PLG as modifiers was evaluated in this study to determine the performance of the modified asphalt binder. The performance of the modified asphalt binders was evaluated in terms of their coatability and bonding characteristics. On the other hand, the physical properties and viscoelastic performances of modified asphalt binders were also assessed and compared with conventional asphalt binder. Based on the conducted laboratory work and analysis, several conclusions can be made and summarized as underneath:

1. The incorporation of bonding promoters does not affect the physical properties as the penetration grade has remained 60/70 PEN. Besides, the softening points of all modified asphalt binders are also not affected by the addition of both bonding promoters. On the other hand, ALM and PLG bonding promoters enhance the ductility and elastic recovery of asphalt binders.
2. Nonetheless, when compared to the control sample, both ALM and PLG additions resulted in comparable production temperatures. The rutting resistance of both ALM and PLG increased, with both ALM and PLG sustaining higher $G^*/\sin\delta$ than the control binder. Only ALM at 0.5%, on the other hand, increased the binder failure temperature of all tested samples.
3. The utilization of both ALM and PLG bonding promoters enhanced the coatability and bonding characteristics of asphalt binders under severe moisture damage conditions.
4. The incorporation of ALM and PLG bonding promoters strengthened the pull-out strength under extreme moisture conditions. In addition, a higher percentage of ALM and PLG incorporation significantly further enhanced the pull-out strength. On the other hand, ALM and PLG bonding promoters are proven to resist shearing force due to higher loading under severe moisture damage conditions.
5. Based on the physical properties of modified asphalt binders, 0.5% ALM is found to achieve better performance. Moreover, the modified asphalt binders possessed higher tensile and shear strength and an increasing percentage of adopted modifiers.

In summary, the binder-aggregate composition is a key parameter in the bonding potential against moisture damage in asphalt mixtures. Nevertheless, the effects of asphalt binder aging and stiffness, aggregate type, and its correlation on bond strength and moisture susceptibility of asphalt binder-aggregate mechanism are also very important to be evaluated and studied. The significant and satisfactory distinctions between failure patterns and bond energies under dry and wet conditions show that changes in the adhesive and cohesive failures in asphalt binder-aggregate compositions are highly dependent on the SFE properties. Therefore, surface free energy (SFE) and failure patterns in the BBS test can be applied in predicting changes in the bond strength of asphalt binder-aggregate systems [17].

Acknowledgements The authors sincerely acknowledge the Ministry of Higher Education Malaysia for the Fundamental Research Grant Scheme with Project Code: FRGS/1/2021/TK01/USM/02/1 that enabled us to conduct this research work. The scholarship provided by the Ministry of Higher Education Malaysia (MOHE) through Hadiah Latihan Persekutuan (HLP) for the professional and managerial officer to pursue a doctoral degree (Ph.D.) is also highly appreciated. Special thanks go to material suppliers for extending their support. The authors also express their appreciation to the technicians of the Highway Engineering Laboratory and School of Materials Engineering at Universiti Sains Malaysia for their continuous assistance.

References

1. Kanitpong, K., Bahia, H.: Relating adesion and cohesion of asphalts to the effect of moisture on laboratory performance of asphalt mixtures. Transp. Res. Rec. **901**, 33–43 (2005)
2. Cihlářová, D., Fencl, I., Cápayová, S., Pospíšil, P.: Use of Adhesion Promoters in Asphalt Mixtures. Slovak J. Civ. Eng. **26**(1), 19–24 (2018)
3. Tan, Y., Guo, M.: Using surface free energy method to study the cohesion and adhesion of asphalt mastic. Constr. Build. Mater. **47**, 254–260 (2013)
4. Moraes, R., Velasquez, R., Bahia, H.: Using bond strength and surface energy to estimate moisture resistance of asphalt-aggregate systems. Constr. Build. Mater. **130**, 156–170 (2017)
5. Kakar, M.R., Hamzah, M.O., Valentin, J.: A review on moisture damages of hot and warm mix asphalt and related investigations. J. Clean. Prod. **99**, 39–58 (2015)
6. Mohd Hasan, M.R., You, Z., Porter, D., Goh, S.W.: Laboratory moisture susceptibility evaluation of WMA under possible field conditions. Constr. Build. Mate. **101**, 57–64 (2015)
7. Chaturabong, P., Bahia, H.U.: Effect of moisture on the cohesion of asphalt mastics and bonding with surface of aggregates. Road Mater. Pavement Des. **19**(3), 741–753 (2018)
8. Moraes, R., Velasquez, R., Bahia, H.U.: Measuring the effect of moisture on asphalt–aggregate bond with the bitumen bond strength test. *Transp. Res. Rec. J. Transp. Res. Board* 2209(1), 70–81 (2011).
9. Canestrari, F., Cardone, F., Graziani, A., Santagata, F.A., Bahia, H.U.: Adhesive and cohesive properties of asphalt-aggregate systems subjected to moisture damage. Road Mater. Pavement Des. **11**, 11–32 (2010)
10. Sheng, N.J., Hasan, M.R.H., Poovaneshvaran, S., Jaya, R.P., Ibrahim, A.H.: Assessments on the performance of asphalt mixtures prepared with adhesion promoters. Int. J. Recent Technol. Eng. **8**(3S3), 332–339 (2019)
11. Xu, G., Wang, H.: Molecular dynamics study of oxidative aging effect on asphalt binder properties. Fuel **188**, 1–10 (2017)
12. Kumbargeri, Y.S., Biligiri, K.P.: understanding aging behaviour of conventional asphalt binders used in India. Transp. Res. Procedia. **17**, 282–290 (2016)
13. Park, D.W., Phan, T.M., Kim, Y.M.: Influence of antistripping additives and rejuvenators on healing performance of moisture-damaged HMA. Adv. Mater. Sci. Eng. 2020 (2020)
14. Qu, X., Liu, Q., Guo, M., Wang, D., Oeser, M.: Study on the effect of aging on physical properties of asphalt binder from a microscale perspective. Constr. Build. Mater. **187**, 718–729 (2018)
15. Aguiar-Moya, J.P., Salazar-Delgado, J., Baldi-Sevilla, A., Leiva-Villacorta, F., Loria-Salazar, L.: Effect of aging on adhesion properties of asphalt mixtures with the use of bitumen bond strength and surface energy measurement tests. Transp. Res. Rec. **2505**, 57–65 (2015)
16. Rossi, C.O., Teltayev, B., Angelico, R.: Adhesion promoters in bituminous road materials a review. Appl. Sci. **7**(5), (2017)
17. Bionghi, R., Shahraki, D., Ameri, M., Karimi, M.M.: Correlation between bond strength and surface free energy parameters of asphalt binder-aggregate system. Constr. Build. Mater. **303**, 124487 (2021)
18. Little, D.N., Allen, D.H., Bhasin, A.: Modeling and design of flexible pavements and materials. Springer International Publishing AG (2018)
19. Abdullah, M.E., Zamhari, K.A., Hainin, M.R., Oluwasola, E.A., Hassan, N.A., Yusoff, N.I.M.: Engineering properties of asphalt binders containing nanoclay and chemical warm-mix asphalt additives. Constr. Build. Mater. **112**, 232–240 (2016)
20. Zhu, C., Xu, G., Zhang, H., Xiao, F., Amirkhanian, S., Wu, C.: Influence of different anti-stripping agents on the rheological properties of asphalt binder at high temperature. Constr. Build. Mater. **164**, 317–325 (2018)
21. Aksoy, A., Şamlioglu, K., Tayfur, S., Özen, H.: Effects of various additives on the moisture damage sensitivity of asphalt mixtures. Constr. Build. Mater. **19**(1), 11–18 (2005)
22. ASTM D5-97.: Standard test method for penetration of bituminous materials. ASTM Stand. (1997)

23. ASTM D36-95.: Standard test method for softening point of bitumen (ring-and-ball apparatus). ASTM Stand. (1995)
24. ASTM D113-17.: Standard test method for ductility of asphalt material. ASTM Int. (2017)
25. Statements, B.: Standard test method for elastic recovery of bituminous materials by Ductilometer 1. Annu. B. ASTM Stand. 1–5 (2008)
26. ASTM D7175-15.: Standard test method for determining the rheological properties of asphalt binder using a dynamic shear rheometer, ASTM International, West Conshohocken, PA. ASTM Stand (2015)
27. Standard, B.: Bituminous mixtures—test methods for hot mix asphalt. Part **3**(1), (2007)
28. ASTM D 3625.: Standard practice for effect of water on bituminous-coated aggregate using. ASTM **96**, 6–7 (2015)
29. ASTM D4541-09.: Standard test method for pull-off strength of coatings using portable adhesion. ASTM Int. 1–16 (2014)
30. Hunter, R.N., Self, A., Read, J., Hobson, E.: The Shell Bitumen Handbook. ICE Publishing, London, UK (2015)
31. Poovaneshvaran, S.: Evaluation Of Crumb Rubber And Latex Modified Asphalt Binders For Pavement Crack Sealant Purpose. Universiti Sains Malaysia, School Of Civil Engineering (2018)
32. Roberts, F.L., Kandahal, P.S., Brown, R.E., Lee, D.Y., Kennedy, T.W.: Hot Mix Asphalt Material-Mixture Design and Construction, 2nd edn. National Asphalt Pavement Association Research and Education Foundation, Lanham, Maryland (1996)
33. Bi, Y., Wu, S., Pei, J., Wen, Y., Li, R.: Correlation analysis between aging behavior and rheological indices of asphalt binder. Constr. Build. Mater. **264**, 120176 (2020)
34. Li, Y., Hao, P., Zhao, C., Ling, J., Wu, T., Li, D., Liu, J., Sun, B.: Anti-rutting performance evaluation of modified asphalt binders: a review. J. Traffic Transp. Eng. **8**(3), 339–355 (2021)
35. Oeser, M., Pellinien, T.: Computational framework for common visco-elastic models in engineering based on the theory of rheology. Comput. Geotech. **42**, 145–156 (2012)

Mechanical Properties of Hot Mix Asphalt Incorporating Coal Fly Ash Filler

Ashraf Abdalla M. Radwan, Mohd Khairul Idham Mohd Satar, Norhidayah Abdul Hassan, and Muhammad Naqiuddin Mohd Warid

Abstract In coal-fired power plants, fly ash accounts for 60–88% of the total combustion residues. Coal fly ash (CFA) has been frequently used in concrete production; however, there are just a few occasions where CFA have been used in the construction of asphalt pavement. The research aims to evaluate CFA's performance as a filler in hot mix asphalt (HMA) mixtures. In this investigation, four CFA contents were used as filler by aggregate weight in the dry method. The resilient modulus (MR), dynamic creep, and wheel tracking tests were used to investigate the performance characteristics of mixtures. Increasing the CFA content in asphalt mixtures improved rutting and fatigue performance compared to a conventional mixture. Thus, it can be concluded that CFA can be used as an alternative filler in HMA.

Keywords Coal fly ash · Resilient modulus · Dynamic creep · Wheel tracking

1 Introduction

Highway paving materials must be of higher quality due to the rapid expansion in traffic demands and increased vehicle loads. The primary responsibility of highway authorities is to produce safe, cost-effective, long-lasting, and smooth pavements capable of carrying anticipated loads. Many specialists, engineers, and researchers are working toward this goal by selecting paving materials that can lessen the intensity and density of distress while also improving the overall performance of asphalt pavements [1, 2]. This would be accomplished by selecting appropriate materials that can improve construction efficiency while also extending pavement service life [3]. According to studies, the strength of hot mix asphalt (HMA) is affected by a variety of parameters, including the type of filler and aggregate used, as well

A. A. M. Radwan · M. K. I. Mohd Satar (✉) · N. Abdul Hassan · M. N. Mohd Warid
Faculty of Civil Engineering, Universiti Teknologi Malaysia, 81310 Skudai, Johor, Malaysia
e-mail: khairulidham@utm.my

A. A. M. Radwan
Department of Civil Engineering, Faculty of Engineering, University of Tripoli, Tripoli, Libya

N. Sabtu (ed.), *Proceedings of AWAM International Conference on Civil Engineering 2022—Volume 2*, Lecture Notes in Civil Engineering 385,
https://doi.org/10.1007/978-981-99-6018-7_5

as the bitumen grade [4]. Among them, filler material significantly impacts HMA properties, particularly those relating to mixture compatibility and aggregate bitumen adhesion [5]. CFA is a combustion by-product that accounts for 60–88% of total coal-fired power plant combustion residues. The utilization of coal fly ash in mineral-asphalt combinations has not been substantially researched, despite the fact that it is frequently used in the construction of concrete. Only a few studies have indicated that fly ash can help asphalt concrete function better [5, 6]. Its annual output is expected to be between 0.75 and 1 billion tons worldwide [3]. The need for affordable power in emerging countries is predicted to expand in the following years, resulting in an increase in this amount [7]. Compared to hydrated lime as a filler replacement in hot mix asphalt, investigations by Mistry and Kumar have demonstrated that fly ash as a filler can provide less deformation and improved strength (HMA) [8]. In this study, the performance metrics of asphalt mixtures with different CFA concentrations were assessed and contrasted with those of a standard mixture. The performance attributes of CFA mixtures will be evaluated through resilient modulus (MR), dynamic creep, and wheel tracking tests.

2 Materials and Methods

2.1 Material

Kemaman Bitumen Company supplied the penetration grade 60/70 asphalt binder (KBC). The softening point, penetration, ductility, density, and viscosity experiments were carried out in the laboratory to analyze the asphalt characteristics. The results are summarized in Table 1.

The coarse and fine aggregates used were crushed rock from the Hanson Quarry in Kulai. Table 2 illustrates the laboratory tests used to evaluate the properties of fine and coarse aggregates. where the results satisfied all of the Malaysia Public Work Department's [9] standards. Figure 1 shows the aggregate distribution of particle size.

Table 1 Asphalt binder properties

Property	Standard	Result
Ductility	ASTM D113	> 100 cm
Penetration	ASTM D5	64.4 mm
Softening point	ASTM D36	51 °C
Density	ASTM D70	1.02 gm/cm^3
Viscosity @ 135 °C	ASTM D4402	3500 mPa s
Viscosity @ 165 °C	ASTM D4402	700 mPa s

Table 2 Properties of the aggregate

Property	Standard	Results
Aggregate crushing value (%)	BS EN 1097-2	17.1
Aggregate impact value (%)	BS EN 1097-2	13.37
Specific gravity (fine aggregate)	ASTM C 128	2.681
Specific gravity (coarse aggregate)	ASTM C 127	2.863
Flakiness (%)	BS EN 933-3	18.4
Elongation (%)	BS 812	22.19
Los Angeles abrasion (%)	AASHTO T96	17.4

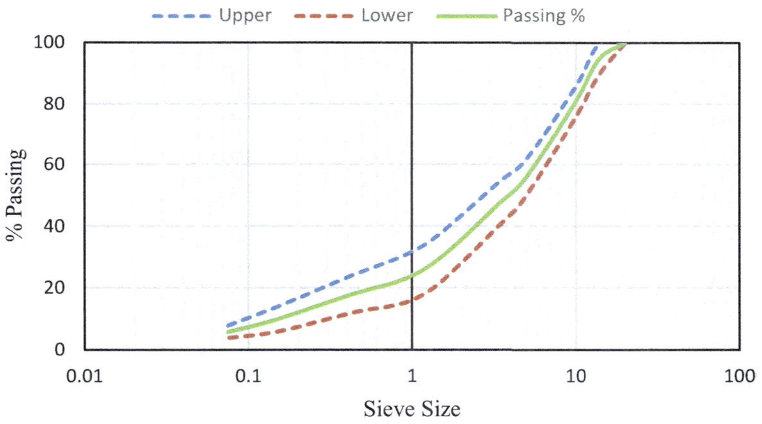

Fig. 1 Aggregate gradation [10]

The traditional mineral fillers utilized in this study were fine aggregate dust, 2% hydrated lime, and coal fly ash (CFA) collected from the Tanjung Bin power plant in Johor. The hydrated lime was used as an anit-stripping agent, according to the Malaysia Public Work Department. According to ASTM C618 [10], the CFA used in this study was classified as type F, where it has a value greater than 70% of the total elements SiO_2 (65.9%), Al_2O_3 (19.7%), and lower than 5% of CaO (3.29%) as in Table 4. The aforementioned fillers were preheated before sieving on a 0.075 mm sieve, with material finer than 0.075 mm being used in this study (Table 3).

2.2 *Samples Preparation*

Four mixes with various CFA concentrations were prepared for this study, as shown in Table 5.

Table 3 CFA chemical component

Component	CFA %	ASTM C618 specification
Al_2O_3	19.7	–
SiO_2	65.9	–
$Al_2O_3 + SiO_2$	85.6	> 70%
Fe_2O_3	5.82	–
TiO_2	1.21	–
Na_2O	1.28	1.5 max
MgO	0.958	–
SO_3	0.322	–
K_2O	1.52	–
CaO	3.29	5 max

Table 4 CFA content in mixtures

Specimens	Mixture A	Mixture B	Mixture C	Mixture D
Hydrated lime	2	2	2	0
Fine aggregate	2	2	0	0
CFA	0	2	4	6

Table 5 Marshall properties of asphalt mixtures

Properties	Mixture				Specifications
	A	B	C	D	
Stability (kN)	15.23	16.85	17.15	18.24	> 8
Stiffness (kN/mm)	4.73	5.40	6.12	7.70	> 2
Flow (mm)	3.22	3.12	2.80	2.37	2–4
VTM %	3.60	3.54	3.23	3.1	3–5
VFA %	75.5	76.1	76.9	78	70–80

The Marshall mix design approach was used to determine the optimum asphalt composition in accordance with ASTM: D6926 [11], with 75 blows applied to each face. To ensure consistency throughout the study for comparability purposes, a 5.05% asphalt component was used for all four mixes. Table 5 lists the results of the measurements made to evaluate stability, stiffness, flow, VTM, and VFA.

2.3 Tests

The experimental tests included in this research were the resilient modulus (MR) test (ASTM D4123) [12], the dynamic creep test (BS DD226) [13], and the wheel tracking test (EN 12697–22) [14]. Three specimens were prepared for each mixture with a diameter of 100 mm in order to conduct the test. The MR and dynamic creep samples were conditioned at the testing temperature (25 °C and 40 °C) and 40 °C respectively for each specimen by using the UTM machine and kept for three hours before the tests. The wheel tracking test was conducted using the double wheel tracker (DWT) equipment that required two molds at a time as show in Fig. 2, each containing two samples with a diameter of 150 mm and a temperature of 50 °C for five hours. Dynamic creep test creep strain slope and creep stiffness modulus were determined using Eqs. 1 and 2.

$$E = \sigma/\varepsilon \tag{2}$$

$$\mathrm{CSS} = (\log \varepsilon 3600 - \log \varepsilon 1200) - (\log 3600 - \log 1200), \tag{2}$$

where

E = creep stiffness modulus (MPa)

σ = applied stress (KPa)

ε = cumulative axil strain at 3600 cycles (mm)

CSS = creep strain slope

Fig. 2 Wheel tracking machine

ε3600 = strain at 3600 cycles

ε1200 = strain at 1200 cycles.

3 Result and Discussion

3.1 *Resilient Modulus*

The average resilient modulus of all mixtures is shown in Fig. 3. It was found that mixtures B, C, and D recorded MR values of 5337, 5543.8, and 5607.7 MPa, respectively, whereas mixture A had MR value of 4436 MPa. As can be seen, adding CFA to the asphalt mixture improves the MR. According to Ali et al. [15], at room temperature, all mixtures had CFA showed an increase in their Mr values by 20%, 24%, and 26% for mixtures B, C, and D, respectively, compared to the control mixture A. The addition of CFA enhanced the stiffness of the mixture, resulting in a higher MR, showing that asphalt mixtures incorporating CFA are less sensitive to fatigue deformation at 25 °C than control mixtures [16]. The MR variation became more vivid as the temperature increased, associated with a decrease in stiffness at 40 °C. The results of MR at 40 °C followed the same pattern as MR at 25 °C. The mixtures recorded 1161.8, 1269, and 1644 MPa for the mixtures B, C, and D, respectively higher than the A mixture. It means that the mixture containing CFA is more resistant to rutting than the control mixture.

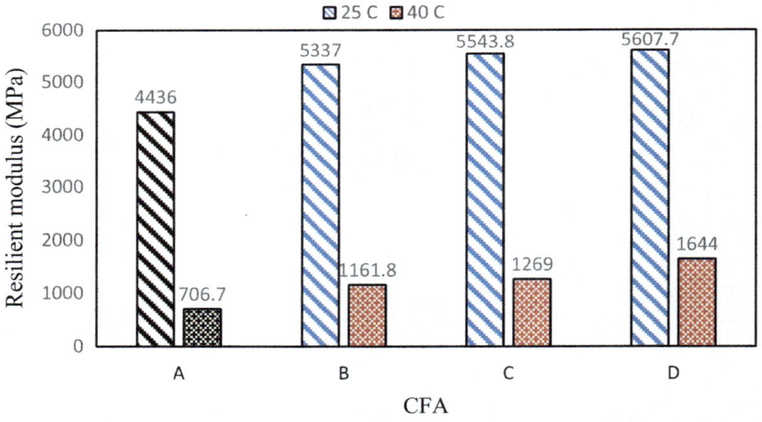

Fig. 3 Result of resilient modulus

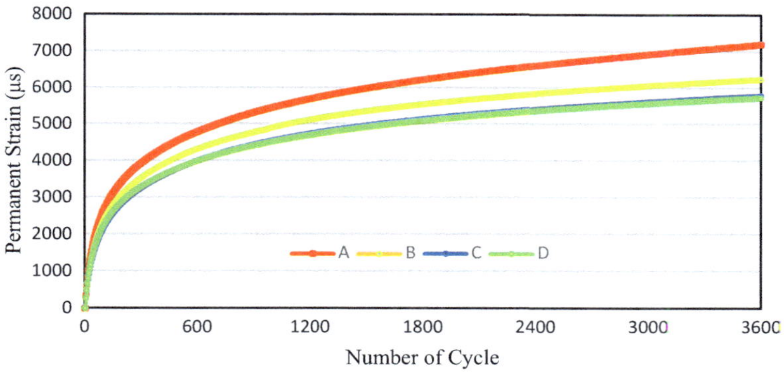

Fig. 4 Permanent strain of CFA mixture

3.2 *Dynamic Creep*

The cumulative permanent strain of every mixture with a load cycle of 3600 is shown in Fig. 4. where the lowest cumulative permanent strain was observed at (5740 µs) in mixture D, followed by mixture C and mixture B with values of (5793.3 µs) and (6253.3 µs), respectively. In comparison, mixture A recorded the highest value (7207 µs). Where increasing CFA content increased the stiffness and thus improved permanent strain. According to Likitlersuang and Chompoorat [17], coal fly ash can improve performance deformation resistance. CSM and CSS were used to determine the asphalt mixture's resistance to permanent deformation. Figure 5a shows that the control mixture A had the greatest CSS value of (0.213). The CSS decreased when CFA was added, as evidenced by mixtures B, C, and D. Mixture B recorded a CSS value of (0.181), mixture C (0.179). and mixture D was the lowest with (0.178). The higher the CSS value, the less resistant the material is to permanent deformation. According to Fig. 5b, mixture D had the largest CSM value (293.3 MPa). Increased creep stiffness indicates more rutting resistance. Mixture C had the highest CSM value of 292.6 MPa, followed by mixture B. (275.7 MPa). Meanwhile, mixture A had the lowest CSM value (200.1 MPa). Fillers can improve the rutting resistance of asphalt concrete, particularly at high temperatures where the asphalt performs softer [17]. In terms of rutting resistance, CFA mixtures performed better than control mixture A due to the inclusion of CFA, which increased the stiffness of mixtures. The dynamic creep test results are consistent with those obtained from the MR test.

3.3 *Wheel Tracking*

The progression of rut depths for asphalt mixture samples is depicted in Fig 6. As observed, mixture D had the least rut depth after 10,000 wheel loading cycles

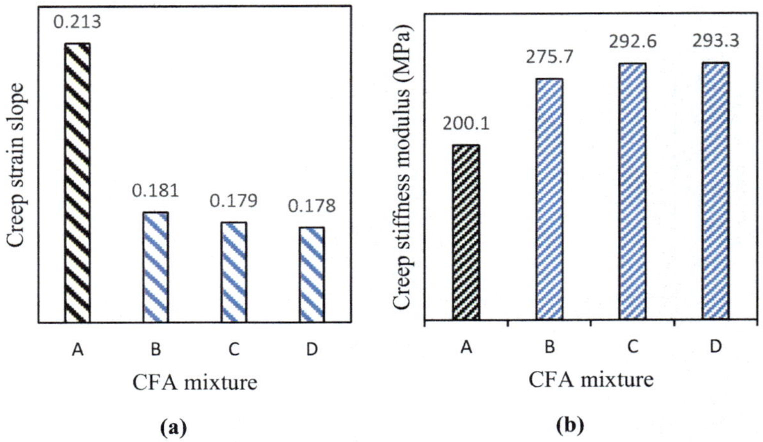

Fig. 5 Graph of **a** CSS of CFA mixture, **b** CSM of CFA mixture

(1.6 mm). The use of filler in HMA has been shown to strengthen the resistance of HMA to rutting deformation [18]. More specifically, mixture C had a rut depth of 1.63 mm, while mixture B had a rut depth of 1.88 mm. Meanwhile, mixture A obtained the greatest rut depth (2.1 mm). These findings indicate that CFA-containing mixtures are more resistant to rutting than the control mixture. The results of the wheel tracking tests are consistent with those of the MR and dynamic creep testing.

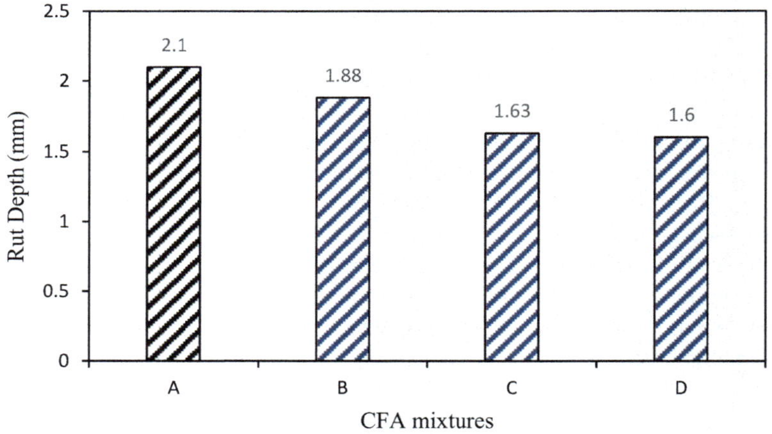

Fig. 6 Wheel tracking

4 Conclusion

The key findings are as follows:

- The addition of CFA increased the mix's resilient modulus at both high and low temperatures.
- CFA mixtures outperformed control mixture A due to the addition of CFA, which increased the stiffness of the mixtures and improved the permanent strain.
- Additionally, these results show that CFA-containing mixtures are less sensitive to rutting than control mixtures.

Acknowledgments Authors wish to thanks Universiti Teknologi Malaysia for supporting this work. In addition, this work was supported by the Ministry of Higher Education under Fundamental Research Grant Scheme (FRGS/1/2019/TK06/UTM/02/1). The author also appreciates the Libyan Government for providing a scholarship through the Ministry of Higher Education and Scientific for supporting this research.

References

1. Mohammed, A., Suhana, B., Nor, K., Ramli, H., Zahid, S., Khan, H.: A systematic review of the utilisation of waste materials as aggregate replacement in stone matrix asphalt mixes. Environ. Sci. Pollut. Res. 35557–35582 (2022). https://doi.org/10.1007/s11356-022-19447-w
2. Radwan, A.A.M., Satar, M.K.I.M., Hassan, N.A., Rogo, K.U.: The influence of coal fly ash on the mechanical properties of hot mix asphalt mixture. IOP Conf. Ser. Earth Environ. Sci. **971**(1), (2022). https://doi.org/10.1088/1755-1315/971/1/012012
3. Tian, Y., et al.: Laboratory investigation on effects of solid waste filler on mechanical properties of porous asphalt mixture. Constr. Build. Mater. **279**, 122436 (2021). https://doi.org/10.1016/j.conbuildmat.2021.122436
4. Dindi, A., Quang, D. V., Vega, L.F., Nashef, E.: Applications of fly ash for CO_2 capture, utilisation, and storage. J. CO_2 Util. **29**(December 2018), 82–102 (2019). https://doi.org/10.1016/j.jcou.2018.11.011
5. Woszuk, A., Bandura, L., Franus, W.: Fly ash as low cost and environmentally friendly filler and its effect on the properties of mix asphalt. J. Clean. Prod. **235**, 493–502 (2019). https://doi.org/10.1016/j.jclepro.2019.06.353
6. Elmagarhe, A., Lu, Q., Alharthai, M., Alamri, M., Elnihum, A.: Performance of porous asphalt mixtures containing recycled concrete aggregate and fly ash. Materials (Basel) **15**(18), 6363 (2022). https://doi.org/10.3390/ma15186363
7. Shanmugan, S., Deepak, V., Nagaraj, J., Jangir, D., Viyagula Jegan, S., Palani, S.: Enhancing the use of coal-fly ash in coarse aggregates concrete. Mater. Today Proc. **30**, 174–182 (2020). https://doi.org/10.1016/j.matpr.2020.05.734
8. Mistry, R., Roy, T.K.: Effect of using fly ash as alternative filler in hot mix asphalt. Perspect. Sci. **8**, 307–309 (2016). https://doi.org/10.1016/j.pisc.2016.04.061
9. JKR: Standard specification for road works. Section 4: flexible pavement. In: Kuala Lumpur: Jabatan Kerja Raya Malaysia, vol. 1–317, pp. S4-58–S4-69 (2008)
10. ASTM C 618-12a: Standard specification for coal fly ash and raw or calcined natural pozzolan for use, pp. 1–5 (2014)
11. ASTM D6926-10: Standard practice for preparation of bituminous specimens using marshall. ASTM Int. 1–6

12. ASTM D4123: Standard test method for indirect tension test for resilient modulus of bituminous. Annu. B. Am. Soc. Test. Mater. ASTM Stand. **82**(Reapproved), 2–5 (1995)
13. BS EN 12697-25: Bituminous mixtures—test methods for hot mix asphalt —part 25: cyclic compression test, vol. 3 (2005)
14. BS EN 12697-22: Standard bituminous mixtures—test methods for hot mix asphalt—part 22 : wheel trakin (2004)
15. Ali, N., Chan, J.S., Simms, S., Bushman, R., Bergan, A.T.: Mechanistic evaluation of fly ash asphalt concrete mixtures. J. Mater. Civ. Eng. **8**(1), 19–25 (1996). https://doi.org/10.1061/(ASCE)0899-1561(1996)8:1(19)
16. Cetin, A.: Evaluation of moisture sensitivity performance of stone mastic asphalt mixes with additional filler: a laboratory comparison of dry and wet mixing methods. Mater. Sci. **28**(3), 369–375 (2022). https://doi.org/10.5755/j02.ms.29552
17. Likitlersuang, S., Chompoorat, T.: Laboratory investigation of the performances of cement and fly ash modified asphalt concrete mixtures. Int. J. Pavement Res. Technol. **9**(5), 337–344 (2016). https://doi.org/10.1016/j.ijprt.2016.08.002
18. Russo, F., et al.: Sustainable asphalt mastics made up recycling waste as filler. J. Environ. Manage. **301**(August 2021), 113826 (2022). https://doi.org/10.1016/j.jenvman.2021.113826

Characterization of Asphalt Emulsion with Different Percentage of Asphalt Content

Ahmad Kamil Arshad, Mohd Izzat Asyraf Mohamad Kamal, Juraidah Ahmad, Ekarizan Shaffie, Noryantizputra Rais, Sanjay Grover, and Xavier Guyot

Abstract Asphalt emulsion is known as an environmentally friendly product to be used in today's paving and conservation efforts. In many road construction, the use of emulsions offer a safer technique than conventional hot asphalt given that the risks of fire, burns, and emissions are averted and the use much less energy. The purpose of this study is to elucidate the effects of asphalt content on the attributes of the emulsion. The test was carried out by using three different percentages of asphalt content on emulsion which are 55%, 60%, and 65%. The tests conducted on asphalt residue are softening point, penetration, ductility, viscosity, and solubility. By using Decision Matrix Analysis, it turned that emulsion with 60% asphalt content is best to be used as it presented the first-rate percentage to develop low penetration, high softening point, and high penetration index.

Keywords Asphalt content · Asphalt emulsion · Physical properties

A. K. Arshad · E. Shaffie
Institute for Infrastructure Engineering and Sustainable Management, Selangor, Malaysia

M. I. A. M. Kamal (✉) · J. Ahmad
Faculty of Civil Engineering, Universiti Teknologi MARA, Selangor, Malaysia
e-mail: asyrafmohamadkamal@gmail.com

N. Rais
Malaysia Institute of Transport, Universiti Teknologi MARA, Selangor, Malaysia

S. Grover
Kemaman Asphalt Company Sdn. Bhd, Selangor, Malaysia

X. Guyot
Tipco Asphalt Public Company Limited, Bangkok, Thailand

N. Sabtu (ed.), *Proceedings of AWAM International Conference on Civil Engineering 2022—Volume 2*, Lecture Notes in Civil Engineering 385,
https://doi.org/10.1007/978-981-99-6018-7_6

1 Introduction

Concerned about global warming, highway industries never lag behind in producing sustainable products replacing existing conventional hot mix asphalt (HMA) by using cold mix asphalt (CMA) which no heating process required hence reducing carbon footprint. Asphalt is a major likely source of urban secondary organic aerosol (SOA) precursors because of its widespread use in the built environment and evidence of reactive precursor emissions [1]. With increasing SOA yields and prolonged SOA generation, emissions increased significantly with modest solar exposure (e.g., 300% for road asphalt) [2]. SOA is a major component of PM2.5 (particulate matter with a diameter less than 2.5 m) and has significant public health implications [3]. A sustainable pavement should maximize the use of natural resources while lowering energy consumption, lowering greenhouse gas emissions, minimizing pollution, increasing health, safety, and risk prevention, and providing a high level of user comfort and safety. The emulsions used for CMA have significantly lower viscosity than asphalt itself and so can be applied at lower temperatures. Emulsions are a safer and more environmentally friendly alternative to hot asphalt in many road construction applications since fire, burns, and emissions are avoided, and the processes consume less energy. Emulsion can be used in different stages of road construction, from rehabilitation to maintenance of pavement structure. For all types of traffic, they offer a wide range of solutions, both in urban and rural areas such as surface dressing, micro-surfacing, grave-emulsion, cold mix asphalt concreter, bond coats, seal coats, and cold recycling and rehabilitation of pavement.

Asphalt emulsion used in CMA can be described as the dispersion of microscopic droplets of one liquid into another. It begins as an oil-in-water emulsion, in which the asphalt is distributed in water with the help of a little amount of an emulsifying agent (known as “soap”). In typical storage circumstances, the substance stabilized the asphalt particles. The emulsions may contain between 40% and 80% asphalt by weight manufactured by using colloid mill. Emulsion can be classified into three types which are water-in-oil (W/O), oil-in-water (O/W), and complex emulsions such as water-in-oil-in- water (W/O/W) as shown in Fig. 1. Oil-in-water (O/W) emulsions are those in which the continuous phase is water and the dispersed (droplet) phase is an “oily” liquid. Water-in-oil (W/O) “inverted” emulsions are those in which the continuous phase is oil and the dispersed phase is water. Emulsions can have a variety of different structures. The dispersion phase of multiple emulsions contains another phase that may or may not have the same composition as the continuous phase [4].

The properties of the asphalt emulsion could be affected by several variables, such as type of emulsifier, amount of the emulsifier used, the pH of the soap solution, and the mixing time. These variables may influence the particle size, residues, viscosity, and storage stability of the asphalt emulsion [5]. Many studies have been done in the past to see how changing the formulation of asphalt emulsion can affect its performance and material qualities. The amount of emulsifier in an emulsion can affect the viscosity of the asphalt emulsion in the lab and the penetration behavior in the field [6]. Furthermore, increasing the emulsifier level resulted in a smaller asphalt

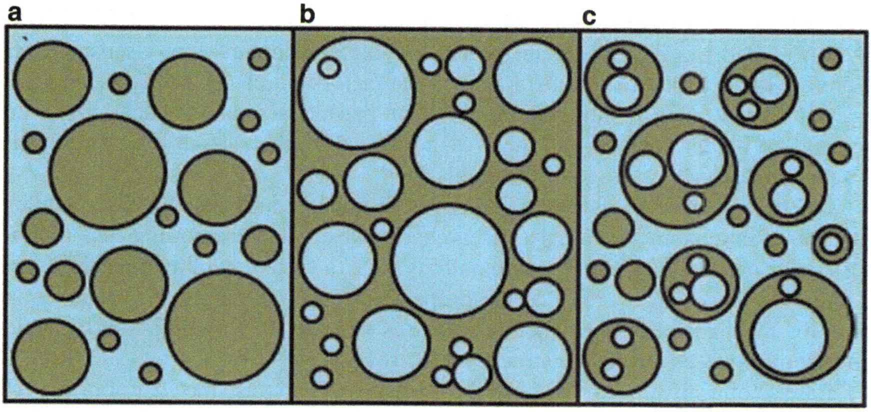

Fig. 1 Types of emulsions: **a** oil-in-water emulsion, **b** water-in-oil emulsion, and **c** complex emulsion

emulsion particle size [7]; the smaller the droplet size in an emulsion, the more stable the emulsion. The viscosity and interfacial tension of asphalt emulsion are affected by the pH of the soap [8].

The emulsion breaks down when it comes into touch with the aggregates, and the mix begins to solidify. Cationic emulsion is utilized in the majority of cases for pavement construction and maintenance. The first asphalt emulsions with an anionic surfactant basis were utilized. However, since the introduction of cationic emulsions, they have all but vanished [9]. Song et al. found that cationic emulsion-prepared mixes had greater strengths and elastic modulus than anionic emulsion-prepared mixes [10]. In the asphalt emulsion sector, cationic emulsifiers are significantly more often utilized than anionic or non-ionic emulsifiers in terms of the surface charge of the hydrophilic end [11]. This is due to its capacity to cling to the most frequent mineral type on Earth, electronegative siliceous solid aggregates [12, 13].

The internal cohesion of the binder is the ability of the binder to withstand shear forces. It depends on several factors, the main one being the viscosity. Cohesion has an optimum value corresponding to a given viscosity range which is dependent on temperature. For instance, below a certain temperature, the binder becomes too hard and brittle when subjected to mechanical force. On the other hand, if the temperature exceeds a certain value, the binder becomes fluids and too soft to be able to withstand the tensile forces. Unlike anionic emulsions, this binder has better wetting characteristics due to the lower hydrophilic nature of the cationic surfactants. On the other hand, cationic emulsion gives significantly higher passive adhesion to the bitumen.

Standard asphalt emulsions are classified as oil-water (O/W) and contain 40–75% asphalt, 0.1–2.5% emulsifier, 25–60% water, and other minor ingredients. The asphalt droplets have a diameter of 0.1–20 μ. This internal water phase has a significant impact on the emulsion's viscosity and, in particular, changes in viscosity during storage [14]. In chip-sealing, low-viscosity emulsion is likely to run off the road, whereas too viscous emulsion may not distribute well over the surface. Viscosity is also related to other important properties like settlement—low-viscosity emulsions are liable to settle during storage. Of the factors which can influence emulsion viscosity, most often mentioned are emulsion particle size and particle size distribution, the binder content, and the presence of salt in the asphalt which can cause viscosity rise during storage of the emulsion.

The performance of asphalt mixtures is frequently influenced by the qualities of asphalt emulsion. Previously, researchers investigated the rutting and fracture resistance of asphalt emulsion combinations under a variety of climatic and material conditions. They discovered that the properties of asphalt emulsions influence not only the mechanical performance of mixtures but also the mixing and compaction energy of mixtures during construction [15, 16].

The use of a solvent can lengthen the curing time and raise the cost of prime coat [17]. As the need for reduced volatile organic compounds (VOCs) has developed in recent years, solvent-free asphalt emulsion has become more popular as a main coat. When opposed to prime coat formulations employing petroleum-based solvents, solvent-free asphalt emulsion generates no VOCs and can dramatically reduce curing time. Furthermore, the adhesion between the base and the asphalt layer that has been treated with asphalt emulsion is stronger than the adhesion between the base and the asphalt layer that has been treated with asphalt cutback [18].

The properties of the binder after extraction from the emulsion were always different from those of the binder before it was emulsified. Therefore, knowledge of the emulsification process and emulsion properties is critical in the above-mentioned elements of bitumen development. The purpose of this study is to expound the characteristics of asphalt emulsion with different percentages of asphalt content to choose the best emulsion for further testing by using Decision Matrix Analysis.

2 Methodology

This study consists of two phases which are sample preparation and material evaluation and characterization.

2.1 *Preparation of Asphalt Emulsion*

The three major chemical components of an emulsion, asphalt, water, and emulsifier are required ingredients in the chemistry and production of paving asphalt emulsions associated with mechanical action of a colloid mill. Other additives may be used to achieve a specific goal [19]. Three primary variables that affect the properties asphalt emulsion are [13]:

(i) Formulation parameters: chemical structure of surfactant and asphalt, as well as temperature
(ii) Composition parameters: as the relative ratio between phases (water-in-oil ratio), and surfactant concentration
(iii) Mechanical factors: affected by emulsification equipment, agitation strength, and emulsion production method.

Taking into account the three variables above, for this study, asphalt with 60/70 PEN grade was used to prepare the emulsion. To make an emulsion of 55% AC (asphalt content), 55% of the emulsion contains asphalt while the remaining 45% is soap solution, while 60:40 for the emulsion of 60% AC and 65:35 for the emulsion of 65% AC. Soap solutions compositions are water, hydrochloric acid (HCl), and emulsifier as shown in Fig. 2. For melting purposes, the asphalt was heated to 135 °C in the oven. The soap solution was created for the water phase, and the temperature was kept between 30 °C and 40 °C. Determination of pH value describes a dedicated method specifically for measuring the pH factor of both anionic and cationic asphalt emulsion. The pH of the soap solution was checked using electronic pH meter and ensured to be at pH 1.7–2.3 to make it cationic. When both the soap solution and the asphalt are ready, the emulsification process begins, the asphalt and soap solution were poured into the colloid mill at the same time for 2–3 min to disperse and homogenize the mixture (Fig. 3).

2.2 *Material Evaluation and Characterization*

A few laboratory tests were conducted on the asphalt emulsion. Table 1 shows the properties tested and method used for the testing. These tests have been selected to verify that the properties of the emulsion are appropriate to be used.

Residue by Evaporation. Evaporation method was chosen over distillation or other methods which relatively involve high-temperature testing because high temperature could lead to damage of original state of asphalt emulsion. According to ASTM D244, this test method is used to assess the percentage of asphalt emulsion residue. 50 g of asphalt emulsion sample was poured into three weighted aluminum beakers with a rod. The beaker was placed in a prepared oven and heated for 2 h at a temperature of 163 °C ± 3 °C. The asphalt emulsion was stirred after 2 h, and the heating process was continued in the oven for another hour at the same temperature. After 3 h of oven heating, the asphalt emulsion now known as asphalt residue was left cool at room temperature and weighed with the rod. The percentage of asphalt residue on each beaker was calculated as follows:

$$\text{Residue, percent} = 2(A - B), \tag{1}$$

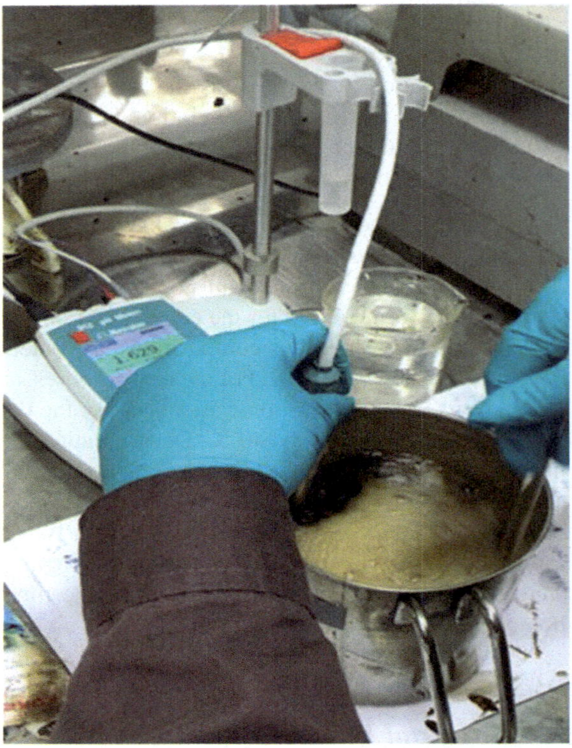

Fig. 2 Preparing soap solution

where

A weight of beaker, rod, and residue (g)
B tare weight of beaker and rod (g)

Softening Point. In this test, the asphalt residue obtained out of the evaporation technique was poured into a brass ring with 5/8 in. diameter and 1/4 in. height. With the device in place, the ring is suspended in a water bath maintained at 5 °C for 15 min after cooling. On the surface of the asphalt residue contained in the ring, steel ball with 3/8 in. diameter is placed. The asphalt residue softens as the temperature rises, and the ball sinks through the ring, carrying the asphalt residue with it. The softening point is supposed to be the temperature at which the asphalt residue taps the bottom plate, despite the fact that the ball's activity is watched and the temperature displayed by the thermometer. To acquire accurate findings, the softening point test procedure must be followed to the letter. Sample preparation, rate of heating, and accuracy of temperature measurement are critical. The recorded temperature is the designated softening point of the bitumen and represents an equi-viscous temperature.

Penetration. The depth to which a standard needle penetrates the sample under standard conditions according to ASTM D5 determines the penetration value. The

Fig. 3 Colloid mill

Table 1 Test conducted on emulsion

Properties	Method
Residue by evaporation	ASTM D244
Softening point	ASTM D5
Penetration @ 25 °C	ASTM D5
Ductility @ 25 °C	ASTM D113
Viscosity at 25 °C	ASTM D244
Solubility in TCE	ASTM D2042

asphalt residue was poured into the sample container in this experiment (diameter 55 mm, 35 mm depth). After cleaning with turpentine or another suitable solvent and drying with a clean cloth, the penetration needle is put into the penetrometer. Lower the needle slowly until the tip lands on the asphalt residue. 25 °C, 50 g, and 5 s were chosen as the temperature, load, and time. The needle was dipped into the asphalt residue for five seconds, and the test was repeated five times on the asphalt residue's surface. Each distance between points should be no less than 10 mm from the container's edge and no less than 10 mm from one point to the next. Before moving on to the following point, make sure the needle you are using is clean. Asphalt binders with high penetration values classified as soft are used for cold climates, while asphalt binders with low penetration values classified as hard are used for warm climates.

Ductility. This test involves pouring asphalt residue into a conventional mold and allowing it to cool to ambient temperature for 35 min. Following the cooling procedure, the asphalt residue-filled mold was placed in a water bath that was kept at the proper temperature for the test, which was usually 25 °C. The sample is placed in the ductility apparatus once the desired temperature has been reached. The device is set up so that one end of the mold is held in place, while the other end is dragged horizontally at a rate of 5 cm/min. The elongation of the asphalt residue's "thread" was noted, and the elongation was sustained until the thread broke. The "ductility" of the asphalt residue is measured in centimeters of elongation traveled.

Viscosity. Viscosity is an essential performance characteristic of asphalt emulsion because it controls the workability and flow properties of emulsified asphalt. The standard test method for viscosity was followed as ASTM D244. Asphalt emulsion passing 850 μm sieve was filled in the viscometer with the cork stopper in position. The viscosity was measured by the time taken (in seconds) from the first drop of asphalt emulsion into receiving flask until it reached the mark on the receiving flask. The time is recorded to the nearest 0.1 s.

Solubility. The solubility test is used to detect contamination in asphalt cement. Presently, solubility in trichloroethylene was used in asphalt cement specifications to replace solubility in carbon disulfide due to the extreme flammability of carbon disulfide. The standard solubility test using trichloroethylene was followed as ASTM D2042. The asphalt residue is dissolved in trichloroethylene and filtered through a glass fiber pad. The insoluble material is washed, dried, and weighed. Specifications for asphalt cements normally require a minimum solubility in trichloroethylene of 99.0%. Total percentage of solubility was calculated as follows:

$$\text{Percent of solubility} = [B - (C - A)]/B, \tag{2}$$

where

A mass of crucible and filter
B mass of sample
C mass of crucible, filter, and insoluble material

3 Results and Discussions

3.1 Residue by Evaporation of Asphalt Emulsion

There are two fundamental motives of having asphalt emulsion recovered. First is to check the quantity of asphalt, or non-water phase within side the asphalt emulsion, and second is to acquire the asphalt residue to be used for further testing. Figure 4 shows the percentage of asphalt residue obtained by the process of evaporation at 163 °C. All types of asphalt emulsion presented residue percentage as percentage of asphalt content in emulsion. This shows that the process of evaporation applied is acceptable.

3.2 Softening Point

The softening point is important for classifying asphalt and determining the consistency of shipments or sources of supply. It also indicates the material's inclination to flow at high temperatures encountered in service. Test results on softening points showed that asphalt residue with 60% of asphalt content emulsion presents the higher temperature followed by 65% of asphalt content and 55% of asphalt content emulsions with 56 °C, 55 °C, and 54 °C, respectively, as in Fig. 5. The higher temperature of softening point represents the harder asphalt but from the results, emulsion with 65% of AC content does not represent the value as it supposed to be which is lower than emulsion with 60% of AC content.

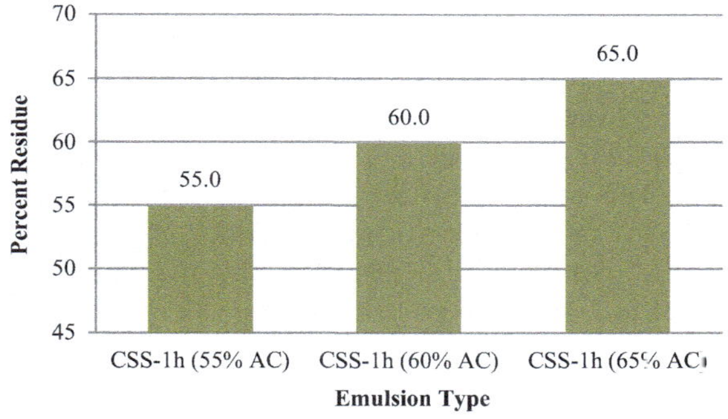

Fig. 4 Percentage residues of evaporated asphalt emulsions

Fig. 5 Softening point test of asphalt residue

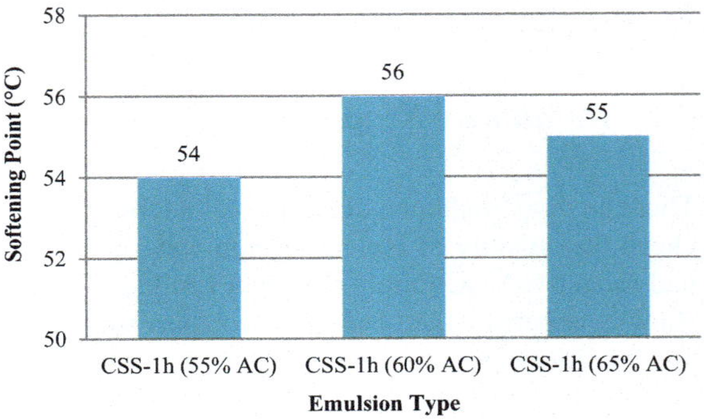

Fig. 6 Penetration test of asphalt residue

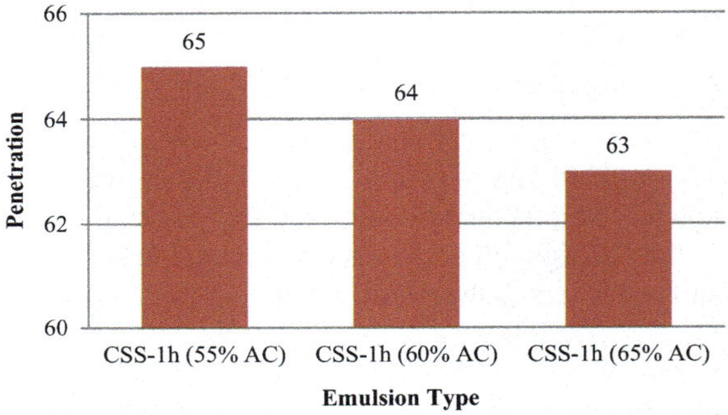

3.3 Penetration

The penetration test is used as a measure of consistency where higher values of penetration indicate softer consistency. From the test, the asphalt residue showed the pattern where the more asphalt contents the lower penetration value. The increment of 5% of asphalt content into the formulation of the asphalt emulsion provided lower penetration value. As shown in Fig. 6, the softest is asphalt emulsion with 55% asphalt content with penetration value 65; followed by asphalt emulsion with 60% asphalt content with penetration value 64; and the hardest is presented by asphalt emulsion with 65% asphalt content with penetration value 63; this trend shown the tendency of strength.

3.4 Penetration Index

The penetration index represents a quantitative measure of the response of asphalt to variation in temperature and Fig. 7 revealed that 60% asphalt content emulsion has the highest penetration index value with 0.83, followed by 65% and 55% of

Fig. 7 Penetration index of asphalt residues

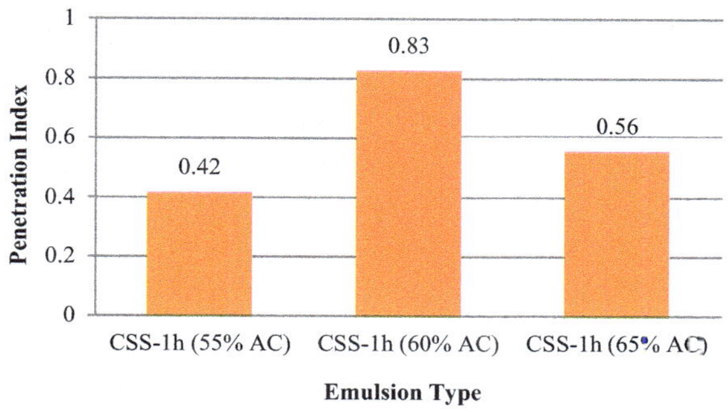

asphalt content emulsion with 0.56 and 0.42, respectively. This tells that the road using emulsion with 60% asphalt content presented better sensitivity to temperature changing and can be indicated the trend on the durability of the road.

3.5 *Ductility*

This test method provides one measure of tensile properties of bituminous materials. As shown in Fig. 8, the value of ductility for all emulsion types tested is more than 150 cm. This indicates that formulation of asphalt emulsion with 55%, 60%, and 65% of asphalt content gives high ductility value and is able to present good adhesive properties and thus serve good performance in service.

Fig. 8 Ductility test of asphalt residue

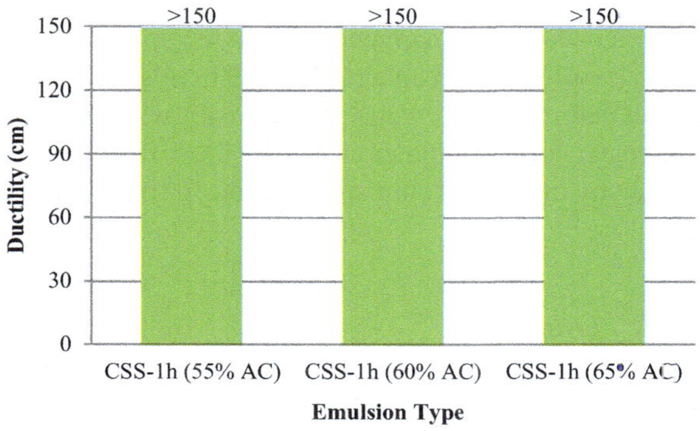

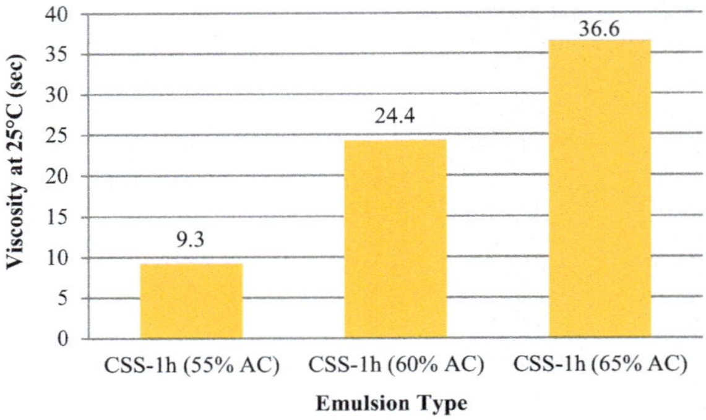

Fig. 9 Viscosity test of asphalt emulsion

3.6 Viscosity

Viscosity is one of major component in grading asphalt emulsion. The viscosity test measures the fluidity of the emulsion, which affects how the asphalt emulsion is pumped and how it flows along the paved road. If the asphalt emulsion is too low, burrs will form when applied, and the aggregates will not be able to maintain a uniform layer. If a thick asphalt emulsion material is not sprayed at right temperatures, it can also cause incorrect spraying and scratches on the carpet. The viscosity of CSS-1 h emulsion with 55%, 60%, and 65% of asphalt content was tested at 25 °C followed as ASTM D244. This test method utilizes the Saybolt Furol viscometer to measure the consistency of an asphalt emulsion. As shown in Fig. 9, it is found that viscosity becomes higher with the increment of asphalt content; this happened due to the concentration of asphalt in emulsion. Asphalt emulsion with 55% of asphalt content presented value of viscosity below minimum specification value which is 20 s. This happened due to the proportion of the mixture which was higher in soap solution percentage. Asphalt emulsion with 60% and 65% of asphalt content exceeded minimum value for viscosity with 24.4 s and 36.6 s, respectively; lower viscosity asphalt emulsion can be used for dense graded aggregate, while higher viscosity of asphalt emulsion is good for open-graded aggregate.

3.7 Solubility

The standard solubility test using trichloroethylene was conducted as ASTM D2042. Specifications for asphalt cement normally require a minimum solubility in trichloroethylene of 99.0%. From Fig. 9, it can tell that the amount of asphalt content affecting the value of solubility, where 65% asphalt content emulsion gives the highest value with 99.84% followed by 60% and 55% asphalt content emulsion with 99.83% and 99.81%, respectively. The solubility test is used to detect contamination in asphalt cement.

Table 2 Score and ranking of asphalt emulsion

	Properties	55% AC	60% AC	65% AC
	Softening point	3	1	2
Properties	Penetration @ 25 °C	3	1	2
	Ductility @ 25 °C	1	1	1
	Viscosity at 25 °C	3	1	2
	Solubility in TCE	3	2	1
	Total score	13	6	8
	Rating	3	1	2

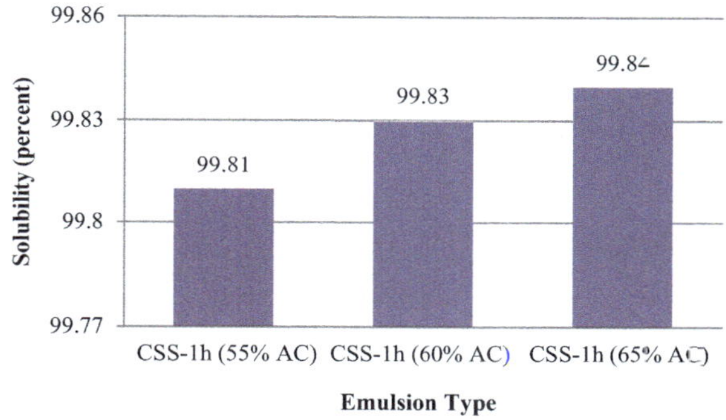

Fig. 10 Solubility test asphalt residue

3.8 *Decision Matrix Analysis*

The use of Decision Matrix Analysis to create a call could be beneficial. It is especially effective when you have a wide range of excellent options to choose from and a lot of different things to consider. This makes it a smart strategy to apply in almost any important decision where there is not a clear and obvious best alternative. Based on above tests, scores for each asphalt emulsion based on its properties are presented in Table 2. Scores 1, 2, and 3 were given to each asphalt emulsion base on their criterion performance with which score 1 for good to 3 for poor. Weighted score was then obtained by multiplying score and weighing factor. The greater score indicates poor performance and smaller score indicates better performance (Fig. 10).

4 Conclusion

Laboratory works have been conducted on asphalt emulsion to check on the effects of percentage of asphalt content in emulsion toward its properties. The purpose of this study is to characterize asphalt emulsion with different percentage of asphalt content. This study investigates the physical properties of emulsion with 55%, 60%, and 65%

of asphalt content hence determining the best emulsion to be used for further testing. Based on experimental physical properties result in this study, the best emulsion to be used on further test is with 60% of asphalt content followed by 65% and 55% asphalt content emulsion; the results show low penetration, high softening point, and high penetration index. This proves that it is important to have the right formulation in producing asphalt emulsion. Too much of soap solution or too much of asphalt content significantly affects the performance of asphalt emulsion. In formulating emulsion, the best emulsifier must be selected to give the desire properties to ensure the long term adhesion to the base and aggregate, but also the best compromise must be made with regard to the composition of the binder so that asphalt emulsion able to perform at its best when applied to the site. It is believed that the objective of this research as stated in the introduction has been met with the completion of this study. There are still areas for improvement that can be consider for further works; different percentage of asphalt content is suggest to be taken into account to describe the diversity of the effects of asphalt formulation on its performance.

References

1. The Freedonia Group: World Asphalt (Asphalt): Industry Study with Forecasts for 2019 and 2024 (Study No. 3351). The Freedonia Group, Inc. (2015)
2. Khare, P., Machesky, J., Soto, R., He, M., Presto, A.A., Gentner, D.R.: Asphalt-related emissions are a major missing nontraditional source of secondary organic aerosol precursors. Sci. Adv. **6**(36) (2020). https://doi.org/10.1126/sciadv.abb9785
3. Khare, P., Gentner, D.R.: Considering the future of anthropogenic gas-phase organic compound emissions and the increasing influence of non-combustion sources on urban air quality. Atmos. Chem. Phys. **18**(8), 5391–5413 (2018). https://doi.org/10.5194/acp-18-5391-2018
4. James, A.: Overview of Asphalt Emulsion. Transportation Research Circular E-C102, pp. 1–15 (2006). https://doi.org/10.17226/23246
5. Kiihnl, L., Braham, A.F.: Developing a particle size specification for asphalt emulsion. Constr. Build. Mater. **293**, 123414 (2021). https://doi.org/10.1016/j.conbuildmat.2021.123414
6. Ouyang, J., Sun, Y., Zarei, S.: Fabrication of solvent-free asphalt emulsion prime with high penetrative ability. Constr. Build. Mater. **230**, 117020 (2020). https://doi.org/10.1016/j.conbuildmat.2019.117020
7. Liu, Z.G., Zong, L., Zhao, L.L., Xie, X.M.: Preparation and storage stability of asphalt emulsions made from modified lignin cationic asphalt emulsifiers. Appl. Mech. Mater. **357–360**, 781–785 (2013). https://doi.org/10.4028/www.scientific.net/amm.357-360.781
8. Cui, D., Pang, J.: The effect of pH on the properties of a cationic asphalt emulsifier. Tenside, Surfactants, Deterg. **54**(5), 386–392 (2017). https://doi.org/10.3139/113.110520
9. Balavoine, G.: Asphalt Emulsions. RGRA, Paris (2008)
10. Song, H., Do, J., Soh, Y.: Feasibility study of asphalt-modified mortars using asphalt emulsion. Construct. Build. Mater. **20**, 332e337 (2006). https://doi.org/10.1016/j.conbuildmat.2005.01.027
11. Read, J., Whiteoak, D.: The Shell Asphalt Handbook, 5th edn. Thomas Telford, London (2003)
12. Mallawarachchi, D.R., Amarasinghe, A.D.U.S., Prashantha, M.A.B.: Suitability of Chitosan as an emulsifier for cationic asphalt emulsions and its behaviour as an additive to asphalt emulsion. Constr. Build. Mater. **102**, 1–6 (2016). https://doi.org/10.1016/j.conbuildmat.2015.10.111
13. Ronald, M., Luis, F.P.: Asphalt emulsions formulation: State-of-the-art and dependency of formulation on emulsions properties. Constr. Build. Mater. **123**, 162–173 (2016). https://doi.org/10.1016/j.conbuildmat.2016.06.129

14. Furlong, S., James, A., Kalinowski, E., Thompson, M.: Water enclosed within the droplets of asphalt emulsions and its relation to viscosity changes during storage. Colloids Surf., A **152**(1–2), 147–153 (1999). https://doi.org/10.1016/s0927-7757(98)00628-1
15. Miljković, M., Radenberg, M.: Characterising the influence of bitumen emulsion on asphalt mixture performance. Mater. Struct. **48**(7), 2195–2210 (2014). https://doi.org/10.1617/s11527-014-0302-y
16. Miljković, M., Radenberg, M., Fang, X., Lura, P.: Influence of emulsifier content on cement hydration and mechanical performance of bitumen emulsion mortar. Mater. Struct. **50**(3) (2017). https://doi.org/10.1617/s11527-017-1052-4
17. Freeman, T.J., Button, J.W., Estakhri, C.K.: Effective prime coats for compacted pavement bases (No. FHWA/TX-10/0-5635-1). Texas Transportation Institute (2010) http://tti.tamu.edu/documents/0-5635-1.pdf
18. Minwu, J., Li, H.Z.: Development and application of high cohesiveness prime coat on cement-stabilized macadam base. J. Chang'an Univ. (Nat. Sci. Ed.) **34**(5), 15–23 (2014)
19. Asphalt Emulsion Manufacturers Association: AEMA Recommended Performance Guidelines, 5th edn (2017)

Full Extraction Asbuton Bitumen Using Different Organic Solvents for Binder Replacement

M. R. A. Maha, M. K. Idham, and M. R. Hainin

Abstract Natural asphalt buton or asbuton can be utilized as alternative binder in asphalt mixture by semi-extraction or fully extraction process. Semi-extraction asbuton or called refinery buton asphalt contained of bitumen and mineral asbuton, while full extraction asbuton by using solvent can maximize the content of bitumen and discarded the mineral. However, usage of different solvent influenced the asbuton bitumen properties. Thus, this study evaluated the effect toluene and turpentine on the full extraction asbuton bitumen properties. Four different percentages of solvent content (1–4%) of both turpentine and toluene were used in this study. The asbuton bitumen was then tested for penetration, softening point, flash point, specific gravity, and thin film oven test. Based on the results, 2% solvent either turpentine or toluene is recommended to produce full extraction asbuton bitumen.

Keywords Full extraction · Asbuton · Turpentine · Toluene · Organic solvent

1 Introduction

Natural buton bitumen or asbuton is natural rock asphalt from Buton Island, North-west Sulawesi, Indonesia. The largest asbuton deposit is located in Buton Regency with a total deposit of 638.2 million tons [1]. Figure 1 shows one of the natural buton quarry located in Indonesia. It can be utilized as alternative binder in asphalt mixture through semi-extraction or fully extraction process. Semi-extraction asbuton or refinery buton asphalt contained bitumen with some mineral. There are many studies were conducted on semi-extraction asbuton. In general, semi-extraction produced 60–70 PEN bitumen together with mineral asbuton. Table 1 shows previous results on semi-extraction asbuton bitumen properties.

M. R. A. Maha · M. K. Idham (✉) · M. R. Hainin
Faculty of Civil Engineering, Universiti Teknologi Malaysia, 81310 Skudai, Johor, Malaysia
e-mail: khairulidham@utm.my

N. Sabtu (ed.), *Proceedings of AWAM International Conference on Civil Engineering 2022—Volume 2*, Lecture Notes in Civil Engineering 385,
https://doi.org/10.1007/978-981-99-6018-7_7

Table 1 Properties of semi-extraction asbuton bitumen by previous studies

No	Type of test	Properties				Requirement	Method
		[2]	[3]	[4]	[5]		
1	Penetration, PEN	41	73.8	60.6	51.6	60–70	SNI 2456:2011
2	Softening point, °C	57	47.25	53.5	–	> 48	SNI 2434:2011
3	Specific gravity	1.090	1.045	1.062	1.062	> 1.0	SNI 2441:2011
4	Flash point, °C	318	340	335	277	> 200	SNI 2433:2011
5	Loss of weight, %	0.019	–	0.77	0.77	> 0.8	SNI 06-2440-1991
6	Penetration after LOH, PEN	75.6	61.3	61.9	62.8	60–70	SNI 2456:2011

Fig. 1 Natural buton bitumen in quarry

Based on Table 1, properties of semi-extraction asbuton bitumen could fulfill the requirement; thus, it can be used to minimize the need for oil-based bitumen obtained from petroleum refinery process [11].

The requirement for the research is used on Indonesian Specs General specification 2018 which is referred to *Standar Nasional Indonesia* (SNI), while 100% bitumen could be obtained from the full extraction asbuton, namely full extraction asbuton bitumen. It can be used as an alternative to oil-based bitumen for asphalt mixture. A solvent is needed in order to maximize the bitumen extraction process (full extraction) without mixing bitumen oil. Petroleum-based solvents including kerosene, naphtha, normal heptane, sulfuric acid, and standard solvent trichloroethylene (TCE) are usually used to produce asbuton bitumen. In addition, organic solvent also can be used for the extraction process. Therefore, this study was conducted to evaluate the properties of asbuton bitumen extracted using organic solvent. It is expected that the usage of toluene and turpentine (organic solvent) will affect the extraction process of asbuton.

2 Materials and Test Methods

2.1 Material

- **Asbuton**
  The natural bitumen available on Buton Island has very large reserves, with asphalt content varying between 10% and 50% with locations spread from Sampolawa Bay to Lawele Bay along 75 km with a width of 27 km plus the Enreke area which belongs to the North Buton district, with deposit amount as shown in Table 2.
- **Solvent**
  In general, any type of materials that can dissolve asphalt can be used to separate bitumen from mineral asbuton. With the increases in crude oil price, the price of petroleum-based solvents also increases. To avoid the dependence of asbuton solvent, prices on crude oil prices asbuton extraction solvents based on organic materials are chosen. Toluene, also known as ethylbenzene or phenyl methanol, is a clear, colorless liquid that is insoluble in water with an odor like paint thinner and smells like benzene. Turpentine is the sap from the pine tree (*Pinus merkusi*) which is then processed into turpentine. Turpentine is used as a raw material for the cosmetic industry, paint oil, a mixture of solvents, antiseptics, camphor, and pharmaceuticals. Toluene and turpentine are carried out to support the full work process of asbuton bitumen extraction with extraction tools through variation percentages of solvent used.

Table 2 Estimated location and deposit of asbuton [7]

No	Region	Bitumen content %	Deposit (million ton)
1	Kabungka	10–20	60
2	Winto	10–20	3.20
3	Winil	10–20	0.60
4	Siantopina	10–20	181.25
5	Olala	10–20	47.089
6	Enreko	10–20	174.725
7	Lawele	20–40	210

2.2 Test Method

Figure 2 shows the flow of the study. First, asbuton is prepared by crushing process before it is placed into the reactor. Asbuton is prepared according to specified temperatures (25 °C, 60 °C, and 80 °C) (Fig. 3). Then, asbuton is mixed with the solvent (turpentine or toluene) and stirred until all the bitumen content stored in the rock pores is completely dissolved.

The bitumen solution is heated to evaporate the solvents, and the solvents are then condensed for reuse (recycling) in the reactor. The residue obtained after all the solvent has evaporated is pure bitumen or full extraction asbuton bitumen.

The rock that has been separated from the bitumen solution is washed several times to ensure that all the bitumen content remaining in the rock pores has completely dissolved. Then the rock is heated to evaporate the turpentine and toluene solvents, after which the solvents are condensed to be reused (recycled) in the reactor. The residue obtained after all the solvent has evaporated is rock that is free of bitumen.

After separated from mineral buton rock, the asbuton bitumen properties tests were carried out to determine the value of penetration, softening point, flash point, specific gravity, and thin film oven test based on variant content of solvent both of turpentine and toluene.

Fig. 2 Full extraction process of asbuton bitumen

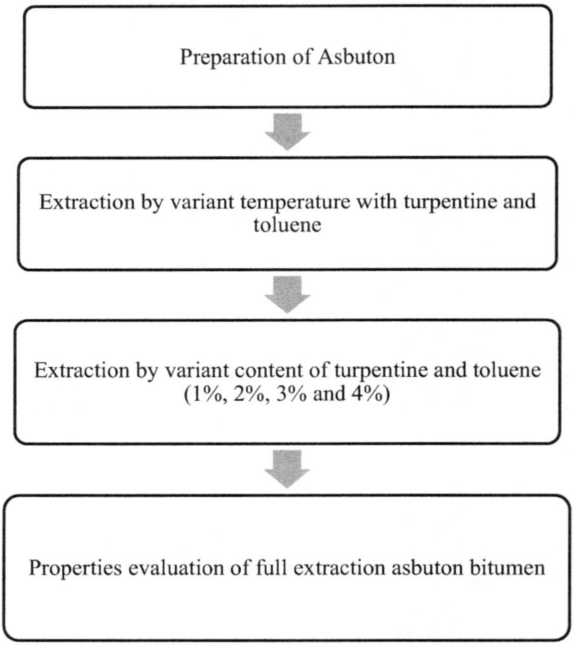

(a) (b)

Fig. 3 **a** Asbuton before for full extraction; **b** full extraction process

3 Results

3.1 Asbuton Bitumen Extraction with Toluene and Turpentine at Variant Temperature

Table 3 shows the result of asbuton bitumen obtained from extraction process at different temperature. It shows that the higher the temperature during the extraction process, the higher the asbuton bitumen could be produced [11]. Based on the preliminary test, even though higher temperature could produce more bitumen, the rest of the extraction process is only carried out at a maximum temperature of 140 °C to avoid aging of the asbuton bitumen.

Table 3 Percentage of asbuton bitumen extracted at variant temperature

Type of Solvent	Percentage asbuton bitumen at temperature			Method
	25 °C	60 °C	80 °C	
Turpentine	10.8	12.1	13.1	SNI 03-3640-1994
Toluene	16.1	17.4	20.2	SNI 03-3640-1994

3.2 *Asbuton Bitumen Properties with Turpentine and Toluene Solvent*

Tables 4 and 5 show the properties of asbuton bitumen extracted with different solvent content. The results show that there is an increase and decrease in bitumen properties with variation of solvent content between turpentine and toluene. Figure 4 shows the penetration, softening point, thin film oven, flash point and specific gravity tests conducted for this study. While Figs. 5, 6 and 7 presented the results.

- **Penetration**
  Figure 5 shows that the penetration value of extraction with 1% turpentine and toluene solvents is smaller than the extraction with the content of 2% and 3%, and slight decrease in the concentration of 4%. It shows that usage of 2% and 3% solvents (turpentine and toluene) is within the range of 60–70 PEN bitumen. The red line indicates the specification area that has been mentioned in the table.

Table 4 Properties asbuton bitumen extracted by turpentine

No	Type of test	Turpentine content (%)				Requirement	Method
		1	2	3	4		
1	Penetration	38.7	62.9	63.2	58.2	60–70	SNI 2456:2011
2	Softening point; °C	32	48.3	47.2	44.2	> 48	SNI 2434:2011
3	Specific gravity	1.071	1.067	1.061	1.054	> 1.0	SNI 2441:2011
4	Flash point; °C	221	233	226	228	> 200	SNI 2433:2011
5	Loss of weight; %	0.0327	0.0191	0.0223	0.0272	> 0.8	SNI 06-2440-1991
6	Penetration after LOH	51.2	61.3	62.8	56.3	60–70	SNI 2456:2011

Table 5 Properties asbuton bitumen extracted by toluene

No	Type of test	Toluene content (%)				Requirement	Method
		1	2	3	4		
1	Penetration	42.3	62.3	64.6	61,2	60–70	SNI 2456:2011
2	Softening point; °C	42.7	48.7	47.4	47.8	> 48	SNI 2434:2011
3	Specific gravity	1.054	1.031	1.028	1.021	> 1.0	SNI 2441:2011
4	Flash point; °C	221	231	226	219	> 200	SNI 2433:2011
5	Loss of weight; %	0.0105	0.0091	0.0106	0.0114	> 0.8	SNI 06–2440-1991
6	Penetration after LOH	41.7	62.2	62.9	60.3	60–70	SNI 2456:2011

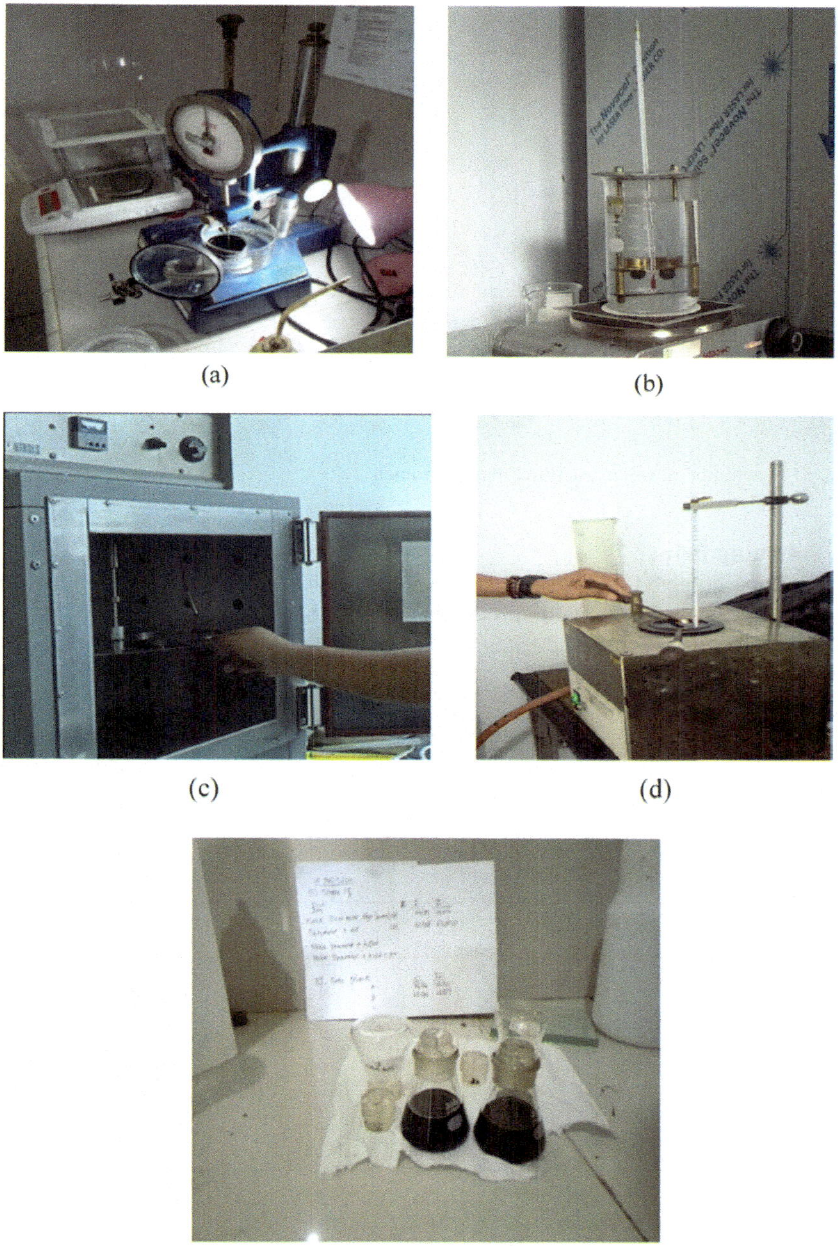

Fig. 4 **a** Penetration test; **b** Softening point test; **c** Thin film oven test; **d** Flash point test; **e** Specific gravity test

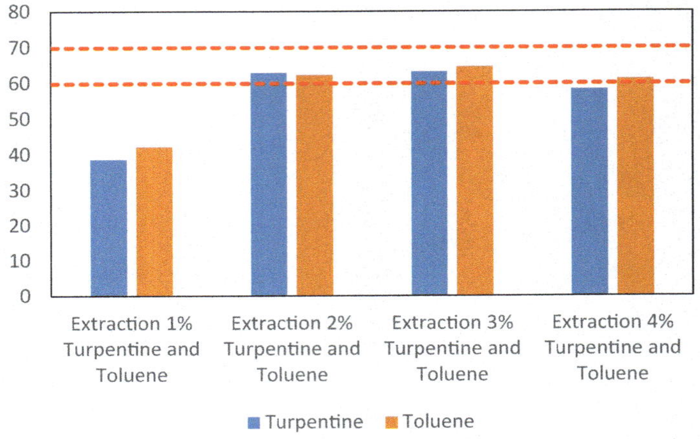

Fig. 5 Penetration of full extraction asbuton bitumen

- **Softening Point**
  Figure 6 depicted that 1% solvent content produced low softening point value. Only 2% solvents content produced comparable temperature to 60–70 PEN bitumen. The red line indicates the specification area that has been mentioned in the table.
- **Flash Point**
  Based on Fig. 7, the most flammable point is experienced by extraction with a solvent content of 1%. While the longest burning is extraction with a solvent content of 2, the red line indicates the specification area that has been mentioned in the table.

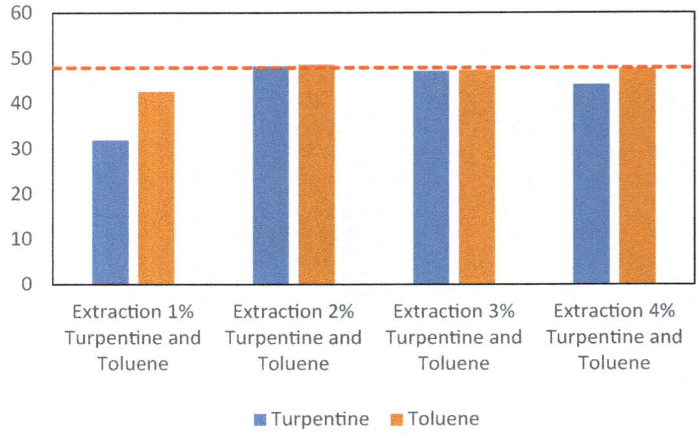

Fig. 6 Softening point of full extraction asbuton bitumen

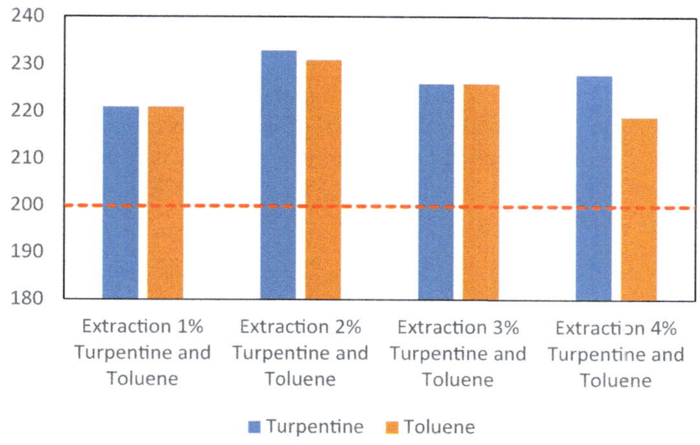

Fig. 7 Flash point of full extraction asbuton bitumen

- **Thin Film Oven**
  The bitumen which is suitable to weather and temperature condition is extraction with a solvent content of 2% both of turpentine and toluene.
- **Specific Gravity**
  The specific gravity of asphalt is important in the design of aggregate and asphalt mixtures. In general, all asbuton bitumen has specific gravity greater than 1.0 with the highest value which is extracted by 1% solvent content.

Based on the results, 2% solvent contents are selected for both turpentine and toluene because it produced better results in terms of softening point and flash point compared to 3% solvent content. Table 6 summarizes the properties of asbuton bitumen extracted with 2% turpentine and toluene.

Table 6 Summary of full extraction asbuton bitumen

No	Type of test	Extracted asbuton bitumen		Method
		Turpentine	Toluene	
1	Penetration	62.9	62.3	SNI 2456:2011
2	Softening point; °C	48.3	48.7	SNI 2434:2011
3	Specific gravity	1.067	1.031	SNI 2441:2011
4	Flash point; °C	233	231	SNI 2433:2011
5	Loss of weight; %	0.0191	0.0091	SNI 06-2440-1991
6	Penetration after LOH	61.3	62.2	SNI 2456:2011

4 Conclusion

Application of 2% turpentine and toluene content as a solvent for full extraction process of asbuton could produce optimum properties of asbuton bitumen. Thus, it is expected that full extraction asbuton bitumen can be used as an alternative of oil-based bitumen in asphalt mixture due to its comparable properties with 60–70 PEN oil-based bitumen.

References

1. Affandi, F.: Characteristics of asbuton granular as bituments for hot mix asphalt. Bandung. J. Road Bridge Res. Center Dev. Road Bridge Ministry Public Work (2008)
2. Affandi, F.: Effect of asbuton semi extraction on stone mastic asphalt. Bandung. J. Road Bridge Res. Center Dev. Road Bridge Ministry Public Work (2010)
3. Damm, K.W., Abraham, J., Butz, T., Hilderbrand, G., Riebeschl, G.: Asphalt flow improvers as, intelligent fillers for hot asphalt—a new chapter in asphalt technology. J. Appl. Asphalt Binder Technol. 36–69 (2002)
4. Departemen Pekerjaan Umum Direktorat Jenderal Bina Marga: Utilization a Asbuton 4th Book of Warm Mis Asphalt with Asbuton Granular. Ministry of Public and Work, Jakarta (2006)
5. Didin: Solvent for Extraction Asbuton. Institute of Technology Bandung, Bandung (2008)
6. Gierhart, D.: Warm mix asphalt—"what is it and how can it benefit?". In: Conference Southeast Pavement Preservation Partnership Meeting. Chinese Society of Pavement Engineering, Jhongli May 1113 (2009)
7. Hamirhan Saodang Msce, Ir.: Highway Construction, 2nd Book. Bandung: Nova (2005)
8. Hasriana, T.: The Testing of Conditionally Asset Buton (Asbuton) By Extraction Using the Sockhlet Method in Pt. Wika Bitumen Southeast Sulawesi. Skripsi/Thesis Undergraduated of University Sultan Alauddin, Makassar, Indonesia (2018)
9. Head of Loka Litbang Asbuton: Strategic plan 2015–2019 asbuton. Bandung, Research and Development for Road and Bridge (2018)
10. Japan Road Association: Manual for Design and Construction of Asphalt Pavement. JRA, Tokyo (1980)
11. Kementerian Pekerjaan Umum: Specification of Warm Mix Asphalt. Direktorat Jenderal Bina Marga, Jakarta (2018)
12. Kurniadji, Nono: Review of additional asbuton in hot mix asphalt from technical and financial side. Bandung, J. Road Bridge Res. Dev. Center Road Bridge **23**(3) (2006)
13. Kurniadji: Refinement instrument model of the asbuton extraction. Bandung, Scientific Manuscripts of Research and Development Center for Road and Bridge, Informatics Publishers (2011)
14. Kurniadji: Asbuton extraction using organic based solvent and water media. Bandung, J. Road Bridge Res. Dev. Center Road Bridge (2014)
15. Setiawan, A., Sanusi, Sujatmiko, G.: Strength and durability of coating on thin hot aspal mixture with retona blend 55. Surakarta, E-J. Civ. Eng. 628 (2014)
16. Sumarji: The use of bitumen asbuton of retona blend type 55 As HRS-B mixed collection material. Yogyakarta J. Eng. **2**(4), 18–24 (2012)
17. Sumiati, Mahmuda, Puryanto: Advantages of premixed asbuton and shell bitumen over concrete asphal mixture (AC-BC). Palembang Polytechnol. **18**(1), 53–64 (2019)
18. Winangsi: Quality analysis of retona blend 55 asphalt materials and oil bitumen. Gorontalo J. Radial **3**(1), 29–34 (2015)

Determination of Optimum Binder Content of Dry-Mixed Crumb Rubber in Stone Mastic Asphalt Using Sandstone Aggregate

Nurul Ariqah Ispal, Lillian Gungat, and K. T. Jeffrey Koh

Abstract The primary objective of this study is to evaluate the optimum binder content of the stone mastic asphalt blending different types of crumb rubber and locally available sandstone aggregate in Sabah, Malaysia. The crumb rubber was dry mixed from 0–3% and bitumen ranging from 5–7% of the aggregate weight, compacted at 150 °C compaction temperature. Gradation of stone mastic asphalt used was following the CR-SMA grading in the Malaysia Standard. The process was adapting the Marshall mix design method, including stability and flow test, and volumetric testing to discover the bulk density and voids in mix, and binder drain down test. In short, engulfing sandstone locally from Sabah in SMA with crumb rubber has some drawbacks finding compared to the past research.

Keywords Sandstone aggregate · Crumb rubber · Dry mix · Stone mastic asphalt

1 Introduction

Asphalt mixtures are made up of aggregates and fillers, bound with bitumen binder. It has good mechanical qualities, smoothness, and durability, and so it is used to build most roadways and highways. Environmental circumstances, traffic loads, and, most importantly, the qualities of the raw materials affect the performance of asphalt mixtures [1], where the mechanical behaviour is influenced by the physical and mineralogical qualities of aggregates. In Sabah, the aggregate widely used in asphalt production is sandstone aggregate; inexorably, Sabah also faces destructive road deterioration issues that cause riding discomfort to the customers for years.

N. A. Ispal · L. Gungat (✉)
Faculty of Engineering, Universiti Malaysia Sabah, Kota Kinabalu, Malaysia
e-mail: lillian@ums.edu.my

N. A. Ispal · K. T. Jeffrey Koh
Astenpave Sdn Bhd, Kota Kinabalu, Malaysia

N. Sabtu (ed.), *Proceedings of AWAM International Conference on Civil Engineering 2022—Volume 2*, Lecture Notes in Civil Engineering 385,
https://doi.org/10.1007/978-981-99-6018-7_8

One of the methods to increase the performance of roads is by modifying the conventional mixture, and in this case, is adapting the stone mastic asphalt (SMA) mixture replacing the dense graded asphalt. SMA is a gap-graded mixture derived around the 1970s in Germany to yield better performance against rutting [2]. It comprises a larger aggregate size that will enhance the loading sustain. However, due to determining factors that affect the structure formation with higher bitumen content, SMA tends to have binder runs off from the aggregate prior to compaction process. This feature and the need to increase the longevity of the mixes have resulted in the introduction of stabilisers to provide better bonding between aggregate and binder [3]. Stabilisers can be originated from many sources, either waste products, synthetic, or natural. Due to the unprecedented rise of waste tyre vehicles that affect environmental pollution worldwide including Sabah, figuring its ability as an additive in asphalt mixtures which purposely to improve the performance of the asphalt pavement, the use of waste tyre rubber is a valid alternate and cheap. It is considered a sustainable technology, "greening asphalt", that might transform wasted residue into a new asphalt mixture substantially resistant to pavement's deformation. Thus, utilising crumb rubber obtained from scrap automobile tyres in SMA, not only cost reduction but also a less ecological impact on keeping the environmental clean and achieving better natural resources balance [4, 5].

In this study, crumb rubber in SMA is a special consideration in attempts to generate economic value. In asphalt technology, there are two methods of using crumb rubber, either as a binder modifier (wet process) or as an aggregate (dry process), in which comparing both, the dry process is a far less popular method as it contributed to inconsistent infield and laboratory performance. However, the dry process is used in this study due to the capability of the Sabah's local premix plant. The primary aim of this study is to investigate the interaction between locally available sandstone aggregate with crumb rubber to obtain optimum bitumen content (OBC) by using the dry process.

2 Experimental Design

2.1 Materials and Sample Preparation

The main materials used in this study were sandstone aggregate, bitumen, quarry dust as filler, and crumb rubber sourced from different types.

The gradation was following the crumb rubber stone mastic asphalt (CR-SMA) in the Malaysia Standard Specification of Public Work Department (PWD) 2019 [6]. Sandstone aggregate and quarry dust were supplied by one of the top blasting quarries in Sabah, with the specification of physical and chemical composition as in Tables 1 and 2, respectively. Penetration grade 60/70 of bituminous binder was used where the physical properties as in Table 3, whereas ASTM Mesh 40 (0.425 mm) crumb rubber was used sourced from truck tyres with composition as depicted in

Table 4. Crumb rubber sourced from truck tyres is chosen due to its composition complying the PWD 2019 standard requirement.

For blending, crumb rubber at 0–3% of the total aggregate weight was dry mixed with aggregate before introducing bitumen at 5–7% of the aggregate weight to the mixture. The mixture was mixed at 160 °C and compacted at 150 °C to adopt site compaction temperature mode, where a 10 °C difference indicates the loss of temperature during transporting period from the premix plant to the construction site. Prior to compaction using the Marshall method at 75 blows per face, the mixture was short term aged for 2 h at compaction temperature. The samples were extruded out of the moulds after 24 h and proceeded for testing.

Table 1 Sandstone aggregate physical properties

Property	Specification	Value	Requirement for road works
Flakiness index (%)	ASTM D4791	13.78	< 20.0
Elongation index (%)	ASTM D4791	0.00	< 20.0
Aggregate impact value (%)	ASTM C125	10.59	< 30.0
Aggregate crushing value (%)	ASTM D5821	23.47	< 30.0
Los Angeles abrasion (%)	ASTM C131-01	23.3	< 40.0
Water absorption—coarse than 10 mm (%)	ASTM D570	1.11	< 2.0
Water absorption—smaller than 9.5 mm (%)	ASTM D570	1.56	< 2.0
Specific gravity of coarse aggregate	ASTM C127	2.56	–
Soundness value of coarse aggregate (%)	ASTM C88	0.94	< 12.0
Soundness value of fine aggregate (%)	ASTM C88	1.80	< 12.0

Table 2 Sandstone aggregate chemical composition

Compound	SiO_2	Al_2O_3	MgO	CaO	K_2O	Fe_2O_3	TiO_2	P_2O_5
Conc unit (%)	71.97	12.90	2.66	2.30	4.24	3.62	0.51	0.61

Note SiO_2 = silicon dioxide; Al_2O_3 = aluminium oxide; MgO = magnesium oxide; CaO = calcium oxide; K_2O = potassium oxide, Fe_2O_3 = iron oxide; TiO_2 = titanium oxide; P_2O_5 = phosphorus pentoxide

Table 3 Bitumen physical properties

Property	Specification	Value	Requirement
Ductility (mm)	ASTM D113-99	129	> 100
Softening point (°C)	ASTM D36-06	51	49–56
Penetration (dmm)	ASTM D5	64	60–70

Table 4 Truck tyre crumb rubber composition

Content	Composition	Requirement in PWD 2019 standard specification
Acetone extract (%)	21.1	< 22
Rubber hydrocarbon (%)	46.6	42–65
Carbon black content (%)	27.08	28–38
Natural rubber (%)	43.85	> 30
Ash content (%)	5.2	< 8.0

Table 5 CR-SMA requirement in PWD 2019

Test	Requirement	OBC determination
VIM	3–6%	4.5%
Bulk density	–	Peak
Stability	Min 6000 N	Peak
Flow	2.0–5.0 mm	3.5%
Binder drain down	Max 0.2%	–

2.2 Experimental Procedure

This study is using the Marshall mix design method due to the aptness of the local premix industry as well as the PWD standard adopted. Experiments were conducted to evaluate the Marshall stability, flow, volumetric properties to determine the OBC. Bitumen drain down test and resilient modulus test also were conducted in the study. Marshall stability measures the capability of compacted asphalt mixture to sustain maximum load until it fails. Marshall flow is to show the vertical deformation when the compacted asphalt mixture sample reaches the maximum load it can sustain. Volumetric properties give findings on bulk density of the mixture and the voids in mix (VIM). The binder drain down test determines the amount of bitumen than drain off in the loose asphalt mixture sample, whereas resilient modulus test is to determine the stiffness of the mixture. All these parameters follow the Malaysian Standard Specification of PWD 2019 to identify the OBC of the CR-SMA following the requirement as in Table 5.

3 Results and Discussion

3.1 Marshall Stability

The Marshall stability results obtained for different percentages of crumb rubber for each bitumen content are shown in Fig. 1. Marshall stability represents the maximum loading that can be sustained at 60 °C testing temperature to stipulate climate in real, with loading rate of 50.8 mm/min, until the sample fails. Studies in Peninsular Malaysia discovered that the addition of crumb rubber in asphalt improves the Marshall stability and quotient [4, 7]. However, this study has come out with the result as shown in Fig. 1 which shows the stability values by differing crumb rubber content in tandem with bitumen content; the value is lower compared with the control sample without any crumb rubber. Though the trend of the Marshall stability value shows anomalies bell-shaped curve and end spike for all mixtures, the control mixture without the addition of crumb rubber is generally higher at all bitumen content. Though past studies stated that crumb rubber has good adhesion to aggregate surface [5, 8, 9], and dry process could increase the stiffness [10], it is indisputable that the previous researches were using different aggregate source with different mineralogy content. Indeed, mineralogy is essential for adhesion quality in asphalt mixture [11]. Apart from the mineralogy concern, the reduction of stiffness in this study is possibly due to the property of crumb rubber that reduces the cohesion of the mixture as a result of deficient interconnection between bitumen and crumb rubber which thus affects the mixture bonding.

Inclusion of crumb rubber, the mixture with the least crumb rubber added at 1% recorded to have the lowest Marshall stability value, followed by the crumb rubber added at 2% and 3%. As the crumb rubber content increases, the stability value increases, which is opposite with the findings by [5]. Increasing stability is due to the elastic behaviour of the crumb rubber that helps in sustaining the load, but alas, too much crumb rubber added will cause the mixture to be agglomerated and losing cohesiveness. In the dry process, usually, crumb rubber is added up to 3% of the total mixture weight [12].

Mashaan found that adding crumb rubber to asphalt mixture improved the Marshall stability as better adhesion was developed [5]. However, in this research,

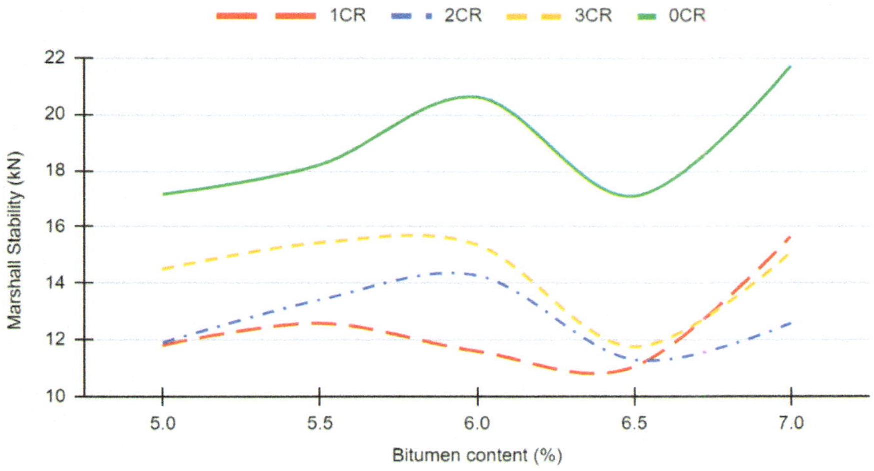

Fig. 1 Marshall stability of CR-SMA

the opposite findings were obtained; thus, adhesion between binder, crumb rubber, and aggregate could be the possible benefactor towards the hitch of the mixture. In general, factors influencing adhesion in asphalt mixes and pavements are aggregate type (surface texture porosity adsorption, mineralogy, moisture and dust), bitumen type, rheology (viscosity), chemical composition, mixture type, air voids, workmanship—segregation, traffic volume, salt, drainage, and climate. One possible factor, the aggregate chemistry, has previously been shown to be the most major contributing component in the adhesion phenomena within bitumen-aggregate systems [13, 14]. A lack of proper aggregate characterisation procedures has led to the contradictory conclusion that moisture susceptibility in hot mix asphalt is positively correlated with aggregate acidity (i.e. SiO_2) and negatively correlated with aggregate basicity (i.e. calcium carbonate ($CaCO_3$) content) [15, 16].

3.2 Marshall Flow

The Marshall flow results obtained for different percentages of crumb rubber for each bitumen content are shown in Fig. 2. The flow value is defined as movement in units of 0.25 mm of the permanent strain in between no load and maximum load applied during the stability test. The depicted result in Fig. 2 shows that despite the amount of crumb rubber added, the flow value increases as the bitumen content increases. This is due to the amount of bitumen added causes aggregate to float within the mixture, which thus enhances the flow value. Based on Fig. 2, an increase in crumb rubber content does not occasionally increase the flow value; however, as the bitumen content increases, the mixture with crumb rubber has higher flow value up to an optimum level. In this study, the increase of flow value with the inclusion of crumb rubber agrees with the past findings [17, 18]. The attributor of high flow value is due to low voids and excessive bitumen content, while vice versa on the contrary. A high flow value may steer to the effect of rutting, whereas a low flow value may cause to the effect of cracking [4, 19].

3.3 Volumetric Analysis

Bulk Density. The bulk density (G_{mb}) of CR-SMA for different percentages of crumb rubber for each bitumen content is shown in Fig. 3. The obtained results indicate characteristics of compaction of different mixtures at a same number of blows and temperature to have a notable effect on the mixture density. The graph shows with the addition of crumb rubber, and the increase of bitumen content has reduced the density. This is explained as the crumb rubber increases the porosity of the matrix by absorbing the bitumen, hence leaving the aggregate with more voids and decreasing the density. Every mixture having various densities is due to viscosity effect of bitumen and crumb rubber interaction. Rubber particles will absorb the maltenes

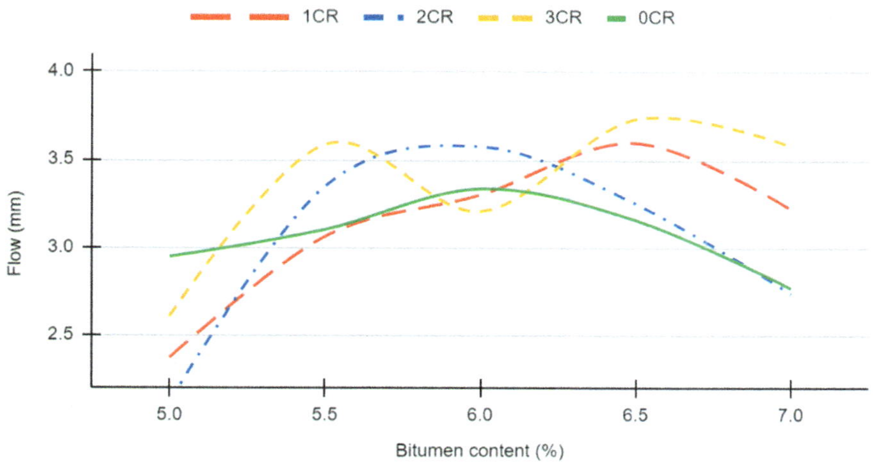

Fig. 2 Marshall flow of CR-SMA

fraction of bitumen and make the residual bitumen have higher asphaltenes content which thus increases the viscosity [20]. With higher viscosity of binder, it makes better resistance during compaction, which consequently reduces the density [21].

Voids in Mix (VIM). The VIM or porosity of CR-SMA for different percentages of crumb rubber for each bitumen content is shown in Fig. 4. This property is essential in the asphalt mixture design as it has a significant effect on the pavement durability. From the graph, all mixtures show the same trend where with higher bitumen content, porosity is reduced, and mixture with crumb rubber leads to a rise in VIM, which

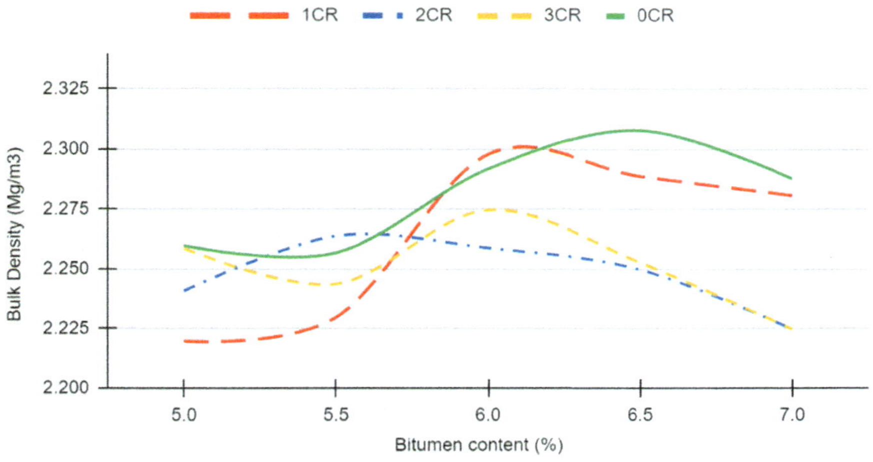

Fig. 3 Bulk density of CR-SMA

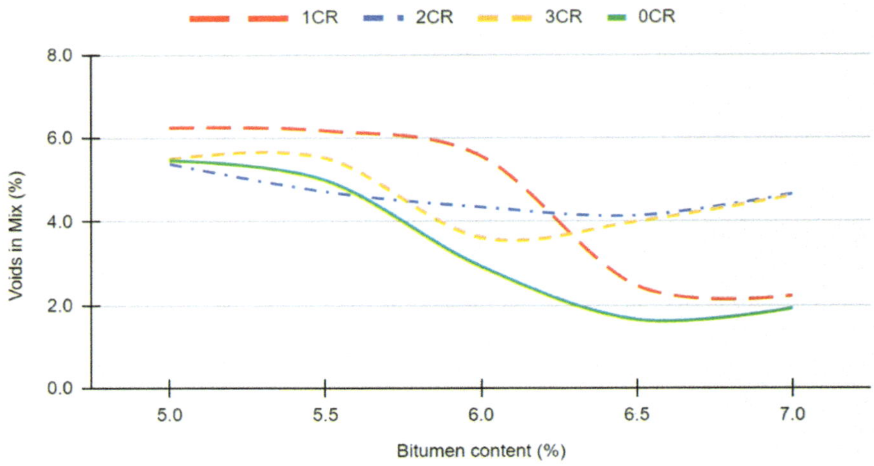

Fig. 4 VIM of CR-SMA

is owing to a lower contacting surfaces between aggregates when crumb rubber content is raised. The high volume of crumb rubber particles absorbs the bitumen needed to embody the aggregate and then fills the spaces between the aggregates. With excessive bitumen content compared to the optimal value of the mixture, VIM started to reduce as much more gaps were filled. In this study, the porosity is varied because although the dry process thoroughly mixed the crumb rubber with aggregate, when bitumen was added, the bitumen-crumb rubber-aggregate interaction was not consistent. When the bituminous mixture has a high porosity, it has many spaces that allow air and water to pass through. Lower porosity, on the other hand, necessitates bleeding, which involves squeezing bitumen from the mixture to the surface.

3.4 Binder Drain Down

Figure 5 represents the binder drain down percentage with the addition of crumb rubber at a different amount. This analysis indicates that the existence of crumb rubber will enhance the bonding between the binder and aggregates, which is very crucial during production, storage, transport, and placement on real site [22]. The bar chart shows improvement on the drain down by adding the crumb rubber up to 2%. However, the drain down percentage elevates when the crumb rubber content is at 3% which has an almost a similar value to the control mixture. This explained that higher crumb rubber content decreases the cohesiveness of bitumen, as the crumb rubber absorbs some of the bitumen fraction, making the bitumen have higher viscosity. Though the mixing was using dry process, it was stated by the previous researchers that the finer rubber particles partially reacted with the bitumen, increasing its viscosity [23]. When rubber comes into contact with bitumen, it absorbs the low

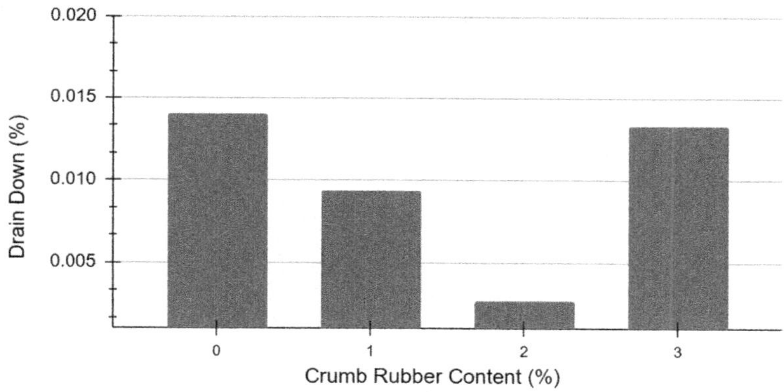

Fig. 5 Binder drain down of CR-SMA

molecular weight, which is maltenes fraction. This leaves the remaining bitumen with a greater part of high molecular weight, which is asphaltenes, which have a and cause the bitumen to have a higher viscosity [23–25]. Thus, at high viscosity, the flow of bitumen is reduced, and this has caused the mixture to have less bonding.

Figure 6 portrays the images of different mixtures after the binder drain down test. Based on the visual, it is vividly shown the difference of binder material to be drained off with different proportions of crumb rubber. For the control sample and mixture with the highest crumb rubber content as in Fig. 6a and b, respectively, bitumen was drained together with the aggregate particles, proving the bonding was not good enough. The trend is also the same with the previous research where crumb rubber addition managed to improve the drain down of the mixture [5].

3.5 *Resilient Modulus*

Figure 7 presents the resilient modulus with the addition of crumb rubber in a different amount prepared at OBC. This analysis was done to determine if the addition of crumb rubber brought any significant changes in the stiffness of the SMA. The bar chart shows a reduction of resilient modulus as the increasing amount of crumb rubber is added to the mixture. These results may be explained by the fact that the strong matrix was generated by homogeneously virgin SMA and only slight rubber particles in the mixes and that this maintained an adhesion between the bitumen and the aggregate. It is possible to draw the conclusion that the rubber particle had a detrimental impact on the cohesiveness [26]. Apart from the nature of crumb rubber that affects the property of mastic in CR-SMA, this study is also using the sandstone type of aggregate, which is different from most of the previous study. Referring to Table 2 shows that the sandstone aggregate used in this study has a high SiO_2 content,

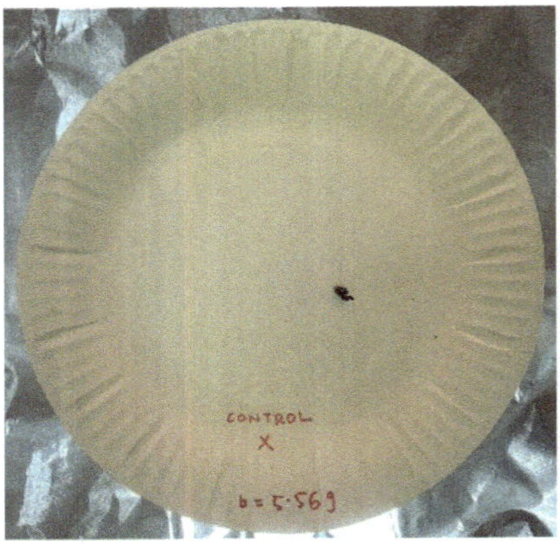

(a) Control

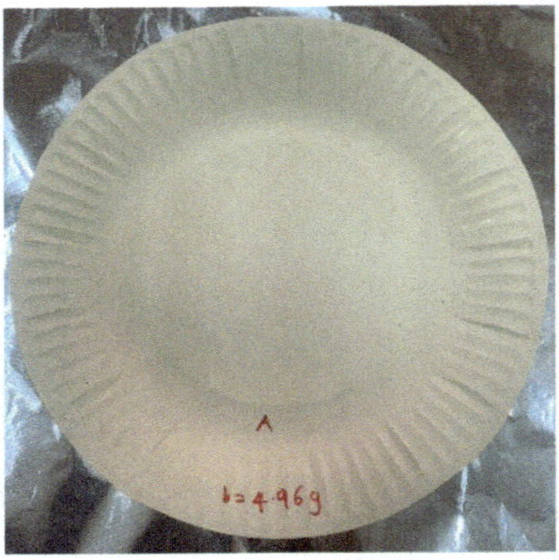

(b) 1% crumb rubber

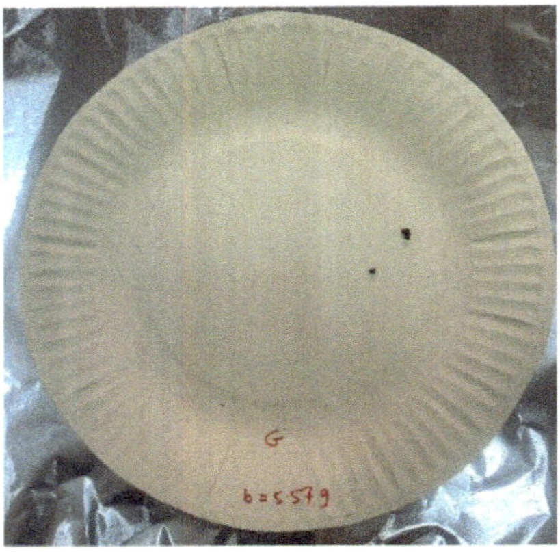

(d) 2% crumb rubber

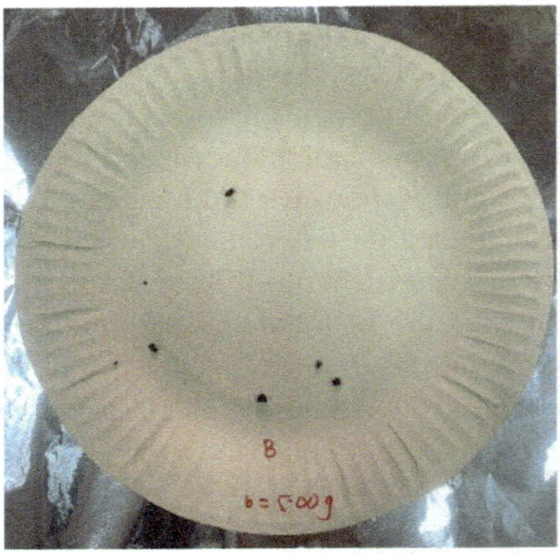

(c) 2% crumb rubber

Fig. 6 Visual observation of CR-SMA binder drain down

which could be the reason for adding a lower resilient modulus value. With high SiO_2 content, aggregate will produce poor adhesion of aggregate-bitumen relation [11].

3.6 Optimum Binder Content (OBC)

Based on the results presented in Sects. 3.1 to 3.5, the specifications to obtain OBC of CR-SMA in accordance to the Malaysia Standard Specification of Public Work Department 2019 as in Table 5, the requisite Marshall stability, Marshall flow, volumetric analysis and binder drain down analysis is adequate to determine the OBC of CR-SMA. The OBC was found at 5.30%, 5.80%, 5.90%, and 5.90% for 0%, 1%,

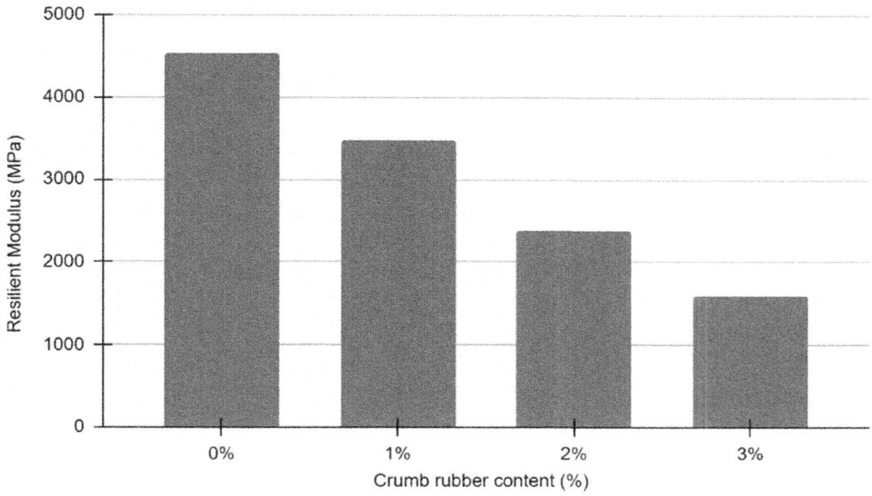

Fig. 7 Resilient modulus of CR-SMA

Table 6 OBC of CR-SMA

Crumb rubber content (%)	Stability (kN)	Flow (mm)	G_{mb} (Mg/m^3)	VIM (%)	Drain down (%)	OBC (%)
0	17.20	3.015	2.256	5.28	0.014	5.30
1	12.50	3.185	2.273	6.00	0.009	5.80
2	13.85	3.605	2.261	4.41	0.003	5.90
3	15.60	3.230	2.272	3.95	0.013	5.90

2%, and 3% of crumb rubber inclusion, respectively, after taking the average of the OBC result for every parameter as in Table 6.

4 Conclusion

The findings of this study indicate that crumb rubber in SMA incorporating the local sandstone aggregate in Sabah, Malaysia, does not improve the stiffness and the ability of the mixture to resist being squeezed out from under a load. Though in loose form mixture at the OBC, the findings give promising results in terms of stability for production, storage, transport, and placement in real site purpose, in a compacted sample, it does not help to provide sturdy structure compared to the virgin SMA mixture. From the results of all tests conducted, conclusions can be drawn from the findings;

(i) Crumb rubber at 2% of the total aggregate weight is recommended to use as it shows the gratifying result of volumetric properties and bitumen drain down; though the Marshall stability has lower compared to control, it gives a double reading of the minimum requirement in the standard.
(ii) It is recommended to study the mineralogy of the sandstone aggregate, as it is the only raw material that differs from the past research incorporating crumb rubber in asphalt. In most cases, a comprehensive chemical analysis of aggregates is not considered when decisions are being made about the selection of aggregates from various sources for the manufacturing of asphalt mixes. This is something that should be brought to attention. In spite of this, it is possible to get an experimental and trustworthy formula for linking the chemical composition of aggregates and the suggested mechanical performance indices if the aggregate type employed in this study is expanded further and comprehensive chemical analysis is performed.

Acknowledgements The authors would like to acknowledge UMSGreat Research Grant (GUG0527-2/2020) from Universiti Malaysia Sabah and Industry Research Grant (LPS2101) from Astenpave Sdn Bhd that funded this study.

References

1. Teurquetil, F., Raju, S.: Chapter 11: choosing asphalts for use in flexible pavement layers. In: The Shell Bitumen Handbook, pp. 261–293. ICE Publishing, London. ISBN: 978-0-7277-5837-8 (2015)
2. Blazejowski, K.: Stone Matrix Asphalt Theory and Practice. CRC Press (2011)
3. Fu, Z., Dang, Y., Guo, B., Huang, Y.: Laboratory investigation on the properties of asphalt mixtures modified with double-adding admixtures and sensitivity analysis. J. Traffic Transp. Eng. (2016)
4. Mashaan, N.S., Ali, A.H., Koting, S., Karim, M.R.: Performance evaluation of crumb rubber modified stone mastic asphalt pavement in Malaysia. Adv. Mater. Sci. Eng. **2013** (2013)
5. Mashaan, N.: Use of Waste Tyre Rubber in Enhancing the Mechanical Properties of Stone Mastic Asphalt Mix. Doctoral dissertation, Universiti Malaya, Kuala Lumpur (2016)
6. PWD: Standard Specification for Road Works (Section 4: Flexible Pavement Adendum Specification for Rubber Modified Asphalt). Cawangan Jalan, Jabatan Kerja Raya Malaysia, Kuala Lumpur, Malaysia (2019)
7. Samsuri, A.: Properties of rubberised bitumen from reclaimed rubber. In: Conference on the Use of Rubberised Bitumen in the Road Construction, pp. 15–23. Malaysia (1997)
8. Sangiorgi, C., Eskandarsefat, S., Tataranni, P., Simone, A., Vignali, V., Lantieri, C., Dondi, G.: A complete laboratory assessment of crumb rubber porous asphalt. Constr. Build. Mater. **132**, 500–507 (2017)
9. Mull, M., Stuart, K., Yehia, A.: Fracture resistance characterization of chemically. J. Mater. Sci. **7**, 557–566 (2002)
10. Navarro, F.M., Gámez, M.C.R.: Influence of crumb rubber on the I direct tensile strength and stiffness modulus of hot bituminous mixes. J. Mater. Civ. Eng. **24**(6), 715–724 (2012)
11. Cala, A., Caro, S., Lleras, M., Rojas-Agramonte, Y.: Impact of the chemical composition of aggregates on the adhesion quality and durability of asphalt-aggregate systems. Constr. Build. Mater. **216**, 661–672 (2019)

12. Hassan, N.A., Airey, G.D., Jaya, R.P., Mashros, N., Aziz, M.M.A.: A review of crumb rubber modification in dry mixed rubberised asphalt mixtures. Jurnal Teknologi **70**(4), 127–134 (2014)
13. Curtis, C.W., Ensley, K., Epps, J.: Fundamental properties of asphalt-aggregate interactions including adhesion and absorption. In: SHRP-A-341. Strategic Highway Research Program, National Research Council, Washington, DC (1993)
14. Curtis, C.W.: Investigation of Asphalt-Aggregate Interactions in Asphalt Pavements Fuel. American Chemical Society, pp. 1292–1297 (1992)
15. Airey, G.D., Collop, A.C., Zoorob, S.E., Elliott, R.C.: The influence of aggregate, filler and bitumen on asphalt mixture moisture damage. Constr. Build. Mater. **22**(9), 2015–2024 (2008)
16. Kakar, M.R., Hamzah, M.O., Valentin, J.: A review on moisture damages of hot and warm mix asphalt and related investigations. J. Clean. Prod. **99**, 39–58 (2015)
17. Mahrez, A.: Properties of rubberised bitumen binder and its effect on the bituminous mix. MSc. Dissertation, Faculty of Engineering, University of Malaya, Malaysia (1999)
18. Kumar, P., Mehndiratta, H.C., Singh, K.L.: Rheological properties of crumb rubber modified bitumen—A lab study. J. Sci. Ind. Res. **68**, 812–816 (2009)
19. Putri, E.E., Idral, M., Makinda, J., Gungat, L.: Marshall properties of porous asphalt with Gondorukem rubber addition. J. Eng. Sci. Technol. **14**(1), 167–180 (2019)
20. Thodesen, C., et al.: Effect of crumb rubber characteristics on crumb rubber modified (CRM) binder viscosity. Constr. Build. Mater. **23**(1), 295–303 (2009)
21. Lee, S.J., Amirkhanian, S.N., Thodesen, C., Shatanawi, K.: Effect of compaction temperature on rubberized asphalt mixes. In: Proceeding Asphalt Rubber, Palm Springs, CA, USA (2006)
22. Bindu, C.S., Beena, K.S.: Influence of additives on the drain down characteristics of stone matrix asphalt mixtures. Int. J. Res. Eng. Technol. **3**(7), 83–88 (2014)
23. Airey, G.D., Rahman, M.M., Collop, A.C.: Absorption of bitumen into crumb rubber using the basket drainage method. Int. J. Pavement Eng. **4**(2), 105–119 (2003)
24. Gawel, I., Stepkowski, R., Czechowski, F.: Molecular interactions between rubber and asphalt. Ind. Eng. Chem. Res. **45**(9), 3044–3049 (2006)
25. Lee, S.J., Amirkhanian, S.N., Shatanawi, K.: Effect of reaction time on physical and chemical properties of rubber-modified binders. In: International Rubber Conference. Compendium of Papers CDRom, Lyon, France (2006)
26. Cetin, A.: Effects of crumb rubber size and concentration on performance of porous asphalt mixtures. Int. J. Polym. Sci. **2013** (2013). https://doi.org/10.1155/2013/789612

Structural Engineering

Non-standard Large-Scale Fire Tests of Structures: A Mini Review

Ibrahim Almeshal, B. H. Abu Bakar, and Bassam A. Tayeh

Abstract Testing of large-scale structural fire has seen a resurgence in recent years, after nearly a century of using the test of standard fire resistance to understand how structural members respond to fires. The regulatory and scientific communities are grappling with a host of issues related to demonstrating adequate structural performance utilizing unrealistic temperature–time curves that are applied on isolated structural members. As a result, non-standard fire testing that is done on a large scale with real fire rather than conventional fires is gaining popularity. Several non-standard, custom-made testing facilities have recently been developed or are almost finished. Over the last three decades, non-standard fire testing has revealed substantial faults in our knowledge of real-world building performance in the face of real-world fire; in most cases, these problems would not have been discovered in tests of conventional furnaces. This study provides a brief overview of necessary non-standard structural fire engineering research conducted on a wide scale in recent decades. It highlights gaps and study requirements based on past research findings and the writers' evaluation of the information. There is also a summary of comparative research needs evaluations that have been conducted or presented in the last ten years. The overall goal is to identify knowledge gaps and guide future studies in structural fire engineering, especially large-scale experimental studies focusing on reinforced concrete structures.

Keywords Non-standard fires · Standard fire testing · Fire resistance

I. Almeshal · B. H. Abu Bakar (✉)
School of Civil Engineering, Engineering Campus, Universiti Sains Malaysia, 14300 Nibong Tebal, Malaysia
e-mail: cebad@usm.my

B. A. Tayeh
Civil Engineering Department, Faculty of Engineering, Islamic University of Gaza, Gaza Strip, Palestine

N. Sabtu (ed.), *Proceedings of AWAM International Conference on Civil Engineering 2022—Volume 2*, Lecture Notes in Civil Engineering 385,
https://doi.org/10.1007/978-981-99-6018-7_9

1 Introduction

One of the most serious environmental risks is fire, and the safety design to mitigate this risk is an essential aspect of civil construction design [1, 2]. Natural catastrophes such as floods and earthquakes and events such as fires can cause structural damage. Every year, fire damage targets about 2.5% of all residential structures worldwide [3]. One of the most severe occurrences that have occurred in Malaysia is fire. Year after year, fires grow, notably in residential structures, with the most significant fire rate among building types [4]. Figures 1 and 2 depict the overall number of fire cases and fires in Malaysian residential structures. Between 2008 and 2018, Malaysia saw 383,621 fires, resulting in RM25 billion in damages [5]. These occurrences push an engineer to think more imaginatively and innovatively while designing a structure to prevent or postpone a fire.

It is crucial to comprehend fire behaviour and the impacts of thermal exposure on structural components in today's fast-changing built environment for the continuous supply of design strategies of fire protection [6–8]. Fire behaviour concepts generated from enclosure fire research have historically affected general structure design considerably [9, 10]. Most fire tests in recent decades have been conducted on single isolated structural elements (i.e. walls, slabs, beams and columns) that have been heated in a furnace. Such experiments do not reflect the behaviour of a whole structure in the event of a fire [11, 12]. Member interaction influences many aspects of structural behaviour. Hence, isolated element testing cannot predict or detect structural continuity. The support and loading conditions of any structural element can be altered by interactions between multiple structural parts in a whole system. This change might result in structural behaviour that is totally different from that predicted by the initial combination of loading and boundary conditions. In reality, structures

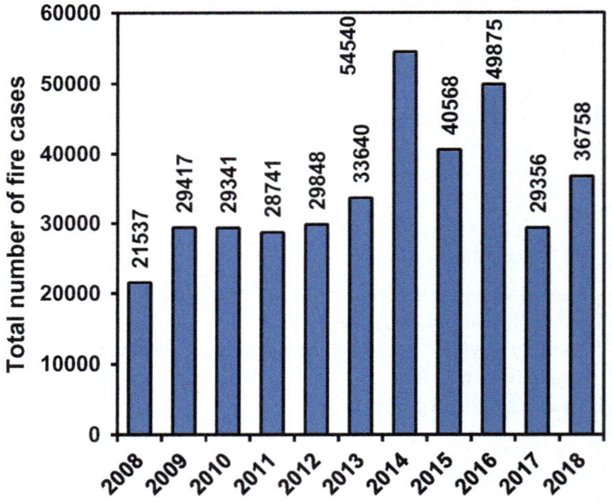

Fig. 1 Fire cases in Malaysia

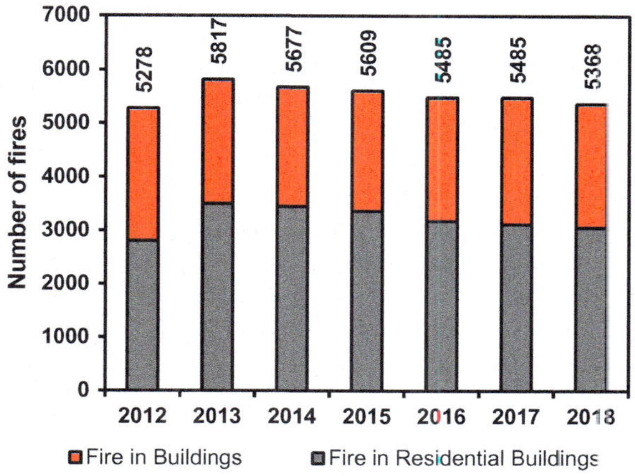

Fig. 2 Fire in buildings and contents

exposed to fire hazards perform significantly better to what is anticipated in standard tests [13]. A natural fire in a fundamental structure does not follow the usual temperature–time heating curve. In fact, many significant parameters in full-scale compartments are not considered in standard fire tests such as structural continuity, boundary conditions, restraint, redistribution and membrane activities. Traditional fire tests utilizing a fire furnace for single members can only reveal local failure of specific members and cannot offer adequate information on the collapse of the entire structure. Individual member tests, as a result, provide little insight into overall building behaviour, needing full-scale fire testing to get this information [14]. After more than a century of demonstrating sufficiency using standard contrived temperature–time curves to test isolated single structural members, the time has come to move forward. Both the scientific and regulatory communities are now using the standard temperature–time curves to determine the fire resistance of single-element on isolated structural components [15]. During the preceding three decades, non-standard fire testing revealed substantial faults in our comprehension of real structural behaviour during real-world flames; in the majority of cases, these defects would not have been identified in typical furnace tests. Fire temperature distribution in big space fires differs from that in small compartment flames, according to full-scale fire studies on large space buildings [16]. Heat transmission and thermal deterioration in assemblies subjected to real flames differ from that in a homogeneous furnace exposure, according to experience [17, 18]. A full-scale experimental series is being conducted to examine and define flames in significant open-plan areas, such as those seen in modern buildings [19]. Several non-standard, custom-built testing facilities have also lately been built or are approaching completion [15].

Buildings that have collapsed due to actual flames (such as the World Trade Centre) [20] and Delft University of Technology's Faculty of Architecture Building [21] demonstrate that structural performance in actual flames is thought to be more complex than in furnace testing. A fire-protection design that is based on furnace testing may not be conservative when the influence of fire-produced temperature gradient is considerable and fire spreading occurs. As a result, testing structures in realistic flames has become a priority in the transition to fire-protection design in structural engineering, and the establishment of improved fire-testing procedures has been a major topic of research. As a result, there is still a considerable need for a better knowledge of the failure processes of structural systems in fire and the improvement of existing design techniques, both pre- and post-fire [22].

It is known that the concrete structures have a complex behaviour in fire. This mini review study aims to present a constructive criticism on the standard fire testing and to provide some of the research that used the non-standard fire furnaces on concrete structures. This paper addresses a discussion on simplified methods that have been utilized effectively for many years for designing concrete members and structures to withstand the impact of real fires.

2 Criticisms of the Standard Fire Test

There are several flaws in standardized structural fire resistance testing [23]. Before giving a review of existing material and previous research on standard large-scale fire, it is worth going over some of the previous reviews that have addressed this or comparable topics:

- Two essays regarding the history of ASTM E119 were written by [24]. The authors confirmed that the ASTM E119 standard was not based on information on building fire intensities. The fire standard was established based on a single variable that determined the intensity levels. While several countries have a viable alternative for the "standard" fire, the US testing technique has remained unchanged since 1918 [24, 25].
- After a workshop at the National Institute of Standards and Technology, Grosshandler [26, 27] gave a complete report on the topic (NIST). The study included several recommendations, including developing novel experimental methods for evaluating the mechanical and thermal characteristics of structure and insulating materials at high temperatures [26]. In a journal paper published in 2003, the same author made numerous recommendations about the standard fire test, including:
  - The current test techniques of structural fire resistance must address (1) structural element reaction up to ultimate failure, (2) statistical failure uncertainty and (3) a choice of classification or system of rating in units other than time, as well as accompanying recommendations.
  - To enable the transition to fire resistance design of performance-based structural, a persistent, multi-national research effort is required [27].
- In 2004, NIST also released a report by Almand et al., which included several recommendations, the most important of which are: (1) standard fire tests do not give information on a component's actual performance in a real fire; (2) no analytical or practical tools are adequate to assess the effect of alternative design

and strategies of fire protection to enlist the help of firefighters; and (3) there are no analytical or practical techniques that are sufficient to assess the efficiency of various design and fire safety measures [28].

- In discussion papers concerning the fire resistance of structures, Kodur et al. said that present structural fire safety measures have major limitations and serious knowledge gaps in the literature. The absence of substantial research efforts in this sector and the dearth of teaching and training programmes at universities are two significant causes for these restrictions [29, 30].
- Beitel and Iwankiw, commissioned by NIST, released a study after investigating 22 multi-storey fires between 1970 and 2002. According to the authors, catastrophic failure modes found in actual structures could not have been predicted using average fire resistance (furnace) testing in most situations. In all situations, structural interactions and connection response were also essential [31].
- A recent study conducted by [15] evaluated structural fire testing on non-standard large-scale compartments and agreed with [32], who indicated that standard fire test is not an accurate tool for making comparison between structures' behaviour and should not be even used to measure the performance of a single member in a real fire [32]. Despite that, the authors agreed with the fire testing industry that standard furnace testing offers benefits such as repeatability and control and is thus helpful for benchmarking and comparison testing [15].
- Gales et al. produced a book on the performance of post-tensioned concrete that was subjected to standard fire test (2015). According to the authors, standard fire testing is unable to recreate a wide range of critical behaviours that may be anticipated and have been observed in genuine unbonded post-tensioned buildings during natural flames [33]
- The standard temperature–time curve for buildings was developed by Gales et al. [34]. According to the author, the standard temperature–time curve (ASTM E119) has not changed appreciably since 1916. Worse, science does not support the heating curve's portrayal of an actual fire [34].

3 What is the Solution to Predict the Fire Resistance of Concrete Structures?

Both actual fire and fundamental structure should be applied through an experimental or numerical model to comprehend a structure's fire resistance fully. That simulation has two parameters: the structure, which is challenging to build due to the high expenses. The majority of prior research on concrete fire behaviour focused on: (1) materials and partial members; (2) reinforced concrete (RC) member/single element; (3) sub-frame assemblies; (4) transiently simulated restrained assemblies; and (5) actual buildings. The second is the sort of fire, which is categorized as follows:

(i) Elevated temperature exposures (transient or steady state): increasing the temperature in linear association and above 100 °C, when hydrothermal reaction starts [35]. This type of fire tests is often used to determine the fire resistance of building materials.
(ii) Standard fires: are time–temperature curves provided by international institutes such as ISO 834, BS 476 and ASTM E119. The test involves exposing a test specimen to a standard fire that is controlled to maintain specific temperatures for a set amount of time [36].
(iii) Equivalent fire severity to a standard fire: is beneficial where standard fire resistance test results are available; however, time equivalent formula may not be accurate beyond the data range for which they were calibrated. It is far more precise to calculate structural fire resistance by using the first principles [37].
(iv) Parametrically fires: the parametric fire model presented in Annex A of Eurocode 1 [38].
(v) Localized fires: Unless there are highly unusual circumstances, every fire in a building begins as a tiny, localized fire. When a fire flashes over, it loses its localized nature. Even a localized fire may have a large influence on a structure. This depends on the relative location of the fire with respect to the structural components and on the type of the structure in general [39].
(vi) Zone model: separating the fire into few numbers of major zones such as the hot smoky gas layer and the plume under the roof. This method is considered simpler compared to the field model. It works to rely on empirical and well-established correlations and equations for smoke and heat transfer between these zones [40].
(vii) Field model fire: is a field of applied mathematical modelling that deals with fire dynamics. To analyse the behaviour of fire, the fundamental laws of fluid mechanics and heat transfer are considered as represented in the laws of conservation mass, momentum and energy. Along with other concepts and equations that support the field modelling method, there is the fundamental basis of the fire field modelling [41].
(viii) Real fire/Non-standard fire: The term "real fire" is often used to describe fires that start within structures, either inadvertently or intentionally [42].

There are two difficulties with concrete in a fire: mechanical property degradation as temperature increases, induced by physicochemical variations in the material during the heating process; and explosive spalling, which leads to material loss, section size reduction and reinforcing steel exposure to high temperatures [35]. In the case of concrete structures, there are three types of failure criteria: (1) structural adequacy (load-bearing capability), (2) structural integrity (flame resistance) and (3) structural insulation (the capability to stop fire from spreading due to an extreme elevation in the temperature of a non-heated face).

The reader can see from the preceding section's analysis of building kinds, fire types and failure criteria that there are numerous possibilities for selecting the sort of fire that can be executed on a complete compartment. The standard fire is an essential time–temperature relationship that academics use for comparison rather than to assess a building's actual performance in a fire. A genuine fire (non-standard) should be carried out on an actual structure (large scale) to fully understand the performance of a fireproof building, which will be described in the next section.

4 Non-standard Large-Scale Fire Test of Reinforced Concrete

Even while key structural parts work well when isolated during normal furnace testing, it is worth mentioning that the concrete buildings, rather than focusing on standardized tests, pioneered what is now known as unique non-standard fire tests on structural "assemblies" of concrete. In the 1950s, for example, early research on beam-slab assemblies and two-way post-tensioned concrete slabs in the USA used a combination of slabs and beams to investigate the two-dimensional response of these assemblies (Troxell 1959). Despite the fact that this early research produced a lot of furnace test data, the current study focuses on more recent endeavours. As a result, these "historical" tests are no longer valid.

Several academics have proposed a large-scale non-standard fire test to address all of the above-mentioned problems, including [15]. However, few non-standard structural large-scale fire tests on RC structures have been conducted owing to the common perception that, cover spalling in the absence of explosive cover, concrete structural components outperform exposed steel parts in conventional furnace testing. As a result, the concrete industry appears to see little value in utilizing time-consuming and expensive testing programmes to examine and demonstrate the potential advantages of considering the overall structure interactions and ability of structure members to carry the design load such as membrane action and load-carrying mechanisms in RC structures during a fire hazard. Because of the evident economic benefits and competitive advantages of non-standard large-scale structural fire testing, the steel construction industry has actively promoted this testing methodology. Despite this, there is a tiny amount of concrete structuration that is significant. Building behaviour in fire is more complex than what present prescriptive standards would indicate, according to available concrete testing. This might have both positive and negative implications for concrete structures in the event of a fire [15].

Concrete constructions offer both potential benefits and hazards in a fire [15]. Many researchers have identified the use of custom-built or modified conventional furnaces to examine unique structural member performance concerns or certain sorts of concrete structures that are difficult to investigate using a normal "component" method. A few notable cases are summarized in Table 1.

Table 1 Non-standard large-scale RC fire tests

Researcher	Material type	Structure/element type	Significant outcome
Van Herberghen [43]	Unbonded post-tensioned RC	Flat floor slabs	• The authors used amended standard furnace for studying the performance of the floor under fire • To ensure a better understanding of post-tensioned RC slabs resistance in fire, numerical models could be used after careful validations
Baily [44]	Cast-in-place RC	Seven storey building 3 × 4 bays and 22.5 m × 30 m in plan	• After 25 min, a maximum temperature of 950 °C in the gas phase was measured before the equipment was turned off (the temperature was assumed to have kept increasing) • Vertical displacements at the building's perimeter were more significant than those near the centre, and there was no trace of a stabilizing plateau • Designers should also consider concrete spalling when selecting fasteners between compartment walls and the soffit of concrete slabs
Kelly and Purkiss [45]	Unbonded post-tensioned RC	Three-span continuous slab strips	• The authors used amended standard furnace for studying the performance of the slab under fire
Bailey and Lennon [46]	Precast RC	Slabs on steel beam flooring systems	• The optimal average temperature in the gas phase was above 1000 °C; during the cooling phase of the fire, the edge units cracked locally; however, this did not result in a loss of overall load-bearing capability • Due to thermal expansion limitations, there was evidence of a lateral compressive strip forming at the ends of the units, which would have increased the units' flexural and perhaps shear capacity

(continued)

Table 1 (continued)

Researcher	Material type	Structure/element type	Significant outcome
[42]	Post-tensioned RC	High-strength columns and slabs	• Adding fibres to a high-strength mix reduces spalling significantly • Spacing the column ties closer together has no effect on how much spalling occurs • Polypropylene fibres should be considered for this sort of construction since a post-tensioned slab containing a single aggregate type spalled poorly • One of the post-tensioned slabs showed no signs of spalling, whereas another spalled on both ends for no apparent reason
Wong and Ng [47]	RC	40 unloaded columns, in an actual building	• Insulation, fire-resistant coating materials, polypropylene fibres, or wire mesh should be used in high-strength column members to prevent spalling. These tests are not applicable to real-world fires in structures. The testing techniques utilized in this study were unusual, even for non-standard structural fire testing
Ring et al. [48]	RC	Four large-scale non-standard fire tests on "frame-like" structures	• There are several advantages of using polypropylene fibres to prevent explosive spalling in a fire

5 ACI 216 and IBC 2021 Methods

The technique utilized by the American Concrete Institute (ACI 216) committee and the Masonry Society (TMS) to find out how fire-resistant concrete members (ACI/TMS 216.1-14(19)) will be presented in this part. Although ASTM E 119 testing is arguably the most dependable method, due to the cost, effort and time required to create and test the assemblies on large scale, this method is impractical and, in many cases, led to unreal results. The methodologies and specifications used in ACI 216.1 were developed based on several previous fire studies that were done between 1958 and 2005, and they are still the most widely used in everyday assessment and design practices. Based on the heat transfer endpoint, the fire resistance of a specific concrete member or full assembly is calculated by determining its equivalent thickness (only the concrete core) and then relating it to the corresponding fire resistance duration in the tables and charts. The correct thickness of slabs and solid walls that have level surfaces is the same as solid wall thickness. ACI 216.1 formulae must be used to estimate the equivalent thickness of walls and slabs with gaps, undulations, ribs, or several layers of different materials (e.g. a sandwich of concrete, insulation and concrete) [49].

ACI 216.1 specifies an analytical method for assessing flexural members' fire resistance. This approach entails calculating the true temperatures of the concrete and the steel reinforcement and then assessing the materials' characteristics at those temperatures. The bottom joyful moment steel, according to the method, will be exposed to high temperatures and start to decay before the top concrete and reinforcement. It enables the member's moment to shift from a weaker, positive zone to a more significant, negative zone. Once the part or assembly has been determined to have sufficient comparable thickness and once the member or assembly has been determined to have sufficient equivalent thickness to satisfy the heat transmission endpoint, it must be determined whether the reinforcing steel has sufficient cover to prevent excessive heat from reducing the yield strength to the point where it can no longer carry the load. The necessary fire rating, aggregate type, restrained or uncontrolled construction and prestressed or unstressed reinforcement all had an impact on the slab cover requirements [50].

All around the world, national building regulations control the fire resistance of different parts and assemblies that form a building structure. Structural frames (consisting of columns and beams), load-bearing walls and floor systems must be able to endure the stresses and strains imposed by fully developed fires while also carrying their dead and superimposed loads for the needed duration. Prescriptive rules for building components are presented in Chap. 7 of the 2021 International Building Code (IBC). This chapter's tables explain various combinations of materials and finishes that fulfil certain fire resistance ratings for a building. The techniques or approaches listed below were specified:

(i) Prescriptive fire resistance: in this method, tables are used to describe fire resistance ratings for using insulating materials such as plaster and unit masonry. The table provides fire resistance periods of these insulating material based on the thickness.
(ii) Calculated fire resistance: in this method, the fire resistance can be calculated through tables and figures which are in compliance with the tables found in ACI 216.1, with the exception of the provisions for the use of high-strength concrete columns found in ACI 216.1.
(iii) Engineering evaluation based on a comparative analysis of the designs of building elements, components or assemblies with fire resistance ratings as indicated by ASTM E119 or UL 263 test procedures.
(iv) Fire resistance designs are documented in approved agency by IBC or approved sources.

6 Conclusion and Research Needs in Structural Fire

- Fundamental techniques for defining fire settings have been utilized in the past, such as assuming homogenous temperatures or using basic temperature–time curves, both of which are inaccurate depictions of a real fire. There is now a need for more thorough research on variations in concrete mixtures, including temporal as well as spatial differences in heating. To understand the holistic behaviour of concrete buildings, additional testing of whole concrete structures in real fire is required, including interconnections between different structural components, and to assist in the validation of sophisticated computer models. Detailed research of concrete building performance in real-world fires can also aid our knowledge of real-world behaviour.
- The behaviour of concrete in fire is currently unknown, and other studies are needed in most themes. The physical performance of concrete degrades in diverse ways based on the concrete mix's details, notably the moisture content and critical climatic variables like maximum fire temperature and duration. It is vital to do a systematic study on the impacts of various heating settings on concrete.
- Linking these finely detailed small-scale behaviours to the performance of whole structures in realistic fires is more difficult. Using exact models to anticipate spalling behaviour remains a severe challenge, despite substantial advances in modelling the mechanical of concrete structures, especially when the basic function of LITS is fully accounted for. Moreover, the capacity to forecast structural interactions that may contribute to failures is lacking.

References

1. Li, G.-Q., Zhang, C., Jiang, J.: A review on fire safety engineering: key issues for high-rise buildings. Int. J. High-Rise Build. **7**, 265–285 (2018)
2. Ni, S., Gernay, T.: Predicting residual deformations in a reinforced concrete building structure after a fire event. Eng. Struct. **202**, 109853 (2020)
3. Božović, M., Mišić, M.: Understanding fire and protecting the buildings. In: Kosanović, S., Klein, T., Konstantinou, T., Radivojević, A., Hildebrand, L. (eds.) Sustainable and Resilient Building Design Approaches, pp. 49–66, Methods Tools, TU Delft Open, Dublin (2018)
4. Khiyon, M.I., Kadir, M.A.A., Zukri, S.N.N.M., Osman, M.H.: Preliminary assessment of fire damaged reinforced concrete structure and steel truss. Malaysian J. Civ. Eng. **29** (2017)
5. Fire and Rescue Department of Malaysia, Annual Report (2016). http://www.bomba.gov.my/index.php/pages/view/746?mid=251
6. Chow, W.K., Zou, G.W., Dong, H., Gao, Y.: Necessity of carrying out full-scale burning tests for post-flashover retail shop fires. Int. J. Eng. Performance-Based Fire Codes **5**, 20–27 (2003)
7. Almeshal, I., Abu Bakar, B.H., Tayeh, B.A.: Behaviour of reinforced concrete walls under fire: a review. Fire Technol. (2022). https://doi.org/10.1007/s10694-022-01240-3
8. El-Sayed, A.A., Fathy, I.N., Tayeh, B.A., Almeshal, I.: Using artificial neural networks for predicting mechanical and radiation shielding properties of different nano-concretes exposed to elevated temperature. Constr. Build. Mater. **324**, 126663 (2022). https://doi.org/10.1016/j.conbuildmat.2022.126663

9. Maluk, C., Linnan, B., Wong, A., Hidalgo, J.P., Torero, J.L., Abecassis-Empis, C., Cowlard, A.: Energy distribution analysis in full-scale open floor plan enclosure fires. Fire Saf. J. **91**, 422–431 (2017). https://doi.org/10.1016/j.firesaf.2017.04.004
10. Mostafa, S.A., Tayeh, B.A., Almeshal, I.: Investigation the properties of sustainable ultra-high-performance basalt fibre self-compacting concrete incorporating nano agricultural waste under normal and elevated temperatures. Case Stud. Constr. Mater. **17**, e01453 (2022). https://doi.org/10.1016/j.cscm.2022.e01453
11. Nagy, N.: Determination of Thermal Properties of Mineral Wool Insulation Materials for Use In Full-Scale Fire Modelling (2020)
12. Udi, U.J., Almeshal, I., Megat Johari, M.A., Mohd Zahid, M.Z.A., Hassan, M.H., Musa, K.A., Albiajawi, M., Abu Bakar, B.H.: Efficiency of high performance fiber reinforced cementitious composites as a retrofit material for fire-damaged concrete. Mater. Today Proc. **61**, 477–486 (2022). https://doi.org/10.1016/j.matpr.2021.12.278
13. Yang, Z.N., Dong, Y.L., Xu, W.J.: Fire tests on two-way concrete slabs in a full-scale multi-storey steel-framed building. Fire Saf. J. **58**, 38–48 (2013). https://doi.org/10.1016/j.firesaf.2013.01.023
14. Roy, K., Lim, J.B.P., Lau, H.H., Yong, P.M., Clifton, G.C., Johnston, R.P.D., Wrzesien, A., Mei, C.C.: Collapse behaviour of a fire engineering designed single-storey cold-formed steel building in severe fires. Thin-Walled Struct. **142**, 340–357 (2019). https://doi.org/10.1016/j.tws.2019.04.046
15. Bisby, L., Gales, J., Maluk, C.: A contemporary review of large-scale non-standard structural fire testing. Fire Sci. Rev. **2**, 1 (2013)
16. Du, Y., Li, G.: A new temperature–time curve for fire-resistance analysis of structures. Fire Saf. J. **54**, 113–120 (2012). https://doi.org/10.1016/j.firesaf.2012.07.004
17. Ellingham, J., DiDomizio, M.J., Weckman, E.J.: A new test methodology for studying the response of walls to real fire environments. Fire Mater. **44**, 323–332 (2020)
18. Didomizio, M.: Experimental study of thermal degradation of fire resisting compartment partitions in fires (2017)
19. Hidalgo, J.P., Cowlard, A., Abecassis-Empis, C., Maluk, C., Majdalani, A.H., Kahrmann, S., Hilditch, R., Krajcovic, M., Torero, J.L.: An experimental study of full-scale open floor plan enclosure fires. Fire Saf. J. **89**, 22–40 (2017). https://doi.org/10.1016/j.firesaf.2017.02.002
20. Shyam-Sunder, S., Gann, R.G., Grosshandler, W.L., Lew, H.S., Bukowski, R.W., Sadek, F., Gayle, F.W., Gross, J.L., McAllister, T.P., Averill, J.D.: Federal Building and Fire Safety Investigation of the World Trade Center Disaster: Final Report on the Collapse of World Trade Center Building **7** (2008)
21. Engelhardt, M.D., Meacham, B., Kodur, V., Kirk, A., Park, H., van Straalen, Ij., Maljaars, J., van Weeren, K., de Feijter, R., Both, K.: Observations from the fire and collapse of the faculty of architecture building, delft university of technology. Struct. Congr. 2013 Bridg. Your Passion Your Prof. 1138–1149 (2013)
22. Zhang, C., Grosshandler, W., Sauca, A., Choe, L.: Design of an ASTM E119 fire environment in a large compartment. Fire Technol. **56**, 1155–1177 (2020)
23. Gales, J., Bisby, L., Gillie, M.: Unbonded post tensioned concrete slabs in fire—part I–experimental response of unbonded tendons under transient localized heating. J. Struct. Fire Eng. **2**, 139–154 (2011). https://doi.org/10.1260/2040-2317.2.3.139
24. Babrauskas, V., Williamson, R.B.: The historical basis of fire resistance testing—Part I. Fire Technol. **14**, 184–194 (1978). https://doi.org/10.1007/BF01983053
25. Babrauskas, V., Williamson, R.B.: The historical basis of fire resistance testing—Part II. Fire Technol. **14**, 304–316 (1978). https://doi.org/10.1007/BF01998390
26. Grosshandler, W.: Fire resistance determination and performance prediction research needs workshop: US Department of Commerce, Technology Administration, National Institute of …. Gaithersburg, MD (2002). https://doi.org/10.6028/NIST.IR.6890
27. Grosshandler, W.L.: The international FORUM of fire research directors: a position paper on evaluation of structural fire resistance. Fire Saf. J. **38**, 645–650 (2003). https://doi.org/10.1016/S0379-7112(03)00068-7

28. Almand, K.H., Phan, L.T., McAllister, T.P., Starnes, M.A., Gross, J.L.: NIST-SFPE workshop for development of a National R&D Roadmap for Structural Fire Safety Design and Retrofit of Structures: Proceedings (2004)
29. Kodur, V.K.R., Garlock, M.E.M., Iwankiw, N.: NISTGCR 07–915–Structures in fire: state of the art, research and training needs. Natl. Inst. Stand. Technol. Gaithersburg, MD USA (2007)
30. Kodur, V.K.R., Garlock, M., Iwankiw, N.: Structures in fire: state-of-the-art, research and training needs. Fire Technol. **48**, 825–839 (2012). https://doi.org/10.1007/s10694-011-0247-4
31. Beitel, J.J., Iwankiw, N.: Analysis of needs and existing capabilities for full-scale fire resistance testing, US Department of Commerce, Technology Administration, National Institute of … (2008)
32. Harmathy, T.Z., Lie, T.T.: Fire test standard in the light of fire research. In: Fire Test Perform, pp. 85–85–13. ASTM International, West Conshohocken, PA 19428-2959 (1970). https://doi.org/10.1520/STP44713S
33. Gales, J., Hartin, K., Bisby, L.: Structural Fire Performance of Contemporary Post-tensioned Concrete Construction. Springer New York, New York, NY (2016). https://doi.org/10.1007/978-1-4939-3280-1
34. Gales, J., Chorlton, B., Jeanneret, C.: The historical narrative of the standard temperature-time heating curve for structures. Fire Technol. **57**, 529–558 (2021). https://doi.org/10.1007/s10694-020-01040-7
35. Khoury, G.A.: Effect of fire on concrete and concrete structures. Prog. Struct. Eng. Mater. **2**, 429–447 (2000)
36. ASTM E119-20, Standard Test Methods for Fire Tests of Building Construction and Materials (2020). https://doi.org/10.1520/E0119-20
37. Thomas, G.C., Buchanan, A.H., Fleischmann, C.M.: Structural fire design: the role of time equivalence. Fire Saf. Sci. **5**, 607–618 (1997)
38. European Committee for Standardization: EN 1991-1-2.Eurocode 1, Actions on structures—Part 1–2: General actions—Actions on structures exposed to fire (2002)
39. Franssen, J.-M., Real, P.V.: Fire Design of Steel Structures: Eurocode 1: Actions on structures; Part 1–2: General actions—Actions on structures exposed to fire; Eurocode 3: Design of steel structures; Part 1–2: General rules—Structural fire design. John Wiley & Sons (2012)
40. HUME, B.T.: Evaluation of Fire Models Summary Report (1992)
41. Yeoh, G.H., Yuen, K.K. (eds.): Field modeling approach. Comput. Fluid Dyn. Fire Eng. 29–133 (2009). https://doi.org/10.1016/B978-0-7506-8589-4.00002-8
42. CCAA, Fire Safety of Concrete Buildings, Cement Concrete and Aggregates Australia (2010). https://www.ccaa.com.au/CCAA/Public_Content/PUBLICATIONS/Technical_Publications/Guides/Guide_to_Fire_Safety_of_Concrete_Buildings.aspx?WebsiteKey=4998d6ce-2791-4962-b1e2-6b717f54a8d3
43. Van Herberghen, P., Van Damme, M.: Fire resistance of post-tensioned continuous flat floor slabs with unbonded tendons, FIP Notes 3–11 (1983)
44. Bailey, C.: Holistic behaviour of concrete buildings in fire. Proc. Inst. Civ. Eng. Struct. Build. **152**, 199–212 (2002). https://doi.org/10.1680/stbu.2002.152.3.199
45. Kelly, F., Purkiss, J.: Reinforced concrete structures in fire: a review of current rules. Struct. Eng. **86**, 33–39 (2008)
46. Bailey, C.G., Lennon, T.: Full-scale fire tests on hollowcore floors. Struct. Eng. **86**, 33–39 (2008)
47. Wong, Y., Ng, Y.: Effects of water quenching on reinforced concrete structures under fire, Hong Kong (2011)
48. Ring, T., Zeiml, M., Lackner, R., Eberhardsteiner, J.: Large scale fire tests on concrete design and results. In: Poster Present. 18th Inter-Institute Seminer Young Res. Sept. Budapest, Hungary (2011)
49. ACI/TMS Comm 216, ACI CODE-216.1-14(19) Code Requirements for Determining Fire Resistance of Concrete and Masonry Construction Assemblies 28 (2019)
50. Bilow, D.N., Kamara, M.E.: Fire and concrete structures. Struct. Congr. 1–10 (2008). https://doi.org/10.1061/41016(314)299

Effect of Hollow Bodies on the Strength and Density of Bubble Concrete

Xiangdong Yan, Pei-Shan Chen, Bashar S. Mohammed, and Baoxin Liu

Abstract Bubble concrete is a new type of lightweight concrete achieved by mixing high-strength hollow bodies into concrete. Different from the stress mechanism of the voided biaxial slab, the hollow bodies are used not only to create multiple cavities in concrete but also to transfer internal stresses. To achieve a better strength of the specimen, this study investigates the arrangement and shape effects of the hollow bodies on the mechanical properties. In addition, a novel hollow body model was proposed to improve the stress distribution to increase strength and reduce density. The density, strength, and stiffness of three types of bubble concrete were investigated through concrete compression experiments. Then, the stress distribution and failure mechanism of the bubble concrete were explored with nonlinear elastoplastic analysis. The results show that the strength and density of bubble concrete with regular arrangement hollow bodies yielded better than the random arrangement. Furthermore, with the new hollow body models, the bubble concrete further increased the strength of concrete to 79.7–85.3% and reduced the density to 80.0–85.0%.

Keywords Bubble concrete · Stress distribution · Hollow body

X. Yan · P.-S. Chen (✉) · B. Liu
Department of Civil and Architectural Engineering, Kyushu Institute of Technology, Kitakyushu 804-8550, Japan
e-mail: chen@civil.kyutech.ac.jp

X. Yan
e-mail: yan.xiangdong760@mail.kyutech.jp

B. S. Mohammed
Department of Civil and Environmental Engineering, University Technology PETRONAS (UTP), 32610 Bandar Seri Iskandar, Malaysia
e-mail: bashar.mohammed@utp.edu.my

N. Sabtu (ed.), *Proceedings of AWAM International Conference on Civil Engineering 2022—Volume 2*, Lecture Notes in Civil Engineering 385,
https://doi.org/10.1007/978-981-99-6018-7_10

1 Introduction

The good thermal insulation and seismic properties of lightweight concrete have made it an increasingly popular building material worldwide. Among the material of lightweight concrete, light aggregate concrete is one of the most concerned research objects [1–5]. Lightweight aggregate concrete is made from expanded shale, clay, or slate materials that are fired in a rotary kiln to form a porous structure. Materials such as clay, perlite, or vermiculite are suitable raw materials for producing lightweight aggregate concrete [6]. However, the porous structure of lightweight aggregate concrete makes it generally unable to sustain a high strength. Thus, lightweight concrete is mainly used for nonstructural elements such as partition walls. As a result, more research work is required to use lightweight concrete for structural elements such as load-bearing structures.

In a previous study, author Pei-Shan Chen proposed a new lightweight concrete inspired by hollow bird bone. It was formed by mixing high-strength hollow bodies into concrete to form new concrete and was named as bubble concrete (hereafter named as BC.) [7–12]. The BC differs from foam concrete and biaxial hollow concrete slabs [13–15], because the hollow bodies are used to create multiple cavities inside the concrete and transfer internal stresses. A schematic diagram of the BC is shown in Fig. 1. By the virtue of hollow bodies, BC has a lower density and is subjected to a low seismic load while maintaining strength simultaneously.

In general, cavities in lightweight concrete reduce the effective cross-sectional area for transferring stresses, thereby initiating internal micro-cracks and reducing the strength of the concrete. However, high-strength hollow bodies in the BC could provide an effective cross-sectional area for transferring stress. In addition, the contact between the concrete and the hollow body can provide adhesion and avoid premature failure.

The shape of the hollow body and the adhesion of the material can significantly affect the strength of BC [12, 16]. Therefore, it is necessary to investigate the effect of different hollow body shapes on the strength of BC. The various types of hollow

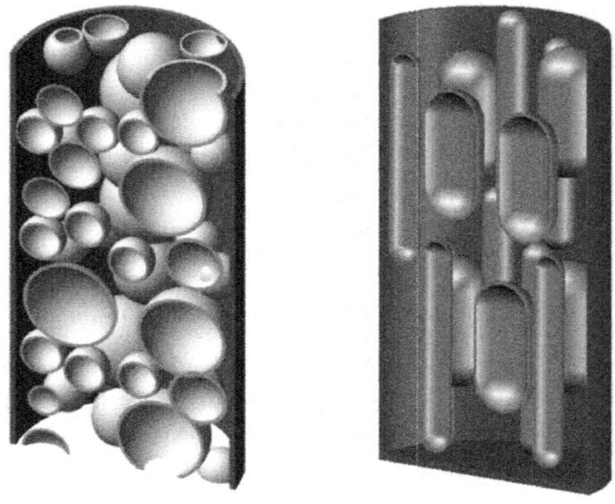

Fig. 1 Bubble concrete contains high-strength hollow bodies

(a) Hollow spheres with surface treatments.

(b) Special stress mechanisms modeling.

Fig. 2 Hollow bodies with different shapes and mechanisms

body models were considered to improve the stress distribution in BC, as shown in Fig. 2. In Fig. 2a, the surface roughening is aimed to increase the adhesion of the hollow body to the concrete, while in Fig. 2a, b special stress mechanism has been proposed to improve the internal stress distribution.

Previous studies found that the strength of the BC with spherical hollow bodies was the most appropriate model, which transfers stresses efficiently. However, the deformation of the spheres would cause the concrete to be squeezed and cracked. Therefore, a new capsule hollow body model was proposed in this study to reduce the area where the concrete is squeezed and finally improve the strength of the specimen.

Because the arrangement and shape of the hollow body will affect the strength and stress path of BC, the study reported in this paper aims to obtain a suitable arrangement and hollow body model to enhance the compressive strength through concrete experiments and nonlinear analysis methods. In this study, the stress distribution of BC under the random arrangement and regular arrangement of the hollow bodies were investigated. Then, the suitable hollow body model was determined based on the results. However, it should be noted that the concrete experiment of capsule hollow body model is still under investigation, so only analytical results were reported in this paper.

2 Randomly Arranged Hollow Spheres

In this section, concrete compression experiments and nonlinear elastoplastic finite element analysis were carried out to study the effect of random arrangement of hollow bodies on the strength of BC. The hollow bodies were randomly mixed into the concrete, and several concrete specimens were made according to the diameter and number of hollow bodies. The experimental and analytical results were compared to investigate the mechanical properties and failure principle of BC.

2.1 Concrete Compression Experiments

Cylindrical concrete specimens with a diameter of 150 mm and a height of 300 mm have been cast and prepared in accordance with the specification of the Japanese Industrial Standard (JIS) [17]. The hollow body is a hollow steel sphere (welded of two hemispheres) made of an SS400 steel plate with a yield point of 235 MPa. Two diameters of 48.6 mm and 60.5 mm with a thickness of 2.3 mm hollow bodies have been used. Five concrete specimens without steel spheres have been prepared as controls.

As shown in Table 1, five groups with three concrete specimens are made in each group according to the diameter and number of steel spheres. Three concrete specimens without steel spheres as control have been prepared using the same concrete mixture. The concrete used in the experiment has a compressive strength of 40 MPa, a maximum aggregate size of 20 mm, a water-reducing agent of 5.4 kg/m^3, and a slump of 21.5 cm. The concrete specimens were filled in three layers, the bottom and sides were vibrated with vibrating rods, and they were cured in water for 28 days before testing.

The density and strength of the specimens are shown in Fig. 3. The average density of concrete without spheres is 2.3 g/cm^3, while the average density of bubbled concrete is 2.2–2.2 g/cm^3. The average strength of the specimens without steel spheres is 43.7 MPa, while the average strength of the BC is 25.2–29.4 MPa. The stress–strain curve of the concrete specimen is shown in Fig. 4. The horizontal axis is compressive strain, and the vertical axis is compressive strength. Symbols G1-1 and G1-2 of Fig. 4 represent the strain measured by the strain gauges on the left and right sides of the concrete specimen. The results show that mixing hollow bodies into concrete have reduced the strength and elastic modulus and further decreased as the number of the sphere increased.

In addition, increasing the diameter of the steel sphere causes a reduction in the strength and elastic modulus of concrete. The failure pattern of BC is shown in Fig. 5. Several cracks occurred in the BC specimen from top to bottom, and the concrete

Table 1 Sphere diameters and the number of spheres

Type	Diameter (mm)	Thickness (mm)	Sphere number	Specimen number
M50-20	48.6	2.3	18	3
M50-40	48.6	2.3	27	3
M60-8	60.5	2.3	8	3
M60-12	60.5	2.3	12	3
M60-16	60.5	2.3	15	3

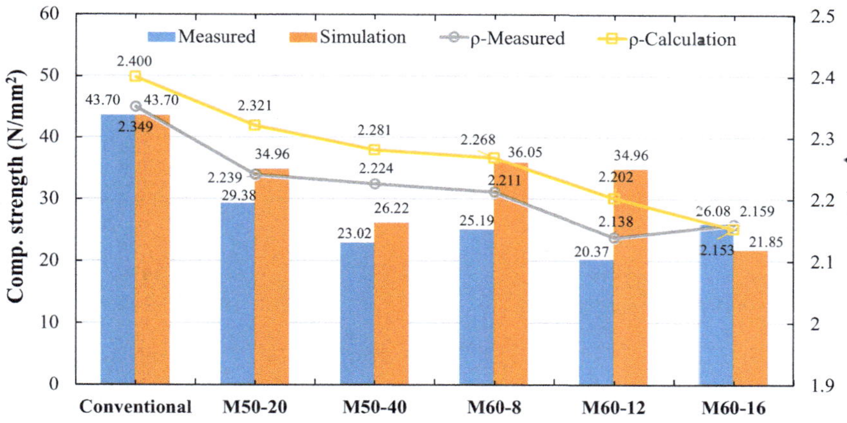

Fig. 3 Density and compressive strength of specimens

severely peeled off the steel spheres. The concrete was crushed, and the specimen failed subsequently. Due to the smooth contact surface between the sphere and the concrete, there is no damage on the sphere surface.

2.2 *Nonlinear Elastoplastic Simulation*

Due to the difficulty in measuring the stress distribution inside the concrete and the sphere, the FEM software ADINA was utilized to perform nonlinear elastoplastic analysis. The dimension and material curves of the concrete and sphere models used for the analysis are shown in Fig. 6. The spheres are randomly arranged in the concrete, and a uniform load is applied on the top. The simulated concrete strength is consistent with the control concrete strength each time. The model is uniformly loaded and increased in ten steps until the concrete failure. To save the time for finite element analysis by CPU, only the middle 1/3 of the BC included the hollow body and the compressive stress–strain curve of the analysis model is calculated as 1/3 of the height.

The stress distribution inside the BC is shown in Fig. 7. When the concrete is loaded in the *Z*-direction, significant stress concentration has occurred on the surface of the sphere, causing the sphere to deform. The deformation of the sphere reduces the stress on the upper and lower sides of the sphere, while the stress in the oblique direction has increased. In addition, the stress around the sphere has been superimposed, and part of the concrete was crushed. Premature concrete damage occurs when the sphere is close to the edge of the concrete.

According to the results of the experiment and the stress distribution diagram of the BC, the main reasons for the specimen failure could be summarized as follows:

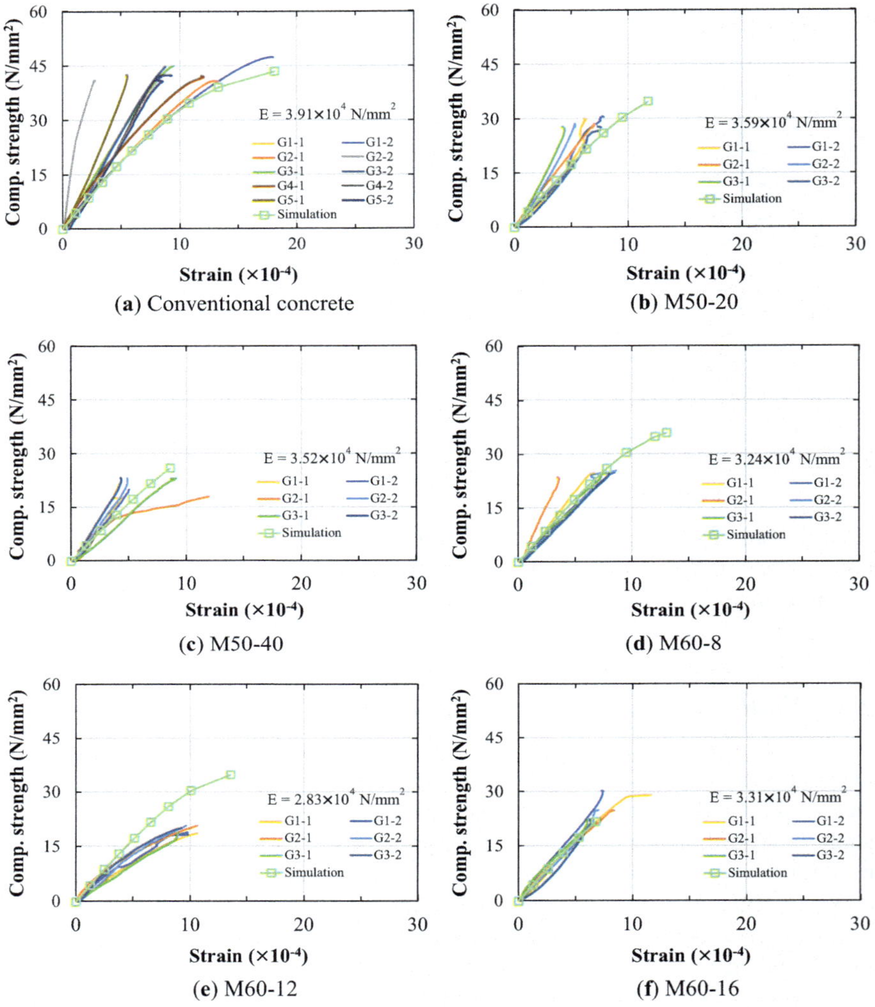

Fig. 4 Stress–strain curve of concrete with spheres. *Notes* Subcaption of each figure **a–f** denotes: multiple spheres, diameter, and sphere number. The average elastic modulus (E) is calculated based on the 1/3 strength of the specimen

(1) Compression is generated in the vertical direction, and the stress in the longitudinal region around the sphere decreases. Conversely, expansion is generated in the oblique direction, and the stress in this region around the sphere increases.

(2) The adhesion stress between the steel sphere and the concrete is insufficient, so the concrete is prone to separate from the steel sphere. Furthermore, the steel sphere close to the edge of the concrete may cause the outer concrete to peel off easily.

Fig. 5 Failure pattern of concrete specimens with randomly arranged hollow spheres

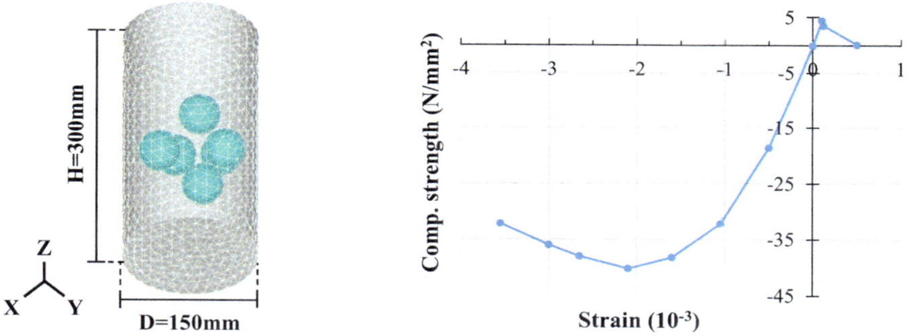

(a) Material curve of concrete specimen.

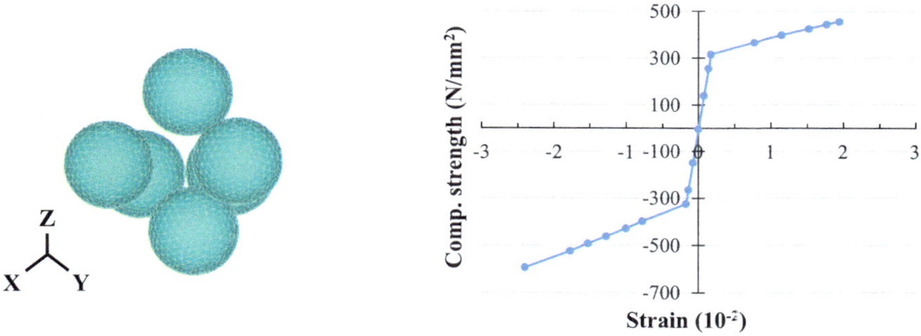

(b) Material curve of steel sphere.

Fig. 6 Dimension and material curves

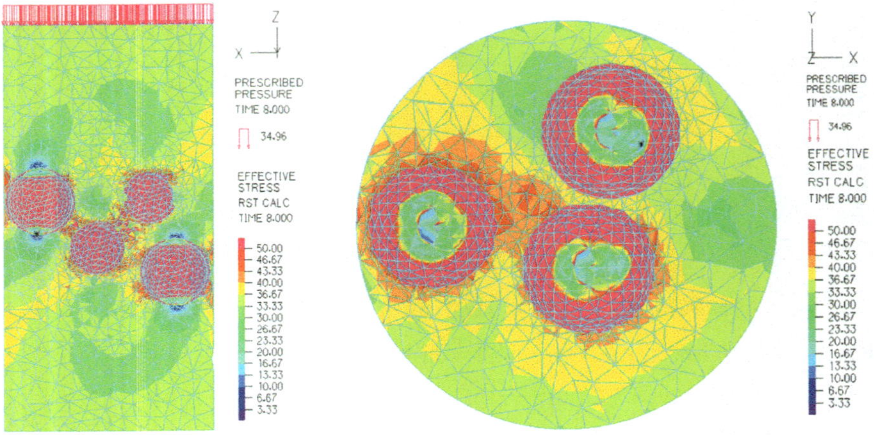

Fig. 7 Internal stress diagram (M50-20)

(3) Steel spheres act as additional aggregates and therefore lead to a reduction in cement paste content, which leads to a decrease in adhesion stress within the specimen. In addition, since the density of steel spheres is lower than cement, they tend to float on the concrete paste when manufacturing concrete.

3 Regularly Arranged Hollow Bodies

The effects of the regular distribution of hollow bodies on the density, compressive strength, and elastic modulus of BC will be reported in this section. The hollow body was fixed by the wire, and concrete compression experiments and nonlinear analysis were performed.

3.1 Concrete Compression Experiments

As shown in Fig. 8a, the hollow bodies have been interconnected with wire to produce a frame and placed in the concrete mold. The dimensions of the concrete specimens were same with previous. The diameter of the spheres was 40 mm with a thickness of 1 mm, and the distance between each neighboring sphere was about 10 mm. Two types of hollow bodies' frames were fabricated for the experiment, as shown in Fig. 8b. Type-A frame was arranged in the same layer of hollow bodies with seven spheres per layer, while the Type-B frame was fixed with six spheres on the outside and one sphere in the middle of the two layers. Three concrete specimens were made for each type, and three specimens without steel spheres were prepared as controls. The strength of precast concrete was 40 MPa, with a maximum aggregate size of

(a) The frame of the hollow bodies. (b) Image of the configuration type.

Fig. 8 Bubble concrete with regularly arranged hollow bodies

15 mm (adjust the direction of the aggregate, an aggregate with a maximum diameter of 15 mm can be placed in a gap with a width of 10 mm). Proportionally mixed with a water-reducing agent of 5.4 kg/m^3, with an average slump flow of 50 cm.

The density and strength of the concrete specimens are shown in Fig. 9. The average density of conventional concrete is 2.37 g/cm^3, while the BC is 1.97–2.00 g/cm^3. The average strength of conventional concrete is 47.73 MPa, while the average strength of BC is 27.54–28.95 MPa. Compared with the experimental results of random arrangement (M50-20, M60-16), this type of BC has a lower concrete density of 6.3–13.3%. The strain curve of concrete is shown in Fig. 10. Symbols G1-1, G1-2; G2-1, G2-2; G3-1, G3-2 are the readings of strain gauges on the left and right sides of the three concrete specimens. Similar to the previous experimental results, incorporating hollow bodies reduces concrete strength and elastic modulus. However, Type B is slightly better than Type A in strength.

The failure pattern of concrete was shown in Fig. 11. When the concrete was subjected to load, cracks were initiated inside the concrete. The concrete peeled off from the hollow bodies, causing stress to increase in the surrounding concrete and

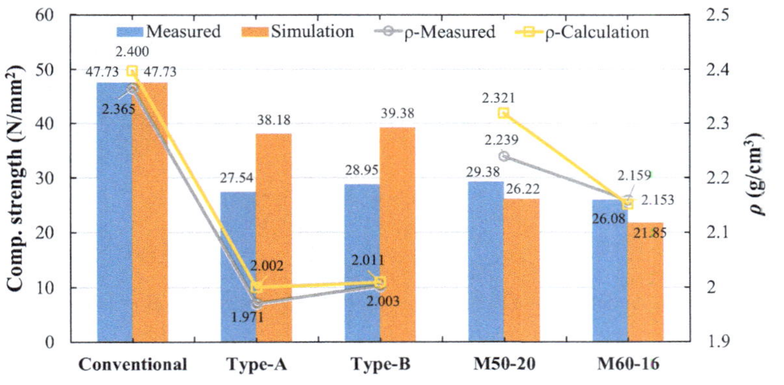

Fig. 9 Density and compressive strength of specimens

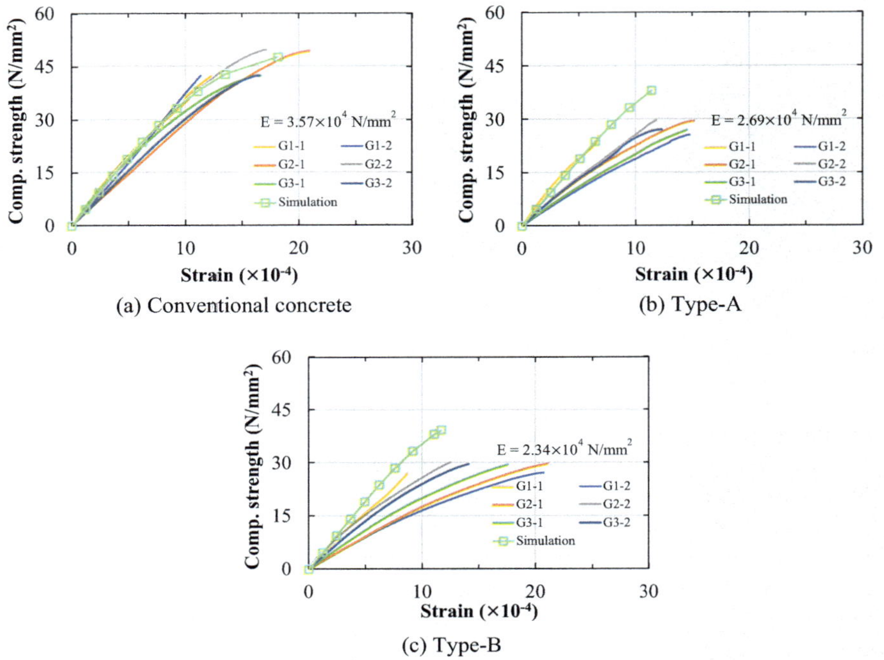

Fig. 10 Stress–strain curve of three type specimens. *Notes* The average elastic modulus (E) is calculated based on 1/3 strength of the specimen

crack. Moreover, the smooth surface of the sphere leads to a weak adhesion to the concrete, which makes the concrete easily damaged after being stressed.

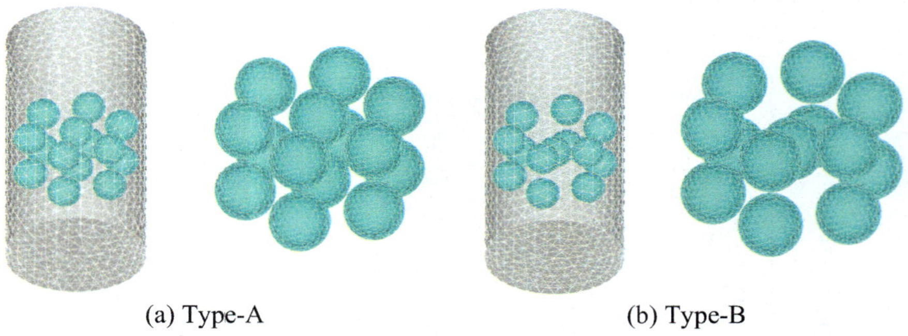

Fig. 11 Analytical model for bubble concrete with regularly arranged hollow bodies

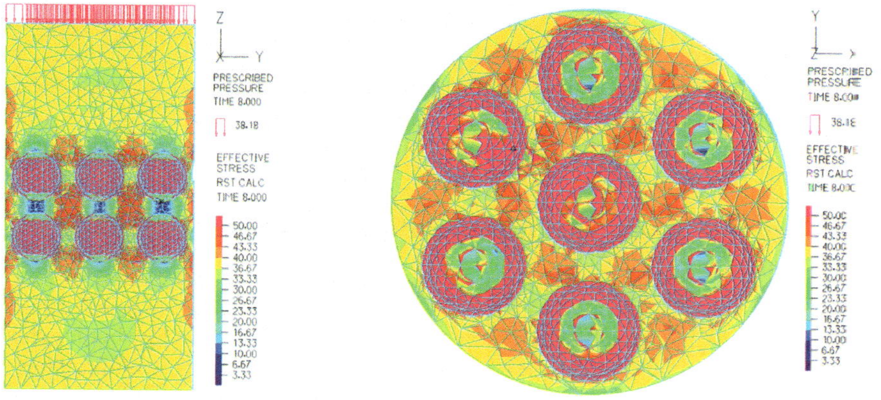

(a) Type-A

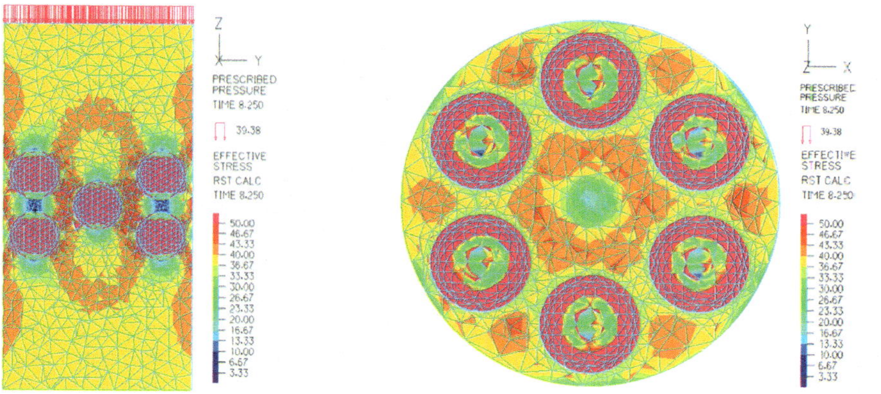

(b) Type-B

Fig. 12 Internal stress diagram of Type-A, Type-B specimen

3.2 Nonlinear Elastoplastic Simulation

The analytical models for the Type A and Type B concrete models are shown in Fig. 11. Since the steel wire is only used to fix the sphere, it has been ignored in the analysis. The stress distribution inside the concrete is shown in Fig. 12. Similar to the previous results, both models produce vertical compression and oblique expansion. However, for Type-B concrete, the concrete stress around the sphere is less than that of Type-A due to the sphere distribution.

Fixing the position of the steel sphere with steel wire can make the stress in the concrete evenly distributed and improve the strength of specimen. However, the regular arrangement may lead to stress concentration in the concrete between the spheres. When the spheres are close to the edge of the specimen, the deformation of the sphere pushes out the partial concrete, causing it to peel off. In addition, the adhesion of spheres to the concrete is insufficient and cannot provide sufficient tensile stress.

4 Nonlinear Elastoplastic Simulation for Hollow Capsule Model

The experimental and analytical results of regularly distributed hollow bodies show that the hollow bodies are deformed inside the concrete specimen. Compression in the vertical direction causes the around stress to decrease, while expansion occurs in the oblique direction, causing the around stress increases. Considering that the hollow body mainly carries the vertical load inside the concrete, the hollow body model has not yet reached the yielding stress. Therefore, the method of merging two adjacent spheres into a capsule model was proposed, as shown in Fig. 13. The capsule model could continue to transmit the vertical stress while reducing the volume of concrete, which leads to a further reduction in the density of the specimen.

The diameter of the capsule remains the same as before, which is 40 mm; the height of the capsule is the height of the merging two spheres, which is 90 mm. Because the hollow body is close to the edge of the concrete will cause the concrete cover to be peeled off. The distance between the hollow body and another hollow body and the concrete edge was adjusted to 7.5 mm, and the loading remained the same as before.

The internal stress diagram and stress–strain curve of concrete are shown in Figs. 14 and 15. The BC with capsule hollow bodies caused a reduction in the stress concentration inside the concrete so that the concrete could carry a higher load, increasing the strength of concrete to 38.96 MPa and reducing the density to

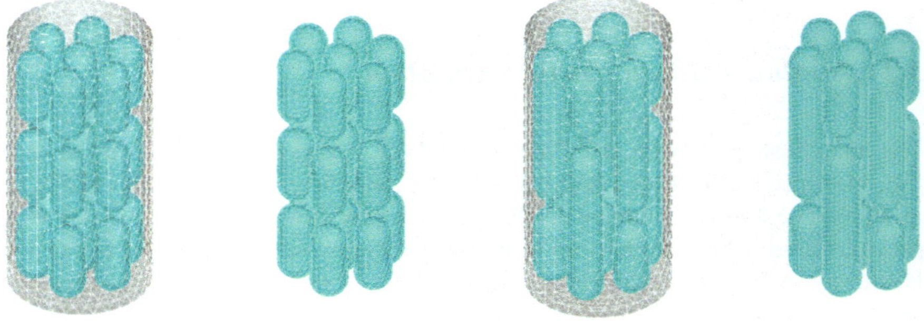

(a) Type-A (b) Type-B

Fig. 13 Analytical model for bubble concrete with capsule models

80%. However, as the stress concentration area still exists, the strength of the specimen is still lower than that of the control. In addition, the material of the hollow body was steel with a density of 7.85 g/cm^3, which causes the density of the BC to be unfavorable. As the concrete experiments of the capsule model are under preparation now, the experimental stress–strain curve used in Fig. 15 is the same as the data in Fig. 10b.

5 Conclusions

Bubble concrete has been developed by mixing high-strength hollow bodies into concrete. Unlike foam concrete and biaxial hollow slabs, hollow bodies mixed into concrete create cavities and transfer stress. Through compression experiment and nonlinear finite element simulation, the effect of hollow sphere arrangement on the strength of BC has been studied.

The hollow bodies are compressed vertically and expanded obliquely for BC with the random arrangement. However, due to the random position of the hollow bodies inside the concrete, the stress distribution may vary with the number of those hollow bodies. The stress is superimposed in the area with many hollow bodies, while the stress is close to the load. When the hollow body is close to the edge of the concrete, it will cause the concrete cover to be peeled off.

For the BC with the regular arrangement, the strength of the concrete was improved, and the density of the concrete was further reduced. However, such an arrangement will cause regular stress concentration inside the concrete, leading to the initiation of cracks, thus reducing the ultimate strength.

For the BC with capsule model. The cylindrical model of hollow bodies improves the stress concentration condition inside the concrete, enhances the strength of the concrete, and further reduces the density of the concrete. In addition, the hollow body material in this experiment is steel, which is dense and cannot lead to a further concrete reduction. Moreover, the surface of the hollow body was polished, resulting in poor adhesion of the hollow body to the concrete. The concrete could not provide sufficient internal tension stress to prevent the external concrete spalling.

The density and strength of the BC with random arrangement hollow spheres, regular arrangement hollow spheres, and regular arrangement spheres are shown in Table 2. The regular arrangement pattern can minimize the density of the concrete and can avoid stress concentration. However, the deformation of the hollow body will squeeze the concrete and cause damage. Thus, the capsule hollow body could reduce the stress concentration area and the density of the concrete.

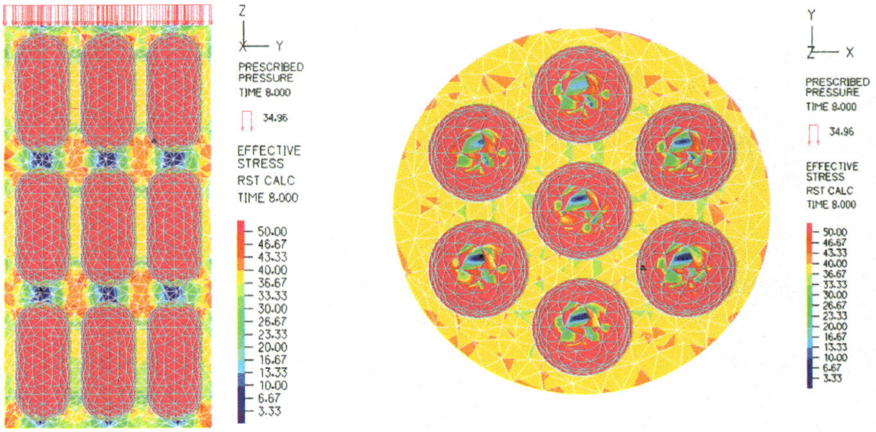

(a) Type-A

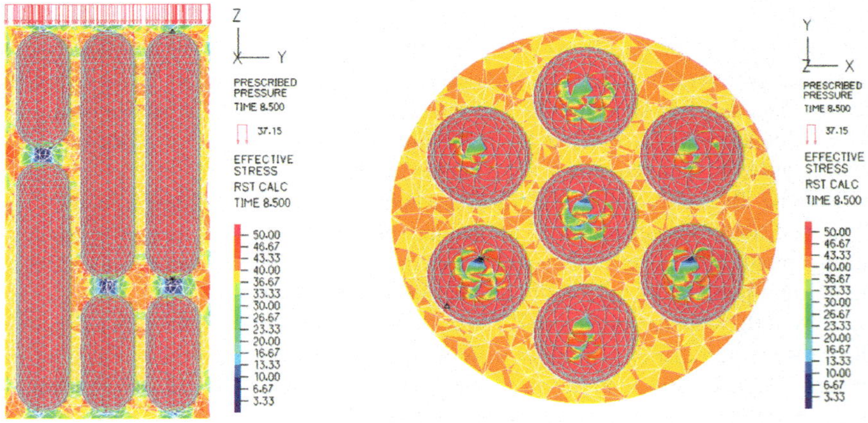

(b) Type-B

Fig. 14 Internal stress diagram of Type-A, Type-B specimen

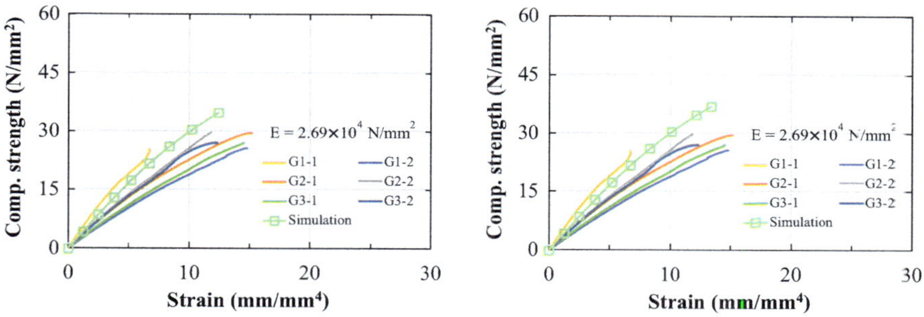

(a) Type-A (b) Type-B

Fig. 15 Stress–strain curve of specimens. *Notes* The average elastic modulus (E) is calculated based on 1/3 strength of the specimen

Table 2 Density and strength of bubble concrete

Type	Density (g/cm^3)	Average strength (N/mm^2)	Simulation strength (N/mm^2)
Random arrangement	2.138–2.239 (91.0–95.3%)	20.372–29.382 (46.6–67.2%)	21.850–36.053 (50.0–82.5%)
Regular arrangement	1.971–2.003 (83.3–84.7%)	27.536–28.954 (57.7–60.7%)	38.184–39.377 (65.0–70.0%)
Capsule model	1.912–2.047 (79.7–85.3%)	–	34.960–37.150 (80.0–85.0%)

Note The percentages in parentheses are ratios based on the control concrete

Appendix

The following are the specific experimental and analytical results of the study. It is necessary to contact the corresponding author Pei-Shan before utilizing the data.

Category	Type	Specimen	Compression strength (N/mm^2)	Weight (g)	Height (mm)	Diameter (mm)
Figure 3	Conventional	1	47.53	12,590	301.0	150.1
		2	41.18	12,480	301.0	150.4
		3	45.06	12,600	302.0	150.2
		4	42.20	12,600	304.0	149.9
		5	42.54	12,540	302.0	150.1
	M50-20	1	30.96	11,970	305.0	149.9
		2	28.84	11,940	301.5	149.9
		3	28.37	11,930	299.0	150.2
	M50-40	1	22.14	11,830	300.0	150.1

(continued)

(continued)

Category	Type	Specimen	Compression strength (N/mm^2)	Weight (g)	Height (mm)	Diameter (mm)
		2	23.62	11,880	301.0	150.1
		3	23.30	11,790	302.0	149.9
	M60-8	1	24.94	11,820	301.0	149.9
		2	25.71	11,840	301.0	150.4
		3	24.92	11,750	302.0	150.2
	M60-12	1	19.11	11,280	301.0	150.2
		2	21.54	11,410	301.0	150.1
		3	20.47	11,500	302.0	149.9
	M60-16	1	30.35	11,660	301.5	150.2
		2	24.97	11,540	303.0	150.1
		3	22.92	11,370	300.0	150.1
Figure 9	Conventional	1	51.13	12,530	300.0	150.0
		2	49.50	12,530	300.0	150.0
		3	42.55	12,560	300.0	150.0
	Type-A	1	25.75	10,400	300.0	150.0
		2	29.66	10,520	300.0	150.0
		3	27.20	10,420	300.0	150.0
	Type-B	1	27.29	10,640	300.0	150.0
		2	29.92	10,490	300.0	150.0
		3	29.65	10,720	300.0	150.0

References

1. Haque, M.N., Al-Khaiat, H., Kayali, O.: Strength and durability of lightweight concrete. Cem. Concr. Compos. **26**, 307–314 (2004). https://doi.org/10.1016/S0958-9465(02)00141-5
2. Min, H.Z., Odd, E.G.: Mechanical properties of high-strength lightweight concrete. Mater. J. **88**, 240–247 (1991)
3. Kan, A., Demirboğa, R.: A novel material for lightweight concrete production. Cem. Concr. Compos. **31**, 489–495 (2009). https://doi.org/10.1016/j.cemconcomp.2009.05.002
4. Lo, T.Y., Tang, W.C., Cui, H.Z.: The effects of aggregate properties on lightweight concrete. Build. Environ. **42**, 3025–3029 (2007). https://doi.org/10.1016/j.buildenv.2005.06.031
5. Abdullah, M.M.A.B., Hussin, K., Bnhussain, M., Ismail, K.N., Yahya, Z., Abdul Razak, R.: Fly Ash-based geopolymer lightweight concrete using foaming agent. Int. J. Mol. Sci. **13**, 7186–7198 (2012). https://doi.org/10.3390/ijms13067186
6. Ducman, V., Mirtič, B.: The applicability of different waste materials for the production of lightweight aggregates. Waste Manag. **29**, 2361–2368 (2009). https://doi.org/10.1016/j.wasman.2009.02.013

7. Chen, P.-S.: A study report on light weight concrete mixed with high strength hollow bubbles. In: Proceeding of AIJ Tohoku Chapter Architectural Research Meeting (Kouzoukei), No. 71, pp. 57–62. Architectural Institute of Japan, Tokyo, Japan (2008)
8. Chen, P.-S.: Bubble concrete. Japan Patent Office, JP2008308395, 25 December 2008. Available online: https://patentfield.com/patents/JP2007181500A. Accessed on 25 December 2008
9. Galilei, G.: Dialogues concerning two new sciences. In: Crew, H., de Salvio, A. (eds.) Macmillan Company, London, England (1638)
10. Evans, H.E., Heiser, J.B.: What's inside: anatomy and physiology. In: Podulka, S., Rohrbaugh, R.W., Bonney, R. (eds.) The Cornell lab of ornithology's handbook of bird biology. Wiley-Blackwell, Hoboken, NJ, USA (2004)
11. Yoshiji, N., Wataru, K., Koji, N.: Failure processes of concrete under triaxial compressive stress. Proc. Jpn. Soc. Civ. Eng. **185**, 31–41 (1971). Available online: https://www.jstage.jst.go.jp/article/jscej1969/1971/185/1971_185_31/_pdf. Accessed on 28 June 2021
12. Yan, X., Chen, P.-S., Al-Fakih, A., Liu, B., Mohammed, B.S., Jin, J.: Experiments and mechanical simulation on bubble concrete: studies on the effects of shape and position of hollow bodies mixed in concrete. Crystals **11**, 858 (2021). https://doi.org/10.3390/cryst11080858
13. Andersen, O., Waag, U., Schneider, L., Stephani, G., Kieback, B.: Novel metallic hollow sphere structures. Adv. Eng. Mater. **4**, 192–195 (2000)
14. Lim, T.-J., Smith, B., McDowell, D.L.: Behavior of a random hollow sphere metal foam. Acta Mater. **50**, 2867–2879 (2002). https://doi.org/10.1016/S1359-6454(02)00111-8
15. Schnellenbach-Held, M., Pfeffer, K.: Punching behavior of biaxial hollow slabs. Cem. Concr. Compos. **24**, 551–556 (2002). https://doi.org/10.1016/S0958-9465(01)00071-3
16. Chen, P.-S., Tsukinaga, Y.: Basic research on the development of light weight concrete mixed with hollow spheres. J. Soc. Mater. Sci. Jpn. **64**, 711–717 (2015). https://doi.org/10.2472/jsms.64.711
17. Japanese Industrial Standard (JIS): JIS A 1108 (2018). Available online: https://webdesk.jsa.or.jp/preview/pre_jis_a_01108_000_000_2018_j_ed10_ch.pdf. Accessed on 28 June 2021

Thermal Performance of Natural Fiber-Reinforced Geopolymer Concrete

Al-Baldawi Maryam Firas, Farah Nora Aznieta Abdul Aziz, Al-Ghazali Noor Abbas, Noor Azline Mohd Nasir, and Nor Azizi Safiee

Abstract Malaysia has a hot tropical climate; hence, the energy required to cool buildings increases daily. Therefore, an attempt was made to produce an eco-friendly construction material with good thermal performance. Geopolymer concrete (GPC) reinforced with two different types of natural fibers, kenaf fiber (KF) or coconut fibers (CFs), was introduced as insulation material. The proposed concrete mixtures were subjected to a non-destructive thermal test using a hotbox apparatus. Three panels of 100 mm thick were fabricated; the first panel consists of GPC as control, the second panel consisted of GPC with 1% KF, and the third panel consists of GPC with 0.75% CF. The composite that achieved the best test performance was selected and used to fabricate different thicknesses, 75 and 50 mm. The results showed that the compressive strength of GPC prepared with KF was 2.4 and 5% higher than that of GPC with CF at 7 and 28 days, respectively. Moreover, the thermal conductivity of GPC prepared with KF was more effective than that with CF which was lower by 3.8%, while compared to GPC without fiber, the thermal conductivity of geopolymer kenaf fiber decreased by 7.1%. In addition, the decrease in the thickness of the panel leads to a decrease in the thermal conductivity by 23% and 42% of 75 mm and 50 mm, respectively. In conclusion, the addition of KF to GPC can improve compressive strength and thermal insulation. Hopefully, this compound provides the industry with an alternative material of high thermal resistance for wall panels.

Keywords Geopolymer concrete · Kenaf fiber · Thermal insulation

A.-B. M. Firas · F. N. A. A. Aziz (✉) · A.-G. N. Abbas · N. A. M. Nasir · N. A. Safiee
Housing Research Centre (HRC), Department of Civil Engineering, Faculty of Engineering, Universiti Putra Malaysia (UPM), 43400 Serdang, Selangor, Malaysia
e-mail: farah@upm.edu.my

N. Sabtu (ed.), *Proceedings of AWAM International Conference on Civil Engineering 2022—Volume 2*, Lecture Notes in Civil Engineering 385,
https://doi.org/10.1007/978-981-99-6018-7_11

1 Introduction

The climate and global warming changes have many impacts on the engineering field and particularly the construction industry. Therefore, many researchers try to find a solution in different ways to decrease temperature effects on buildings and geopolymer is one of the alternative materials. Geopolymers possess distinctive properties such as high strength and resistance to high temperatures, i.e., a good heat insulator. The production of one ton of geopolymer binders produces 0.18 tons less carbon dioxide (CO_2) from the burning of carbon fuel compared to one ton of CO_2 from Portland cement [1]. Geopolymer concrete is an alternative material to normal concrete (made of Portland cement), it is an eco-friendly material consisting of source materials rich in silica and alumina such as fly ash, ground granulated blast furnace slag (GGBS) and metakaolin GPC possess high compressive strength at M14 of sodium hydroxide up to 82 MPa [2]. Utami reported that the thermal conductivity of geopolymer composites was ranging between 0.19 and 0.82 W/m C with an average of 0.421 W/m C, which enabled them to be used as insulator materials. However, although geopolymer composites are similar to OPC in terms of their favorable features, geopolymer binders exhibit brittle behavior and have low flexural and tensile strengths. Utilizing short fibers in the mix is one common and practical technique to increase flexural strength and toughness while reducing the brittleness of the binders [3].

The addition of fibers in brittle matrices is an efficient approach for reducing crack propagation and thereby improving the fracture toughness of the matrix when subjected to load [4]. Different types of fibers have been used as reinforcement in cementitious and geopolymer composites to enhance their flexural capacity and toughness like steel, polypropylene, basalt, sisal, jute, wood, bamboo, coconut, and asbestos [5]. Wang et al. worked on natural fiber-reinforced fly ash geopolymer mortar high calcium. The study showed two kinds of natural geopolymer composites. The authors used coconut and sisal fiber, with a content of 0.5–1.0%. The mechanical, physical, and thermal characteristics of fiber-reinforced geopolymer composites have been examined and compared to the composites prepared with glass fiber and plain geopolymer without fibers. The findings showed that the addendum of natural fibers (sisal/coconut fibers) as reinforcement substances has resulted in serious enhancements in flexural and tensile strength comparable to glass fibers utilization, while there is an initiating decline in dry density, workability, and compressive strength [6]. Irshidat et al. reported that the addition of polypropylene in the range of 0.05–0.2% reduced the thermal conductivity of geopolymer composites. The reduction increased with fiber content due to the ability of the fibers to entrain air or because of the PP fibers' lower heat conductivity value as compared to the geopolymer matrix [7].

Recently, the inclusion of natural fibers within the geopolymer compound had many benefits because it is an eco-friendly and lightweight material and is effective in thermal insulation [8]. Kenaf fiber is a well-known natural fiber. It consists of cellulose, hemicellulose, lignin, and various compounds that dissolve in water. Lignin, hemicelluloses, and cellulose were the main compounds [9]. Abirami et al. [10] used 0.25–1% of various types of natural fibers (kenaf and sisal fibers) in the cement composite. The results indicated that adding both fibers improved the compressive and tensile strength by increasing about 6.5% and 12.7%, respectively. Using 1% of sisal or kenaf fibers in the composite obtained the highest mechanical properties than other fiber dosages. Moreover, kenaf fiber-reinforced composite concrete (KFRCC) achieved better compressive strength than using sisal fibers in the composite. Abbas et al. demonstrated that KFRCC showed higher ductility by delaying the formation of cracks under applied compression and flexure loads than the normal composite. There was a large amount of cement matrix and aggregates spalling in the normal sample. Kenaf fibers can act as bridges across microcracks caused by stress, preventing them from diffusing in composites. Depending on their properties, KF prevents the cracks to transfer to the whole composite [11].

Coconut fiber is also a well-known natural fiber used to reinforce geopolymer with excellent properties as reported by abbass et al. The inclusion of coconut fiber in geopolymer matrix improves the properties of the composite, and the tensile strength is enhanced by increasing from 3.6 to 3.89 MPa; however, the inclusion of higher amount of fiber increased the voids content in the matrix [12]. The above-mentioned literature showed that geopolymer composites had good thermal performance and fire resistance. In addition, a positive effect of microfibers on the mechanical properties of the geopolymer composites has been described. Reinforcing geopolymer binders with fibers has been capable to support higher resistance against the crack formation, hence defeating their brittleness problem. However, the current study focuses on the thermal performance of geopolymer with natural fibers such as kenaf and coconut. Additional information from this study will provide better thermal conductivity materials to the construction industry.

2 Experiments

2.1 Materials

The materials used in this research are fly ash (FA), ground granulated blast slag (GGBS) as the binder, alkaline activator, coarse and fine aggregate, kenaf, and coconut fibers. The combination of FA and GGBS was used to produce the geopolymer binder because FA's reactivity is insufficient for alkaline activators to activate it at room temperature, and activation of FA needs a curing temperature of 60–85 °C. Under ambient temperature, curing GGBS enhanced the properties of FA-based GPC, due to the CaO that exists in the GGBS components. Therefore, the use of FA/GGBS-based GPC can conserve energy and provide excellent mechanical properties [13]. The fly ash was used as 60% of the total binder and it was obtained from a coal power plant Green cement company, Malaysia, which satisfied the ASTM C618-15 [14] requirements for low calcium fly ash (class F). On the other hand, the

GGBS was utilized as 40% of the binder. It was designated according to ASTM C 989 [15].

A combination of sodium hydroxide (SH) and sodium silicate (SS) was used as the alkaline solution to activate FA and GGBS. Sodium hydroxide purity is 97–99%. Finally, coarse and fine aggregates are used in producing geopolymer concrete. The river sand was tested for specific gravity and water absorption according to ASTM C128-07 [16] and sieved ASTM C33-03 [17] to the standard specifications. The specific gravity is 2.52 kg/m^3 with an average of about 1.68 water absorption for fine aggregate, while the specific gravity is 2.65 kg/m^3 with an average of about 0.88 water absorption for crushed granite coarse aggregate. In this work, the maximum size of sand retained sieved which was 4.75 mm and coarse 9.5 mm was used. The concrete mixtures were designed with a target slump of higher than 50mm to achieve acceptable workability and compressive strength of 40 MPa at 28 days.

Kenaf fibers were supplied from Lembaga Kenaf Dan Tembakau Negara (LKTN), Malaysia. The fibers were subjected to the alkaline treatment to enhance the fiber properties before being used in the GPC. The fibers were immersed in 6% NaOH solution for 3 h and then washed with water and immersed in the water for 24 h. After 24 h, the fibers were kept to drying at the ambient temperature. The treatment was conducted by following the previous studies [18]. After the treatment, the fibers were cut to a length of 20 mm and added to the GPC mixtures in a content of 1.0% volume fraction [19]. The other type of natural fiber used in this research is coconut fiber, which was also pre-treated by using the alkaline treatment before being used in the concrete mixtures. After the treatment, the fibers were cut to a length of 30 mm and added to the mixtures in a content of 0.75% volume fraction. The length and content of the fibers were chosen by following the study of [20].

2.2 *Mix Proportion*

The mix proportion of geopolymer concrete was designed as shown in Table 1.

2.2.1 Preparation of Specimen

In this research, all specimens were prepared from 60% of FA and 40% of GGBS mixed with an alkaline solution which consists of sodium silicate and sodium hydroxide at a ratio of 0.55. The sodium hydroxide molarity used in this study was 14 M. It is important to prepare alkali solution before 24 h of each mixture to ensure that all ingredients are dissolved. The dry compounds: coarse aggregate, fine aggregate FA, and GGBS were mixed in an electric mixer for 3 min. Then, the alkaline solution was added and mixed for 2 min. After that, the natural fiber was added in small batches and mixed for another 3 min to obtain a homogeneous mixture. Then, the fresh mixes were poured into clean molds and then covered until the test day. Three identical cubes were prepared for the compressive strength test and five panels

Table 1 Mix proportion of specimens (Kg/m^3)

GPC mix (Kg/m^3)	FA	GGBS	F. Agg	C. Agg	Ratio S/B	SS	SH	Fiber content	Dimension (mm) ($L \times W \times t$)
(P0) GPC	261	174	710	1065	0.55	170.8	68.3	–	500 × 500 × 100
(P2) GPC + KF	261	174	710	1065	0.55	170.8	68.3	1%	500 × 500 × 100
(P3) GPC + CF	261	174	710	1065	0.55	170.8	68.3	0.75%	500 × 500 × 100
(P4) GPC + KF	261	174	710	1065	0.55	170.8	68.3	1%	500 × 500 × 75
(P5) GPC + KF	261	174	710	1065	0.55	170.8	68.3	1%	500 × 500 × 50

[a] *S* Solution, *B* Binder, *FA* Fly ash, *F. Agg.* Fine aggregate, *C. Agg.* Coarse aggregate

for the thermal conductivity test for three mixes of geopolymer concrete (GPC), geopolymer concrete with kenaf fiber (GPC + KF) and geopolymer concrete with coconut fiber (GPC + CF) with different panels thickness 100, 75, 50 mm according to appropriate insulation thickness in Malaysia is in the range of 18–126 mm [21].

2.3 Tests Procedures

The compression test was carried out according to BS EN12390-3 [22] standard. In total, 18 cubes were tested for compressive strength at 7 and 28 days of ambient curing. The thermal test was carried out using hotbox test apparatus according to ASTM1363-97 [23]. The hotbox test is the method to determine the thermal performance for homogeneous and nonhomogeneous specimens at steady-state conditions for small samples. Hotbox apparatus consists of two boxes one for the metering chamber and the other climate chamber as shown in Fig. 1. The apparatus consists of fans and heating sources as shown in Sections X1 and X2, which contain an internal opening for measuring samples with sizes 500 mm × 500 mm × 100 mm. The box has been surrounded with an isolation material to prevent the heat flux from the hot chamber to the cold chamber to decrease the losses. The samples are placed on the interior opening of the hotbox device where five samples are used with different dimensions and different materials (P0) Control panel geopolymer concrete, (P2) Panel with 100 mm geopolymer with kenaf fiber, (P3) Panel with 100 mm geopolymer with coconut fiber, (P4) Panel with 75 mm geopolymer with kenaf fiber, (P5) Panel with 50 mm geopolymer with kenaf fiber, the sensors are placed in different places on both hot and cold surfaces of the sample using a thermocouple to monitor the temperature during the test, in addition, the temperature of the cold room should be set by operating air condition before starting the test, and the device is connected to a computer to collect and analyze data to obtain the accurate results of thermal conductivity (K) (W/m K). Values are measured to evaluate the thermal performance of the samples and to obtain results. Thus, the desired results are obtained from this test to obtain an appropriate environment for residential buildings in a hot climate by using a data logger to calculate the thermal conductivity of the samples studied through the general equation stated in Eq. 3. There are three samples prepared with each being square wall panels of 500 × 500 with 100 mm thickness. Each sample is subjected to the test with repetitive of 3. The geopolymer mixes are made with an addition of 1% by volume of kenaf fiber or with an addition of 0.75% by volume of coconut fibers. Then, the best fiber mix which is kenaf fiber casts on panels with 50 and 75 mm thickness to study their effect on the thermal performance of the geopolymer fiber concrete.

$$U = Q/(A(T_2 - T_0))\text{in}\left(\text{w}/^\circ\text{C m}^2\right), \tag{1}$$

where

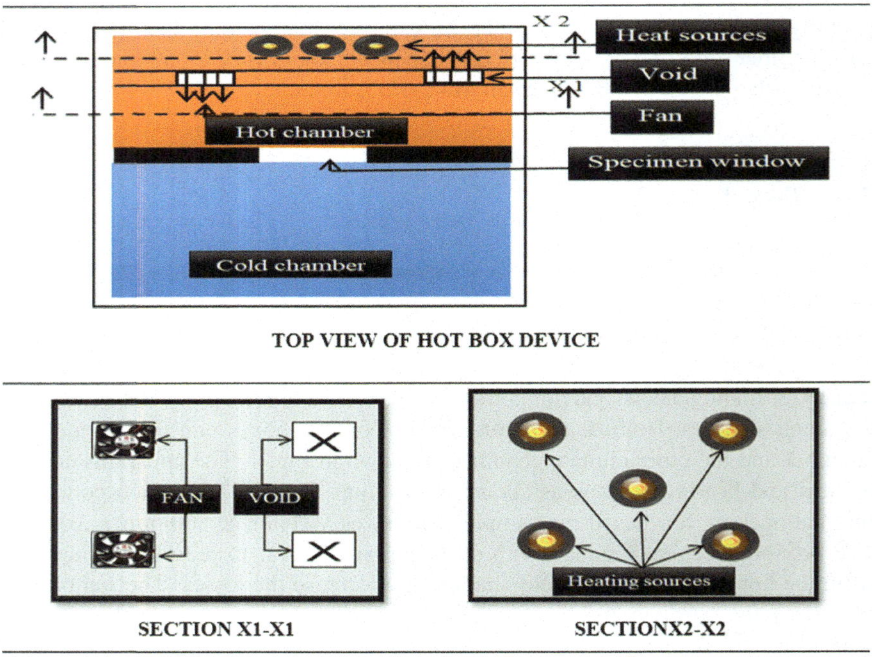

Fig. 1 Scheme view of hotbox apparatus

U: thermal transmittance; Q: represents heat flow rate (W) an exposure area perpendicular to heat flow (m^2), T_2: temperature of the hot side with Kelvin (K), T_0 temperature of cold side (K). After calculating the U-value, the thermal resistance and conductivity were determined using the following equations:

Thermal resistance (R):

$$R = 1/U \text{ in}\left(°\text{C m}^2/\text{w}\right). \tag{2}$$

Thermal conductivity (K):

$$K = d/R \text{ in}(\text{w/m k}), \tag{3}$$

where

d: the specimen thickness (m).

2.3.1 Environmental Temperature

The environmental temperature was recorded using hotbox apparatus. Before the test, the temperature in both the hot chamber and climate chamber is necessary to reach a steady-state condition between 1000 and 1400 min as mentioned in ASTM C1363-97 [23]. The temperature in the cold chamber is at constant of approximately 17–18 °C and in the hot chamber was 70–77 °C. Both temperatures are within limits of between −40 and 85 °C. Figure 2 shows the steady condition which was maintained in the apparatus until 1440 min (24 h), which fulfilled the standard requirements.

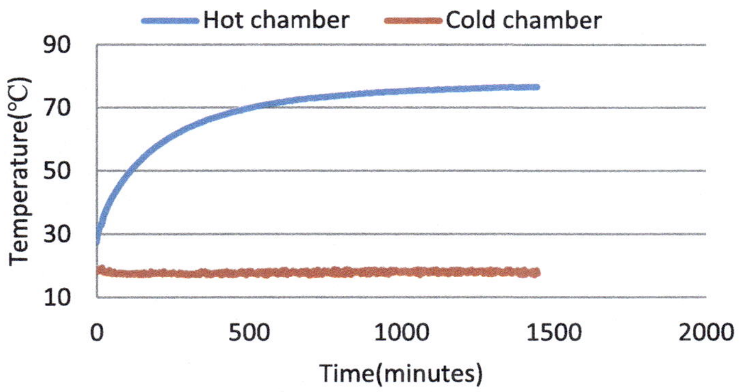

Fig. 2 Average hot and cold air temperature profiles in the hotbox apparatus

3 Results and Discussion

3.1 Compressive Strength

The compressive strength is a very important characteristic of concrete. It is to confirm whether the concrete fits its designated purposes and to determine the concrete quality. In this study, the compressive strength of the GPC control sample and geopolymer with natural fiber was tested by a compressive test machine using a specimen of 7 and 28 days, and the results are presented in Fig. 3. The average compressive strength at 7 days for the GPC control mix is 47.5 MPa and at 28 days is 69.8 MPa. It is observed that by increasing the curing time, the compressive strength increased by 46.9%. Besides, the addition of natural fibers decreased the compressive strength at both testing ages. The compressive strength of geopolymer concrete with kenaf fiber (GPC + KF) is 45.9 MPa at 7 days and 60.7 MPa at 28 days. This shows that the addition of kenaf fibers decreased the compressive strength of GPC by 3.3% at 7 days and decreased by 13% at 28 days. On the other hand, the compressive strength for GPC mixed with coconut fiber (GPC + CF) at 7 days is 44.8 MPa and at 28 days is 57.6 MPa. The reduction in the strength of GPC with the addition of natural fiber could be attributed to the increased voids that formed in concrete [24]. It can be also observed that the compressive strength of GPC mix with kenaf fiber was better than the compressive strength of GPC mix with coconut fiber by 2.4% at 7 days and by 5.1% at 28 days.

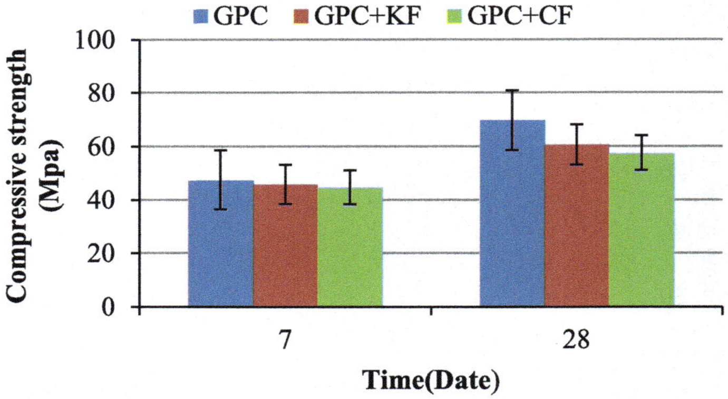

Fig. 3 Compressive strength for geopolymer concrete with kenaf and coconut fibers

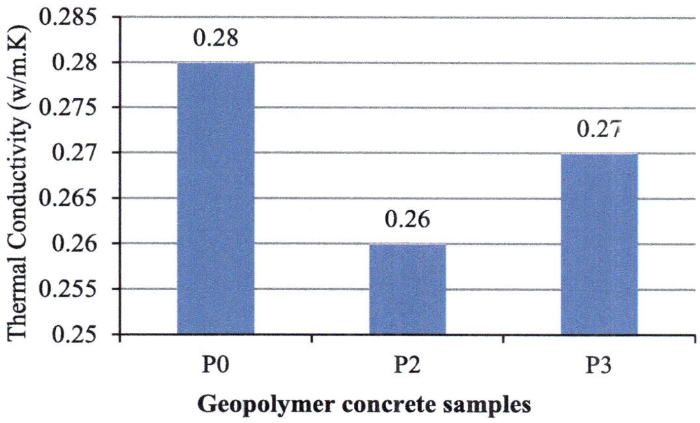

Fig. 4 Thermal conductivity of geopolymer with kenaf (P2) and coconut (P3) fibers

3.2 Effect of Natural Fiber on Thermal Performance

The thermal performance of geopolymer concrete is presented in Fig. 4. As shown, the plain GPC (P0) allows direct heat transfer and has attained the highest thermal conductivity of 0.28 W/m K. However, the addition of natural fibers (kenaf or coconut) slightly decreased the thermal conductivity to 0.26 and 0.27 W/m K for the composites prepared with kenaf fiber (P2) and that prepared with coconut fiber (P3), respectively. This indicates an improvement of about 7.1 and 3.6% for P2 and P3, compared to P0 of the same thickness. This shows that using kenaf fiber with geopolymer concrete achieved a lower thermal conductivity than coconut fiber. This may be due to the hollow and cellular nature of kenaf [25] that traps the air in voids and prevents the heat to transmit from the heat source to the material. The results of the current study showed improvement in heat conductivity as compared to works reported by Aprilia et al. [26] and Sample et al. [27] of geopolymer concrete. However, both kenaf and coconut fibers are used for thermal isolation purposes as they are less than 1 W/m K [28]. Moreover, the thermal conductivity for 1% of kenaf fiber with geopolymer concrete reported excellent results around 0.28 W/m K compared with the thermal conductivity of 1% of kenaf fiber around 0.93 W/m K, and this proved that geopolymer concrete can be a good insulator for buildings [25]. Comparing it with a brick wall, the geopolymer concrete 0.28 W/m K is lower by 65.5% than the brick wall of 0.811 W/m K as reported by Balaji et al. [29].

3.3 Effect of Different Thicknesses on Thermal Performance

The thickness changes of wall panels have an important influence on thermal performance. The results as shown in Fig. 5 are based on Eq. (3). Panels P4 and P5 are

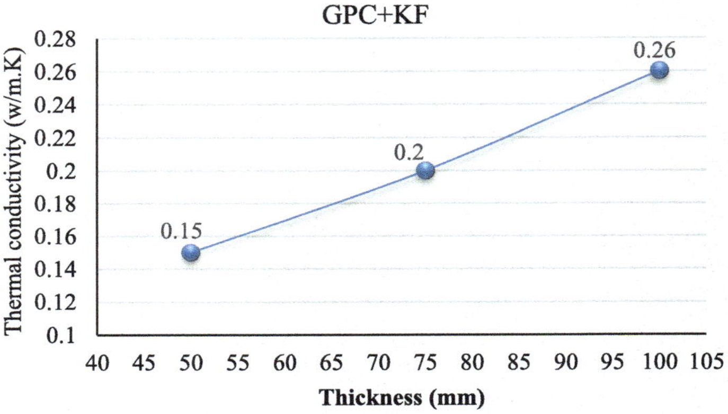

Fig. 5 Relationship between thermal conductivity and thickness of panels

the same materials that consist of geopolymer concrete mixed with kenaf fiber with different thicknesses are 75 and 50 mm. Panels P4 and P5 have thermal conductivity around 0.2 and 0.15 W/m K, respectively. The panel P4 with 75 mm thickness showed an improvement by 28.6%, while the panel P5 with a thickness 50 mm showed an improvement by 46.4% compared with the P0 with a thickness of 100 mm (no fiber specimen). If compared to P2 , a 100mm thickness specimen of GPC with KF, panels P4 improved about 23.1% for 75 mm thickness and P5 of 42.3% for 50 mm thickness. When reducing the thickness of the wall, the thermal conductivity value decreases according to Fourier's law as shown in Fig. 5.

4 Conclusion

This study is an attempt to achieve a better thermal performance of walls using an eco-friendly material made of geopolymer and natural fibers. The two fibers, kenaf and coconut, are compared by using the hotbox method. The following conclusion can be drawn:

1. The addition of fibers reduced the compressive strength of geopolymer, but the kenaf fiber exhibited lesser reduction than the coconut fiber. However, it is insignificant as both are in a range of 60 MPa at 28 days and 40 MPa at 7 days.
2. The addition of natural fibers with geopolymer concrete as an insulating material achieved a great reduction of the heat transmitted through the walls. This is proven by the thermal test, where the geopolymer concrete resistance is 0.28 W/m K.
3. Kenaf fibers have proven more effective in thermal insulation than coconut fiber, where the thermal conductivity of kenaf fibers blended with geopolymer concrete was 0.26 W/m K.

4. The reduced thickness of the wall causes a lower value of the thermal conductivity of the panel.

Acknowledgements The authors gratefully acknowledge the financial support for this research by the Ministry of Education, Malaysia, under the Fundamental Research Grant Scheme (FRGS/1/2020/TKO/UPM/02/32) with Vot no: 5540372 for research work entitled 'An investigation of characterization and parametric effect of kenaf bast fiber in the properties of geopolymer kenaf reinforced concrete'. The authors also gratefully acknowledge Ms Ernaleza Mahsum in the materials' lab of Faculty of Engineering, University Putra Malaysia UPM and the Excel test company.

References

1. Szczypinski, M.M., Hájková, P., Kovacic, V., Bakalova, T., Volesky, L., Louda, P.: Mechanical properties of geopolymer foam at high temperature. Sci. Eng. Compos. Mater. **27**(1), 129–138 (2020)
2. Mortar, N.A.M., Kamarudin, H., Rafiza, R.A., Meor, T.A.F., Rosnita, M.: Compressive strength of fly ash geopolymer concrete by varying sodium hydroxide molarity and aggregate to binder ratio. In: IOP Conference Series: Materials Science and Engineering, vol. 864, no. 1, p. 012037. IOP Publishing (2020)
3. Ganesh, A.C., Muthukannan, M.: Development of high performance sustainable optimized fiber reinforced geopolymer concrete and prediction of compressive strength. J. Clean. Prod. **282**, 124543 (2021)
4. Guo, X., Yang, J.: Intrinsic properties and micro-crack characteristics of ultra-high toughness fly ash/steel slag based geopolymer. Constr. Build. Mater. **230**, 116965 (2020)
5. Ranjbar, N., Zhang, M.: Fiber-reinforced geopolymer composites: a review. Cement Concr. Compos. **107**, 103498 (2020)
6. Wang, Y., Zheng, T., Zheng, X., Liu, Y., Darkwa, J., Zhou, G.: Thermo-mechanical and moisture absorption properties of fly ash-based lightweight geopolymer concrete reinforced by polypropylene fibers. Constr. Build. Mater. **251**, 118960 (2020)
7. Irshidat, M.R., Al-Nuaimi, N., Rabie, M.: Thermal behavior and post-heating fracture characteristics of polypropylene microfiber-reinforced geopolymer binders. Constr. Build. Mater. **332**, 127310 (2022)
8. Lotfi, A., Li, H., Dao, D.V., Prusty, G.: Natural fiber–reinforced composites: a review on material, manufacturing, and machinability. J. Thermoplast. Compos. Mater. **34**(2), 238–284 (2021)
9. Ramesh, P., Durga Prasad, B., Narayana, K.L.: Characterization of kenaf fiber and its composites: a review. J. Reinf. Plast. Compos. **37**(11), 731–737 (2018)
10. Abirami, R., Sangeetha, S.P., Nadeemmishab, K., Vaseem, P.Y., Sad, K.S.: Experimental behaviour of sisal and kenaf fibre reinforced concrete. In: AIP Conference Proceedings, vol. 2271, no. 1, p. 030023). AIP Publishing LLC (2020)
11. Abbas, A.G.N., Aziz, F.N.A.A., Abdan, K., Mohd Nasir, N.A., Norizan, M.N.: Kenaf fibre reinforced cementitious composites. Fibers **10**(1), 3 (2022)
12. Abbass, M., Singh, D., Singh, G.: Properties of hybrid geopolymer concrete prepared using rice husk ash, fly ash and GGBS with coconut fiber. Mater. Today: Proc. **45**, 4964–4970 (2021)
13. Zhang, P., Gao, Z., Wang, J., Guo, J., Hu, S., Ling, Y.: Properties of fresh and hardened fly ash/slag based geopolymer concrete: a review. J. Clean. Prod. **270**, 122389 (2020)
14. ASTM C618: Standard specification for coal fly ash and raw or calcined natural pozzolan for use in concrete. ASTM International, West Conshohocken, PA (2015)

15. ASTM, C., 989: Standard specification for ground granulated blast-furnace slag for use in concrete and mortars. In: Annual Book of ASTM Standards, vol. 4: pp. 15–38.4 (1994)
16. ASTM, C. 127: Standard test method for specific gravity and absorption of coarse aggregate. In: Annual Book of ASTM Standards (2007)
17. ASTM, C: Standard specification for concrete aggregates. American Society for Testing and Materials, Philadelphia, PA (2003)
18. Ibrahim, M.I., Hassan, M.Z., Dolah, R., Yusoff, M.Z.M., Salit, M.S.: Tensile behaviour for mercerization of single kenaf fiber. Malays. J. Fundam. Appl. Sci **14**, 437–439 (2018)
19. Abbas, A.G.N., Aznieta, A.A.F.N., Khalina, A., Azline, M.N.N.: Effects of fibers length on the properties of kenaf fibers-reinforced geopolymer concrete. In: AIP Conference Proceedings, vol. 2447, no. 1, p. 030001. AIP Publishing LLC (2021)
20. Alfianuz Zura binti aiful bahri 188686: Mechanical properties of geopolymer concrete with addition coconut fiber. Final year project thesis Bachelor of Civil Engineering with Honours University Putra Malaysia (2020)
21. Basrawi, F., Ibrahim, H., Taib, M.Y., Lee, G.C.: Optimum thickness of wall insulations and their thermal performance for buildings in Malaysian climate. Int. J. Autom. Mech. Eng. **8**, 1207 (2013)
22. BS EN12390—3 Testing Hardened Concrete—Part 3: Compressive Strength of Test Specimens. British Standards Institution, Her Majesty Stationery Office, London, United Kingdom (2009)
23. ASTM: C1363-97: Standard Test Method for the Thermal Performance of Building Assemblies by Means of a Hot Box Apparatus 1 (1997)
24. Abbas, A.G.N., Aziz, F.N.A.A., Abdan, K., Nasir, N.A.M., Huseien, G.F.: A state-of-the-art review on fibre-reinforced geopolymer composites. Constr. Build. Mater. **330**, 127187 (2022)
25. Ismail, A.S., Jawaid, M., Naveen, J.: Void content, tensile, vibration and acoustic properties of kenaf/bamboo fiber reinforced epoxy hybrid composites. Materials **12**(13), 2094 (2019)
26. Utami, F.A.R., Triwiyono, A., Agustini, N.K.A., Perdana, I.: Thermal conductivity of geopolymer with polypropylene fiber. In: IOP Conference Series: Materials Science and Engineering, vol. 742, no. 1, p. 012031. IOP Publishing (2020)
27. Sampe, F., Triwiyono, A., Agustini, N.K.A., Perdana, I.: Thermal Properties of geopolymers with silica fume. In: IOP Conference Series: Materials Science and Engineering, vol. 742, no. 1, p. 012032. IOP Publishing (2020)
28. Jalilluddin, A.M., Ayop, S.M., Kamaruddin, K.: Evaluation on the thermal conductivity of sand-cement blocks with Kenaf Fiber. In: Advanced Materials Research, vol. 626, pp. 485–489. Trans Tech Publications Ltd. (2013)
29. Balaji, N.C., Mani, M., Venkatarama Reddy, B.V.: Thermal performance of the building walls. In: Preprints of the 1st IBPSA Italy Conference Free University of Bozen-Bolzano, vol. 346, pp. 1–7 (2013)

Performance of Natural Fibre-Reinforced Geopolymer Composites Exposed to Sulphuric Acid

Al-Ghazali Noor Abbas, Farah Nora Aznieta Abdul Aziz, Khalina Abdan, and Noor Azline Mohd Nasir

Abstract Increasing interest in using waste materials and by-products from different industries as renewable sources of construction materials is mainly for environmental concerns and sustainability. Geopolymer and natural fibres are examples of industrial wastes and by-products used. However, their acceptability in the industry is relatively low due to insufficient research on their durability performance. This paper presents an investigation into the durability of kenaf fibre-reinforced geopolymer composites (KFRGCs) when exposed to a 5% solution of sulphuric acid. The main parameters evaluated were the changes in the visual appearance, weight, and compressive strength. Geopolymer specimens' reinforced composites with a 1% volume fraction of different lengths of kenaf fibres (20, 30, and 40 mm) were produced to evaluate the impact of the fibre addition and fibre length on the performance of geopolymer composites. The results showed that the KFRGC in the acidic environment demonstrated a reasonable level of resistance to the attack as compared to plain geopolymer concrete (GPC). Fibres' length effects show shorter fibres which have better resistance than longer fibres. These findings display the positive impact of natural fibres on the performance of GPC in an acidic environment.

Keywords Sustainability · Geopolymer composites · Natural fibres · Kenaf fibres · Durability · Acid attack

A.-G. N. Abbas · F. N. A. A. Aziz (✉) · N. A. M. Nasir
Housing Research Centre (HRC), Department of Civil Engineering, Faculty of Engineering, University Putra Malaysia UPM, 43400 Serdang, Selangor, Malaysia
e-mail: farah@upm.edu.my

K. Abdan
Department of Biological and Agricultural Engineering, Faculty of Engineering, Institute of Tropical and Forest Products (INTROP), Universiti Putra Malaysia, University Putra Malaysia UPM, 43400 Serdang, Selangor, Malaysia

N. Sabtu (ed.), *Proceedings of AWAM International Conference on Civil Engineering 2022—Volume 2*, Lecture Notes in Civil Engineering 385,
https://doi.org/10.1007/978-981-99-6018-7_12

1 Introduction

The high public understanding of the environment and the protection of non-renewable petroleum-based materials have contributed to the widespread use of natural resources for industrial applications. Besides, increasing interest among consumers in sustainable and environmentally friendly materials encourages retailers and manufacturers to invest in the production of sustainable materials at reasonable costs to mitigate the effects of global warming [1, 2].

Geopolymers are inorganic polymers formed through the chemical reaction between aluminosilicate sources derived either from geological minerals like metakaolin (MK) or from industrial by-products such as the ground granulated blast furnace slag (GGBS), fly ash (FA), palm oil fuel ash (POFA) with an alkaline solution like sodium silicate and sodium hydroxide through the geopolymerization process that results in polymer chains and cross-linked networks of comparable or greater strength than that of OPC [3–5]. On the other hand, natural fibres are abundant, lightweight, and cost-effective materials [6, 7]. The composites based on natural fibres and geopolymers could produce green materials for the construction and building industry.

Previous studies have focused on the influence of the properties of natural fibres like cotton, hemp, flax, and coir on the properties of geopolymer composites [8–11]. The authors found that using the appropriate length and volume fraction of short natural fibres was an effective solution to overcome the brittleness of GPC [12]. Hence, these findings pave the way to use kenaf fibre to get more details on the properties of geopolymers reinforced by natural fibres.

Kenaf fibre (KF) is a well-known natural fibre in Malaysia that has been used in biocomposite in the local industry. Malaysian Government agencies like the National Kenaf and Tobacco Board are undertaking intensive research on the kenaf plant. Moreover, Malaysia is towards eco-friendly, with the goal of reducing the worldwide use of non-degradable materials, particularly for domestic applications. About RM25 million has been allocated under the Ninth Malaysia Plan for the kenaf crop [13]. Kenaf fibres have good mechanical properties and require low production energy with environmental benefits, which include bio-renewability, biodegradability, non-hazardous nature, and low pollutant emission [14]. The biodegradability of KF contributes significantly to a sustainable environment, while its low cost and good performance address manufacturers' economic concerns [15–17]. These fibres have great potential for replacing synthetic fibres like glass fibre [18].

The major obstacles to the usage of natural fibres as reinforcing materials are the weak compatibility between the fibre and the matrix in addition to the inherent high absorption of moisture, which could lead to dimensional changes in the fibre that could result in micro-cracking of the composites and deterioration of the mechanical properties. Surface modification by using alkaline treatment is an effective method for minimizing the water-absorbing chemical components like the hemicelluloses, lignin, and impurities such as oils and wax from the fibre surface and increasing the fibre surface roughness of KF, which improves the interface bonding between fibre and matrix [19, 20].

The use of geopolymer in concrete can enhance the durability characteristics, and the fibres demonstrate suitable behaviour in this regard [21]. The decision to utilize sulphuric acid among the other acid types was based on its practical applicability because the most commonly used concrete buildings, like the food processing industries, mining, and wastewater canals, are in touch with sulphuric acid. This paper evaluates the performance of geopolymer concrete containing 1% KF with three different lengths of 20, 30, and 40 mm under the sulphuric acid attack [21].

2 Experimental Programme

2.1 Materials

Materials used to prepare the GPC and KFRGC are the binder, alkaline solution, fine and coarse aggregates, water, and KF. The binder is mainly consisting of FA (Class F) according to ASTM C 618-12a [22] and GGBS. The alkaline solution is comprised a mixture of sodium hydroxide and sodium silicate. These are used to activate the alumina and silica in the binder. Analytical grade sodium hydroxide (98% purity collected from bt science Sdn. Bhd Malaysia) was received as pellets; the pellets were dissolved in water to prepare a sodium hydroxide solution with a 14 M concentration. The solution was prepared 24 h before mixing and left in the lab to be cool. Analytical grade sodium silicate solution was also used to prepare the alkaline activator.

2.2 Kenaf Fibres (KF)

KF is known for its high cellulose content, and thereby, it is biodegradable. Besides, the absence of silica content enables the fibres to be nonabrasive [23]. The fibres were extracted through the water retting process from the bast of the plant. The length of the fibres in the market is between 3.5 and 5 m. The fibres were treated to remove the surface impurities and increase the surface roughness, thus enhancing the interfacial bonding between the fibres and the geopolymer matrix. The alkaline treatment method was chosen to treat the fibres before their inclusion in the geopolymer composites.

The KF was treated in the laboratory by immersing the fibres in sodium hydroxide solution of 6% concentration for 3 h. The sodium hydroxide concentration and the immersion time were chosen based on previous studies [17, 24]. After 3 h, the fibres were then washed and kept in water for 24 h. After 24 h, the fibres were dried at ambient temperature. Afterwards, the specimens were ready for testing.

Table 1 Properties of KF

Diameter (μm)	76
Tensile strength (Mpa)	242
Specific gravity (kg/m^3)	1450

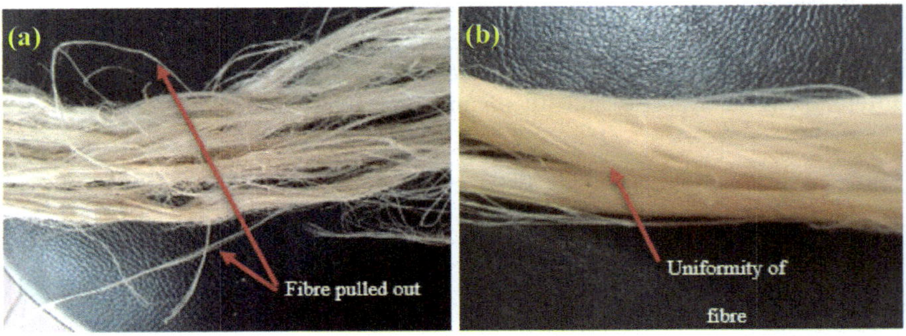

Fig. 1 Surface appearance of KF, **a** untreated KF, **b** treated KF

After the treatment, KF's diameter, tensile strength, and density were evaluated, and the results are presented in Table 1. The physical appearance of kenaf fibre before and after treatment is shown in Fig. 1. Overall observations show that the physical appearance of the fibres was greatly altered by the chemical treatment. In terms of colour, treated fibre shows a different colour after treatment. The colour of the fibre becomes more yellowish after the treatment. The fibres were cut into three different lengths of 20 mm, 30 mm, and 40 mm and added to the geopolymer mixtures.

2.3 Mix Proportions

The mix proportions are shown in Table 2. Three different KFRGC mixes besides the control mix without fibres were prepared. KF was cut to three different lengths of 20, 30, and 40 mm and used in the mixtures at a 1% dosage. The ratio of sodium silicate (SS) to sodium hydroxide (SH) was fixed at 2.5 for all mixtures.

2.4 Mixing, Casting, and Curing

To get a homogeneous mixture, the fine and coarse aggregates were dry mixed for two minutes before the binder materials were added and mixed. Afterwards, KF was added and mixed for another 3 min. Lastly, the alkaline activator was included and mixed for 3 min. Then, fresh mixtures were cast in the moulds in three levels, with

Table 2 Mix proportions (kg/m^3)

Mix	FA	GGBS	Aggregates		Solution		L^a (mm)	V_f^b (%)
			Sand	10 mm	SS	SH		
GPC (Ref)	261	174	704.7	1065.7	170.8	68.3	0	0
KFRGC1	261	174	704.7	1065.7	170.8	68.3	20	1
KFRGC 2	261	174	704.7	1065.7	170.8	68.3	30	1
KFRGC 3	261	174	704.7	1065.7	170.8	68.3	40	1

[a] L: Fibres' length
[b] V_f: Fibres' volume fraction

each layer being consolidated for about one minute on a vibration table to permit air escape. After casting, the specimens were left for 24 h at room temperature (27 ± 1.5) °C under a relative humidity of 75%. The specimens were then removed from the mould and stored in the laboratory until the testing date.

2.5 Testing

For the entire experimental work, a total of 60 geopolymer cubes with a 100 mm size were cast. In the first stage, the air-cured compressive strength of KFRGC and GPC was determined by using 100 mm cube specimens in line with the procedures laid down by BS 12390-3 [25] at 7, 28, and 56 days.

The chemical immersion experiment was used to evaluate the performance of GPC and KFRGC against the sulphuric acid attack. After 28 days of curing, the specimens were cleaned with a cloth to remove any impurities from their surface and weighed to the nearest 0.1 g as the initial weight. The specimens were then immersed in 5% H_2SO_4 solutions for 180 and 365 days, respectively, and subsequent weights were recorded during the exposure period. To keep the pH constant throughout the study period, the acid solution was replaced every two months. After 180 and 365 days of exposure to the acid, the performance of each mixture was evaluated based on its visual appearance, weight loss, and strength loss as follows:

Visual appearance: all the samples were visually inspected to determine the degree of deterioration. The colour, texture, edges, size, and geometry of the sample were observed throughout the evaluation.

Weight loss: to determine the variation in weight, the weight loss percentage for each mix was calculated by using Eq. (1).

$$\text{Weight loss} = \frac{W_2 - W_1}{W_1} \times 100, \tag{1}$$

where

W_1 = the initial weight of specimen before immersion in the acid solution (gr), W_2 = the weight of specimen after immersion in acid the solution (gr).

Strength loss: on the other hand, the degradation of geopolymer samples was also examined by determining the strength loss factor (SLF). The SLF is expressed in percentage and calculated using Eq. 2 [26].

$$\text{SLF} = \frac{F_{\text{ca}} - F_{\text{cs}}}{F_{\text{ca}}} \times 100, \quad (2)$$

where

SLF = the percentage of loss in geopolymer's compressive strength after exposure to the acid attack, F_{ca} = the average compressive strength of air-cured geopolymer, and F_{cs} = the average compressive strength of geopolymer immersed in the chemical solution.

3 Results and Discussion

3.1 Density

The density of geopolymer composites depends largely on the unit weight of the constituent materials employed in the mix. Figure 2 shows the density of the various geopolymer mixes. KFRGC exhibited lower density compared to plain geopolymer. Besides, increasing the length of the fibre caused a higher reduction in density. This is due to the higher porosity resulting from the entrapped air within the composite during the mixing of fibres. Furthermore, because KF has a lower density than GPC, the final density of KFRGC becomes lower.

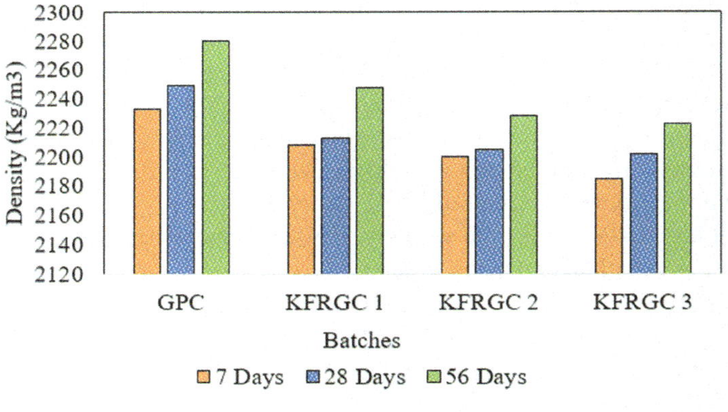

Fig. 2 Density of plain geopolymer (GPC) and KFRGC prepared with different fibre lengths

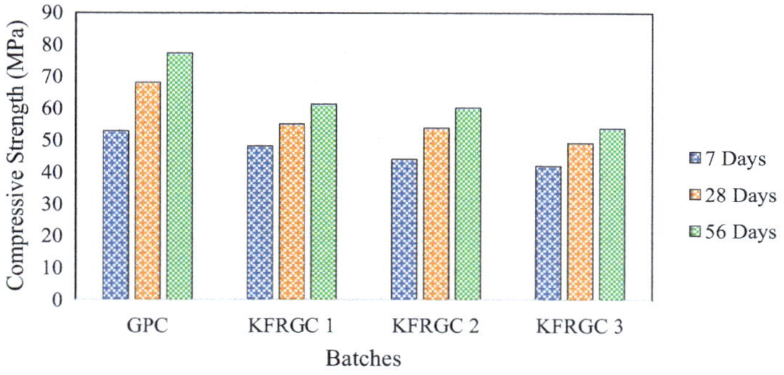

Fig. 3 Compressive strength of plain geopolymer concrete (GPC) and KFRGC prepared with different fibre lengths

3.2 Air-Cured Compressive Strength

The compressive strength values of the GPC and KFRGC are illustrated in Fig. 3. The incorporation of KF decreased the compressive strength by about 19–27% less than the plain geopolymer at 28 days. The reduction is due to the porosity of the composite, as has been shown in the density. The KFRGC containing short fibres (20 and 30 mm) shows better performance compared to longer fibres (40 mm). Longer fibres' ineffectiveness in increasing compressive strength can be ascribed to two distinct phenomena. First, if the fibres are too long, problems can arise during mixing the geopolymer. Fibre clusters develop as a result, which can result in increased porosity. Secondly, the orientation of the fibres inside the hardened KFRGC can also play a vital role in the composite's strength. The adverse impact of fibre length on the compressive strength was also reported by Xu et al. [27], Vilaplana et al. [28], and Novais et al. [29] when the authors used polyvinyl alcohol fibres, carbon fibres, and glass fibres, respectively.

3.3 Acid Attack

Visual Appearance

The appearance of geopolymer specimens after 365 days of exposure to H_2SO_4 is presented in Fig. 4. The control sample showed the highest degree of deterioration with spalling and line cracks at the edges of cubes, as in Fig. 4a. In terms of KFRGC, increasing the length of the fibres from 20 to 30 mm improved the composite's resistance and reduced the deterioration degree with smaller cracks and erosion, as in Fig. 4b, c. However, the further increase in the length of the fibres to 40 mm showed an adverse effect with more spalling, as shown in Fig. 4d.

Fig. 4 Appearance of geopolymer composites exposed to an H_2SO_4 solution for 365 days: **a** plain GPC; **b** KFRGC 1; **c** KFRGC 2; **d** KFRGC 3

Weight Loss

The weight loss percentage was also evaluated after 180 and 365 days of immersion in the acid solution, as shown in Fig. 5. KFRGC samples achieved lower weight loss than the GPC sample (without fibres), showing the attributes of the fibres to improve composite materials' abrasion resistance [30]. Increasing the fibre length from 20 to 30 mm caused an insignificant increase in the weight loss rate of about 2% after 365 days of exposure. However, increasing the fibre length to 40 mm increased the weight loss of KFRGC by about 10%, showing that the porosity due to long fibres is being too large. Hence, it cannot improve the performance of the geopolymer.

Compressive Strength Loss Factor (SLF)

As the weight of the samples decreases, it is expected that the strength will follow after the immersion in solution for 180 and 365 days. The longer the immersion period, the higher the strength reduction. The deterioration was attributed to the aluminosilicate polymer depolymerization in acidic environments and the development of zeolites, which resulted in a significant loss of strength [31]. The reduction of compressive strength of plain GPC and KFRGC mixtures due to acid attack was expressed in the form of a strength loss factor (SLF), and the results are illustrated in Fig. 6. The SLF values of plain GPC were higher than that of KFRGC. However, the SLF of KFRGC highly depends on the length of the fibre. The SLF of the composites decreased with increasing the length from 20 to 30 mm; however, a further increase

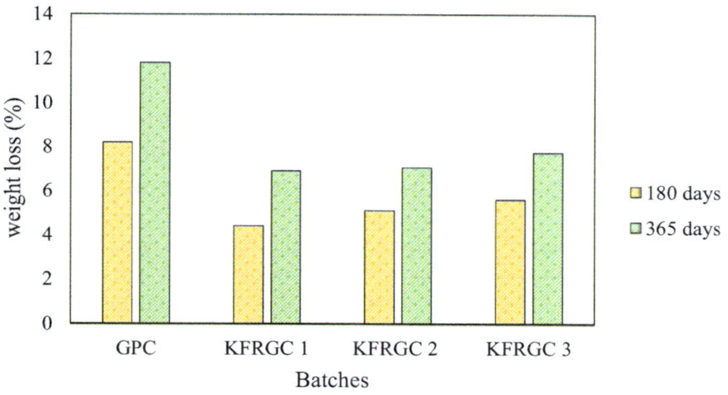

Fig. 5 Effects of KF fibres' length on the weight variation of the geopolymer after 180 and 365 days of exposure to the acid attack

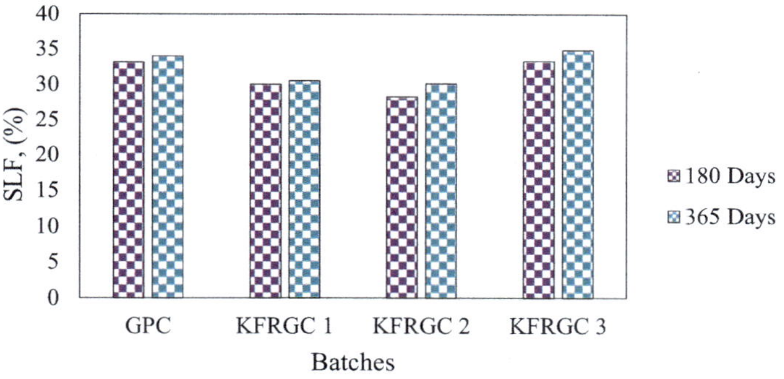

Fig. 6 Strength loss factor of plain GPC and KFRGC containing different fibre lengths exposed to the H_2SO_4 solution

in the length to 40 mm increased the SLF by 15% after 365 days. These findings reveal that the composites with longer lengths lost more strength. This could be attributed to the higher porosity and the distribution of the fibres inside the samples. This demonstrates that increasing the length of fibres to 40 mm will not be able to mitigate acid-induced damage in KFRGC.

4 Conclusion

In this study, the sulphuric acid resistance of plain geopolymer and kenaf fibre-reinforced geopolymer composites was studied. KFRGC exposed to acidic attack exhibited a reasonable level of resistance to the acid environment. The KFRGC reinforced with short lengths of fibres showed better results compared to the composites containing long fibres, and the SLF was 10 and 14.5% lower than the plain GPC for the samples with 20 mm and 30 mm, respectively. Increasing the length of the fibres from 20 to 30 mm showed insignificant change in terms of the weight and strength loss percentage. However, increasing the length of the fibres to 40 mm led to adverse results. Therefore, in terms of durability, KFRGC with 20 mm or 30 mm fibre length shows good performance and potential when exposed to an acidic environment.

Acknowledgements The authors gratefully acknowledge the financial support for this research by the Ministry of Education, Malaysia, under the Fundamental Research Grant Scheme (FRGS/ 1/2020/TKO/UPM/02/32) with Vot no: 5540372 for research work entitled 'An investigation of characterization and parametric effect of kenaf bast fibre in the properties of geopolymer kenaf reinforced concrete'. The authors also gratefully acknowledge the assistant engineer Mr Fairus Ismail in the materials lab of the Faculty of Engineering, University Putra Malaysia UPM.

References

1. Menatalah, M.K.A., Kotop, A., El-Feky, M.S., Alharbi, Y.A, Abadel, A.A., Binyahy, A.S.: Engineering properties of geopolymer concrete incorporating hybrid nano-materials. Ain Shams Eng. J. 2090–4479 (2021)
2. Zhou, Y., Fan, M., Chen, L.: Interface and bonding mechanisms of plant fibre composites: an overview. Compos. Part B: Eng. (2016)
3. Ganesh, A.C., Muthukannan, M.: Development of high performance sustainable optimized fiber reinforced geopolymer concrete and prediction of compressive strength. J. Clean. Prod. **282**, 124543 (2021)
4. Farhan, N.A., Sheikh, M.N., Hadi, M.N.S.: Behaviour of ambient cured steel fibre reinforced geopolymer concrete columns under axial and flexural loads. Structures **15**(July), 184–195 (2018)
5. Abbas, A.G.N., Aziz, F.N.A.A., Abdan, K., Nasir, N.A.M.: Mechanical properties of natural fibre reinforced geopolymer composites: a review. Sci. Technol. **30**(3), 2053–2069 (2022)
6. Castañeda, R.A.D., Silva, G., Salirrosas, J., Kim, S., Bertolotti, B., Nakamatsu, J.: Production of a lightweight masonry block using alkaline activated natural pozzolana and natural fibers. Constr. Build. Mater. **253**, 119143 (2020)
7. Onuaguluchi, O., Banthia, N.: Plant-based natural fibre reinforced cement composites: a review. Cem. Concr. Compos. **68**, 96–108 (2016)
8. Silva, G., Kim, S., Aguilar, R., Nakamatsu, J.: Natural fibers as reinforcement additives for geopolymers—A review of potential eco-friendly applications to the construction industry. Sustain. Mater. Technol. **23**, e00132 (2020)
9. Poletanovic, B., Dragas, J., Ignjatovic, I., Komljenovic, M., Merta, I.: Physical and mechanical properties of hemp fibre reinforced alkali-activated fly ash and fly ash/slag mortars. Constr. Build. Mater. **259**, 119677 (2020)
10. Assaedi, H., Shaikh, F.U.A., Low, I.M.: Effect of nanoclay on durability and mechanical properties of flax fabric reinforced geopolymer composites. J. Asian Ceram. Soc. **5**(1), 62–70 (2017)
11. Korniejenko, K., Fr, E., Pytlak, E., Adamski, M.: Mechanical properties of geopolymer composites reinforced with natural fibers. Procedia Eng. **151**, 388–393 (2016)
12. Abbas, A.G.N., Aziz, F.N.A.A., Abdan, K., Nasir, N.A.M., Huseien, G.F.: A state-of-the-art review on fibre-reinforced geopolymer composites. Constr. Build. Mater. **330**(January), 127187 (2022)

13. Meon, M.S., Othman, M.F., Husain, H., Remeli, M.F., Syawal, M.S.M.: Improving tensile properties of kenaf fibers treated with sodium hydroxide. Procedia Eng. **41**(Iris), 1587–1592 (2012)
14. Abbas, A.G.N., Aziz, F.N.A.A., Abdan, K., Nasir, N.A.M., Norizan, M.N.: Kenaf fibre reinforced cementitious composites. Fibers **10**, 1–24 (2022)
15. Esmaeilpourshirvani, N., Taghavighalesari, A.: Improvement of the engineering behavior of sand-clay mixtures using kenaf fiber reinforcement. Transp. Geotech. **19**(January), 1–8 (2019)
16. Elsaid, A., Dawood, M., Seracino, R., Bobko, C.: Mechanical properties of kenaf fiber reinforced concrete. Constr. Build. Mater. **25**(4), 1991–2001 (2011)
17. Azzmi, N.M., Yatim, J.M., Hamid, H.A., Aziz, A.A., Adole, A.M.: Mechanical properties of pulverized fuel ash concrete kenaf fibrous. MATEC Web Conference, vol. 05007 (2018)
18. Pirmohammad, S., Mengharpey, M.H.: Influence of natural fibers on fracture strength of WMA (warm mix asphalt) concretes using a new fracture test specimen. Constr. Build. Mater. **251**, 118927 (2020)
19. Hamidon, M.H.: Effects of fibre treatment on mechanical properties of kenaf fibre reinforced composites: a review. Integr. Med. Res. **8**(3), 3327–3337 (2019)
20. Muhammad, A., Rashidi, A., Wahit, M.: Alkaline treatment on kenaf fiber to be incorporated in unsaturated polyester. J. Eng. Appl. Sci. January, 1819–6608 (2016)
21. Charkhtab Moghaddam, S., Madandoust, R., Jamshidi, M., Nikbin, I.M.: Mechanical properties of fly ash-based geopolymer concrete with crumb rubber and steel fiber under ambient and sulfuric acid conditions. Constr. Build. Mater. **281**, 122571 (2021)
22. ASTM C618—12a: Standard Specification for Coal Fly Ash and Raw or Calcined Natural Pozzolan for Use, pp. 1–5 (2014)
23. Akil, H.M., Omar, M.F., Mazuki, A.A.M., Safiee, S., Ishak, Z.A.M., Bakar, A.A.: Kenaf fiber reinforced composites: a review. Mater. Des. **32**(8–9), 4107–4121 (2011)
24. Ikhwan, M., Zaki, M., Dolah, R., Mohamed, M.Z.: Tensile behaviour for mercerization of single kenaf fiber. Malays. J. Fundam. Appl. Sci. **14**(4), 437–439 (2018)
25. BS EN 12390-3: Testing hardened concrete—Part 3 : compressive strength of test specimens (2009)
26. Mohammadhosseini, H., Alyousef, R., Ngian, S.P., Tahir, M.: Waste polypropylene food tray fibers and palm oil fuel ash. Crystals, vol. 11 (2021)
27. Xu, F., Deng, X., Peng, C., Zhu, J., Chen, J.: Mix design and flexural toughness of PVA fiber reinforced fly ash-geopolymer composites. Constr. Build. Mater. **150**, 179–189 (2017)
28. Vilaplana, J.L., Baeza, F.J., Galao, O., Alcocel, E.G., Zornoza, E., Garcés, P.: Mechanical properties of alkali activated blast furnace slag pastes reinforced with carbon fibers. Constr. Build. Mater. **116**, 63–71 (2016)
29. Novais, R.M., Carvalheiras, J., Seabra, M.P., Pullar, R.C., Labrincha, J.A.: Effective mechanical reinforcement of inorganic polymers using glass fibre waste. J. Clean. Prod. **166**, 343–349 (2017)
30. Ganesan, N., Abraham, R., Deepa Raj, S.: Durability characteristics of steel fibre reinforced geopolymer concrete. Constr. Build. Mater. **93**, 471–476 (2015)
31. Bakharev, T.: Resistance of geopolymer materials to acid attack. Cem. Concr. Res. **35**(4), 658–670 (2005)

Experimental Evaluation of Bond Properties in Textile-Reinforced Concrete by Digital Image Correlation

Sanjay Gokul Venigalla, Abu Bakar Nabilah, Noor Azline Mohd Nasir, Nor Azizi Safiee, and Farah Nora Aznieta Abd Aziz

Abstract The advancement in construction technology leads to the development of sustainable structural systems with a motive of minimal natural resources usage, lightweight, and slender with added economic benefits. To attain this, the studies are evolved to alter partially conventional reinforcement with short fibers. However, the nonuniform distribution of fibers and the difficulty to achieve targeted slender sections resulted in the development of continuous fiber concrete, also termed textile-reinforced concrete (TRC). Due to its non-metallic property, concrete cover can be reduced, which allows for slender and lightweight structural systems. Even though TRC possesses vast possibilities in terms of its usage, the bond–slip relationship is very difficult to identify experimentally, and design standards are still in the research stage. This study aims to determine bond–slip at interface experimentally and explain how the bond length influences the overall performance of the textile-reinforced composite. The bond performance of TRC was assessed on tensile specimens, utilizing double-sided pullout tests with different anchorage lengths. The anchorage length is limited by creating two-sided notches, making sure that only one fiber is tested. The samples are reinforced with one layer of carbon textile coated with epoxy. The bond effect is measured in the form of applied load versus the displacement using the digital image correlation (DIC) technique. Multiple images

---

S. G. Venigalla · A. B. Nabilah (✉) · N. A. M. Nasir · N. A. Safiee · F. N. A. A. Aziz
Department of Civil Engineering, Faculty of Engineering, Universiti Putra Malaysia, 43400 Serdang, Malaysia
e-mail: nabilah@upm.edu.my

S. G. Venigalla
e-mail: gs54602@student.upm.edu.my

N. A. M. Nasir
e-mail: nazline@upm.edu.my

N. A. Safiee
e-mail: norazizi@upm.edu.my

F. N. A. A. Aziz
e-mail: farah@upm.edu.my

N. Sabtu (ed.), *Proceedings of AWAM International Conference on Civil Engineering 2022—Volume 2*, Lecture Notes in Civil Engineering 385,
https://doi.org/10.1007/978-981-99-6018-7_13

are collected and analyzed using GOM correlate tool, and a detailed bond behavior is visualized with the increase in the loading. The failure due to rupture and pullout are observed in experiments and pullout load–slip curves are obtained. The results were compared with local bond stress–slip interface models, where shorter bond length experimental results are in good agreement.

Keywords Textile-reinforced concrete · Digital image co-relation · Pullout test

1 Introduction

In recent times, many structures are constructed by some form of pre-fabrication elements, which offers in the rise of using innovative materials and adopting new construction techniques. In this process, many studies evolved in replacing conventional reinforcement with short fibers which allow for the production of lightweight, durable, and slender sections. But the disadvantages of using chopped fibers are the random orientation and nonuniform distribution of fibers over the cross-section, reducing the effectiveness. To overcome this issue, continuous fibers (meshed textile) are used, creating textile-reinforced concrete (TRC) [1]. Textile-reinforced concrete is composed of fine-grained concrete with multi-axial textile fabrics. Textile-based composites are in use to retrofit existing structures [2] and also to fabricate structural elements as permanent formwork [3]. The enhanced properties of TRC like resistance to fire and increased flexibility encourage it to be used as an alternative material to fiber-reinforced polymer (FRP) for strengthening existing structures.

Over a decade, distinct collaborative research was carried out in the Europe and USA on the properties of TRC from material to structural behaviors, in terms of tensile and flexural strengths [4, 5]. The specimens are fabricated in the form of plates and are subjected to tension or bending. The results are majorly dependent on the fiber type, aspect ratio, and mortar mix. In TRC elements, the interfacial bond behavior of textile yarn in cementitious matrix plays a major role in influencing the global structural behavior [6]. Further, the characterization of bond behavior is a critical input parameter for numerical models to showcase the structural response of TRC. To explain the bond failure mechanism, fiber pullout test methods are developed [7]. A considerable amount of test setups is developed to examine the bond characteristics of textile in concrete. In general, one-sided and double-sided pullout tests are used for TRC. In a one-sided pullout test [8], a group of testing yarns is bundled with epoxy at one end and the other end is fabricated inside a concrete block. Similar to conventional concrete, the pullout tests are carried on unmanufactured single yarn where the obtained results do not replicate a single fiber pullout. Lorenz and Ortlepp [9] developed a double-sided single fiber pullout test setup with unsymmetrical clamping lengths, where the bond strength of the textile fiber in the concrete is determined using a double-sided notch. The bond behavior of textile with short anchorage length is identified experimentally. To overcome the effect of anchorage length criteria, Williams Portal et al. [10] developed a double-sided pullout test setup and characterized the effect of the bond at different anchorages. A 1D model is developed for short embedded length textile and compared to experimental results and bond–slip curves are plotted. The same bond–slip curve for short embedded lengths is used for all anchorage lengths to determine bond behavior. Thus, a comprehensive study is required to explain the actual behavior of textile at any anchorage length. The present study explains the phenomenon of textile bonds in cement mortar at different anchorages. To explain this, a single fiber pullout test setup is established based on Peled et al. [11] and load-displacements are measured using the digital image correlation method (DIC). DIC is a non-destructive/non-contact technique that uses images to record the requisite output. A series of images will be acquired at different loading intervals and compared with the reference image [12].

2 Experimental Research

The present work is part of evaluating the bond behavior of textile reinforcement in concrete by experimental work. The experiments are conducted to identify an appropriate mix for TRC and pullout response of textile in the mortar at different bond lengths is identified using digital image correlation (DIC) technique. The experimental research carried out is shown in Fig. 1.

2.1 Material Description

In this work, a fine grain concrete with a maximum grain size of 1.18 mm was developed based on Du et al. [13] with the target strength at 28 days of 45 MPa. The mortar mix comprises 68.5% Type I Portland cement@42.5 N, 25% fly ash, and 6.5% micro silica with a W/B of 0.35. As an admixture, a naphthalene-based superplasticizer was used at a dosage of 0.8% of binder content to achieve high workability. To choose a suitable water-binder ratio for TRC, flow tests (see Fig. 2) were conducted on fresh mortar with reference to ASTM C230. The complete details of mix proportion are tabulated in Table 1. Finally, after casting and curing at 25 °C (room temperature), compression tests were conducted on cube specimens of size 50 mm × 50 mm × 50 mm in accordance with BS EN:1015-11. The mean compressive strength of mortar cubes at the age of 28 days is 47 MPa.

The textile reinforcement used in this work was a commercially available carbon fiber with multiple yarns placed in two orthogonal directions. The most commonly used textile is carbon because of its high strength and durability. The fiber has a mesh configuration of the size of 5 mm × 5 mm with a yarn fineness of 2200 tex. Each yarn consists of 12,000 filaments knitted in weft and wrap direction, with each filament measuring 7 μm in diameter. Further geometry and mechanical properties of the fiber are listed in Table 2. To avoid tensile weakening and penetration of mortar through filaments, the textile yarns were coated with epoxy resin. The epoxy used

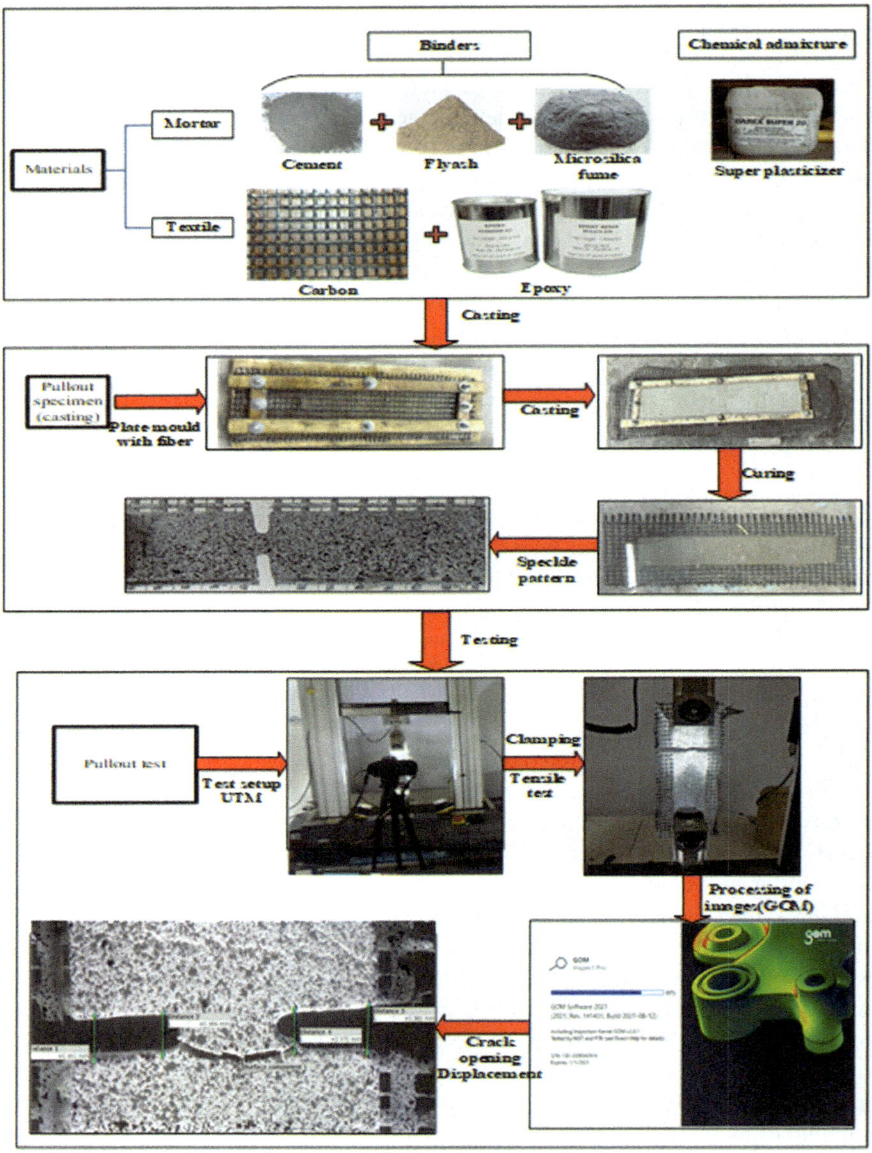

Fig. 1 Process flow diagram

Fig. 2 Flow test

Table 1 Mortar mix proportions

Cement (kg/m³)	Fly ash (kg/m³)	Micro-silica (kg/m³)	Fine aggregate kg/m³)		Water (kg/m³)	Superplasticizer	Mortar flow (mm)
			0–0.6 (mm)	0.6–1.18 (mm)			
582.96	212.76	55.31	390.06	791.94	297.86	0.8% of binder	155

is EPIKOTE Resin 828 produced from bisphenol A resin and epichlorohydrin with polyamide resin 651 hardener. The ratio of epoxy to hardener is 5:2 as recommended by the manufacturer. The fiber is coated with epoxy and cured at room temperature (25 °C) for 48 hours.

Table 2 Specifications of carbon textile

Property	Units	Value
Each yarn	Nos	12,000 filaments
Mesh size (0°/90°)	mm x mm	5 × 5
Tensile strength	GPa	4.5
Elastic modulus	GPa	230
Density	g/cm³	1.79
Diameter	μm	7
Elongation	%	1.9

2.2 Pullout Test Specimen Details

The dimension of tensile specimens was designed based on the recommendations of RILEM TC 232-TDT [11]. The test specimens were fabricated with rectangular cross-sections with a measuring size of 300 mm length, a width of 50 mm, and 12 mm thick. The mortar is placed in two layers, first, a layer of 6 mm thick mortar was poured and carbon textile was placed and fastened by a formwork, followed by placing the second layer of mortar with 6 mm thickness. The tensile specimens were demolded after 24 hours and cured at room temperature for 28 days. Before the date of testing, uneven concrete edges were removed in the load application area and the cross-section of the specimen was checked at different points with an accuracy of 0.2 mm.

2.3 Experimental Setup

The pullout test setup was chosen based on uniaxial double-sided tests in accordance with RILEM recommendations [11]. The tests were conducted, making sure that only one yarn per sample is tested. Two asymmetric anchorage lengths were chosen, such as 40 mm and 30 mm in the upper part of the sample. The anchorage lengths were limited by making a saw cut underneath the clamping length. A notch was created via two saw cuts, to ensure the yarn fail at a predetermined breaking point. The details of the test specimen are shown in Fig. 3. The pullout tests were carried out by varying bond lengths to observe the mode of failures such as rupture or deboning failures.

The pullout tests were conducted on a displacement-controlled universal testing machine with a capacity of 5 kN (see Fig. 4). The specimens were clamped on both

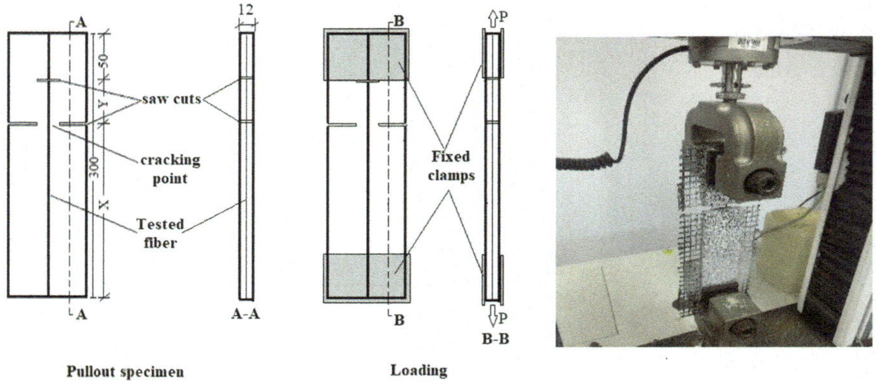

Fig. 3 Details of pullout specimen

sides with a clamping length of 50 mm. To avoid the effect of clamping load on the specimen, galvanized iron (GI) sheets with a thickness of 1 mm were glued on both faces using epoxy. The rate of loading applied on the specimen was 1 mm/min. During the test, the displacements were measured using the digital image correlation (DIC) technique.

A series of images were taken at different loading intervals and compared with the reference image. The differences in images can be recognized by the mode of the speckle pattern. After the specimen was clamped in the universal testing machine, a camera with a resolution of 24.2 megapixels was used to capture the images. The camera was set perpendicular to the sample at a horizontal distance of 32.5 cm. The details of camera inputs used in recording the images are shown in Table 3. The images recorded were analyzed using GOM correlate software.

Fig. 4 Tensile testing in the universal testing machine with DIC

Table 3 Camera inputs for digital image correlation method

Settings	Input
Mode	AV
Shutter speed	1/40
Pixel size	
Width	6000 pixels
Height	4000 pixels
Focal length	55 mm
Exposure time	1/100 s
Aperture	f/7.1

2.4 Local Bond–slip Interface

The load–slip curves obtained from pullout tests are compared to theoretical equations proposed by *fib Bulletin 55: Model Code 2010* for bonded FRP rebars [14]. In the proposed model, the local bond–slip curve shape is a function of geometry, material property, and stress distribution at the interface. In the present study, at the first bond–slip function was evaluated from experimental data (pullout load vs. slip). The bond stress is calculated with respect to load and bond length using the general bond strength equation (Eq. 1).

$$\tau = \frac{P}{\pi dl}, \tag{1}$$

where P is load, d is the diameter of fiber, and L is embedded length (30 mm and 40 mm), and curves are plotted for bond stress versus slip (Fig. 6).

To compare the experimental results, *the fib* local bond stress–slip model is used, where the bond stress is calculated as a function of relative slip according to the following equations.

$$\tau = \tau_{\mathrm{m}}\left(\frac{S}{S_1}\right)^{\alpha} \quad \text{for } 0 \le S \le S_1, \tag{2}$$

$$\tau = \tau_{\mathrm{m}} - \tau_{\mathrm{m}} p(S - S_1)/S_1 \quad \text{for } S_1 \le S \le S_2. \tag{3}$$

Assuming the pullout failure, τ represents the bond stress, τ_{m} is the maximum bond stress, and S is the slip of fiber. The values of S_1 and S_2 are based on experimental results. The bond–slip relation is explained theoretically in two phases [Eqs. (2) and (3)]. The α and p represent the correction factors for theoretical equations. Figure 5 replicates the analytical bond stress–slip relationship for embedded fiber.

Fig. 5 Analytical bond stress–slip curve for fiber [14]

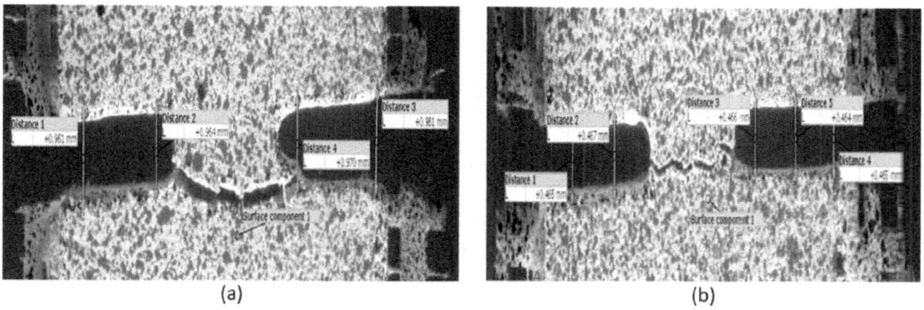

Fig. 6 Crack-opening displacement measurement using digital image correlation analysis for an embedded length of **a** 30 mm, **b** 40 mm

The ascending part illustrates the perfect bonding between fiber and mortar (elastic phase), whereas the descending part represents the decrease of bond strength due to shear (damage evolution debonding phase). The correction factor α is calculated based on the area under ascending part (A_τ) and p is calculated based on maximum bond stress to elastic slip.

3 Results and Discussions

3.1 Experimental Results

The recorded images were analyzed, and the surface components were created on the specimen. The parameters used for surface components were fact size of 19 pixels and point distance of 16 pixels with standard computation. Another two points were selected to measure the displacement and the average distance was calculated for both anchorage length specimens of 30 and 40 mm. For visualization of bond failure, the samples with an embedded length of 30 and 40 mm are shown in Fig. 6. The average crack-opening displacements of 0.93 mm and 0.47 mm were recorded at the maximum load, respectively.

By using displacements from DIC, the average pullout load versus displacement graphs of carbon fiber are plotted as shown in Fig. 7. During pullout load, the restraining forces acting from mortar to fiber are compressive and tension forces exerted by fiber. In this process, the circumferential stresses led to fracture once the mortar reached its maximum tensile strength. A pullout failure was observed for the specimens with short embedded length (30 mm) and rupture was the cause of failure for larger lengths (40 mm). The pullout failure occurred due to the reduced adhesive bond at interface up to pre-peak, followed by slip, due to destruction of the adhesive bond between fiber and mortar. On the other hand, the rupture failure was due to the immediate loss of textile capacity after the peak load was attained. The average

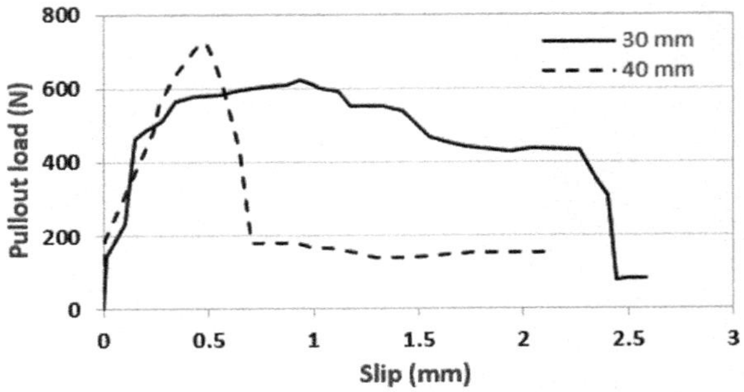

Fig. 7 Pullout load versus slip curve for different anchorage lengths

maximum pullout force corresponding to short and long embedded length specimens was 625.5 N and 726.2 N, respectively.

For the 30 mm anchorage length specimens, it is observed that the load versus displacement curve is linear up to 483.9 N, due to the linear elastic behavior of the composite. Afterward, a slight drop in the load is observed due to the elastic softening at the interface. The pullout load increases to a maximum load of 625.5 N, where pullout of fiber was initiated (start of elastic deboning), and the load capacity drops gradually which resembles the slip due to loss of friction. For the specimen with an anchorage length of 40 mm, the maximum pullout load is recorded as 726.2 N with similar initial stiffness as the previous sample. The load capacity drops immediately which resembles the rupture at the interface (extended elastic softening phase).

3.2 Local Bond Stress–slip Model Results

The theoretical bond stress–slip curves developed using fib model (Sect. 2.4) are shown in Fig. 8. The maximum bond stress for 30 mm and 40 mm is 3.32 MPa and 2.89 MPa, respectively. The curves are plotted for bond versus slip considering experimental results and local bond–slip interface models. The interface models are calibrated using the α and p parameters. The correction factor α is calculated based on the area under the experimental bond–slip curves in the elastic phase (α is 0.12 for 30 mm and 0.63 for 40 mm). The parameter p is used in plotting curve in debonding phase, calculated from maximum bond stress and slip (p is 0.2 for 30 mm and 0.12 for 40 mm).

From the results, it is identified that the local bond–slip curves almost replicate the experimental results in the elastic and deboning phase for 30 mm (short), whereas in the case of fiber with a bond length of 40 mm, the curves are reasonably good correlation only in the elastic phase. In post-peak behavior, due to rupture of fiber, there is a sudden drop in stiffness of the experimental plotted curve. The fib model curve followed the same pattern as 30 mm (short) because of linear material assumption.

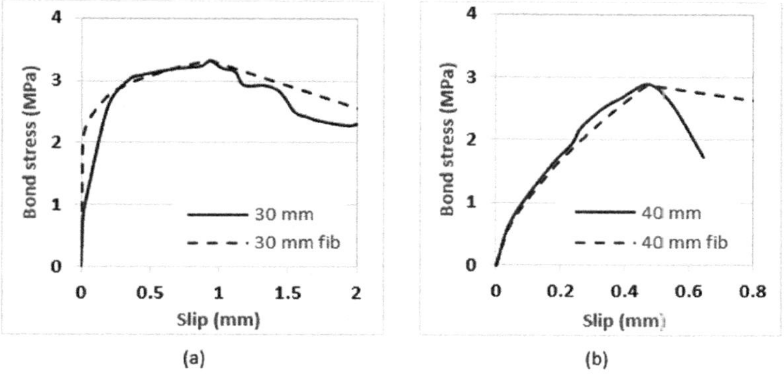

Fig. 8 Local bond–slip results versus experimental results. **a** 30 mm, **b** 40 mm bond lengths

4 Conclusions

The bond properties of carbon textile-reinforced concrete were identified using double-sided pullout tests. Firstly, a suitable mortar mix is identified which satisfies the workability and strength required for textile-reinforced concrete. In addition, a tensile specimen is designed, satisfying the RILEM report guidelines. In addition, the digital image correlation method is used to record the pullout response.

The bond failure is recorded due to pullout and rupture, depending on the anchorage length provided. This is due to the restraining forces leading to fracture at the interface. For the specimens with short anchorage lengths, pullout failure was observed, while rupture failure was noticed in specimens with long anchorage lengths. The loss of bond is noticed due to the loss of adhesion between the textile and cement matrix. Further, the results are compared with a theoretical model where the samples with shorter bond lengths replicate the theoretical model, whereas as longer bond length samples, the theoretical curve is nonlinear, due to linear material assumption.

Acknowledgements The authors would like to thank *Kementerian Pendidikan Tinggi Malaysia* for the financial support of this research work under the Fundamental Research Grant Scheme, FRGS/1/2018/TK01/UPM/02/3.

References

1. Naaman, A.E.: Thin TRC products. In: Textile Fibre Composites in Civil Engineering, no. 60, pp. 413–439. Elsevier (2016)
2. Kouris, L.A.S., Triantafillou, T.C.: State-of-the-art on strengthening of masonry structures with textile reinforced mortar (TRM). Constr. Build. Mater. **188**, 1221–1233 (2018). https://doi.org/10.1016/j.conbuildmat.2018.08.039

3. Daskiran, M.M., Daskiran, E.G., Gencoglu, M.: Seismic performance of PVA textile cementitious composites used as permanent formwork in full-scale circular RC columns. In: Structures, vol. 36, no. December 2021, pp. 445–462 (2022). https://doi.org/10.1016/j.istruc.2021.12.006.
4. Yin, S., Jing, L., Yin, M., Wang, B.: Mechanical properties of textile reinforced concrete under chloride wet-dry and freeze-thaw cycle environments. Cem. Concr. Compos. **96**(August 2017), 118–127 (2019). https://doi.org/10.1016/j.cemconcomp.2018.11.020
5. Trochoutsou, N., Di Benedetti, M., Pilakoutas, K., Guadagnini, M.: Mechanical characterisation of flax and jute textile-reinforced mortars. Constr. Build. Mater. **271**, 121564 (2021). https://doi.org/10.1016/j.conbuildmat.2020.121564
6. Rampini, M.C., Zani, G., Colombo, M., di Prisco, M.: Mechanical behaviour of TRC composites: Experimental and analytical approaches. Appl. Sci. **9**(7) (2019). https://doi.org/10.3390/app9071492
7. Tekle, B.H., Messerer, D., Holschemacher, K.: Bond induced concrete splitting failure in textile-reinforced fine-grained concrete. Constr. Build. Mater. **303**(August, 2021). https://doi.org/10.1016/j.conbuildmat.2021.124503
8. Banholzer, B.: Bond of a strand in a cementitious matrix. Mater. Struct. Constr. **39**(294), 1015–1028 (2006). https://doi.org/10.1617/s11527-006-9115-y
9. Lorenz, E., Ortlepp, R.: Bond behavior of textile reinforcements - development of a pull-out test and modeling of the respective bond versus slip relation. In: RILEM Bookseries, no. June, pp. 479–486 (2011). https://doi.org/10.1007/978-94-007-2436-5
10. Williams Portal, N., Fernandez Perez, I., Nyholm Thrane, L., Lundgren, K.: Pull-out of textile reinforcement in concrete. Constr. Build. Mater. **71**, 63–71 (2014). https://doi.org/10.1016/j.conbuildmat.2014.08.014
11. Peled, A., Mobasher, B., Hegger, J.: Recommendation of RILEM TC 232-TDT: test methods and design of textile reinforced concrete. Mater. Struct. **49**(12), 4923–4927 (2016). https://doi.org/10.1617/s11527-016-0839-z
12. Vervloet, j., et al.: Validation of a numerical bending model for sandwich beams with textile-reinforced cement faces by means of digital image correlation. Appl. Sci. **9**(6) (2019). https://doi.org/10.3390/app9061253
13. Du, Y., Zhang, X., Zhou, F., Zhu, D., Zhang, M., Pan, W.: Flexural behavior of basalt textile-reinforced concrete. Constr. Build. Mater. **183**, 7–21 (2018). https://doi.org/10.1016/j.conbuildmat.2018.06.165
14. Grossetti, M.: International federation for structural concrete (fib), fib Bulletin 55: model code 2010 (2010)

Characterization of Spent Bleaching Earth Ash (SBEA) and Assessing Its Use as Partial Cement Replacement in Cement Mortar

Andrew Lim, Abdul Karim Bin Mirasa, and Hidayati Bte Asrah

Abstract Spent bleaching earth ash (SBEA) originates from spent bleaching earth (SBE), which has been shown to possess pozzolanic properties and has been considered for use as cement replacement in various applications such as concrete and masonry. With large amounts of SBEA being produced annually and often treated as hazardous waste, it would be more sustainable to re-use it an application that could reduce the demand on our environment, such as the cement industry investigated in this paper. This paper characterized SBEA in its untreated form and found that it can be classified as a Class N natural pozzolan as defined by ASTM C618 in terms of chemical composition, fineness and water requirement. At the same time, its 7- and 28-day strength activity index both exceeded 75%, further demonstrating its good pozzolanic reactivity. Finally, an assessment of its early (3- and 7-day) and late (28- and 56-day) compressive strengths as well as its flexural strength further reinforces the theory that SBEA could be effectively used as partial replacement of cement in cement mortar applications.

Keywords Spent bleaching earth · Ash · Cement replacement · Natural pozzolan · Compressive strength

A. Lim (✉) · A. K. B. Mirasa · H. B. Asrah
Faculty of Engineering, Universiti Malaysia Sabah, 88400 Sabah, Malaysia
e-mail: andrew.lim@ums.edu.my

A. K. B. Mirasa
e-mail: akmirasa@ums.edu.my

H. B. Asrah
e-mail: hidayati@ums.edu.my

N. Sabtu (ed.), *Proceedings of AWAM International Conference on Civil Engineering 2022—Volume 2*, Lecture Notes in Civil Engineering 385,
https://doi.org/10.1007/978-981-99-6018-7_14

1 Introduction

Spent bleaching earth ash (SBEA), otherwise also known as eco-processed pozzolan (EPP), is a waste by-product from the oil refining industry. It originates from highly adsorbent types of clay known as bleaching earth which is used to remove pigmentation from unrefined palm oil. The bleaching earth, after being used for this purpose, would contain residual oil which was extracted using the Soxhlet extraction technique to be repurposed into bio-diesel [1]. The remaining earth, then known as spent bleaching earth (SBE), was burnt up as fuel in boilers and the resulting by-product is SBE ash (SBEA). Hence, SBEA typically comes supplied in fine powdery form and has a greyish brown in tone.

SBEA is typically treated as hazardous waste due to its high residual oil content and thus is difficult and expensive to dispose safely. The global production of SBEA exceeded 5.75 million tonnes in 2020 alone [2]. This increases the need to find an alternative use for SBEA rather relegating it to the landfill.

Just as its precursor material, SBE, having shown good pozzolanic reactivity [3], SBEA too may contain sufficient silica oxides to exhibit pozzolanic behaviour and this means that it could potentially be used as cement replacement in various applications. However, there are currently very few work carried out with relation to replacement of cement with cement, with limited research on establishing its strength activity index [4] and applications in foamed concrete [5].

This paper will characterize SBEA samples, then go on to investigate the feasibility of using it as partial cement replacement in cement mortar. The characterization of SBEA will include establishing the physical, chemical properties and its mineralogy phase. This is followed determination of the pozzolanic reactivity of SBEA using the strength activity index and finally the actual compressive and flexural strength of cement mortar samples containing SBEA.

2 Methodology

The materials, its preparation and the experimental procedure used in this research are detailed in this section. All experimental works were carried out at the laboratories located on Faculty of Engineering, Universiti Malaysia Sabah, Malaysia, except for the XRF and LPSA analyses which were done at Universiti Teknologi Malaysia laboratory services.

2.1 Material Preparation

The SBEA used in this research was supplied in bags of 5 kg by Eco-Oils Sdn Bhd based in Lahad Datu region of East Malaysia. The material originated as spent bleaching earth that has previously been used to remove pigmentation from palm oil and then burnt as fuel in the boiler. All SBEA samples were oven dried prior to use.

The Ordinary Portland cement (OPC) used is CEM II conforming to BS EN 197 which is also manufactured in Sabah in East Malaysia by Cement Industries Sabah. The OPC is used as is without any further treatment.

The fine aggregate used in the cement mortar consists of washed and oven-dried river sand from the Kota Belud region of East Malaysia, sieved past 600 μm and retained on 150 μm.

Finally, for the water in the mix, tap water from the concrete laboratory of Faculty of Engineering, Universiti Malaysia Sabah, was used.

2.2 *Characterization*

The SBEA samples underwent several types of characterization tests; namely X-ray fluorescence (XRF) to determine its chemical composition, X-ray diffraction (XRD) to determine its level of crystallinity, laser particle size analysis (LPSA) to determine particle size distribution and specific surface area, and finally the specific gravity was also determined using the pycnometer.

2.3 *Mix Design and Compressive Test*

Compressive strengths are determined using 50 mm × 50 mm × 50 mm cement mortar mixes. The cement mortar mixes were designed in accordance with ASTM C109. Each mix is consisted of three sets of replicates and tested at 3, 7, 38 and 56 days after casting. The control mix was based on ASTM C109, and its mix proportions are detailed in Table 1. There are also seven sets of test mixes containing SBEAs to replace OPC at varying levels of replacement: starting from 10 to 70% at 10% intervals. Cast samples de-moulded after a day are then cured in saturated lime water until the day of testing. In addition to compressive strength, the results were also used in determining the strength activity index of the samples which can be used to interpret the pozzolanic reactivity of SBEA in cement.

Compressive strength testing was carried out at the respective cured age using the universal testing machine (UTM) located at the heavy structures laboratory at the Faculty of Engineering, Universiti Malaysia Sabah.

Table 1 Mix design proportions for six no. of control mortar cubes

Material	Amount
OPC	500 g
Graded sand	1375 g
Water	242 mL (w/c 0.485)

Table 2 Chemical composition of materials (in % weight)

Material	SiO_2	Al_2O_3	Fe_2O_3	CaO	SO_3	LOI
SBEA	49	11	10	11	1	3
OPC	16	4	3	69	3	2
Sand	65	12	5	4	0	3

3 Results and Discussions

3.1 Chemical Properties

The chemical properties for SBEA, OPC and sand used in this research are summarized in Table 2. SBEA, in this instance, contains over 70% of the required oxides ($SiO_2 + Al_2O_3 + Fe_2O_3$) whilst having a SO_3 and LOI content of under 4.0 and 10.0%, respectively, satisfying the limits set by ASTM C618 for a Class N natural pozzolan.

Naturally, OPC contained high amounts of CaO which was expected.

3.2 Morphology

The mineral phases of SBEA are shown in Fig. 1 and are characterized by 39 peaks. The major crystalline phase detected was quartz that exists as SiO_2 and reflected at $2\theta = 20.84°$, $26.65°$ and $50.13°$. Minor crystalline phases include corundum (Al_2O_3), hematite (Fe_2O_3) and quicklime (CaO). The high level of SiO_2 corresponds with the chemical composition in Table 2 as well as past research into similar material [6]. The level of crystallinity was estimated to be 31.4% through evaluation of the areas under the peaks.

3.3 Physical Properties

The SBEA used in this research came in very fine, greyish brown tone and is smooth to the touch (Fig. 2) similar to OPC. Colour-wise, it is darker compared to OPC.

The physical properties (particle sizes, specific surface area and specific gravity) are listed in Table 3. In terms of particle size distribution, SBEA is overall coarser, more than twice as large compared to OPC (Fig. 3). Because of this, SBEA tends to have lower specific surface area as well. However, SBEA particles still fine enough that 33.3% are retained on a 45 μm sieve, which is just within the 34% limit set by ASTM C618 for Class N natural pozzolan. Weight-wise, SBEA is lighter than OPC, coming in at about 2/3 of its the specific gravity.

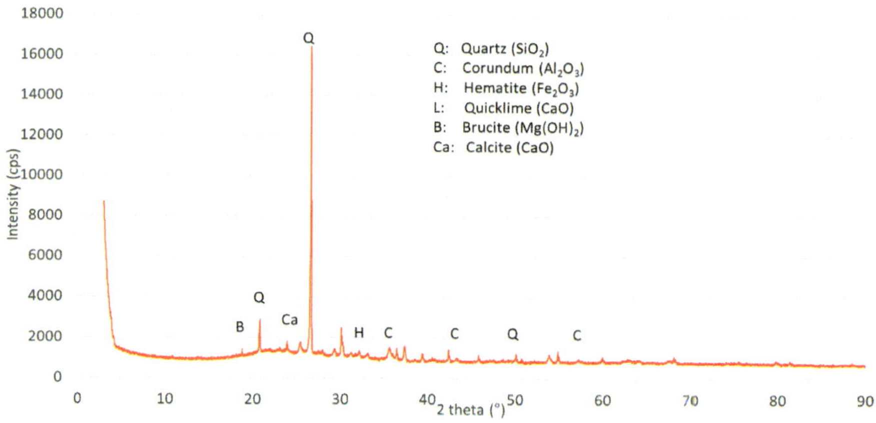

Fig. 1 X-ray diffraction (XRD) pattern for SBEA (truncated to 2 theta <70°)

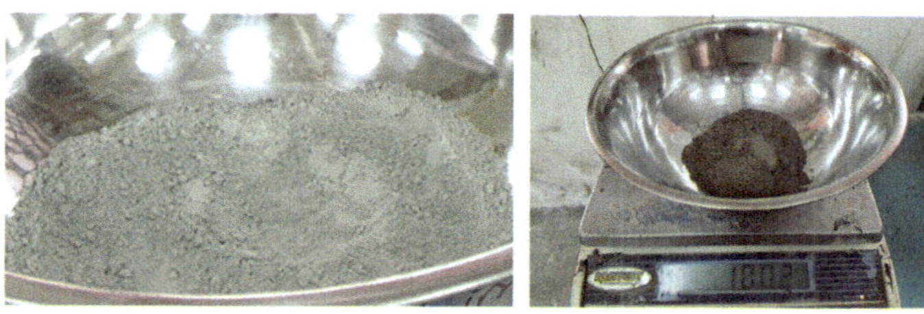

Fig. 2 OPC (left) and SBEA (right)

Table 3 Physical properties of materials

Material	Particle size (μm)			Specific surface area (m^2/kg)	Specific gravity
	d_{10}	d_{50}	d_{90}		
SBEA	8	36	100	347	2
OPC	3	16	48	753	3
Sand (unsieved)	272	345	1202	–	–

SBEA under the microscope reveals a mix of large proportions of angular-shaped particles with few spherical ones interspersed in between (Fig. 4). When zoomed in at 1000× magnification, the individual particles reveal pores on the particle surface and voids inside. These pores explain the lower specific gravity of SBEA compared to OPC [7]. There also appears to be substantial agglomeration occurring, evident from the magnified image showing smaller particles adhering onto the surfaces of larger ones.

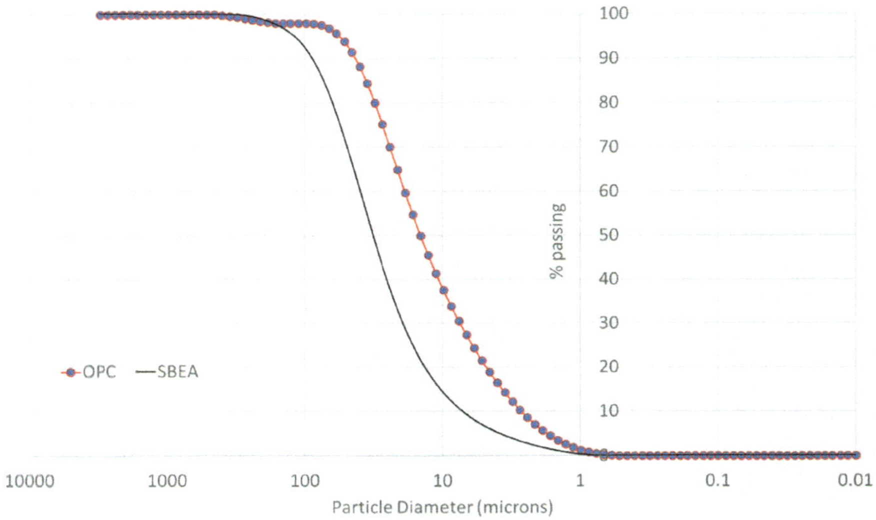

Fig. 3 Particle size distributions for SBEA and OPC

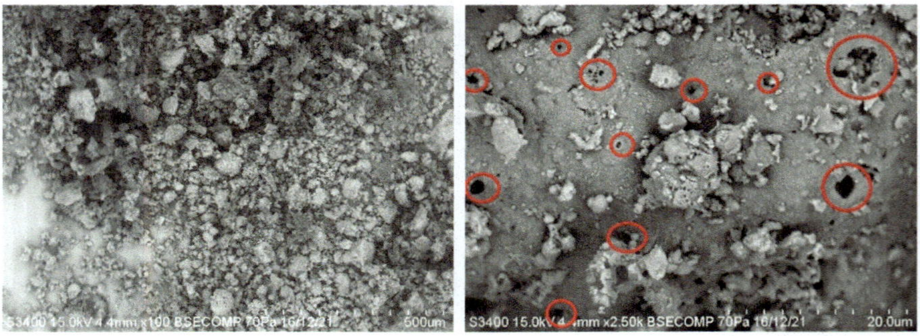

Fig. 4 SBEA particles under SEM at ×100 magnification (left) and ×1000 magnification (right)

3.4 *Water Requirement/Flow-Ability*

It is commonly understood that use of pozzolan in cement mortar will result in loss of flow-ability and increase in water requirement in the mortar samples [8]. The same phenomenon has been observed in the use of SBEA in cement mortars where increasing amounts of water are required to maintain the same workability for increasing levels of replacement (Fig. 5). At 20% replacement, the water requirement is 20 mL on top of the amount needed for the control sample, which is 108% of the control sample's content. However, this still falls within the limit of 115% set by ASTM C618. The increase is relatively linear up to about 50% where it began to spike sharply.

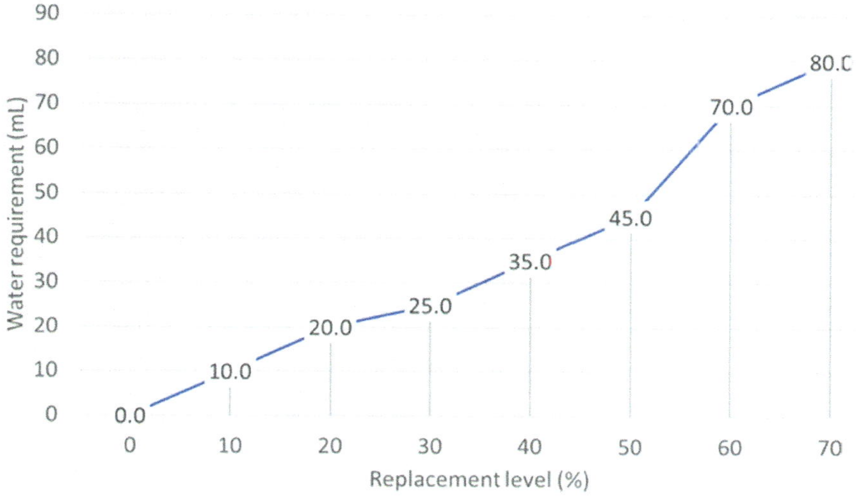

Fig. 5 Water requirement in cement mortar containing SBEA

This increase in water requirement can be explained by the existence of pores on the surface of SBEA particles which were as large as 2 μm (Fig. 4). These pores are capable of attracting water via capillary action and holding them within [9].

3.5 *Pozzolanic Reactivity Assessment Using Strength Activity Index (SAI)*

The strength activity index (Fig. 6) method was used to assess the pozzolanic reactivity of SBEA in cement mortar. At 7 days, the test sample achieved 84.8% of the control sample strength, but with further curing, the strength exceeded past that of the control samples at 106.8%. This was in contrast to past research which found that the SBEA samples showed higher early age strength compared to the 28-day strength.

Despite this, the samples, at both ages, demonstrated strength exceeding that of the target criteria of 75% as specified by ASTM C618 and thus met one of the reactivity criteria for pozzolans.

3.6 *Compressive Strength at 3, 7, 28 and 56 Days*

The compressive strengths of cement mortar samples containing SBEA substitution for all four curing ages (3, 7, 28 and 56 days) are shown in Fig. 7.

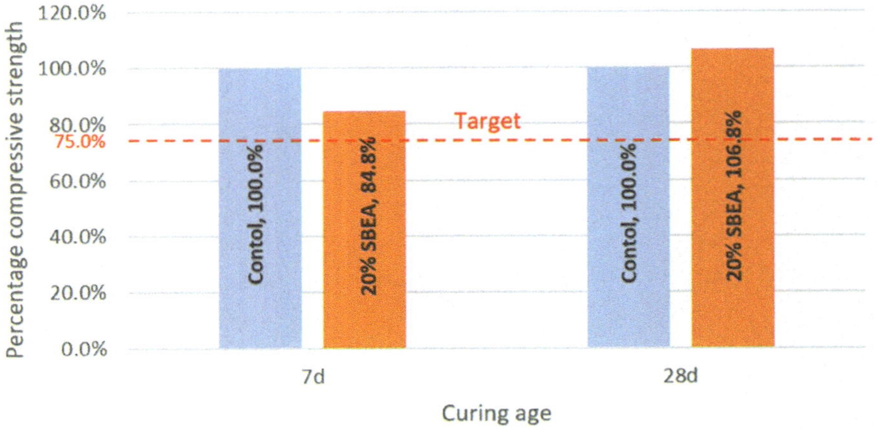

Fig. 6 Strength activity index at 7 and 28 days

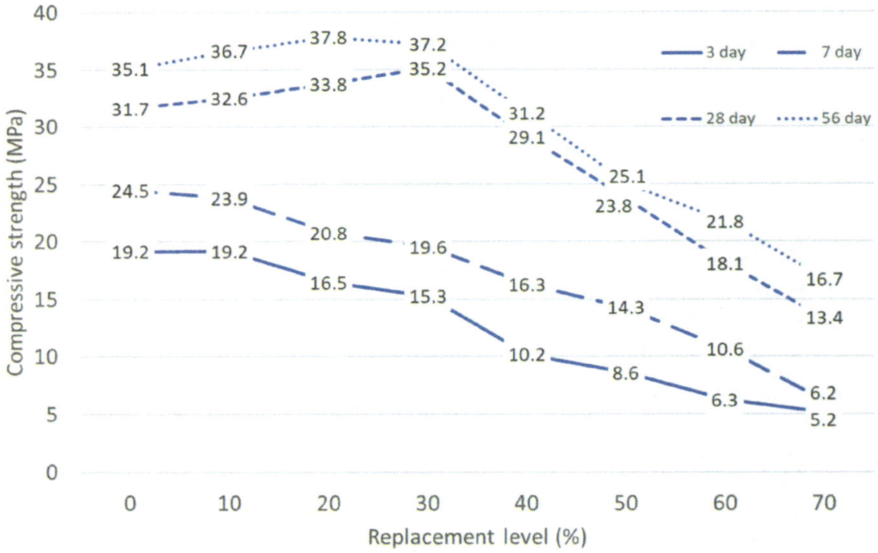

Fig. 7 Compressive strength trend of cement mortar test samples

As expected with the use of natural pozzolans, partial replacement of cement with SBEA in cement mortar saw significant reduction in its early age compressive strengths. Both 3 and 7-day compressive strength tests revealed that any addition of SBEA resulted in loss of compressive strength, and whilst this loss is insignificant at lower levels of replacement, it became markedly worse over replacement levels of over 10%. This lack of early age strength was due to insufficient portlandite (CH) being produced to initiate pozzolanic reaction at initially stages [10].

At 28 days, however, the use of SBEA began to exhibit significant gains as compressive strength increases with level of replacement up to a peak of 30%. Past the peak, compressive strength quickly declines below the control sample mix at 40% and beyond. This additional strength can be attributed to the formation of additional calcium/calcium alumina silicate hydrates (C-S-H/C-A-S-H) from the pozzolanic reaction between the silica and alumina oxides in SBEA with portlandite. Past research has also attributed the additional strength to 'filler' effect that the finer pozzolan particle has in packing the microstructure of the mortar [11], but this cannot be substantiated here as SBEA particles were found to be coarser than OPC.

The 56-day compressive strength trend is identical to the 28-day profile with compressive strength improving with increasing levels of replacement until it plateaued off between 20 and 30% and then reduces drastically at 40% replacement. This continual development in strength is likely the result of continuous pozzolanic reaction that is occurring until either all the silica/alumina oxides in SBEA or portlandite from the hydration process have been depleted.

Both the 7 and 28-day compressive strengths are comparable to results from past research which showed improvement strength up to 20% SBEA replacement in cement mortar [12].

3.7 *Flexural Strength at 28 Days*

The trend for 28-day flexural strength appears similar to that of the compressive strength for the same age (Fig. 8). There was a gradual increase in strength at lower levels of replacement up to a peak of 20%. Thereafter, there was a steep decline in the strength with further addition of SBEA past 20% replacement.

The trends' profile of both the 28-day flexural and compressive strength was similar but with only differing peak locations.

4 Concluding Remarks

1. SBEA contains the requisite chemical composition to qualify as natural pozzolan in accordance with ASTM C618. Use of SBEA as cement replacement also achieved strength in excess of the required 75% of the control sample strength, as required by ASTM C618.
2. Mineralogically, about 31% of it is crystallized with quartz forming the bulk of these, followed by hematite, bauxite and quicklime.

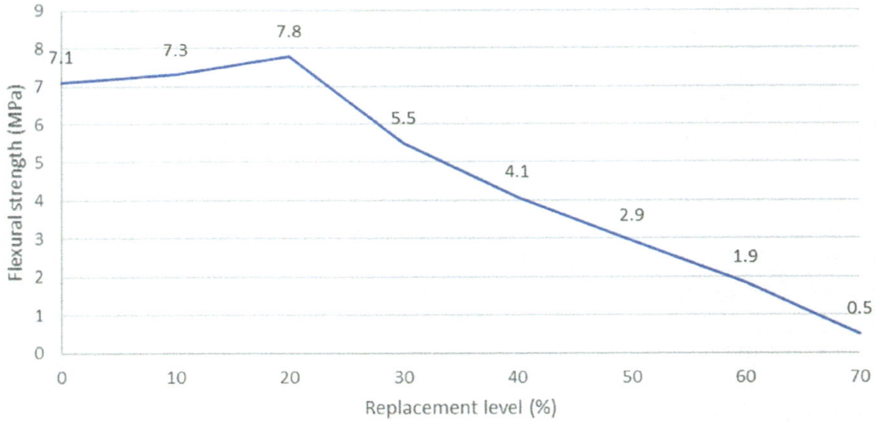

Fig. 8 Flexural strength development of cement mortar test samples at 28 days

3. Physically, SBEA is supplied in dark grey powdery form that is a little coarser than OPC but at the same time, also weighs less. Despite this, with a percentage retained on a 45 μm sieve under 34%, it also satisfies the fineness requirement of ASTM C618.
4. As expected from past research, use of pozzolans such as SBEA increases water requirement in cement mortar samples. However, this increase is within the acceptable limits of ASTM C618.
5. The compressive strength closely reflects those from past research involving use of pozzolan where poor early age performance was observed, but improvement was noticed later at 28 and 56 days.
6. The flexural strength of cement mortar containing SBEA was assessed only at 28 days of curing, but noticeable improvement was also observed for low level of improvements of 20% and below.

Acknowledgements The authors wish to thank the Faculty of Engineering as well as the Green Materials & Advanced Technology Research Unit (GMACT) of Universiti Malaysia Sabah for their support in this research endeavour. The authors also wish to acknowledge the assistance from Mr. Jonathan Chung of Cement Industries (Sabah) Sdn. Bhd. for supplying the cement needed for this research project.

References

1. Loh, S.K., Choo, Y.M., Ma, A.N.: Residual oil from spent bleaching earth (SBE) for biodiesel and biolubricant applications. MPOB Inf. Ser. 2–5 (2007)
2. Vegetable oil prices on an upward trend (2020)

3. Nalobile, P., Wachira, J.M., Thiong'o, J.K., Marangu, J.M.: Pyroprocessing and the optimum mix ratio of rice husks, broken bricks and spent bleaching earth to make pozzolanic cement. Heliyon **5**(9), e02443 (2019)
4. Rahman, R.F.A., Asrah, H., Rizalman, A.N., Mirasa, A.K.: Characterization of eco-processed pozzolan as pozzolanic material. In: International Conference on Environmental Sustainability and Resource Security (IC-ENSURES), 2019, pp. 89–93 (2019)
5. Rokiah, O., Khairunisa, M., Youventharan, D., Arif, S.M.: Effect of processed spent bleaching earth content on the compressive strength of foamed concrete. IOP Conf. Ser. Earth Environ. Sci. **244**(1) (2019)
6. Rahman, R.F.A., Asrah, H., Rizalman, A.N., Mirasa, A.K., Rajak, M.A.A.: Study of eco-processed pozzolan characterization as partial replacement of cement. J. Environ. Treat. Tech. **8**(3), 967–970 (2020)
7. Tangchirapat, W., Saeting, T., Jaturapitakkul, C., Kiattikomol, K., Siripanichgorn, A.: Use of waste ash from palm oil industry in concrete. Waste Manag. **27**(1), 81–88 (2007)
8. Kaid, N., Cyr, M., Khelafi, H.: Characterization of an algerian natural pozzolan for its use in eco-efficient cement. Int. J. Civ. Eng. **13**(4A), 444–454 (2015)
9. Zhang, C., Lu, N.: Peering into pores: what happens when water meets soil? EOS (2019). Available https://eos.org/editors-vox/peering-into-pores-what-happens-when-water-meets-soil
10. Abdelli, K., Tahlaiti, M., Belarbi, R., Oudjit, M.N.: Influence of the pozzolanic reactivity of the Blast Furnace Slag (BFS) and metakaolin on mortars. Energy Procedia **139**, 224–229 (2017)
11. Fernandez, R., Martirena, F., Scrivener, K.L.: The origin of the pozzolanic activity of calcined clay minerals: a comparison between kaolinite, illite and montmorillonite. Cem. Concr. Res. **41**(1), 113–122 (2011)
12. Yunus, E., Asrah, H., Rizalman, A.N.: Compressive strength of eco-processed pozzolan concrete under chloride and sulphate exposure. **1**(1), 1–9 (2019)

An Innovative Shear Connector for Composite Beam with Cold-Formed Steel

Ahmed N. M. Almassri, Mahmood Md Tahir, and Arizu Sulaiman

Abstract In steel construction, it is well established that composite construction is very popular in the construction of buildings and bridges with hot-rolled steel (HRS) and headed stud connectors. Due to the composite action of concrete and steel, composite beams are more cost-effective to be constructed in buildings than conventional hot-rolled steel beams. This is due to shallower beam depth, very stiff and significantly reduces the steel's weight. The increasing in demand of residential and housing construction necessitates the development of low-cost, simple, and rapid construction systems to meet the growing demand for housing. As a result, using cold-formed steel sections as an alternative to the conventional hot-rolled steel sections could contribute to better cost-effective construction to the total cost of the building and could lead to better sustainable construction requirements. Typically, the composite action is accomplished through the use of conventional headed shear stud connectors. However, the use of headed studs is impractical for cold-formed steel (CFS) sections due to their extremely thin section, which results in welding difficulty. The behaviour of novel shear connectors used in cold-formed steel–concrete composite beams comprised lipped C-channels with profiled metal decking was investigated experimentally in this study. The purpose of the experimental work is to ascertain the maximum resistance of the proposed shear connectors. Two push-out test specimens in accordance with Eurocode 4 were prepared and tested. It was discovered that the strength capacity of a shear connector is highly dependent on its deformation and the crushing of concrete. The ductility of the shear connectors tested met the Eurocode 4 requirements for ductility. The proposed shear connectors are found to be suitable for construction of composite beams with CFS because they are stronger than standard headed studs.

Keywords Shear connector · Push-out test · Cold-formed steel · Composite beam

---

A. N. M. Almassri (✉)
Faculty of Engineering, School of Civil Engineering, Universiti Teknologi Malaysia, Johor Bahru, Malaysia
e-mail: nmaahmed2@graduate.utm.my

M. M. Tahir · A. Sulaiman
School of Civil Engineering, Universiti Teknologi Malaysia, Johor Bahru, Malaysia

N. Sabtu (ed.), *Proceedings of AWAM International Conference on Civil Engineering 2022—Volume 2*, Lecture Notes in Civil Engineering 385,
https://doi.org/10.1007/978-981-99-6018-7_15

1 Introduction

CFS sections, also known as cold-formed steel, are lightweight materials that are produced by cold-forming, or the process of bending a flat sheet of steel into a defined shape at room temperature [1–3]. It is capable of supporting a load that is greater than the flat sheet itself. Because of the higher structural performance they offer, they are useful for the building industry [3, 4]. The most frequent types of CFS sections are channel-lipped C and Z sections. Their depths can range from around 51–305 mm, while their thicknesses may vary from 1.2 to 6.4 mm [3, 5]. Both hot-rolled steel (HRS) and cold-formed steel (CFS) are widely known and popular forms of steel that are utilised widely in the construction of buildings. However, engineers are more familiar with the HRS steel material than any other type of steel. Cold-formed steel sections are excellent than hot-rolled steel sections in several ways, including their ease of handling, rapid production and transportation, high strength-to-weight ratio, cost-effectiveness, and composition of material that is entirely recycled [5–7]. Because of these qualities, cold-formed steel is a cost-effective option for residential and commercial construction. In spite of this, the use of beams made from cold-formed steel sections (CFS) as major structural components in buildings is still debatable. This is due to the thin sections of the beams, which result in insufficient strength and stiffness. In recent years, a topic of worry for researchers has been the possibility of creating composite sections built of CFS beams and concrete as an answer to CFS thinness [8]. The shear connection that exists between the steel section and the concrete slab is an essential element in the composite beam's capacity to maintain its structural performance. This connection is made using mechanical shear connectors, which not only allow for the transmission of loads from the concrete to the steel but also provide resistance to vertical uplift forces at the point where the steel and concrete come into touch with one another [9, 10]. To provide shear resistance, shear connectors are usually welded to the top flange of a steel beam and embedded in the concrete slab. The most common type of the shear connectors is headed stud, which are commonly used for hot-rolled beams, are not suitable for CFS beams because their section is very thin, making welding of shear studs impractical [11]. Although other types of connectors, such as self-drilling screws, are viable alternatives, the implementation of these type of connectors may not strong enough to develop composite action; therefore, additional research into strength and stiffness of a new type of shear connectors being suitable for CFS composite constructions is proposed in this paper.

Several academics have investigated the use of cold-formed steel in composite sections with concrete, and their findings have been published. According to the findings of a researcher, the cost of manufacturing concrete beams can be reduced by replacing traditional reinforcing steel with cold-formed steel lipped channel sections of an equal cross-sectional area [5, 7, 12, 13].

2 Push-Out Test

2.1 Description of the Push-Out Test Specimens

Two push test specimens were created in accordance with Eurocode 4 recommendations [14]. Holes were created to act as shear connectors and were incorporated into the top flange beam embedded in the slab. Cold-formed lipped channel steel sections, a metal deck, and concrete slabs were used to construct the specimens. Two 660 mm long cold-formed steel lipped C channel sections were oriented back-to-back to create an I-section steel beam as in Fig. 1. Specimen No. 2 was created using pre-fabricated holes with a diameter of 30 mm that were drilled into cold-formed steel sections to act as a shear connector as shown in Fig. 2. In specimen No. 2, four shear connectors (holes) were incorporated into each slab. Eurocode 4 recommends this arrangement to allow for load redistribution. Two 660 mm long, 800 mm wide, and 125 mm thick concrete slabs were used in conjunction with a perpendicular metal deck. A single layer of fabricated reinforcement mesh (A142) with a diameter of 5.7 mm and a spacing of 200 mm in both directions was installed with a nominal concrete cover of 25 mm to prevent surface cracking and longitudinal shear. The horizontal casting of the concrete slabs for the push test specimens was recommended by the standard (EN1994-1-1, 2004). As a result, the first concrete slab labelled as specimen "A" was cast prior to the second concrete slab labelled as specimen "B." The concrete slab "A" with the beam was cast on the first day of the first phase. The concrete slab "B" was then cast for the same specimens. Compressive strength of concrete was determined using six cubes and six cylinders at the time of the push test specimen and after 28 days. The push test specimens were painted white to facilitate visual inspection of cracks. The push test specimens included a 100 mm recess between the bottom of the concrete slab and the lower end of the cold-formed steel beam to allow for slippage during testing. Typical test specimens with their dimensions and details are presented in Figs. 3, 4 and 5.

The longitudinal spacing between connectors (pre-drilled holes) was kept constant at 330 mm, while the transverse spacing was preserved at 100 mm. The sections utilised were cold-formed lipped channel steel with a minimum yield strength of 500 MPa, a nominal thickness of 3 mm, and a depth of 300 mm. All specimens were cast in C25 concrete (25 N/mm^2). A metal deck dubbed "AJIYA PEVA 50 profiled steel sheet" was introduced. It has a nominal thickness of 1 mm and minimum yield strength of 550 MPa (Table 1).

2.2 Test Setup and Instrumentation

The test setup is shown in Fig. 6. Each push-out specimen was positioned appropriately over an 800 mm × 800 mm × 100 mm thick steel piece so that the concrete slabs could be correctly positioned. The lower half of the steel beam, which has a

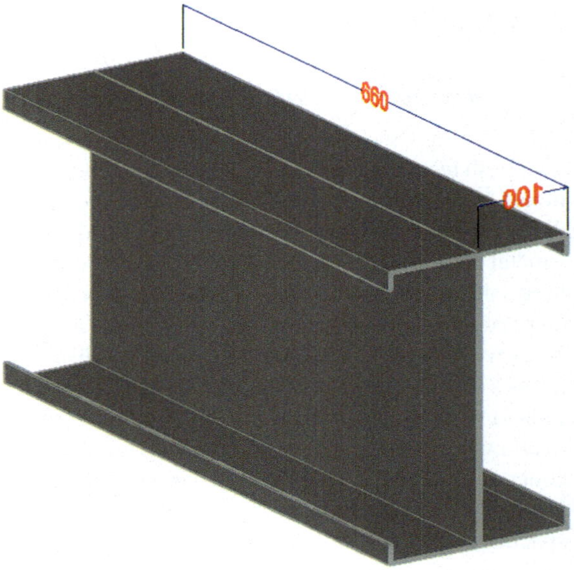

Fig. 1 Cold-formed steel lipped C channel beam fabrication for specimen without holes

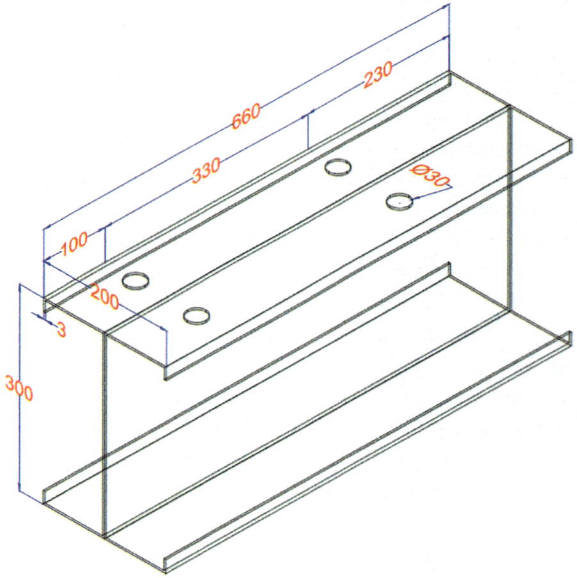

Fig. 2 Cold-formed steel lipped C channel beam fabrication for specimen with holes

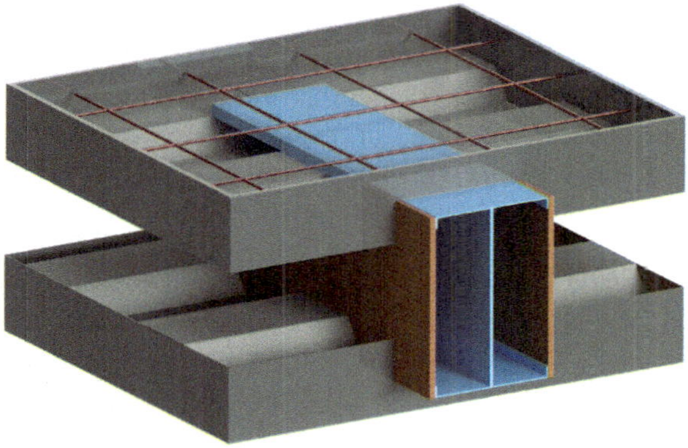

Fig. 3 Full view of specimen no. 1 where the beam is oriented back-to-back without holes

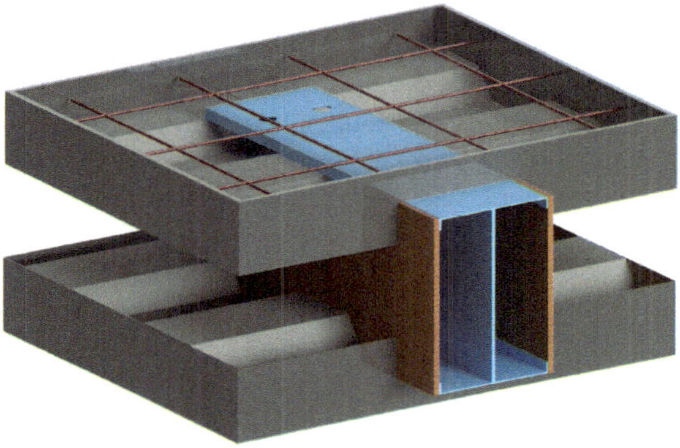

Fig. 4 Full view of specimen no. 2 where the beam is oriented back-to-back without holes

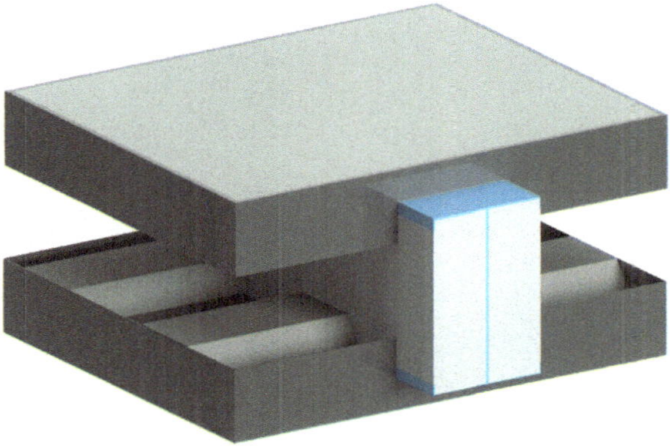

Fig. 5 Specimens after casting

Table 1 Shear connection type of the CFS beam

Specimen no.	The orientation of CFS beam	Shear connector type
1	Back-to-back	Embedded CFS beam without holes
2	Back-to-back	Embedded CFS beam with holes

load cell capacity of 2000 kN, was loaded by means of a DARTEC Jack machine. The load cell capacity is measured in kilonewtons. In order to determine the amount of slippage that occurred in the vertical direction between the concrete slab and the steel beam, each push-out specimen was fitted with four linear variable displacement transducers (LVDTs). These transducers were positioned as follows: two on either side of the steel beam web and the other two on the upper surface of the concrete slabs. The load was measured in "kN" and the displacement was measured in "mm" using the LVDTs and the load cell, which were all connected to a data logger for data collecting. The procedure of testing was carried out in accordance with Eurocode 4. The load was imposed at a steady rate of up to 40% of the specimen's expected failure load. In order to guarantee that the specimen was in an equilibrium state, the loading was cycled 25 times (loading and unloading) between 5 and 40% of the estimated failure load. After the cyclic loading sequence, the load was then applied at a constant rate until failure. When there was a decrease of 20% of the maximum load that could be placed on the specimen or when the specimen was unable to withstand any additional load, the loading was terminated. Approximately forty-five minutes had passed for each specimen before it was deemed a failure. The values of the slip that occurred between the steel beam and the concrete were recorded at every single load increment. The average displacement value is plotted against the load.

Fig. 6 Setup of push-out test

Table 2 Results for cylinder and cube tests for concrete material

Curing time	Cylinder compressive strength, f_{ck} (MPa)	Cube compressive strength, f_{cu} (MPa)	Splitting tensile stress, f_{ct} (MPa)	Modulus of elasticity, E (MPa)
Concrete @ 28 days	22	28.7	2.2	27,230

3 Material Properties

CFS channel-lipped section with a depth of 300 mm, a width of 100 mm, a lipped depth of 20 mm, and a thickness of 3 mm; metal deck; steel rebar shear connector of 500 N/mm^2 grade; welded wire fabric mesh A142 of 5.7 mm thick spaced 200 mm × 200 mm; and concrete grade of 25 N/mm^2 are the materials used in this study. After the concrete is cast for the push-out specimens, concrete cylinders (100 mm in diameter 200 mm in length) are prepared during each pouring. These concrete cylinders are tested for compressive strength on 28 days. Table 2 shows the results of the compressive strength tests. The tensile strength and the elastic modulus of the concrete are obtained according to the procedure proposed by Eurocode 2. The properties of the steel materials used in these specimens are prepared for tensile test in accordance with BS EN ISO 6892-1:2009 (2009).

4 Results

Table 3 summarises the results of the push-out tests and the failure modes. The load–slip relationships of tested specimens are depicted in Figs. 7 and 8. All specimens developed cracks at varying rates depending on the type of shear connector used before reaching their ultimate load levels. A maximum load close to 490 kN was recorded for specimen No. 2. The failure mode observed in all push test specimens except those that sheared off or buckled is identical and could be attributed to concrete cracking followed by concrete crushing longitudinally along the length of the beam and transversely across its width as illustrated in Fig. 9. The cracks were first noticed in slab "B" and then in slab "A." This is because concrete slab "B" is cast after concrete slab "A" and thus has a lower compressive strength. During the advanced stage of loading, longitudinal cracks were observed running through the slabs' centres. The reason for these longitudinal cracks is that as applied load increases, the longitudinal shear force transmitted from the concrete slab to the steel beam via shear connectors tends to split the slab longitudinally. Transverse separation between concrete slab and steel beam was also observed in some specimens, such as those with embedded steel rebar from the top shear connector's specimen, as illustrated in Fig. 9. It is worth noting that during the test, no sign of buckling was observed on the flange or web of the steel beam. After dismantling the specimens to investigate and visualise

Table 3 Push-out test results

Specimen no.	Total P_u (kN)	P_u per connector (kN)	δu (mm)	First crack @ (kN)	Failure mode
1	412.6	No shear connectors used	5.02	125	Transverses and longitudinal cracks followed by concrete crushing
2	486.6	60.83	9.1	185	Transverses and longitudinal cracks followed by concrete crushing

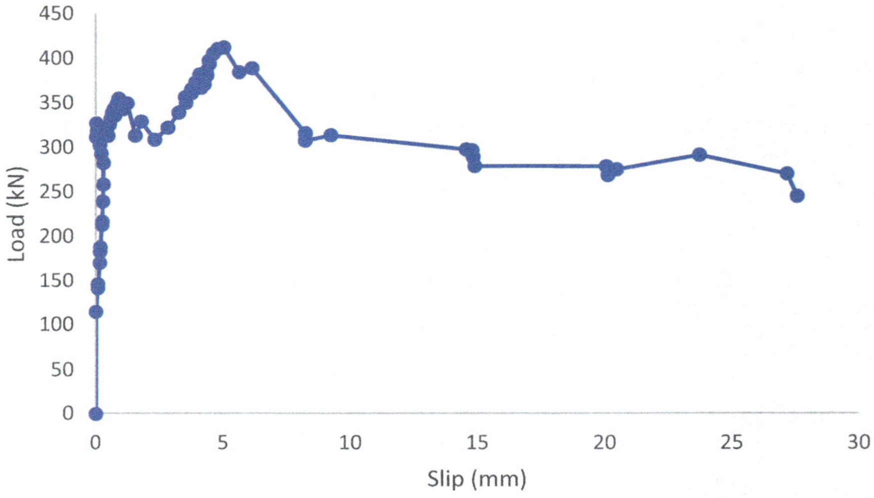

Fig. 7 Load–slip curve for specimen no. 1

the CFS and shear connector status, it was found that there was no deformation of the CFS and shear connector embedded in the concrete slab.

5 Discussion

The embedded back-to-back CFS beam is available in two configurations: the first is "without pre-drilled holes," and the second is "with pre-drilled holes." CFS beam is embedded "back-to-back" in the concrete slab, which allows for full contact throughout the whole width of the CFS beam, resulting in an extremely strong connection between the concrete slab and the CFS beam. Thus, the embedded portion of the CFS beam resisted the applied load along the whole length of the beam section

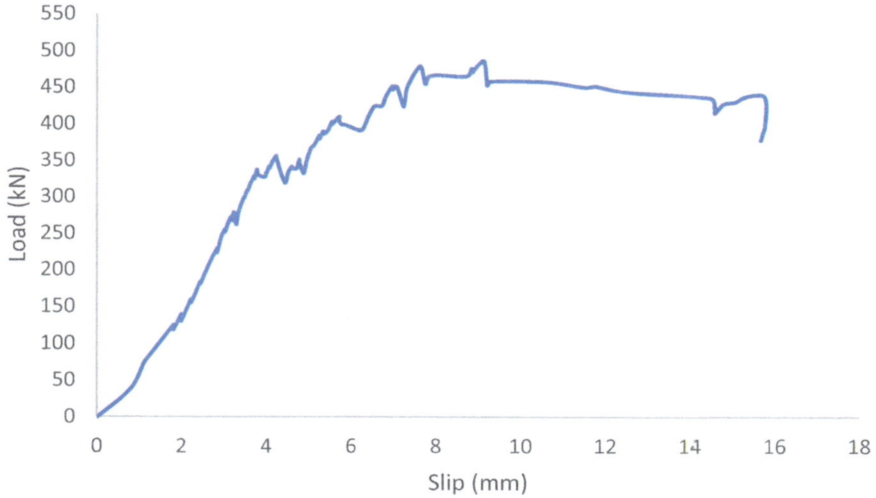

Fig. 8 Load–slip curve for specimen no. 2

Fig. 9 Longitudinal and transverse cracks of back-to-back specimens

resulted in a high loading capacity. At the advanced stage of the loading, the longitudinal and transverse cracks were developed followed by concrete crushing, without any deformation to CFS section. However, when the concrete is crushed, the bonding between the concrete slab and CFS beam is lost, which shows that higher concrete grade could develop higher load capacity. Concrete grade in this configuration is found to govern the design load, not the CFS section deformation. The embedded back-to-back CFS beam with holes achieved the highest loading capacity which achieved approximately 490 kN compared to the embedded back-to-back CFS beam without holes' configuration which achieved 412.6 kN. This could be due to the surface area of CFS exposed to a concrete slab which is larger than back-to-back CFS beam without holes, which shows that the pre-drilled holes effectively work as a shear connector. This proves that using pre-drilled holes with the embedded back-to-back beam can be an effective alternative, easy fabricated, time-saving, easy handling, and very cheap to the traditional shear connector types which are time-consuming, difficult to fabricate, expensive, and not suitable to CFS beam.

References

1. Hancock, G.J., Murray, T., Ellifrit, D.S.: Cold-Formed Steel Structures to the AISI Specification. CRC Press (2001)
2. Tahir, M.M., et al.: Finite element analysis of flush end-plate connections connected to column web. Int. J. Steel Struct. **11**(3), 247 (2011)
3. Lawan, M.M., Tahir, M.M., Osman, M.H.: Composite construction of cold-formed steel (CFS) section with high strength bolted shear connector. Jurnal Teknologi **77**(16) (2015)
4. Yu, W.-W., LaBoube, R.A., Chen, H.: Cold-Formed Steel Design. Wiley (2019).
5. Bamaga, S.O., et al.: Feasibility of developing composite action between concrete and cold-formed steel beam. J. Central South Univ. **20**(12), 3689–3696 (2013)
6. Lakkavalli, B.S., Liu, Y.: Experimental study of composite cold-formed steel C-section floor joists. J. Constr. Steel Res. **62**(10), 995–1006 (2006)
7. Bamaga, S.O., et al.: Structural behaviour of cold-formed steel of double C-lipped channel sections integrated with concrete slabs as composite beams. Latin Am. J. Solids Struct. **16**(5) (2019)
8. Bamaga, S.O., et al.: Push-out tests on three innovative shear connectors for composite cold-formed steel concrete beams. Constr. Build. Mater. **223**, 288–298 (2019)
9. Mazoz, A., Benanane, A., Titoum, M.: Push-out tests on a new shear connector of I-shape. Int. J. Steel Struct. **13**(3), 519–528 (2013)
10. Titoum, M., et al.: Experimental study and finite element modelling of push-out tests on a new shear connector of I-shape, pp. 487–506 (2016)
11. Shi, Y., et al.: The flexural behavior of cold-formed steel composite beams. Eng. Struct. **218** (2020)
12. Nguyen, R.P.: Thin-walled, cold-formed steel composite beams. J. Struct. Eng. **117**(10), 2936–2952 (1991)
13. Dar, A.R.: An experimental study on the flexural behavior of cold-formed steel composite beams. Mater. Today: Proc. **27**, 340–343 (2020)
14. Standardization, E.C.F.: Eurocode 4: Design of Composite Steel and Concrete Structures. European Committee for Standardization, Brussels (2004)

Approximate Equation for Estimating Global Buckling Load of Single-Layer Cylindrical Space Frames

Baoxin Liu, Pei-Shan Chen, Jialiang Jin, and Xiangdong Yan

Abstract This study proposes an approximate equation to evaluate the global buckling load of a single-layer cylindrical space frame without buckling analysis. In this research, large amounts of linear buckling analysis are carried out with various geometric parameters, and an imaginary stiffness G is proposed to represent the overall stiffness of the whole structure. Then, according to the influence of geometric parameters on buckling load factors and imaginary stiffness G, the authors derive an approximate equation based on regression analysis. Finally, the authors prove that the proposed approximate equations can evaluate the global buckling load in a high precision range.

Keywords Approximate equation · Imaginary stiffness · Buckling analysis · Regression analysis

1 Introduction

As a principal problem for the structural system of the single-layer space frames, estimating its global buckling capacity has aroused wide concern for many years [1, 2]. Currently, two main analysis methods are proposed to evaluate the global buckling capacity for single-layer space frames: continuum shell analogy [3] and finite element analysis (FEA) [4]. The continuum shell analogy is an approximate

B. Liu · X. Yan
Department of Civil Engineering and Architecture, Kyushu Institute of Technology, Kitakyushu 804-8550, Japan
e-mail: yan.xiangdong760@mail.kyutech.jp

P.-S. Chen (✉)
Graduate School of Engineering, Kyushu Institute of Technology, Kitakyushu 804-8550, Japan
e-mail: chen@civil.kyutech.ac.jp

J. Jin
Graduate School of Environmental Studies, Nagoya University, Nagoya 464-8601, Japan

N. Sabtu (ed.), *Proceedings of AWAM International Conference on Civil Engineering 2022—Volume 2*, Lecture Notes in Civil Engineering 385,
https://doi.org/10.1007/978-981-99-6018-7_16

theory that transforms a space frame into an equivalent continuum shell and then estimates its buckling load based on conventional shell theory [5]. However, it has the limit of calculation accuracy and structural forms. With the development of numerically structural analysis for discrete systems, FEA is fully utilized [6]. But FEA even requires professional researchers to deeply understand global buckling analysis, not to mention ordinary engineers and designers. Moreover, the FEA software for analyzing the global buckling load is not widespread, especially that processing nonlinear buckling analysis.

Based on this background, the authors propose a simple method to approximately evaluate the global buckling load for single-layer space frames without complex buckling analysis. Unlike the previous research methods that evaluate the buckling load with mechanical analysis, this study adopts regression analysis to find the relationship between the buckling load and design parameters and consequently derive the approximate equations from evaluating the buckling load. For such a purpose, an imaginary stiffness G is introduced to express the overall stiffness of the whole structure. The imaginary stiffness G is related to the maximum displacement $d_{\max}$ gained by general elastic linear structural analysis, so that the designers can estimate the buckling load easily by the proposed approximate equation during the structural design. However, as the first stage of the present study, the authors only target part of geometric parameters for single-layer cylindrical space frames with the orthogonal grid pattern, and show the process of approximate equations derivation in this paper.

2 Geometrical Parameters and Analytical Models

As well known, the buckling load factor λ depends on design parameters like span L, angle θ, structural stiffness, etc. [6, 7]. This study divides the design parameters into geometric parameters and overall stiffness. As shown in Fig. 1, the geometric parameters of a single-layer space frame are introduced as span L_x and L_y, angle θ (related to Rise h), mesh-density n_x and n_y, the diameter of member sections Ø.

The former researches based on the continuum shell analogy defined the structural stiffness of a single-layer space with the member sections, configuration grids,

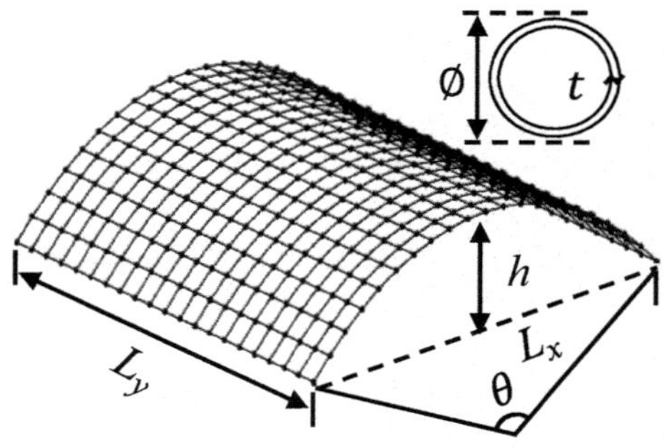

Fig. 1 Geometric parameters

Fig. 2 Maximum displacement

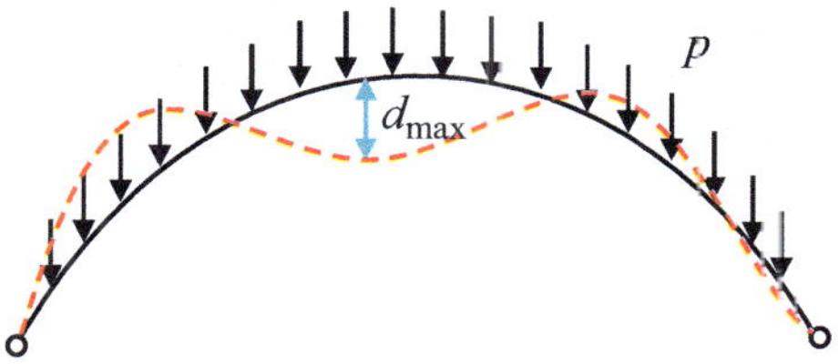

etc. [3, 8]. However, such methods cannot be used in this study to evaluate the overall stiffness, because the continuum shell analogy is not employed in this study. According to the statistical analysis results based on general elastic linear structural analysis, the authors investigated the deformation on angles, curvatures and displacements, and found that only the maximum displacement d_{max} (shown in Fig. 2) was monotonously changed with various design parameters. Consequently, an imaginary stiffness G related to the d_{max}. is proposed with consideration of span L_x and L_y, mesh-density n_x and n_y, vertical load p acting on unit area. For the imaginary stiffness G expressed as Eq. 1, the value of pL_xL_y can be considered as F (total load), while the d_{max}/n_xn_y can be assumed as δ. (distribution of d_{max} on each grid).

$$G = \frac{F}{\delta} = \frac{pL_xL_y}{d_{max}/n_xn_y} \times 10^{-11}\ (\text{N/mm}). \tag{1}$$

This study aims to find one or a series of approximate equations to evaluate the global buckling load factor λ, thus, large amounts of linear buckling analysis are carried out with various geometric parameters. The number and process of analysis cases for each geometric parameter are shown in Fig. 3 [9]. As the first stage of this study, the span L_i. is set as $L_x = L_y$ and mesh-density n_i. is set as $n_x = n_y$. In the practical structure design, the member sections of a space frame may be designed in different shapes and dimensions. The researches [10, 11] show that the relationship of the mathematical statistics between the buckling load and design parameters just appears slight difference if replace all the member sections with a uniform member section. Therefore, for each analysis case, all the diameters Ø of member sections are assumed as the same, which is changed from 80 to 200 mm, the thickness is constant $t = 6$ mm. The member material is assumed as SS400 with Young's modulus of $E = 2.05 \times 10^5$ N/mm^2. The uniform distribution vertical load acting on every joint is 1 KN/m^2. The boundary condition is four-edge pin-supporting.

3 Basic Principle

As mentioned before, to determine the influence of geometric parameters on buckling load factors λ, large amounts of linear buckling analysis and regression analysis are carried out. A buckling analysis program CPS, developed by author Chen, is applied

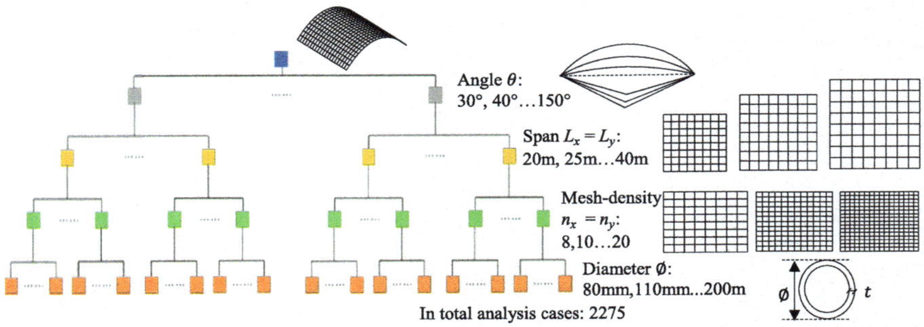

Fig. 3 Hierarchical tree for illustrating the number and process of analysis cases

to perform buckling analysis by automatically displaying the buckling modes and simulating the deformation process. The regression analysis based on TableCurve 3-D is used to forecast the relationship between the geometric parameters and buckling load factors λ.

3.1 Linear Buckling Analysis

The previous researches [5, 12] show that linear buckling analysis is an excellent approximate method to find the buckling loads, but it can neither trace the load–displacement curve nor analyze the post-buckling behaviors. However, the load–displacement curve and post-buckling behaviors are not necessary for this study. In general, linear buckling analysis can be formulated as following eigenvalue problem [13]

$$\left(\boldsymbol{K}_{\mathrm{L}} + \lambda^{(i)}\boldsymbol{K}_{\mathrm{G}}\right)\boldsymbol{U}^{(i)} = \boldsymbol{0}. \tag{2}$$

where $\boldsymbol{K}_{\mathrm{L}}$. and $\boldsymbol{K}_{\mathrm{G}}$. are the matrices representing the linear and geometric stiffness, respectively, $\lambda^{(i)}$ is the linear buckling load factor associated with the i-th buckling mode $\boldsymbol{U}^{(i)}$. In this study, the authors use the first eigenvalue value λ as buckling load factor.

3.2 Regression Analysis

As the core part of this study, regression analysis is used to estimate the causal relationship between the design parameters and buckling load factors, such that the relationship can be expressed as approximate equations. In order to derive the approximate equations with high precision, mathematical regression models, such as linear regression, logistic regression and polynomial regression are applied. The precision of the mathematical regression models is judged by the coefficient of determination R^2, which is defined as follows:

$$R^2 = 1 - \frac{\sum_i (\lambda_i - f_i)^2}{\sum_i (\lambda_i - \overline{\lambda})^2} \quad (3)$$

where λ_i is the analysis result of the buckling load factors, f_i indicates the approximate value of the buckling load factors gained by mathematical regression models and $\overline{\lambda}$ is the average of λ_i. For the best case, the modeled values exactly match the λ_i, which results in R^2 almost closing to 1 [14].

3.3 *Process for Deriving the Equations*

As shown in Fig. 4, according to the influence of geometric parameters on λ and G, the process to derive the approximate equation is separated into two routes. Route-1 first builds the foundational analysis model with the geometric parameters which affect λ and G apparently. Then regression analysis is employed to find the best mathematical regression models according to the foundational analysis model, such that the foundational approximate equations λ_{f} can be derived with high precision. In the route-2, effect factors are determined to adjust the foundational approximate equations λ_{f}. The effect factors are statistical results according to the relationship between the foundational analysis models and geometric parameters (with less or no influence on λ and G). As a result, the approximate equations can be expressed as $\lambda_{\mathrm{app}} = f(\lambda_{\mathrm{f}}, \text{effect factors})$.

4 Derivation of the Approximate Equations

This section will derive the approximate equations based on the proposed principles. Due to the limitation of the paper space, the authors only report and illustrate the necessary parts to detailedly demonstrate the process of deriving the approximate equations. Because the influence of the member section can be reflected by imaginary stiffness G according to the past research [10, 11], the diameter Ø of the member section in this study is not applied to the equations.

4.1 *Foundational Approximate Equations*

The influence of angle θ, span L and mesh-density n on buckling load factor λ and imaginary stiffness G is shown in Fig. 5. It can be found that only the trends of θ on

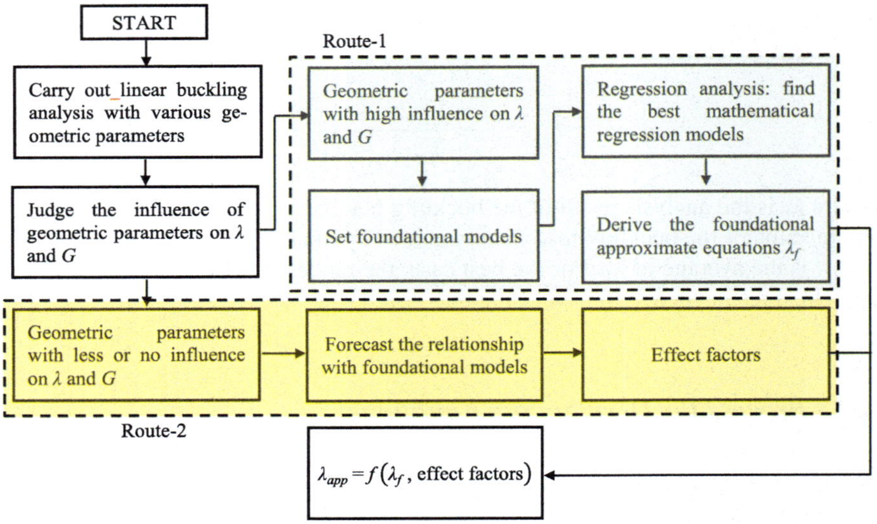

Fig. 4 Process of deriving equations

λ and G are non-monotonic changed, and the peak value for each Ø in (a) and (d) is also different. Accordingly, angle θ has an apparent influence on λ and G, while the influence of span L and mesh-density n is not obvious. In addition, in order to show the influence of span L and mesh-density n more intuitively, the authors combine all the analysis results of each L and n in 3-D graph (Figs. 6 and 7). Figure 6 shows that the changing trends of λ and G on different L are almost identical when $n = 8$. Similarly, Fig. 7 shows that the changing trends of λ and G on each n are close to each other when $L = 20$ m. Therefore, the authors directly build the relationship among λ, θ and G, and set the span $L = 20$ m with mesh-density $n = 8$ as the foundational analysis models.

Because of considering the practical design and improving the accuracy of the approximate equations, the authors reduced the range of angles to 60°–120°, where the distribution of results for the foundational analysis models is shown in Figs. 8 and 9. In order to derive the foundational approximate equations λ_f with high precision, the authors analyzed a series of mathematical regression models and calculated their coefficient of determination R^2 as given in Table 1. It can be found that the R^2 of polynomial regression (No. 1) is the most appropriate. Consequently, based on polynomial regression, the approximate foundational equation is derived as Eq. 4 with a satisfaction condition that the vertical uniform distribution load is set as 1 KN/m^2.

$$\begin{aligned}\lambda_f = f(\theta/10, G) &= 4.744 - 1.485(\theta/10) + 0.142(\theta/10)^2 \\ &- 0.004283(\theta/10)^3 - 0.5497G + 0.005038G^2 \\ &- 1.92E - 06G^3 + 0.1352(\theta/10)G\end{aligned}$$

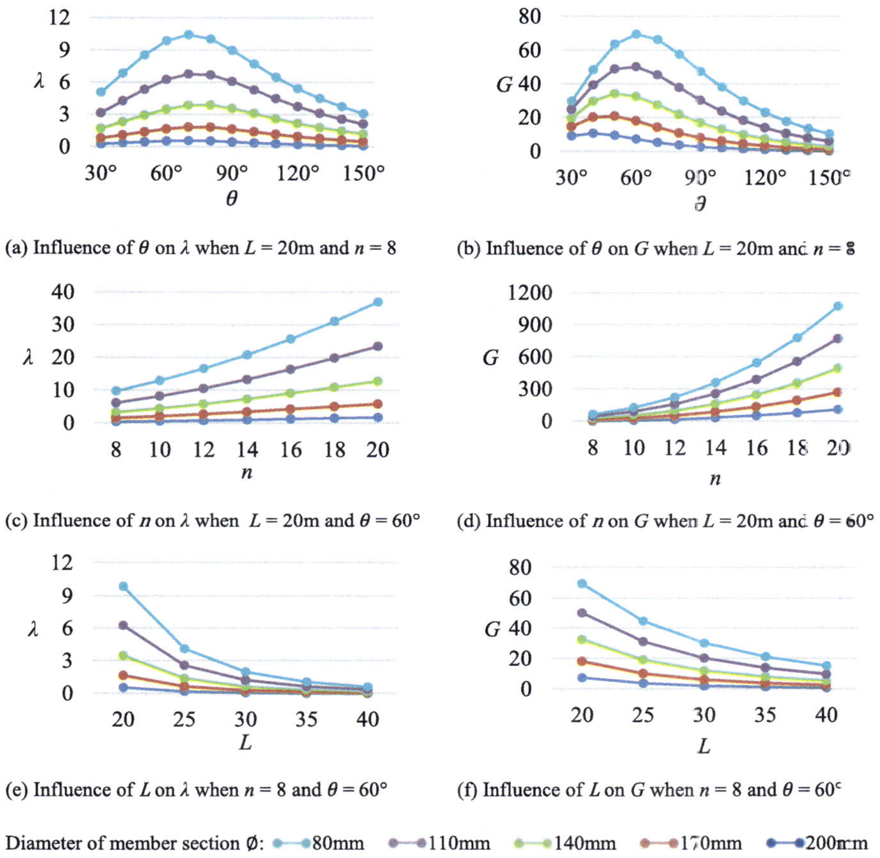

(a) Influence of θ on λ when $L = 20$m and $n = 8$

(b) Influence of θ on G when $L = 20$m and $n = 8$

(c) Influence of n on λ when $L = 20$m and $\theta = 60°$

(d) Influence of n on G when $L = 20$m and $\theta = 60°$

(e) Influence of L on λ when $n = 8$ and $\theta = 60°$

(f) Influence of L on G when $n = 8$ and $\theta = 60°$

Fig. 5 Influence of geometric parameters on buckling load λ and imaginary stiffness G

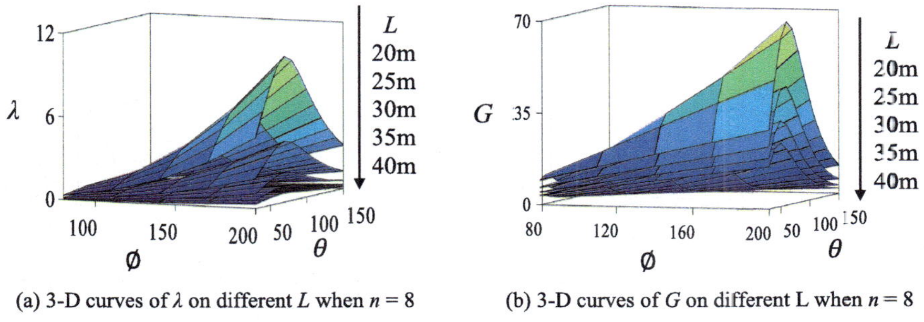

(a) 3-D curves of λ on different L when $n = 8$

(b) 3-D curves of G on different L when $n = 8$

Fig. 6 3-D curves of analysis result for λ and G on different span L

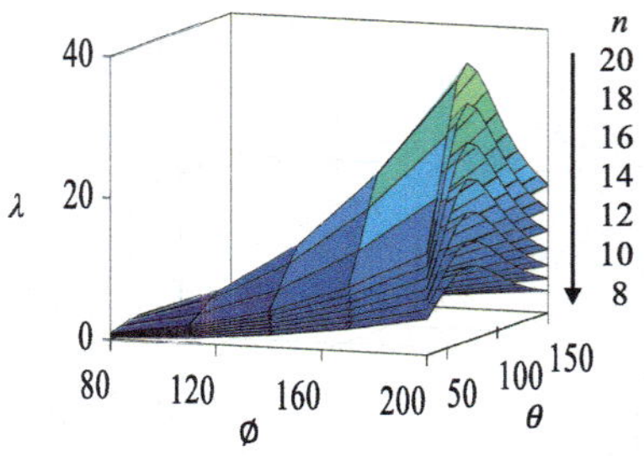

(a) 3-D curves of different n on λ with $L = 20\text{m}$

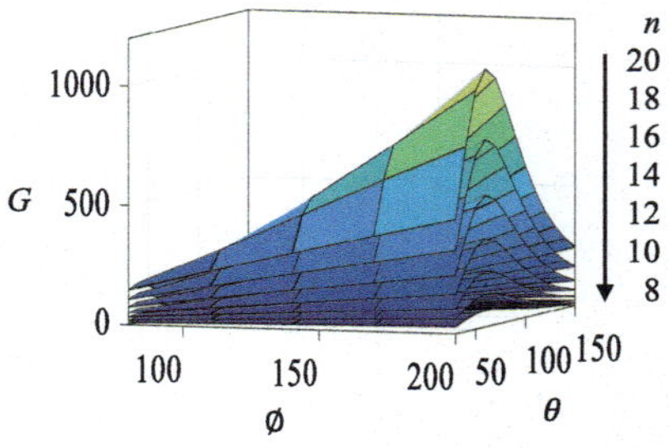

(b) 3-D curves of different n on G with $L = 20\text{m}$

Fig. 7 3-D curves of analysis result for different mesh-density n on λ and G

$$- 0.005352(\theta/10)^2 G - 0.0006438(\theta/10)G^2 \tag{4}$$

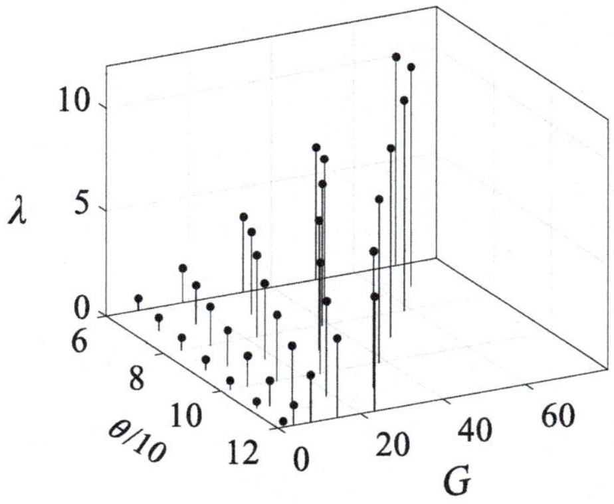

Fig. 8 Distribution of foundational analysis models

Fig. 9 Polynomial regression models

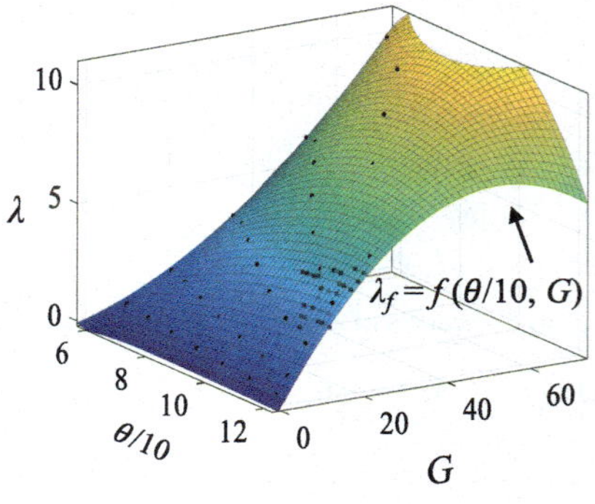

Table 1 Analysis result of mathematical regression models

No.	Equations of mathematical regression models	R^2
1	$\lambda = a + bx + cx^2 + dx^3 + ey + fy^2 + gy^3 + hxy + ix^2y + jxy^2$	0.99971
2	$\lambda = (a + cx + ey + gx^2 + iy^2 + kxy)/(1 + bx + dy + x^2 + hy^2 + jxy)$	0.99964
3	$\lambda = a + b/x + cy + d/x^2 + ey^2 + fy/x + g/x^3 + hy^3 + iy^2/x + jy/x^2$	0.99933
4	$\lambda = (a + b\ln x + cy + dy^2 + ey^3)/(1 + f\ln x + g(\ln x)^2 + hy + iy^2)$	0.99894
5	$\lambda = (a + bx + cy + dy^2 + ey^3)/(1 + fx + gx^2 + hy + iy^2)$	0.99734

$x = \theta/10$, $y = G$

4.2 Effect Factors

In the last section, the foundational approximate equations λ_f. have been gained, but it can be only used as the foundational analysis models with span $L = 20$ m and mesh-density $n = 8$. Thus, to build the relationship between λ_f and spans, mesh-densities, the authors set the result of span $L = 20$ m and mesh-density $n = 8$ as the foundational models, respectively, and then investigate the relationship between the result of foundational models with the other span and mesh-density on λ and G as shown in Fig. 10. It's easy to find that the magnification of each span on $L = 20$ m and each mesh-density on $n = 8$ are almost constantly in a line, which means that t influence of span and mesh-density on λ and G is constant, and almost don't accept the influence from the other geometric parameters such as angle and diameters.

Therefore, the authors take the average of these magnifications and propose four effect factors for spans and mesh-densities as follows:

$$\text{Attenuation factors of } \lambda \text{ for span } L \qquad \alpha_{\lambda L} = \frac{\lambda_{Li}}{\lambda_{20}} \tag{5}$$

$$\text{Attenuation factors of } G \text{ for span } L \qquad \alpha_{GL} = \frac{G_{Li}}{G_{20}} \tag{6}$$

$$\text{Enhancement factor of } \lambda \text{ for mesh-density } n \qquad \beta_{\lambda n} = \frac{\lambda_{ni}}{\lambda_8} \tag{7}$$

$$\text{Enhancement factor of } G \text{ for mesh-density } n \qquad \beta_{Gn} = \frac{G_{ni}}{G_8} \tag{8}$$

where λ_{20} and G_{20} are the analysis results of span $L = 20$ m, while λ_8 and G_8 are the analysis results of mesh-density $n = 8$. As a result, the approximate equations can be expressed as Eq. 10. The values of effect factors $\alpha_{\lambda L}$, α_{GL}, $\beta_{\lambda n}$, β_{Gn} are given in Tables 2 and 3. Here, G should be changed as $\alpha_{GL}\beta_{Gn}G$.

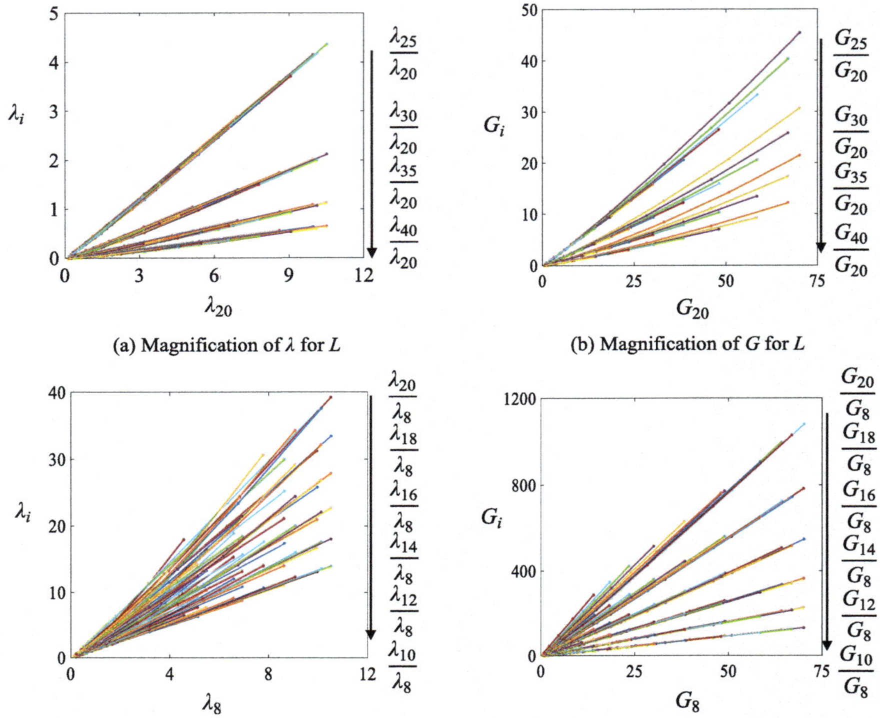

(a) Magnification of λ for L (b) Magnification of G for L

(c) Magnification of λ for n (d) Magnification of G for n

Fig. 10 Relationship of magnification between each span on 20 m and each mesh-density on 8

$$G \to \alpha_{GL}\beta_{Gn}G \tag{9}$$

$$\lambda_{app} = \alpha_{\lambda L}\beta_{\lambda n}\lambda_f(\theta, \alpha_{GL}\beta_{Gn}G) \tag{10}$$

Table 2 Attenuation factors of λ and G on span L

Span L	20	25	30	35	40
$\alpha_{\lambda L}$	1	0.369	0.167	0.083	0.045
α_{GL}	1	0.532	0.321	0.210	0.143

Table 3 Enhancement factor of λ and G on mesh-density n

Mesh-density n	8	10	12	14	16	18	20
$\beta_{\lambda n}$	1	1.294	1.622	1.986	2.385	2.818	3.284
β_{Gn}	1	1.891	3.245	5.188	7.866	11.446	16.119

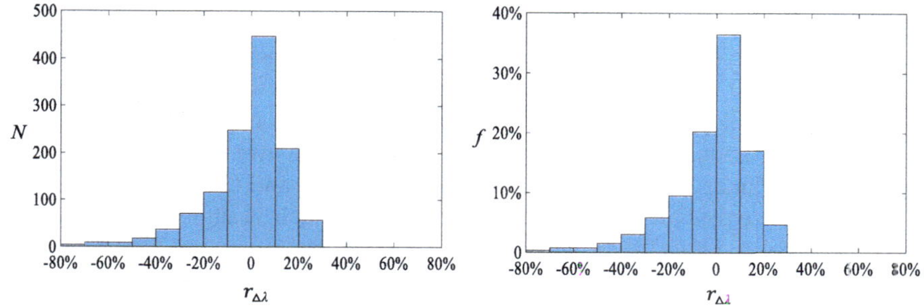

(a) Rate of difference $r_{\Delta\lambda}$ with the quantity N (b) Rate of difference $r_{\Delta\lambda}$ with frequency f

Fig. 11 Distribution for the rate of difference $r_{\Delta\lambda}$

5 The Effectiveness Analysis

Though this study tries to guarantee the high precision for the approximate foundational equations based on regression analysis, some minor difference still exists. In addition, out of considering the briefness of the approximate equations, the authors propose the effect factors by the average of the magnifications, which could result in a big difference. Therefore, to examine the accuracy of the proposed approximate equations, the authors calculate the approximate buckling load factors λ_{app} through Eq. 10, and compare them with the results from linear buckling analysis λ_{an}. As a result, the distribution of difference is illustrated by the distribution histograms as shown in Fig. 11, where the $r_{\Delta\lambda}$ is the rate of difference and expressed as $r_{\Delta\lambda} = (\lambda_{app} - \lambda_{an})/\lambda_{an} = \Delta\ \lambda/\lambda_{an}$, N shown in Fig. 11a is the quantity of $r_{\Delta\lambda}$, while f shown in Fig. 11b is the frequency of $r_{\Delta\lambda}$ and $\sum f = 1$.

From Fig. 11, it can be found that there exists some $r_{\Delta\lambda}$ even up to the -80%, the reason is that the λ_{an} is very small so that the difference $\Delta\ \lambda$ almost close to the λ_{an}. Moreover, the frequency f of $r_{\Delta\lambda} = -80\%$ is very small, which means that the difference of $r_{\Delta\lambda} = -80\%$ does not affect the accuracy of the approximate equations. In addition, the quantity N and frequency f almost concentrates in the range of $r_{\Delta\lambda} = \pm 20\%$, i.e., the proposed equation is enough to evaluate the global buckling load for the single-layer space frames.

6 Conclusions

This paper investigated the influence of geometric parameters on buckling load factor λ and imaginary stiffness G, and consequently proposed an approximate equation to evaluate the global buckling load for the single-layer cylindrical space frames. The significant findings obtained from this study are summarized as follows:

1. The influence of angle θ, span L and mesh-density n is very clear. Both the λ and G will firstly increase and then decrease with the increase of angle, and will decrease with the increase of span L, while increase with the increase of mesh-density n. In addition, the magnifications of each span on $L = 20$ m for λ and G are almost a constant value, so as each mesh-density over $n = 8$, which means that the influence of span L and mesh-density n on λ and G almost don't accept the influence from the other geometric parameters such as angle θ and diameters of member sections Ø.
2. The approximate equation proposed in this paper consists of two parts which are foundational equations and effect factors. The authors only prove the high precision for the foundational equations based on the regression analysis, while the effect factors just are taken by average values which could lead to some big difference, but from the rate of difference $r_{\Delta\lambda}$, it can be found that the frequency f is almost concentrated in the range of $r_{\Delta\lambda} = \pm 20\%$, which means that the envelope equations are valid enough to evaluate the buckling load for the single-layer space frames studied in this paper.

However, though the influence of span L and mesh-density n is demonstrated in this paper, the case of span $L_x \neq L_y$ and mesh-density $n_x \neq n_y$ is not considered. Moreover, structural design (considering the shapes and dimensions of member section) and the initial imperfection are not considered yet. In addition, the authors will extend this research to the other shapes and boundary conditions of space frames in the future.

Acknowledgements The author LIU would like to show his sincere gratitude to ASSURAN International Scholarship Foundation and Kyutech Fund Scholarship for supporting him to focus on this study.

References

1. Chilton, J.: Space Grid Structures. Architectural Press, Great Britain (2000)
2. Bulenda, T., Knippers, J.: Stability of grid shells. Comput. Struct. **79**, 1161–1174 (2001)
3. Wright, D.T.: Membrane forces and buckling in reticulated shells. J. Struct. Div. **21**, 175–201 (1965)
4. Karadeniz, H.: Stochastic Analysis of Offshore Steel Structures—An Analytical Appraisal. Springer, London (2013)
5. Kato, S., Yamashita, T.: Evaluation of elasto-plastic buckling strength of two-way grid shells using continuum analogy. Int. J. Space Struct. **17**(4), 249–261 (2002)
6. Altuna-Zugasti, A.M., Lopez-Arancibia, A., Puente, I.: Influence of geometrical and structural parameters on the behaviour of squared plan-form single-layer structures. J. Constr. Steel Res. **72**, 219–226 (2012)
7. Chen, P.S., Mamoru, K.: Optimization for maximum buckling load of a lattice space frame with nonlinear sensitivity analysis. Int. J. Space Struct. **21**(2), 111–118 (2006)
8. Makowski, Z.S.: Analysis Design and Construction of Braced Domes. Elsevier Applied Science Publishers, New York (1985)

9. Malek, S., Wierzbicki, T., Ochsendorf, J.: Buckling of spherical cap gridshells: a numerical and analytical study revisiting the concept of the equivalent continuum. Eng. Struct. **75**, 288–298 (2014)
10. Liu, B., Chen, P.S.: Buckling characters and approximate buckling load factors of 1.5-layer cylindrical space frames with crossing-units. In: Architectural Institute of Japan/Kyushu Branch, No. 59, pp. 221–224 (2020)
11. Liu, B., Chen, P.S.: Approximate evaluation of global buckling load of single-layer and 1.5-layer cylindrical space frames. In: 2020 Annual Conference of Architecture Institute of Japan, pp. 747–748 (2020)
12. Chair of Working Group 8: Draft Guide to Buckling Load Evaluation of Metal Reticulated Roof Structures. IASS WG 8 for Metal Spatial Structures, Japan (2014)
13. Li, Y.: The analysis of buckling finite element of single layer spherical lattice shell. J. Jiaying Univ. (Nat. Sc.) **23**(6), 99–101 (2005)
14. Wang, N., Guan, Q., Sun, Y., Wang, B., Ma, Y., Shao, W., Li, H.: Predicting the spatial pollution of soil heavy metals by using the distance determination coefficient method. Sci. Total Environ. **799**, 149452 (2021)

Palm Oil Clinker as Coarse and Fine Aggregates in Lightweight Concrete

Nor Fazlin Zamri, Roslli Noor Mohamed, Dinie Awalluddin, and Shariwati Mansor

Abstract This study presents the investigation of the properties of lightweight concrete (LWC) with the palm oil clinker (POC) as the lightweight aggregates. The investigation began with the characterization of the properties of the POC to ensure its suitability in the production of LWC. The fresh state and mechanical strength of concrete was then compared between normal concrete (NC) and LWC. In comparison with NC, LWC had a maximum decrement of about 25%, 37%, and 22% for compressive, splitting, and flexural tensile strengths, respectively. It has been demonstrated that the properties of the aggregates used in concrete production have a direct influence on the workability and strength of concrete.

Keywords Palm oil clinker · Lightweight concrete · Physical properties · Mechanical properties

1 Introduction

Hardened concrete with a density of less than 2000 kg/m^3 [1] can be classified as lightweight concrete (LWC). As the aggregate is the main constituent that contributes to the density of the concrete, different types of materials are being investigated to replace the normal weight aggregates. These include expanding clay [2], lightweight expanded clay aggregate (LECA) [3], and agricultural waste such as oil palm shells (OPS) [4–6], palm oil clinker (POC), [7, 8] and coconut shell [9].

Furthermore, Malaysia accounts for approximately 58% of global palm oil production [10]. However, the palm oil industry also produces abundant waste, including empty fruit bundles (EFBs), mesocarp fibre, and palm oil shells. The electricity in palm oil mills was generated through steam from the combustion of the mesocarp and POS [11]. Following the combustion process, the POC is formed and then deposited in the landfill. POC with a porous characteristic and an irregular shape, offers a

N. F. Zamri · R. N. Mohamed (✉) · D. Awalluddin · S. Mansor
Faculty of Civil Engineering, Universiti Teknologi Malaysia, Johor Bahru, Johor, Malaysia
e-mail: roslli@utm.my

N. Sabtu (ed.), *Proceedings of AWAM International Conference on Civil Engineering 2022—Volume 2*, Lecture Notes in Civil Engineering 385,
https://doi.org/10.1007/978-981-99-6018-7_17

good lightweight characteristic to be used as a lightweight aggregate. Therefore, by taking advantage of this free and abundant waste that is available in Malaysia, and reusing it in the construction industry, it leads to substantial savings in material and construction costs, as well as contributing to a positive impact on the environment.

In conjunction with that, the use of POC has gained popularity in the production of LWC. For example, Ahmmad et al. [7] investigated the use of POC to replace POS as coarse aggregate, with the variation in percentage of replacement ranging from 25 to 100%. Concrete with POC aggregate showed a better performance in terms of strength, with a 43% increment when compared with the concrete with POS aggregate. It is proved that POC has great potential as a coarse aggregate when compared with other industrial waste products. The reduction in strength when using POC as coarse aggregate has been reported by numerous studies [12, 13], and it was expected due to the porous characteristics of the POC. Other than the production of vibrated concrete, the use of POC as coarse aggregate in the production of self-compacting concrete has been investigated by Owi et al. [8]. It is proved that the use of POC still ensured the fresh concrete exhibited the good characteristics of self-compacting concrete. Coarse aggregate from POC has engineering potential and economic advantages. The purpose of this study is to further explore the usage of POC as fine and coarse aggregates in the manufacture of LWC.

2 Methodology

The properties of two types of concrete (i.e. NC and LWC) were investigated in the present study. These concretes were evaluated to determine the effect of fully replacing conventional aggregates with the POC. The properties of the aggregates used were also investigated in order to support the relationship between the properties of aggregate and the properties of concrete in hardened state.

2.1 Preparation of Materials

Ordinary Portland cement (OPC) with a grade of CEM 1 42.5R was used for both types of concrete. The palm oil clinker (POC) in the form of a porous lump was collected from Alaf Palm Oil Mill, Kota Tinggi, Johor as shown in Fig. 1a. The large chunks of POC were crushed into smaller sizes by using hammers and crusher machines as shown in Fig. 1b, c. The crushed POC was then sieved using a sieve shaker and divided into two categories: (i) coarse aggregate (with a particle size of between 4.75 mm and 9.5 mm) and (ii) fine aggregate (passing 4.75 mm) as shown in Fig. 2. As shown in Fig. 3, grading curves demonstrated that material was well-graded as specified in BS EN 12620 [14]. On the other hand, normal weight aggregates, such as crushed granite and river sand were used as the aggregate in the production of NC. All the aggregates are in saturated surface dry condition (SSD).

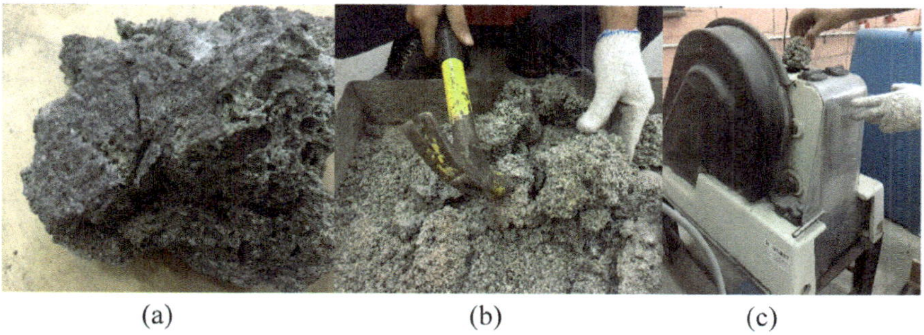

Fig. 1 Preparation of aggregates: **a** Raw POC, **b** Raw POC is crushed into smaller size using hammer and **c** crusher machine

Fig. 2 **a** POC coarse aggregate and **b** POC fine aggregates

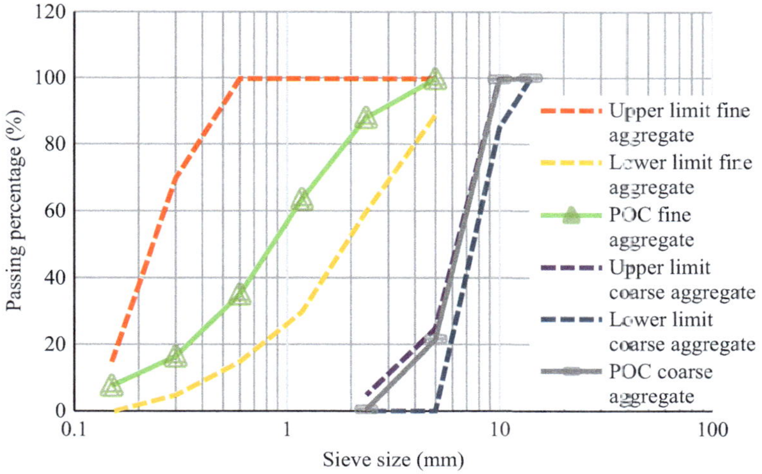

Fig. 3 Grading curves for POC coarse and fine aggregates

Table 1 Constituent of the NC and LWC mix

Types of concrete	LWC	NC
Cement (kg/m^3)	500	483.4
Water (kg/m^3)	250	241.7
Coarse aggregate (kg/m^3)	520	701.3
Fine aggregate (kg/m^3)	630	929.7
Water-cement ratio	0.5	0.5

The POC also tested for its physical and mechanical properties such as specific gravity [15], water absorption [15], bulk density [16], void content [16], aggregate crushing value (ACV), and aggregate impact value (AIV) to support the relationship between the properties of aggregate and the properties of concrete in hardened state.

2.2 Concrete Mix Proportion and Test Procedures

NC and LWC concrete mix proportions were designed using ACI to achieve a compressive strength of 30 N/mm^2 at the age of 28 days. As there is no established standard method for achieving a satisfactory mix design, trial studies need to be conducted to find the suitable method and proportion of the concrete mix. The main issue is that the lack of cohesiveness in the cement paste contributes to poor workability of fresh concrete and honeycombs on the hardened concrete surface. This is an unavoidable issue due to the porous structure of the POC's high moisture and void content. The POC has a higher tendency to absorb water during the casting process. The selected mix proportion is given in Table 1. For comparison purposes, normal concrete (NC) with a similar designated strength was also cast. This comparison was performed to demonstrate how the physical and mechanical properties of aggregates used in concrete affect the overall properties of the concrete.

2.3 Preparation of Specimens

For each concrete, a total of 27 specimens were cast to test its hardened concrete strength at 7, 14, and 28 days. Cube specimens of 100 mm in size were cast to test for compressive strength [17]; cylindrical specimens of 100 mm in diameter and 200 mm in height for splitting tensile strength [18]; and prismatic specimens of 100 × 100 × 500 mm long for flexural tensile strength [19]. All of the hardened concrete specimens were tested after the water curing process following BS EN 12390-2 [20].

Table 2 Comparison of properties between POC and crushed granite

Types of aggregate	POC	Crushed granite [8]
Specific gravity (OD)	1.80	2.61
Specific gravity (SSD)	1.90	2.65
Water absorption (%)	5.5	1.6
Loose bulk density (kg/m^3)	818.2	1611.8
Compacted bulk density (kg/m^3)	949.2	1750.2
Void content (%)	54.5	38.2
ACV (%)	45.8	24.3
AIV (%)	46.0	25.6

3 Result and Discussion

3.1 Physical Properties of POC

Table 2 listed the properties of the POC and crushed granite. The specific gravity in the oven-dried (OD) and saturated surface dry (SSD) states of the POC is 1.80 and 1.90, respectively, which is lower than the value of the crushed granite. The lightness of the coarse POC was primarily due to its cellular structure, with a high number of voids present within the particles contributing to the light density of the aggregate. This porous characteristic also contributed to the high percentage of water absorption and void content. As a result, the high number of pores allows more water to penetrate its structure, and as a result, more water can be retained when compared with crushed granite.

The lower value in the loose and compacted bulk density of the POC is due to the lightness and porousness of the POC. It also shows that the packing of the POC is less dense than the crushed granite, which has solid and dense features. Furthermore, the value of the loose bulk density of the POC is still within the limit of the lightweight aggregate suggested by BS EN 13055 [1], so it is qualified to be used as the lightweight aggregate. The value of the ACV and AIV of the POC was almost two times higher than the crushed granite. This indicates that the POC has a low load-bearing capacity when compared with the crushed granite. This low load-bearing capacity is due to the porous structure of the POC.

3.2 Fresh Properties of LWC

The irregular and angular shape of the POC aggregate used in LWC resulted in a lower slump value of 68 mm when compared with the NC, which had a slump value of 118 mm as shown in Fig. 4. It is also known that the POC aggregate has a high porosity and water absorption, thus, it requires high water content to achieve the

(a) (b)

Fig. 4 Slump test for **a** LWC and **b** NC

targeted slump. In some cases, the addition of a superplasticizer can also help to increase the workability of LWC. However, the slump value of LWC was acceptable and within the targeted range of 60–180 mm.

The irregular shape of POC has produced a harsher and less cohesive fresh concrete mix when compared with the NC. This is because the POC aggregate has a larger surface area that needs to be coated by cement paste when compared with the granite and sand river. Hence, less cement paste is available for lubrication. It is worth mentioning that good workability can only be achieved by balancing the consistency and cohesiveness of the concrete mix.

3.3 Mechanical Properties of LWC

Density. Generally, the average density of NC was around 2400 kg/m^3. However, the use of 100% of POC as aggregate in the LWC mix can reduce the concrete density up to a maximum value of 1967 kg/m^3 in SSD condition as shown in Fig. 5; thus, this concrete mix can be categorized as lightweight concrete [1]. The specific gravity of the POC had contributed to the reduction of the density of the LWC, since it was approximately 28.3% lower than the crushed granite used in NC. In addition to that, the density of the POC was also lower than the crushed granite. The density of the constituent components, particularly the aggregate, had an impact on the density of the hardened concrete. Abutaha et al. [13] also found that using POC as a substitute for crushed granite as coarse aggregate reduced concrete density by about 15%, which supported the finding of this study.

Fig. 5 Densities of the hardened concrete

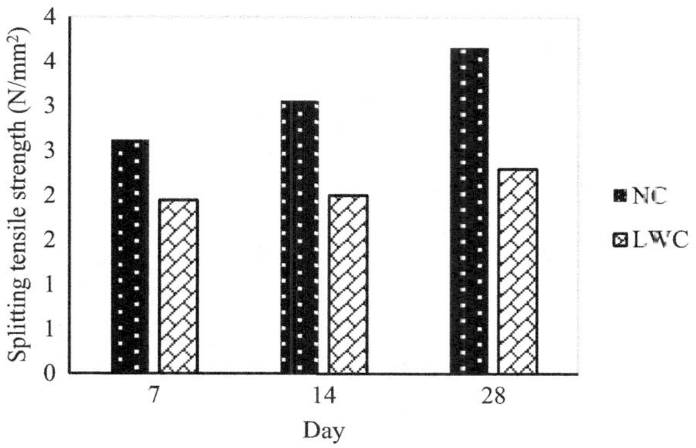

Compressive strength. Figure 6 shows the strength of LWC and NC cubic specimens under compression load. At the age of 28 days, LWC specimens recorded the highest strength of 29.56 N/mm^2, which was about 25.3% lower than NC. This is because the POC had low bearing capacity when compared with the normal weight aggregate. This is proven through its high ACV value. Furthermore, the high void content in POC also contributed to a reduction in the LWC matrix strength. Kanadasan and Razak [12], Abutaha et al. [13] and Shafigh et al. [5] have reported that the concrete with less density has less strength when compared with the normal weight concrete. This study also shows a similar trend, as illustrated in Fig. 7.

Splitting and Flexural tensile strength. Figures 8 and 9 illustrate the comparison of the splitting and flexural tensile strengths between LWC and NC, respectively. The splitting and flexural tensile strength of the LWC were 37.1% and 21.7% lower than the NC, respectively. This maximum decrement in concrete strength was recorded at the age of 28 days. It is proven that the properties of POC have negatively affected the strength of concrete in its hardened state. However, the tensile strength of cylindrical

Fig. 6 Cubic compressive strength of LWC and NC

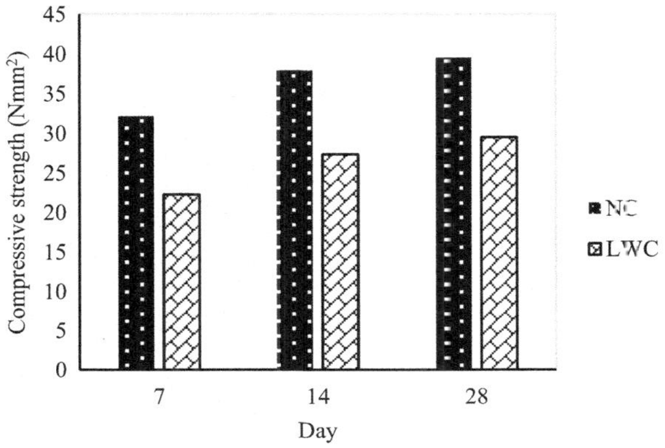

LWC specimens was acceptable for structural application as it meets the minimum strength requirement of 2.1 N/mm^2 [21].

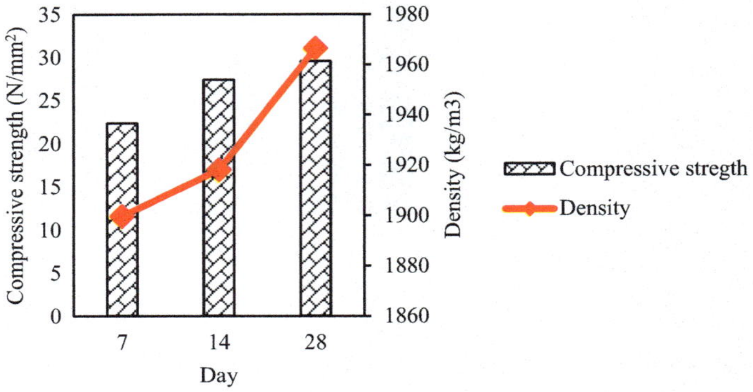

Fig. 7 Relationship of compressive strength and density of LWC

Fig. 8 Splitting tensile strength of cylindrical specimens

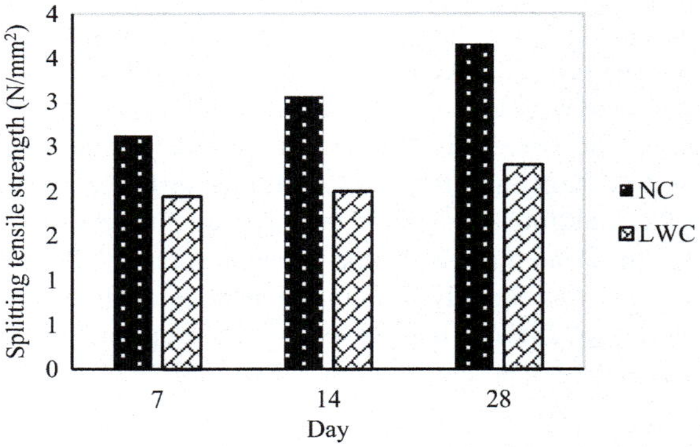

Fig. 9 Flexural tensile strength of prismatic specimens

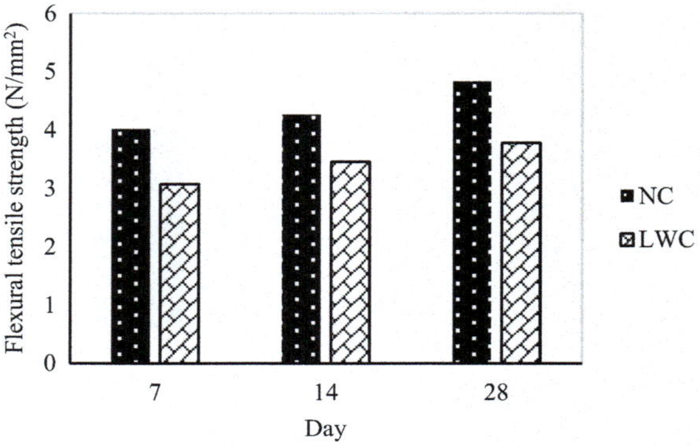

Fig. 10 Cross-section of NC specimens

Fig. 11 Cross-section of LWC specimens

Failure mode. Figures 10 and 11 show the cross-section of NC and LWC specimens after the failure occurred. As can be visually observed in NC specimens, the failure occurred due to an aggregate-cement paste bonding failure. In contrast to the NC, where failure occurred in the interfacial zone, the rupture in LWC specimens occurred in the lightweight POC aggregate. The crack was initiated by the weaker strength of the POC aggregate and these cracks later propagated through the interface between aggregates and mortar.

4 Conclusion

A thorough investigation and comparison of the lightweight concrete using POC aggregates and the normal concrete with normal weight aggregates have been conducted. The following conclusion can be drawn from this study:

- POC has a cellular structure with a high number of voids present within the particles, which contributes to the light density of the aggregate. However, this porous characteristic has resulted to weaker strength aggregate of POC when compared with the normal weight aggregate.
- LWC has comparable properties in both fresh and hardened properties, even though the density of LWC is much lower compared with those of NC.
- The porous characteristics of POC also contributed to the decrement in LWC, with 25%, 37%, and 22% decrements recorded for compressive, splitting tensile, and flexural strength when compared with the NC at the ages of 28 days, respectively.
- This LWC concrete made from POC aggregates has the potential to be used in construction industries where this waste can be fully utilized and reduce the cost of materials needed. Overall, this study has advanced our understanding of POC's characteristics and produced a finished product that may find value in the building industry.

Acknowledgements The authors would like to thank Universiti Teknologi, Malaysia for funding this study through the UTM Encouragement Research Grant No. Q.J130000.2651.17J54.

References

1. BS EN 13055: Lightweight aggregates—Part 1: lightweight aggregates for concrete, mortar and grout, Br. Stand. Inst (2016)
2. Ibrahim, N.M., Salehuddin, S., Amat, R.C., Rahim, N.L., Izhar, T.N.T.: Performance of lightweight foamed concrete with waste clay brick as coarse aggregate. APCBEE Proc. **5**, 497–501 (2013)
3. Scotta, R., Giorgi, P.: Comparative cyclic tests of exterior flat slab–column connections in normal concrete and fiber-reinforced lightweight aggregate concrete. Mater. Struct. Constr. **49**, 4049–4067 (2016)
4. Mannan, M.A., Alexander, J., Ganapathy, C., Teo, D.C.L.: Quality improvement of oil palm shell (OPS) as coarse aggregate in lightweight concrete. Build. Environ. **41**, 1239–1242 (2006)
5. Shafigh, P., Jumaat, M.Z., Mahmud, H.: Oil palm shell as a lightweight aggregate for production high strength lightweight concrete. Constr. Build. Mater. **25**, 1848–1853 (2011)
6. Muthusamy, K., Zamri, N.A.: Mechanical properties of oil palm shell lightweight aggregate concrete containing palm oil fuel ash as partial cement replacement. KSCE J. Civ. Eng. **20**, 1473–1481 (2016)
7. Ahmmad, R., Jumaat, M.Z., Alengaram, U.J., Bahri, S., Rehman, M.A., Bin Hashim, H.: Performance evaluation of palm oil clinker as coarse aggregate in high strength lightweight concrete. J. Clean. Prod. **112**, 566–574 (2016)
8. Owi, S.F., Mohamed, R.N., Mohamed, A.: Fresh and hardened properties of self-compacting lightweight concrete using coarse palm oil clinker **29**, 205–220 (2017)
9. Gunasekaran, K., Annadurai, R., Kumar, P.S.: Plastic shrinkage and deflection characteristics of coconut shell concrete slab. Constr. Build. Mater. **43**, 203–207 (2013)
10. Ahmad, Z., Ibrahim, A.: Drying shrinkage characteristics of concrete reinforced with oil palm trunk fiber. Int. J. Eng. Sci. Technol. **2**, 1441–1450 (2010)
11. Vijaya, S., Ma, A.N., Choo, Y.M., Nik Meriam, N.S.: Life cycle inventory of the production of crude palm oil—a gate to gate case study of 12 palm oil mills. J. Oil Palm Res. **20**, 484–494 (2008)

12. Kanadasan, J., Razak, H.A.: Mix design for self-compacting palm oil clinker concrete based on particle packing. Mater. Des. **56**, 9–19 (2014)
13. Abutaha, F., Abdul Razak, H., Kanadasan, J.: Effect of palm oil clinker (POC) aggregates on fresh and hardened properties of concrete. Constr. Build. Mater. **112**, 416–423 (2016)
14. BS EN 12620: Aggregates for Concrete (2013)
15. ASTM C127: Standard Test Method for Density, Relative Density (Specific Gravity), and Absorption of Coarse Aggregate (2004)
16. ASTM C29/C29M-17a: Standard Test Method for Bulk Density ("Unit Weight") and Voids in Aggregate, ASTM International, West Conshohocken, PA, Annu. B. ASTM Stand (2017), pp 1–7
17. BS EN 12390-3: Testing Hardened Concrete—Part 3: Compressive Strength of Test Specimens, British Standards Institution (2009)
18. BS EN 12390-6: Testing Hardened Concrete—Part 6: Tensile Splitting Strength of Test Specimens (2009)
19. BS EN 12390-5: Testing Hardened Concrete. Part 5: Flexural Strength of Test Specimens (2009)
20. BS EN 12390-2: Testing Hardened Concrete—Part 2: Making and Curing Specimens for Strength Test (2009)
21. ASTM C330: Standard Specification for Lightweight Aggregates for Structural Concrete (2009)

Dynamic Properties and Seismic Performance of 1.5-Layer Space Frames with Lap-Units whilst Considering Vibration of Lower Joints

Jialiang Jin, Pei-Shan Chen, Baoxin Liu, and Xiangdong Yan

Abstract A new structural system called 1.5-layer space frame is proposed as the third member of the family of space frames, which may provide new design possibilities. The 1.5-layer space frames with lap-units are susceptible to deformation due to key part rotation about the axis of its upper chords. Therefore, it is necessary to capture the dynamic behaviors of the lower joints of the lap-units. Considering the vibration of lower joints, the main dynamic properties of four typical structures, including natural frequency and vibration mode shape are first examined, and the seismic performances are then assessed within the linear elastic range under selected earthquake excitations. To provide helpful information for practical design, the investigation on the relationship between structural properties and geometrical/design parameters is also conducted. As a result, the seismic response of the structures is significant when the depth-grid ratio (the ratio of web member and upper chord length) is greater than 0.5.

Keywords 1.5-layer space frame · Lap-unit · Lower joint · Dynamic property · Seismic performances · Geometrical parameters

J. Jin
Graduate School of Environmental Studies, Nagoya University, Nagoya 464-8601, Japan

P.-S. Chen (✉)
Graduate School of Engineering, Kyushu Institute of Technology, Kitakyushu 804-8550, Japan
e-mail: chen@civil.kyutech.ac.jp

B. Liu · X. Yan
Department of Civil Engineering and Architecture, Kyushu Institute of Technology, Kitakyushu 804-8550, Japan
e-mail: yan.xiangdong760@mail.kyutech.jp

N. Sabtu (ed.), *Proceedings of AWAM International Conference on Civil Engineering 2022—Volume 2*, Lecture Notes in Civil Engineering 385,
https://doi.org/10.1007/978-981-99-6018-7_18

1 Introduction

Over the decades, a growing interest in space frame systems has been witnessed worldwide. Single-layer and double-layer space frames have been extensively employed in large-span structures because of their significant structural potentials, such as lighter weight, speedy construction, economy, and aesthetics. In recent years, the search for new structural systems to increase the variety and interest of design and accommodate large-span areas has always been the main objective of architects and engineers. For this purpose, Chen developed a new structural system, namely the 1.5-layer space [1], as the third member of the family of space frame structures. The morphological conception of the 1.5-layer space frame was inspired by the geometrical principle of an elaborate ancient wooden bridge, which is called Hongqiao [2–4].

The 1.5-layer space frame is a bar-linked structure composed of upper chords, web members, and diagonal members, which may provide a higher buckling capacity than a single-layer space frame and achieve a lighter weight than a double-layer space frame [5, 6]. In other words, the 1.5-layer space frame has advantages in structural stability, economic, constructability, and form design compared with conventional single-layer and double-layer space frames. Therefore, the 1.5-layer space frame could be a moderate choice between a single-layer and double-layer space frame, and this study responds to and helps implement the major sustainable development goals (SDGs) adopted by the United Nations agenda. Focusing on this study, Chen [7–9] has proposed various assembling methods of 1.5-layer space frames such as lap-units, crossing-units, triple-connecting units, crossing-lap units, polygon pyramids, traditional Chinese and Japanese patterns, etc., to discover new design possibilities. The assembling method of lap-unit is one of the first ideas to compose the 1.5-layer space frames. The assembling method with lap-unit is to set one end of a key part onto the central joint of an adjacent key part so that the key parts may support mutually like that shown in Fig. 1. As early as 2008, a structure similar to the assembling method of lap-units has been built as an archeological museum in Bibracte Burgundy (Fig. 2), France, which was designed by Paul Andreu and Bemard Vaudeville and Simon Aubry [10]. The completion of this museum is sufficient to prove the feasibility of the lap-units in the actual structure.

In the past decades, to promote the further development of large-span space structures, the seismic behavior of the single-layer and double-layer space frames has

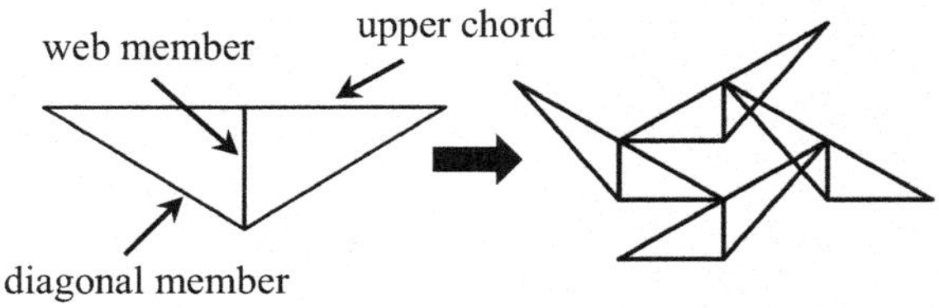

Fig. 1 Lap-unit (left: a key part, right: a lap-unit)

Fig. 2 Structure of the archeological museum in France [11]

been the subject of some research interest, e.g., Cedrón and Elghazouli [12] investigated the main considerations related to the seismic design and evaluation of single-layer steel cylindrical lattice shells, and then provided recommendations for practical application. He et al. [13–15] presented a horizontal movement retractable reticulated mega-structure with single-layer latticed-intersected-cylindrical-shell substructures and then analyzed their natural vibration properties and seismic performances. Nie et al. [16] conducted a series of shaking table tests to examine the seismic behavior of the double-layer space frame structure with a lower supporting structure. In addition, Kato et al. [17, 18] conducted extensive researches on the seismic behavior of space frames, such as the effect of the vertical component of the earthquake on single-layer reticular domes. Moreover, Kunieda [19] provided numerical results of the intensity and distribution of response accelerations of shells subjected to the Kobe earthquake. They have contributed greatly to the development of the space frame system.

The dynamic behavior and structural integrity of frame structures are significantly impacted by the connecting joints [20, 21]. The joints, such as welds or bolts are more vulnerable to damage and essential for overall safety of structures comparing with other members [22]. Damage to the joints of space frame structures has also been observed in the past earthquakes, e.g., the 1995 Kobe earthquake (Japan) and the 2013 Lushan earthquake (China) [23, 24]. Studies on the joints of space frame structures have been taken seriously by several researchers in recent years. The concepts of how to consider the slip effects of joints when dealing with large space frame structures were presented by Mähr et al. [25]. They illustrated the importance and benefits of this holistic analysis approach with the example of the Louvre Abu Dhabi Dome, where the effects of joints were considered in the overall modeling. In a recent study, Hou et al. [22] proposed a two-step damage detection method for joints and bar elements and demonstrated the efficiency of this method by utilizing an experimental 3D space structure with semi-rigid connections. They also performed effective numerical simulations to explore the damage in joints and members of the space structure.

In recent years, much work has been done to develop and improve the property of the 1.5-layer space frame system. For instance, the early research on 1.5-layer space frames has developed the joint technologies and performed stability analysis and buckling path tracking [7–9]. Additionally, the buckling characteristics of 1.5-layer space frames with lap-units and crossing-units, including buckling modes, buckling loads, and equilibrium path have been investigated in detail [5, 26].

Nevertheless, few studies have been conducted on the seismic performances of the 1.5-layer space frames. The previous study by Jin et al. [5] demonstrated that the 1.5-layer space frames with lap-units are prone to deformation due to key part rotation about the axis of its upper chords (Fig. 1). Consequently, it becomes imperative to explore the dynamic behaviors of the lower joints of the lap-units for the development of the 1.5-layer space frames with lap-units. Jin et al. [5] also found that different joint forms have a significant impact on the phenomenon of key part rotation (Fig. 1). For convenience, conventional approaches typically presumptively assume that the joints are either pinned or rigid. Pinched joints normally have some rotational stiffness, whereas rigid joints usually exhibit a certain amount of flexibility. In this paper, these two types of joints are applied to the cylindrical and domed numerical models, respectively. The primary focus of this paper is to investigate the natural vibration property and seismic performance in detail through basic parametric analysis, and to provide helpful information for the missing aspects of practical seismic design of the 1.5-layer space frames with lap-units on the basis of the consideration of the vibration of lower joints. With consideration of the vibration of lower joints, this paper modeled four structures with different geometric configurations and then examined the seismic response of the lower joints of lap-units. Simultaneously, the impact of various geometrical parameters on the dynamic properties and seismic performances of the structures was confirmed to provide recommendations on conducting seismic design for 1.5-layer space frames.

2 Analytical Methods and Models

Aided with the simulation software Automatic Dynamic Incremental Nonlinear Analysis (ADINA), eigenvalue analysis and linear dynamic analysis method are adopted to investigate the natural vibration property and seismic performance of the 1.5-layer space frames with lap-units in this paper. Eigenvalue analysis based on non-damping free vibration [27] is first employed to capture the dynamic behavior of the innovative structures in consideration of the vibration of the lower joints. In earthquake engineering, there are four main kinds of methods for the seismic analysis of a structure, including linear static analysis, linear dynamic analysis, nonlinear static analysis, and nonlinear dynamic analysis. This research is implemented within an elastic range so that the linear time history analysis method [28, 29] is chosen. Though the calculation process of the time history method is complicated and time-consuming, it can accurately reflect the entire response process of the structure under the action of an earthquake and it is theoretically applicable to any complex structure. The cylindrical and domed structures with orthogonal and diagonal grids are considered based on above analysis methods. The key geometrical/design parameters are introduced as open angle θ and depth-grid ratio *D/G*, which significantly influence the architectural and engineering design [30].

2.1 Geometrical Configuration

In this study, four numerical models, i.e., the Cylindrical structures with Lap-units composed of Orthogonal grids (named hereinafter as the CLO-model), the Cylindrical structures with Lap-units composed of Diagonal grids (the CLD-model), the Domed structures with Lap-units composed of Orthogonal grids (the DLO-model), the Domed structures with Lap-units composed of Diagonal grids (the DLD-model), are chosen for the analysis, as shown in Fig. 3. Previous works [5] found that the lower part of the lap-unit is easily liable to degrees of freedom (rotation about the upper chords) in the horizontal direction, especially for the domed models. Therefore, all upper chords and web members are assumed as beam elements to avoid the rotation deformation of the partial key part. All diagonal members are assumed as truss elements for cylindrical models, beam elements for domed models. The detailed settings of elements and joints are shown in Fig. 3 and Table 1. The 1.5-layer space frame is defined by its span $L_X = L_Y$, open angle θ ($\theta_X = \theta_Y$ for domed models), grid number $N_X = N_Y$, and length of the web member D, from which we can calculate the rise h and the radius R. In this study, all analytical models are summed as roofs with the same planar dimension in X and Y-direction. The detailed geometrical parameter designs are displayed in Table 1.

The member material is assumed as steel pipes of SS400 with a constant thickness of 6 mm, and Young's modulus is 2.05 × 105 N/mm^2 and Poisson's ratio is 0.3. The diameter of each member is determined based on long-term allowable stress calculations, which is the preliminary structural design for confirming the safety of the building structures. The minimum diameter is set as 45 mm and increased by a certain change of 5 mm. Long-term allowable stress calculations for structural design have been carried out through an automatic allowable stress calculation program written by Fortran and Batch. The stress design factor is set to $\sigma/f \leq 0.9$ during the trial design in the present study. The self-weight of the space frame and the weight of various accessories, such as cladding, supported lighting, heat, and ventilation equipment, comprise the total dead load. Since the live load is small in a large-span spatial

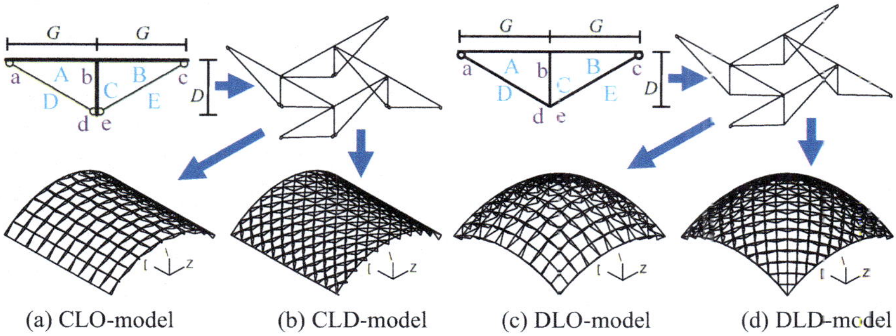

Fig. 3 Structural types of 1.5-layer space frames with lap-units

Table 1 Structural information of 1.5-layer space frames with lap-units

Model	$L_X = L_Y$ (m) [a]	θ (°) [b]	*D/G* [c]	Beam element	Truss element	Rigid joint	Pin joint
CLO-model	20–80 (10–40 grids)	30–120°	0.1–0.6	**A, B, C**	**D, E**	***b***	***a, c, d, e***
CLD-model	20–80 (8–32 grids)	30–120°	0.1–0.6	**A, B, C**	**D, E**	***b***	***a, c, d, e***
DLO-model	20–80 (10–40 grids)	30–120°	0.1–0.6	**A, B, C, D, E**	—	***b, d, e***	***a, c***
DLD-model	20–80 (8–32 grids)	30–120°	0.1–0.6	**A, B, C, D, E**	—	***b, d, e***	***a, c***

[a] $L_X = L_Y$ (m) is the span in X-direction and Y-direction of the structure
[b] θ (°) denotes the open angle of the structure
[c] *D/G* denotes the depth-grid ratio (the ratio of web member length and upper chord length)
Note **A, B, C, D, E** denote members of the key part; ***a, b, c, d*** denote joints of the key part

structure, the long-term load is equivalent to a surface uniformly distributed load, corresponding to the superimposed dead load. This surface load is then multiplied by the tributary area of each joint, and the load is applied to the structure as vertical point loads in the Z-direction on the joints. These loads contribute to the mass of the structure when performing dynamic analyses, where the masses are typically modeled as lumped at the joints as well. During the dynamic analysis, lumped mass is set up at every joint, and the lumped mass at the lower joints is approximately calculated as half of the mass of the web and diagonal members, as shown in Fig. 6.

The boundary conditions considered in this study correspond to those deemed to be found most commonly in practical design. The analysis is carried out in the boundary condition of four-edge supporting, which is assumed as the upper pin supporting with all translational movements restrained, but free to rotate in any direction. In addition, the ends of the upper chords are released except for the rotation about the axes of the upper chords, as referenced to [5]. Previous researches [7–9] have proved that the 1.5-layer space frames directly composed of the basic units are not an extremely stable structure. If some stiffening or support members are added to their edges, the modified 1.5-layer space frames can meet the requirement of structural stabilities. Therefore, the analytical models in this study utilize strengthened members for the edges of the structures to guarantee their stability and easy support. In addition, the diagonal members (upper diagonal members) in the surface of the upper chords are added to resist the earthquake force, as shown in Fig. 6 All stiffening members are assumed as truss elements.

Table 2 List of input ground motion records

No	Recording time	Input waves	PGA (gal)	Input time (s)	Time step (s)
1	1940	EL Centro NS	255.0	53.0	0.02
2	1952	Taft EW	248.0	54.2	0.02
3	1995	JMA Kobe NS	165.2	81.9	0.02
4	1968	Hachinohe NS	166.0	35.9	0.01

2.2 *Ground Motion Inputs*

It is unclear what kind of earthquake will occur, and since each ground motion has its waveform characteristics, it is necessary to use multiple input ground motions in the time history response analysis. In this study, four ground motions widely applied in seismic design in Japan (EL Centro NS, Taft EW, JMA Kobe NS, and Hachinohe NS) [31] given in Table 2 are used to execute the linear time history response analysis. The waveforms and acceleration response spectrums of input ground motions are indicated in Figs. 4 and 5. Rayleigh damping was used for the time history analysis, and the damping ratio is set as $h = 2\%$. Since the analysis is performed in the elastic range and deals with the medium earthquake level, the peak ground velocity (PGV) is assumed as 25 kine.

2.3 *Analytical Models*

Based on the above methods, the ultimate models of cylindrical and domed 1.5-layer space frames with lap-units composed of orthogonal and diagonal grids can be established (Fig. 6). The geometrical parameters are introduced as the open angle of $\theta = 30°–120°$ and the depth-grid ratio of $D/G = 0.1–0.6$. Four ground motions exhibited in Table 2 are adopted for the time history response analysis. The input direction of the ground motion is assumed to be X, Y, and 45°-direction, as shown in Fig. 6.

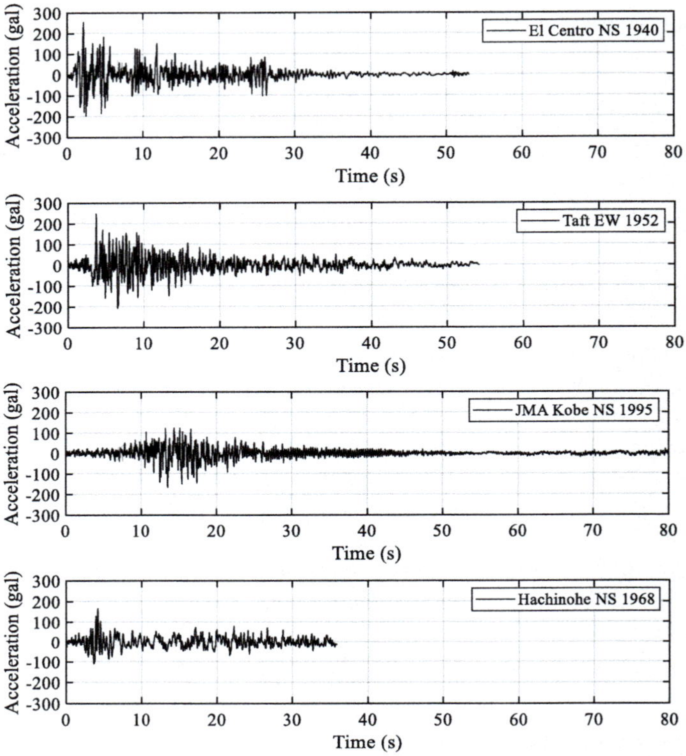

Fig. 4 Waveforms of input ground motions

Fig. 5 Acceleration response spectrums

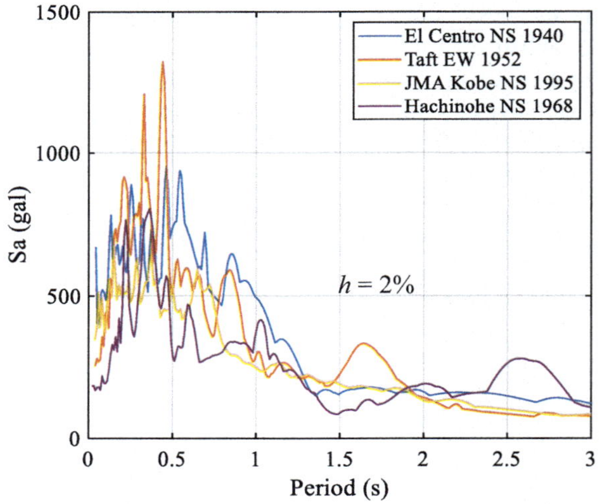

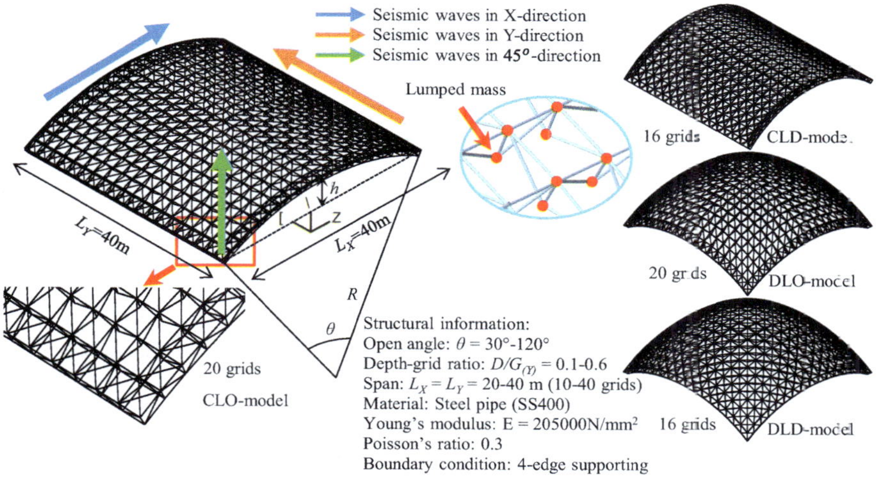

Fig. 6 Analytical models

3 Natural Vibration Properties

Many external factors affect the seismic performance of a space frame, such as seismic intensity and site conditions. However, the most important influence is the natural vibration characteristics of the structure, which should be studied first. This study adopts the eigenvalue analysis method, which is an analysis method for determining the vibration properties of a structure including the natural frequency and vibration mode. It is well-known that the fundamental frequency is the most important index of the dynamic property of a structure. Consequently, in this section, the fundamental frequency is taken as the index to investigate the influence of geometrical parameters (open angle θ and depth-grid ratio *D/G*) on the natural vibration property of the 1.5-layer space frames with lap-units.

Suppose that, the span of the 1.5-layer space frames with lap-units is 40 × 40 m (20 × 20 grids for CLO and DLO-models, 16 × 16 grids for CLD and DLD-models), the open angle is $\theta = 30°–120°$, and the depth-grid ratio is $D/G = 0.1–0.6$. The fundamental frequencies and associated first vibration mode shapes of all the analytical models are displayed in Fig. 7. From Fig. 7, the fundamental frequencies of the domed models are larger than those of the cylindrical models. The fundamental frequency increases with the increase in the open angle and the depth-grid ratio, which implies the increase of their integral rigidity. However, for the models with high open angles and depth-grid ratios, the fundamental frequency is relatively dense or has a downward trend, primarily because the web members are too long to affect the overall stability of the structures, particularly for domed models. Additionally, the structures of different grid combinations (orthogonal and diagonal grids) show similar analysis results, while the geometry of the structure has a greater impact on the results.

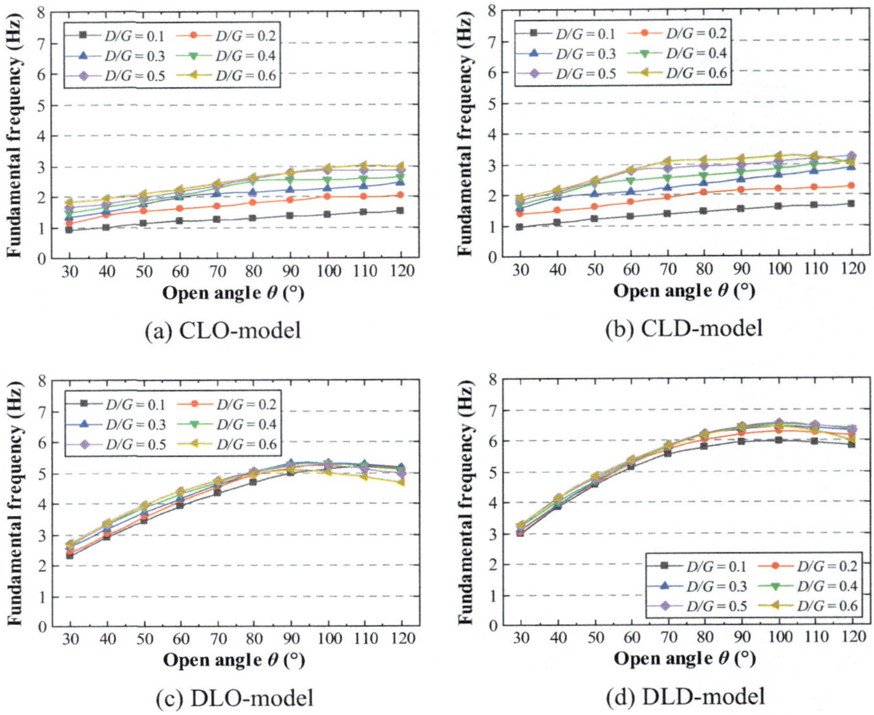

Fig. 7 Fundamental frequency versus open angle and depth-grid ratio

The associated vibration mode shapes of the four structures are shown in Fig. 8. Note that the letters a–j represent the first vibration modes. Observing the vibration modes shows that some key parts appear in rotational deformation about their upper chords instead of overall vibration for the cases of high open angles and depth-grid ratios. This may be generally interpreted as follows: if the open angle and the depth-grid ratio reach larger values, considering the lumped masses at the lower joints, the vibration modes in which the lower end of the web members swings laterally becomes predominant due to the excessively high rise and long web members of the structures.

4 Seismic Performances

The seismic response such as the peak acceleration response of the upper and lower joints, as well as the inter-layer deformation angle of the web members, are used as the response index for the study to investigate the seismic performances of the structures and the dynamic behaviors of the lower joints of the lap-units. Consistent

Eigen modes		θ (°)									
		30 → 40 → … → 110 → 120									
	0.1	a	a	a	c	c	c	c	c	c	c
	0.2	b	b	a	a	a	d	d	d	c	c
$\frac{D}{G}$	0.3	b	b	b	b	a	d	d	d	d	d
	0.4	b	b	b	b	b	b	d	d	d	d
	0.5	b	b	b	b	b	b	b	d	d	d
	0.6	b	b	b	b	b	b	b	e	e	e

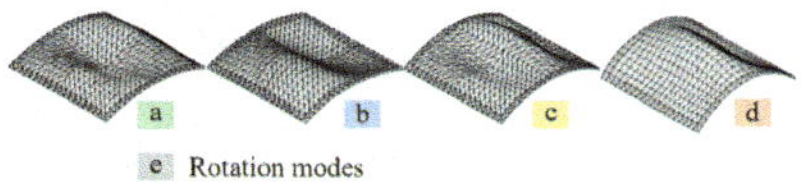

(a) CLO-model

Eigen modes		θ (°)									
		30 → 40 → … → 110 → 120									
	0.1	a	a	c	c	c	c	c	d	d	d
	0.2	a	a	a	a	a	a	c	c	c	c
$\frac{D}{G}$	0.3	b	b	a	a	a	a	a	a	a	c
	0.4	b	b	b	a	a	a	a	a	a	a
	0.5	b	b	b	b	a	a	a	a	a	a
	0.6	b	b	b	b	a	a	a	a	e	e

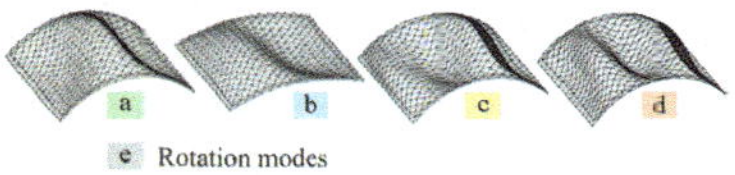

(b) CLD-model

Eigen modes		θ (°)									
		30 → 40 → … → 110 → 120									
	0.1	a	a	c	c	c	c	d	d	d	d
$\frac{D}{G}$	0.2	a	a	a	a	a	a	e	e	e	e
	0.3	b	a	a	a	a	a	e	e	e	e
	0.4	b	b	b	a	a	a	e	e	e	e
	0.5	b	b	b	b	b	b	e	f	f	f
	0.6	b	b	b	b	b	b	f	f	f	f

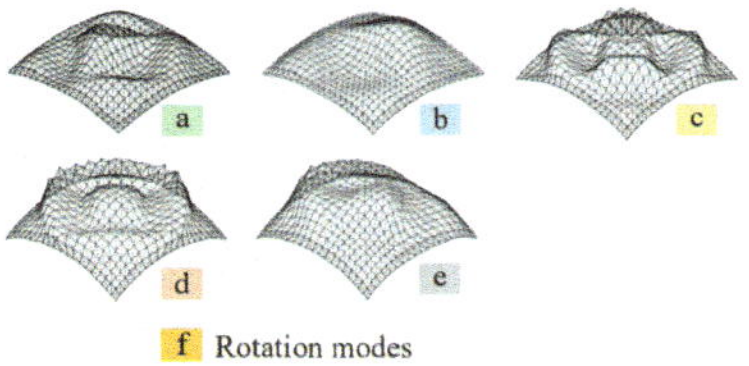

(c) DLO-model

Eigen modes		θ (°)									
		30 → 40 → … → 110 → 120									
	0.1	a	a	e	f	f	f	f	f	f	f
$\frac{D}{G}$	0.2	b	d	a	a	g	g	g	h	h	h
	0.3	b	d	d	d	d	d	g	g	g	g
	0.4	c	d	d	d	d	d	d	g	g	g
	0.5	c	b	d	d	d	d	d	g	g	j
	0.6	c	c	d	d	d	d	d	j	j	j

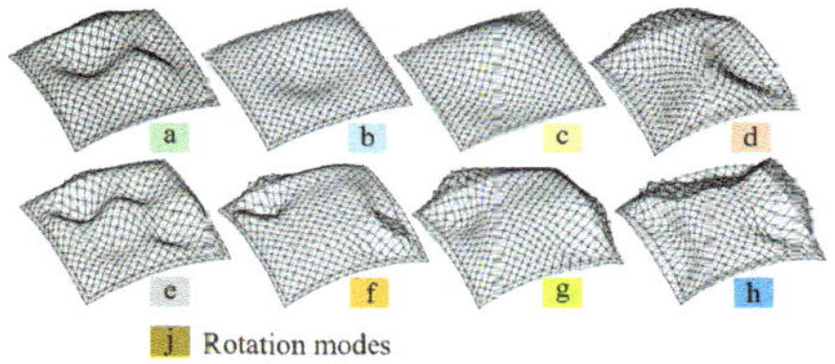

(d) DLD-model

Fig. 8 First eigen vibration mode versus open angle and depth-grid ratio

with the natural vibration analysis, the effect of the open angle and depth-grid ratio on the seismic performances of the structures is investigated.

4.1 Acceleration and Response Spectrum

The acceleration responses and their response spectrums reflecting the characteristics of the response waveform are examined first to investigate the seismic performances of the structures. The models are presumed to have the same structural planar dimension of $L_X = L_Y = 40$ m, an open angle of 90°, and a depth-grid ratio of 0.3. The following describes the results of inputting the ground motions in the *X*-direction.

Figures 9 and 10 show the acceleration responses and response spectrums of the upper and lower joints for the CLO-model in ground motions input direction (*X*-direction) and vertical direction (*Z*-direction) where the seismic response is the largest. Comparing with Fig. 9, it can be found that the acceleration response of the lower joint of the structure is larger than that of the upper joint, especially in the direction of the ground motions input. The other three models show the similar results as the CLO-model, and the analysis results of the lower joints are shown in Figs. 11, 12 and 13. For the CLO-model, CLD-model and DLO-model, the EL Centro and Taft waves cause greater response, while for the DLD-model, the longer-period JMA Kobe wave has a greater response. In addition, the results show that in the case of such a relatively small open angle and depth-grid ratio, the responses of the lower joints of the structures in the vertical direction are more obvious. The acceleration response spectrums are in the form of "single peaks", except for the "double peak" in Figs. 9d and 10d. Furthermore, the predominant periods corresponding to the responses in the X and Z-directions of the upper and lower joints are very close.

Similarly, the structures with the open angle of $\theta = 30°–120°$ and depth-grid ratio of $D/G = 0.1–0.6$ are investigated, with the same planar dimensions of 40×40 m. Using the selected ground motions input along the X, Y, and 45°-direction, a series of time history analyses corresponding to the first natural periods are carried out. Then, the peak acceleration responses in the horizontal and vertical directions of the upper

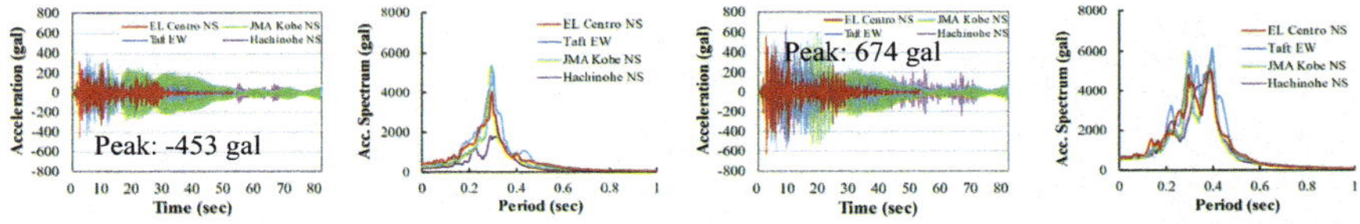

(a) Acc. in X-dir. (b) Acc. spectrum (c) Acc. in Z-dir. (d) Acc. spectrum

Fig. 9 Peak acceleration of upper joint for CLO-model

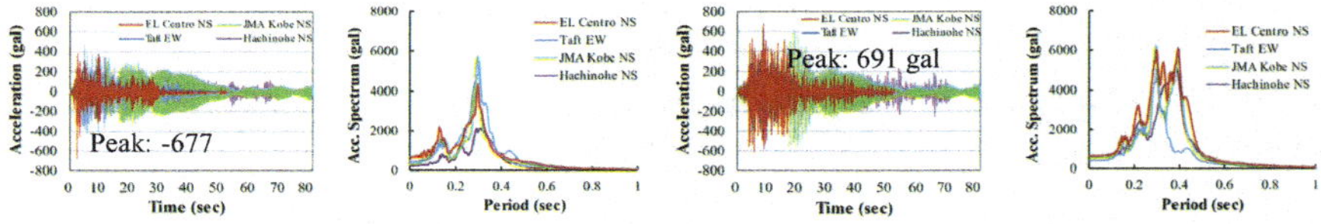

(a) Acc. in X-dir. (b) Acc. spectrum (c) Acc. in Z-dir. (d) Acc. spectrum

Fig. 10 Peak acceleration of lower joint for CLO-model

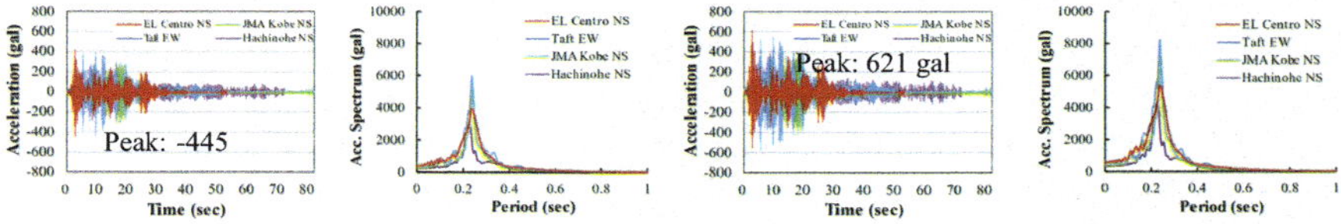

(a) Acc. in X-dir. (b) Acc. spectrum (c) Acc. in Z-dir. (d) Acc. spectrum

Fig. 11 Peak acceleration of lower joint for CLD-model

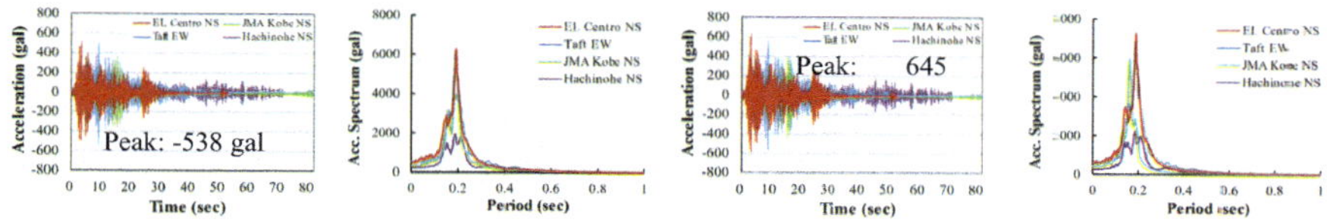

(a) Acc. in X-dir. (b) Acc. spectrum (c) Acc. in Z-dir. (d) Acc. spectrum

Fig. 12 Peak acceleration of lower joint for DLO-model

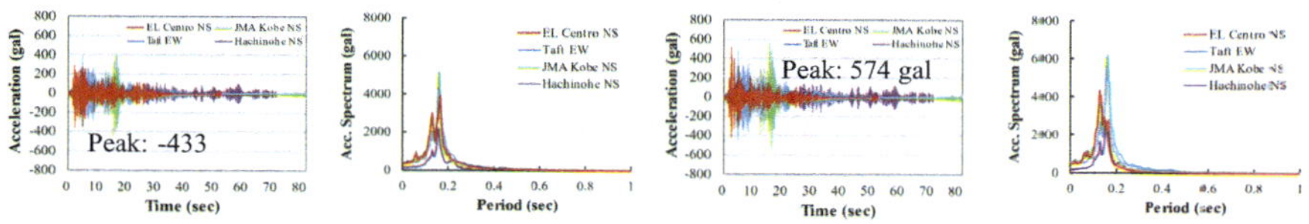

(a) Acc. in X-dir. (b) Acc. spectrum (c) Acc. in Z-dir. (d) Acc. spectrum

Fig. 13 Peak acceleration of lower joint for DLD-model

and lower joints are obtained. Additionally, the maximum inter-layer deformation angle of the web members is examined, which may provide guiding suggestions for the practical design. The results summarized below are the maximum values of the four input ground motions.

4.2 Peak Acceleration Response

The distribution of the peak values of acceleration answers for the ground motions input along the *X*-direction is similar to that of values input along the 45°-direction and is higher than the values input along the *Y*-direction. Due to space limitations, only the analysis results of *X*-direction inputting are reported as shown in Figs. 14, 15, 16, and 17. According to Figs. 14, 15, 16 and Figs. 17a, c, the peak values of the acceleration in the *X*-direction of lower joints are much larger than those of upper joints due to the lumped masses of lower joints. For the *Z*-direction, the distribution of the peak acceleration responses of lower joints is similar to that of upper joints. In addition, the peak acceleration response increases with the increase in the open angle, particularly the response in the *X*-direction increases significantly. For models with lower open angles, the depth-grid ratio seems to have little effect on the peak acceleration besides the response of the lower joints in the *X*-direction. From Figs. 14c, 15c, 16c and 17c, as the open angle increases, the model with a larger depth-grid ratio may reach a much larger response value due to the long length of web members, which should be noted in practical design. In addition, for the upper joints, it is found that the peak values of the acceleration responses in the *Z*-direction

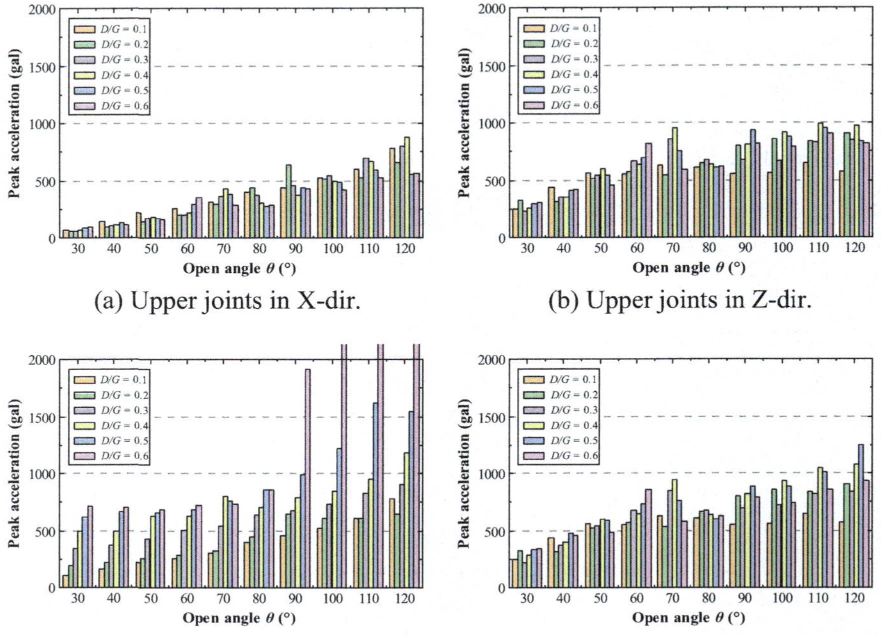

(a) Upper joints in X-dir.

(b) Upper joints in Z-dir.

(c) Lower joints in X-dir.

(d) Lower joints in Z-dir.

Fig. 14 Peak acceleration for CLO-model when ground motions input along X-direction

are bigger than those in the X-direction. The lower joints of the analytical models with $\theta = 90°–120°$ and $D/G = 0.5–0.6$ may obtain higher values of the acceleration in the X-direction, particularly for the cylindrical models.

4.3 Maximum Inter-layer Deformation Angle

In this research, the inter-layer deformation angle of the web member is expressed as following (see Fig. 18):

$$\Delta S = \sqrt{(x_l - x_u)^2 + (y_l - y_u)^2 + (z_l - z_u)^2} \quad (1)$$

$$\alpha \approx \Delta S / D \quad (2)$$

in which ΔS is the inter-layer deformation between the upper and lower joints, D is the length of web members (distance between layers), α indicates the inter-layer

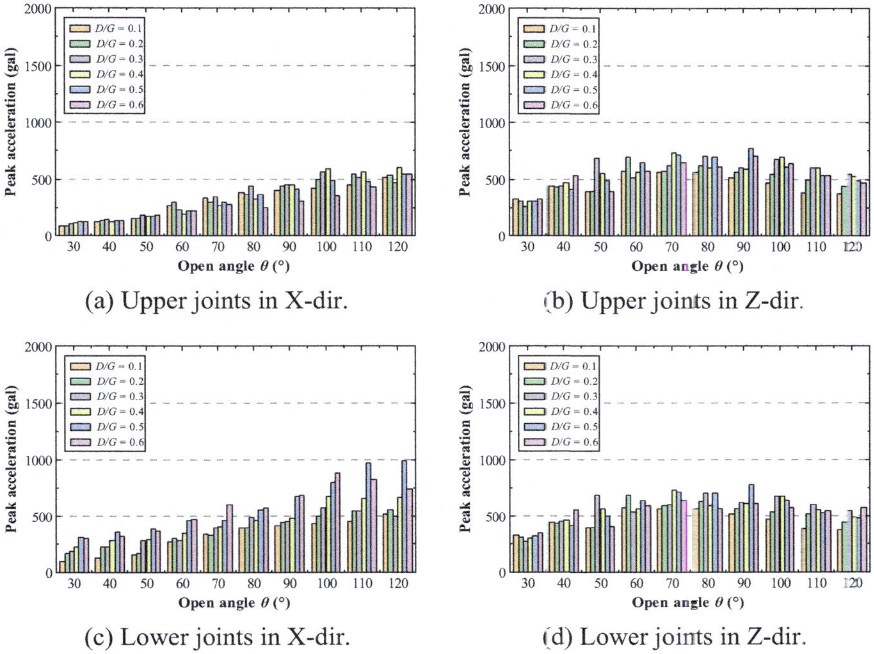

(a) Upper joints in X-dir.

(b) Upper joints in Z-dir.

(c) Lower joints in X-dir.

(d) Lower joints in Z-dir.

Fig. 15 Peak acceleration for CLD-model when ground motions input along *X*-direction

deformation angle of web members, and x_i, y_i, z_i are the displacements of the upper and lower joints. Under the selected ground motions, the influence of the open angle, the depth-grid ratio, and structural span on the maximum inter-layer deformation angle of the web member is investigated in this section. The selected analytical models are the same as the model for peak acceleration analysis.

The distribution of the maximum inter-layer deformation angle for the ground motions input along the *X*-direction is similar to that of values input along the 45°-direction and is higher than the *Y*-direction. Figure 19 shows the variation of the maximum inter-layer deformation angles with the open angle and the depth-grid ratio when the ground motions input along the *X*-direction. The results show that the maximum inter-layer deformation angle of the cylindrical 1.5-layer space frames is more significant than that of the domed one. For the CLO-model and CLD-model, at a lower depth-grid ratio, with the increase of the open angle, the maximum inter-layer deformation angle slowly decreases. While for the model with a higher depth-grid ratio, as the open angle increases, the maximum inter-layer deformation angle sharply increases because of the long web members of the structure. In the case of a low open angle, the maximum inter-layer deformation angle is not so large and the depth-grid ratio has little influence on it. However, the maximum inter-layer deformation angle

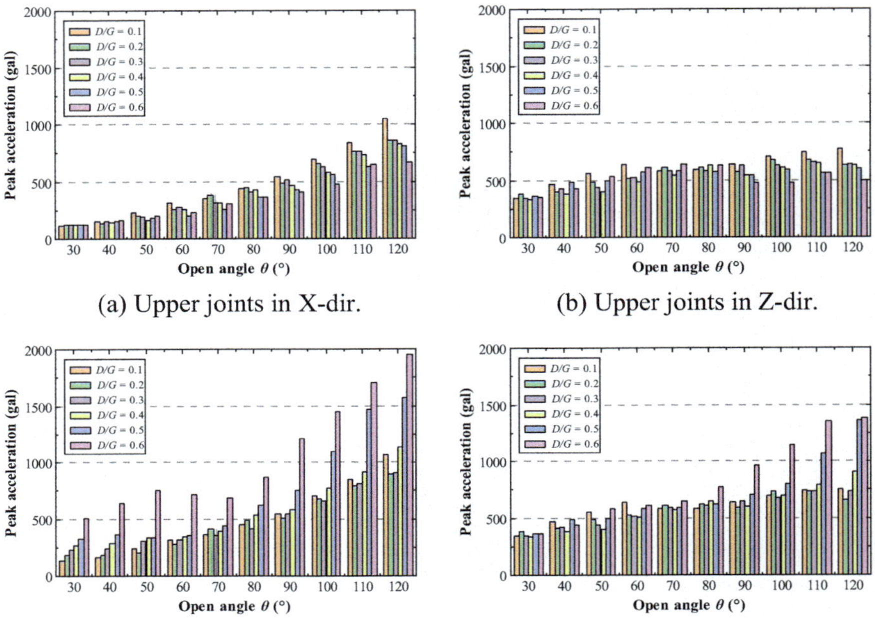

(a) Upper joints in X-dir.

(b) Upper joints in Z-dir.

(c) Lower joints in X-dir.

(d) Lower joints in Z-dir.

Fig. 16 Peak acceleration for DLO-model when ground motions input along *X*-direction

reaches a bigger value in the case of $\theta \geq 90°$ and $D/G \geq 0.5$, especially for the CLO-model. Accordingly, to obtain smaller inter-layer deformation angles and stability of lower joints against seismic loading, it is not recommended to design such a structure with $D/G \geq 0.5$ for practical construction.

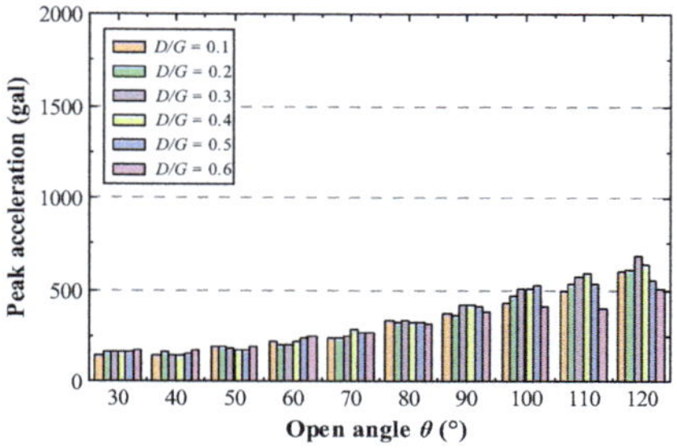

(a) Upper joints in X-dir.

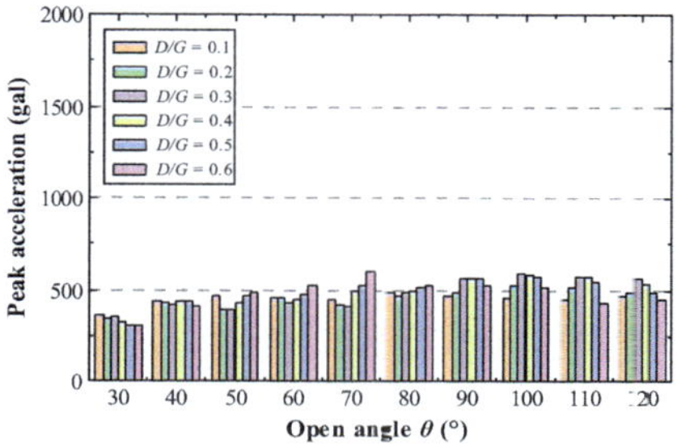

(b) Upper joints in Z-dir.

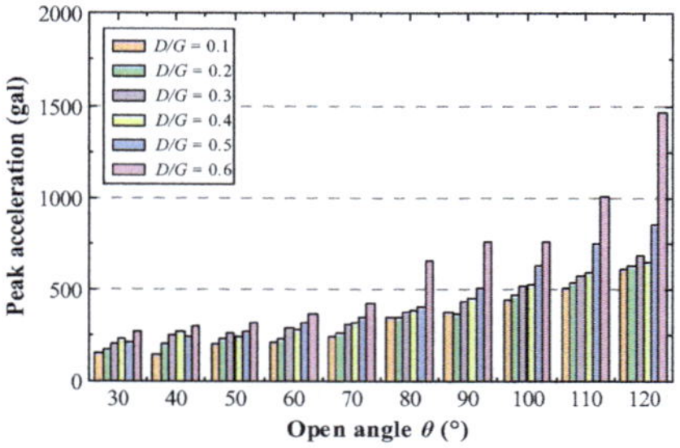

(c) Lower joints in X-dir.

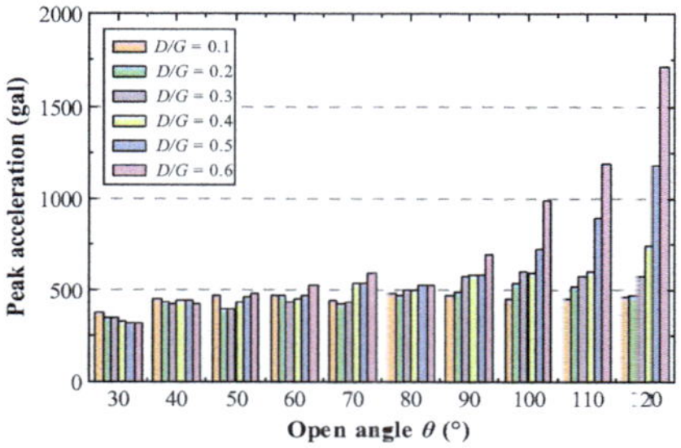

(d) Lower joints in Z-dir.

Fig. 17 Peak acceleration for DLD-model when ground motions input along X-direction

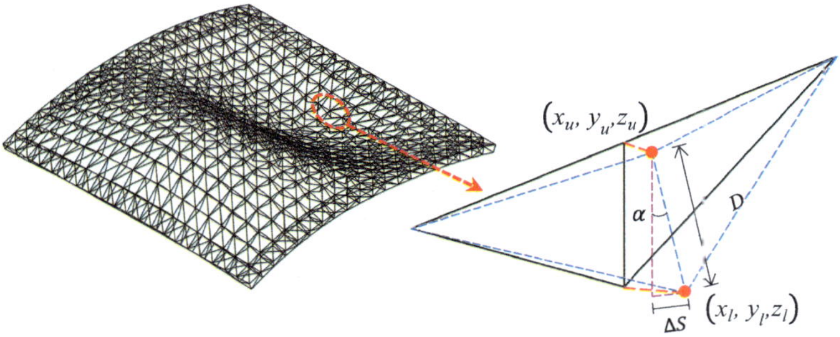

Fig. 18 Definition of inter-layer deformation angle α

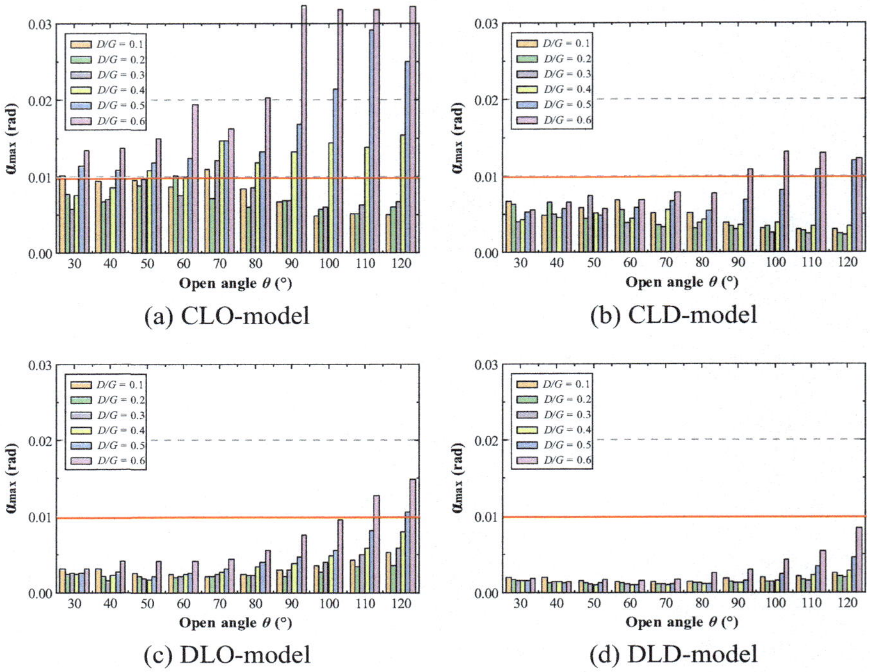

Fig. 19 Max. inter-layer deformation angle of web members (motion input along X-dir.)

5 Conclusion

In this research, eigenvalue analysis and elastic time history analysis were carried out to investigate the natural vibration properties and seismic performances of the 1.5-layer space frames with lap-units considering the vibration of lower joints. A parametric study of the natural vibration properties and seismic performances of four structures was conducted, and the design recommendations of seismic parameters for a quick and accurate assessment of 1.5-layer space frames with lap-units were proposed, which could provide useful information for practical structural design. The major findings obtained from this study are summarized as follows.

1. The fundamental frequency increases with the increase in the open angle and the depth-grid ratio at a low depth-grid ratio. However, at a high open angle and depth-grid ratio, the fundamental frequency is relatively dense or downward. Observing the first mode shapes shows that some key parts appear rotational deformation about their upper chords instead of overall vibration for the cases of high open angles and depth-grid ratios.
2. The EL Centro and Taft waves cause greater seismic response. The domed models exhibit notably more stable performance in the overall stiffness and seismic response.

3. The distribution of the peak acceleration response and the maximum inter-layer deformation angle for the ground motions input along the *X*-direction is similar to that of values input along the 45°-direction and is higher than the values input along the *Y*-direction. The peak accelerations in the horizontal direction of the lower joint are much larger than that of the upper joint, which shows that the lumped masses of lower joints have a significant influence on seismic response and should be considered as much as possible in the numerical simulation.
4. The peak acceleration response increases with the increase in the open angle, particularly the response in the *X*-direction increases significantly. At lower open angles, the depth-grid ratio seems to have little effect on the peak acceleration besides the response of the lower joints in the *X*-direction. The model with a larger open angle and depth-grid ratio may reach a much larger response value due to the long length of web members.
5. For the models with a higher depth-grid ratio, as the open angle increases, the maximum inter-layer deformation angle sharply increases because of the long web members of the structures, especially for the CLO-model. To obtain smaller inter-layer deformation angles and stability of lower joints against seismic loading, it is not recommended to design such a structure with $D/G \geq 0.5$ for practical construction.

The new structural system, 1.5-layer space frame, has advantages in form design, which could be designed into planar or curved shapes through different assembling methods such as lap-units, crossing-units, triple-connecting units, crossing-lap units, polygon pyramids, traditional Japanese pattern, etc. However, in this paper, only the dynamic mechanical properties of 1.5-layer space frames with lap-units were evaluated. The authors will extend this research to the other types of 1.5-layer space frames in the future. Moreover, dynamic analysis in the elasto-plastic range and detailed analysis of joint failure mechanisms should be a subject of further research.

References

1. Chen, P.S.: A report on the innovation of 1.5-layer space frames. In: Proceeding of IABSE IASS, London (2011)
2. Chen, P.S.: Configuration and structural principle of an ancient Chinese Wood Bridge Hongqiao. In: Proceeding of IABSE IASS, Venice, 2007
3. Chen, P.S.: A study report on an ancient wood bridge Hongqiao. Structural Engineering International. 2008; pp. 84–87.
4. Chen, P.S.: A study on the geometrical configuration of an ancient wooden bridge in Qingming Shanghe Tu. In: Proceeding of IABSE IASS, Shanghai, 2010
5. Jin, J.L., Qian, X.X., Chen, P.S.: Study on buckling characteristics of cylindrical and domed 1.5-layer space frames with lap-units. J. Struct. Eng. AIJ. **67B**, 251–260 (2021)
6. Zhu, M.L, Noguchi, S., Chen, P.S.: Structural merits of 1.5-layer space frames with crossing-units, colloquium analysis and generation of structural shapes and systems. AIJ (2017) USB
7. Chen, P.S.: A basic study on the geometrical configuration and design versatility of 1.5-layer space frames. J. Struct. Eng., AIJ. **60B**, 153–158 (2014) (in Japanese)

8. Chen, P.S.: On the configuration of 1.5-layer space frames. In: Proceedings of IABSE IASS, Tokyo, pp. 26–30 (2016)
9. Chen, P.S.: Research on the configuration system of the 1.5-layer space frames. J. Struct. Eng., AIJ. **62B**, 93–98 (2016)
10. Simon, G., Simon, A., Bernard, V.: Nexorade or reciprocal frame system applied to the design and construction of a 850 m^2 archaeological shelter. Int. J. Space Struct. **26**(4), 303–311 (2011)
11. https://www.tess.fr/en/projet/excavation-shelter. Accessed 2021.10
12. Cedrón, F., Elghazouli, A.Y.: Assessment and design considerations for single layer cylindrical lattice shells subjected to seismic loading. Structures **163**, 940–960 (2019)
13. He, Y.J., Quan, Y., Zhou, X.H., et al.: Research on static and stability property of retractable reticulated mega-structure. J. Hunan Univ. Nat. Sci. **36**(4), 24–28 (2009). (in Chinese)
14. He, Y.J., Zhou, X.H., Liu, J.: Dynamic property and seismic performance of square-based membranous latticed shells. Soil Dyn. Earthq. Eng. **43**, 25–32 (2012)
15. He, Y.J., Zhou, X.H., Wang, J.X.: Seismic performance of cylindrical latticed-intersected-three-dimensional-beam-system reticulated mega-structure with single-layer latticed-intersected-cylindrical-shell substructures. Soil Dyn. Earthq. Eng. **57**, 15–24 (2014)
16. Nie, G.B., Zhang, C.X., Dai, J.W.: Seismic behavior study on a space double-layer lattice structure with a lower supporting structure. Adv. Mech. Eng. **10**(11), 1–11 (2018)
17. Kato, S., Murata, M.: Dynamic elasto-plastic buckling simulation system for single layer reticular domes with semi-rigid connections under multiple loadings. Int. J. Space Struct. **12**(3&4), 161–172 (1997)
18. Ishikawa, K., Kato, S.: Dynamic buckling behaviour of single and double layer latticed domes due to vertical earthquake motions. In: Proceeding of IABSE IASS, Guildford, vol. 1, pp. 466–475 (1993)
19. Kunieda, H.: Earthquake response of roof shells. Int. J. Space Struct. **12**(3&4), 149–159 (1997)
20. Nethercot, D.A.: Frame structures, global performance, static and stability behavior, general report. J. Constr. Steel Res. **55**(1–3), 109–124 (2000)
21. Jaspart, J.P.: General report: Session on connections. J. Constr. Steel Res. **55**(1–3), 69–89 (2000)
22. Hou, R., Beck, J.L., Zhou, X., Xia, Y.: Structural damage detection of space frame structures with semi-rigid connections. Eng. Struct. **235**, 112029 (2021)
23. Steel Committee of Kinki Branch, the Architectural Institute of Japan: Reconnaissance Report on Damage to Steel Building Structures Observed from the 1995 Hyogoken-Nanbu (Hanshin/Awaji) Earthquake, 1995
24. Nie, G.B., Dai, J.W., Zhang, C.X., et al.: Failure patterns of large span space structures in Lushan earthquake and numerical simulation. Chin. Civil Eng. J. **48**, 1–6 (2015)
25. Mähr, T., Espinha, M., Ziegler, R.: Global structural modelling of space frame structures with consideration of joint behaviour: an integral analysis approach. Berlin CE/papers. 2017, 1(2–3), pp. 590–598
26. Chen, P.S., Noguchi, S.: Buckling features of 1.5-layer space frames with crossing units. In: Proceedings of IABSE IASS, Germany, pp. 25–28 (2017)
27. Dozio, L., Ricciardi, M.: Free vibration analysis of ribbed plates by a combined analytical-numerical method. J. Sound Vib. **319**(1–2), 681–697 (2009)
28. Kolousey, V.: Dynamics in Engineering Structures. Butterworth, London (1983)
29. Alwar, R.S., Sekhar Reddy, B.: The dynamic analysis of circular plates and shallow spherical shells. J. Sound Vib. **70**(4), 467–473 (1980)
30. Lan, T.T.: Space Frame Structures. Structural Engineering Handbook (1999)
31. Jin, J.L., Chen, P.S.: Dynamic mechanical properties of cylindrical 1.5-layer space frames with inclined lap-units. In: 2021 Summaries of technical papers of annual meeting, AIJ. 2021. pp. 687–688

Study on the Concrete Strength Development with Power Plant Waste as an Aggregate Replacement

Muhammad Nor Syahrul Zaimi, Nur Farhayu Ariffin, Sharifah Maszura Syed Mohsin, Abdul Muiz Hasim, and Nurul Natasha Nasrudin

Abstract Concrete is the most widely utilized construction material due to its design adaptability, availability, and cost effectiveness. Concurrently, the production of power plant waste had become one the major issues in the disposal process that can cause harm toward environmental and economic consequences. The aims of this paper is to compare the strength development of concrete incorporate coal bottom ash (CBA) and fly ash (FA) as a partial and full replacement in concrete. A variety of mix designs were developed based on the varied CBA and FA replacements. CBA is used to replace 50% of fine and coarse aggregate content for partial replacement, and 100% of aggregate content as fine and coarse material for a complete replacement. The FA was utilized as a 20% substitute in cement. The test pieces are manufactured in 100 mm × 100 mm × 100 mm cubes, and all samples are wet cured for a set duration before the compressive strength and water absorption tests. Result shows that the maximum potential of coal ash can be used as 100% replacement. The findings also show that the concrete integration of CBA and FA exhibits a remarkable strength development when compared with the normal concrete. According to the findings of this study, using ground CBA as a material in concrete improves its resilience to harsh environments.

Keywords Coal ash · Strength development · Water absorption

M. N. S. Zaimi · N. F. Ariffin (✉) · S. M. Syed Mohsin · A. M. Hasim · N. N. Nasrudin
Faculty of Civil Engineering Technology, Universiti Malaysia Pahang Al-Sultan Abdullah, Lebuhraya Tun Razak, 26300 Gambang, Pahang, Malaysia
e-mail: farhayu@ump.edu.my

S. M. Syed Mohsin
e-mail: sharifahmaszura@ump.edu.my

N. Sabtu (ed.), *Proceedings of AWAM International Conference on Civil Engineering 2022—Volume 2*, Lecture Notes in Civil Engineering 385,
https://doi.org/10.1007/978-981-99-6018-7_19

1 Introduction

Building construction is the science and art of assembling and erecting buildings, especially those used for shelter. Concrete, steel, wood, and other materials are among those used in construction. Despite, concrete is the most useful material because it is easy to be handled and operated. In the nineteenth century, concrete was mainly used as structural elements in construction industry [1]. The cementitious materials in concrete buildings have this ability that may evolve throughout time [2]. Nowadays, building is an important part of industrial culture, a reflection of its diversity and complexity, as well as a measure of the mastery of natural forces capable of creating a wide range of built environments to meet the diverse needs of society.

The strength of concrete determines its quality [3]. It is important to analyze the environment, the placement process, the ingredient used, and the local weather to estimate performance and duration for the concrete serviceability during its life cycle. The concrete strength also is an important criterion to determine whether it will be accepted or rejected for use in construction. The quality of materials used such as cement, aggregate, sand and water [4, 5], compaction during pouring, age of concrete during production and testing, and others are important variables that affect concrete strength. Generally, the mix ingredient of concrete as per mentioned contributes to a high rate of natural resource consumption, which can lead to depletion [6]. Depletion of natural resources has a number of undesirable effects [7].

The function of the material, such as the use of high-quality cement improves the bond and strength. On the other hand, the curing period that starts from production date also influences the concrete strength due to the hydration process that is used for strength development. Previous study reveals that the size, shape, hardness, texture, and other properties of coarse aggregates varied widely depending on the location of quarried or dredged from riverbeds [8]. For example, when the amount of surface and ground water evaporates due to direct sunlight, the natural aggregate such as river sand is taken as a surcharge to dry the river bed. Rivers and estuaries deepen when sand is lost from the riverbed, as does the width of estuaries and coastal bays [9]. Even materials from the same quarry or pit, and the type of rock used, can differ drastically. Crushed stone from hard rocks like granite and limestone, or river gravels, are typically coarse aggregates used for concrete manufacturing. Barbara [10] mentions that the common condition of coarse aggregate are irregular crushed stones or naturally occurring round gravels used to make concrete (round aggregates). Groundwater and surface water are the primary effects, and contours are deformed; the drainage system is disrupted, and the natural profile and surface relief are interrupted [11]. Natural resources are extremely important in developing countries. Therefore, to minimize the negative impacts of resource depletion, we need to protect natural resources. Researching a substitute material that can be used in concrete is one way to prevent depletion. In a previous study, garbage was used as a concrete substitute to replace natural resources [12].

In the context of waste management, waste refers to the unwanted or useless stuff generated by human activity, which can take many forms. There are diverse disposal and management options for liquid, solid, and gaseous waste. There are categories of solid waste which are hazardous and non-hazardous material. The coal bottom ash and fly ash are hazardous solid wastes that are generated at power plants and must be disposed of safely [13]. Landfill cannot sustain the disposal process due to the high volume of coal waste. Another option is to utilize waste to the construction industry as concrete replacement material.

Coal waste contains coal, dirt, and rock (mine waste), which is the potentially hazardous residue of prior coal burning. In the novel fluidized bed burners, the majority of waste coal is burnt in its natural state. There are different types of by-products that are generated after the combustion process, namely FA, boiler slag, CBA, clinker and other forms of coal ash [14], with FA and CBA being the most common waste materials. In a power plant, the boiler, furnace, or incinerator, CBA is part of the noncombustible combustion residue. It's dark in color and accumulated at the bottom of the boiler, accounting for roughly 20% of total coal ash. CBA typically consists of large in size particles. FA is identified to be used as a cement addition at concrete mixes to increase the concrete's workability and strength [15]. CBA refers to the bottom ash that forms at the bottom of a boiler and is heavier and more difficult to move by combustion gases than fly ash. Bottom ash as a building material has attracted the attentions of a number of researchers. It offers many advantages to industry as it is inexpensive, has a low fill density that can be used as a replacement for concrete materials, making it a particularly sustainable resource [16]. As a result, using coal waste as a substitute for natural resources in concrete can help to prevent resource depletion and CBA disposal. This is owing to the enormous annual production of coal waste in any country that uses raw coal to generate electricity. For instance, recently, there is lack of research on the high amount of CBA as a replacement in concrete. Since there are many researchers that utilized the CBA and FA as partial replacements for concrete, the production of waste cannot be solved by the waste produced and this research can help in accelerating the reduction of coal waste. The industrial player, such as engineer, architecture, contractor would be able to implement this coal waste into concrete production after sufficient knowledge had been provided by researcher. As a consequence, residents will be urged to advocate coal waste disposal alternatives such as ponds and landfills.

2 Experimental Work

2.1 Material

Properties of Material

In this investigation, ordinary Portland cement used was the Orang Kuat brand. The river sand was employed as fine aggregate. Then all the ingredients were mixed and cured using clean tap water. To obtain coarse aggregate size, the aggregate was crushed using the jaw crusher equipment. For this study, the coal ash was collected at the Tanjung Bin Power Plant in Johor, Malaysia. The physical and chemical composition of CBA and FA are illustrated in Fig. 1, Tables 1, and 2, respectively.

In this study, the river sand as fine aggregate and crushed limestone as coarse aggregate were utilized. The coarse aggregate has a range in coarse particle size of 10–5 mm and all fine aggregate particles passing through size of 4.75 mm [17]. The

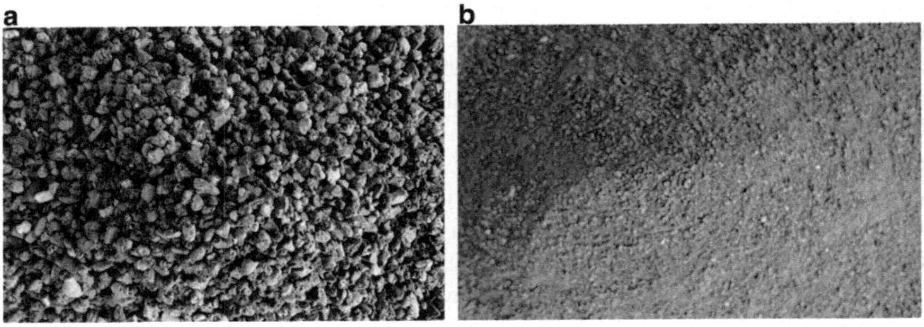

Fig. 1 **a** Coarse coal bottom ash (CBA), **b** fine coal bottom ash (FBA)

Table 1 Physical properties of mix material

Description	Stone	Sand	CBA	FBA
Shape	Angular	Angular	Porous	Fine
Size (mm)	5 mm >	4.75 m <	5 mm >	4.75 m <
Specific gravity	2.51	2.81	2.21	3.2
Density (kg/m^3)	1600	1223.8	641.03	1000
Moisture content (%)	1	4.50	40	25–45

Table 2 Chemical properties of coal ash, fine aggregate, and cement

Formula	CBA	FA	Sand	Cement
SiO_2	33.7	57.13	51	20.9
Al_2O_3	12.9	34.48	6.83	4.8
CaO	6.34	0.32	0.48	65.4
Fe_2O_3	6.98	3.46	0.32	3.4
MgO	0.65	–	–	1.3
TiO_2	0.89	0.48	0.58	–
K_2O	1.19	3.67	0.4	0.4
SO_3	0.9	–	–	2.7
P_2O_5	0.3	0.33	–	–
Na_2O	0.59	–	–	–
BaO	0.22	–	–	–

river sand has a specific gravity of 2.81, and the density of sand is 1223.78 kg/m^3. The moisture content of fine aggregate is 4.5%. The specific gravity of crushed limestone was 2.51 with the density of 1600 kg/m^3, and 1.3% moisture content. The color of CBA varies from brown to dark gray, whereas the color of OPC is gray. The differences in color between the two materials could be due to differences in raw materials, production techniques, and firing temperatures. For CBA, the specific gravity was 2.21, and a density of 641.03 kg/m^3 is lower than coarse and fine aggregates due to its porous form. CBA has a greater moisture content than regular concrete. Singh et al. [18] mentioned that CBA has a greater water absorption of 30% compared with the percentage of natural aggregate.

Table 2 gives the chemical compositions of CBA, FA, sand, and cement. The analysis of the CBA showed that the ratio of $SiO_2 + Al_2O_3 + Fe_2O_3$ was 53.58% at minimum of 50% which corresponds to a pozzolanic class C material according to ASTM C618 [19]. According to Athirah [20], CBA was categorized as a class F ash having a $SiO_2 + Al_2O_3 + Fe_2O_3$ composition greater than 70%. Kurama et al. and Mangi et al. [21, 22] also observed that the amount of $SiO_2 + Al_2O_3$ and Fe_2O_3 is classified as a type-F ash according to ASTM C 618 [19] if it reaches more than 80 percent in the composition. The CBA was crushed and divided to two different particle size gradations, namely coarse (5–10 mm) and fine (below sieve 4.74 mm). The FA has the high chemical content of $SiO_3 + Al_2O_3 + CaO$, which is 91.93%, and consists of a small amount of pozzolanic reactive materials, which account for 5.32% CaO. Sand consists of unconsolidated granular components such as rock fragments, mineral particles, and marine sediments. In sand, the silicate minerals and granular silicate rock particles make up the majority of it [23], because quartz (sand), also known as silicon dioxide, is the most prevalent mineral in fine aggregate when silicon and oxygen are combined and has the greatest chemical composition of 51% SiO_2 [24]. The pozzolanic reactivity of CBA is represented by the granular condition of CBA particles and the percentage content of SiO_2 [25].

2.2 *Sample Preparation and Mix Proportion Design*

Table 3 gives the mix design of the four different composite designs used in this study. CBA was replaced with 100 and 50% with coarse and fine aggregates, respectively, while FA was replaced by 20% with cement. The mill is used to grind coal ash to the best alternative size for sample preparation. The surface of the sample is ground flat before testing, and fresh concrete is poured into the mold according to the cubic size of 100 mm × 100 mm × 100 mm. Specimens were cured at room temperature for an entire day and a damp bag was applied over the specimens. Samples are cured for 28 days after removal from the mold for optimum strength. The composite design used for this survey is given in Table 3.

Table 3 Mix design

Description	Cement (kg/m^3)	Fly ash (20%)	Coarse aggregate (kg/m^3)		Fine aggregate (kg/m^3)	
			Granite	CBA	SAND	FBA
Control	440	0	780	0	881	0
0% CBA 100% FBA		88	780	0	0	910
100% CBA 0% FBA			0	770	881	0
100% CBA 100% FBA			0	770	0	910

2.3 Experimental Program

The attributes of raw material and cured concrete, including sieve analysis, compressive strength, and water absorption were tested in the different concrete mixes, which were separated into four designs and three cube samples for each design. The next sections go through the details of each testing procedure.

For the gradation of coarse aggregate, fine aggregate, coarse CBA, and fine CBA, a sieve analysis test of raw material was performed. To determine the optimal concrete strength, the coarse aggregate was regulated to be 5–10 cm and the fine aggregate was regulated to be less than 2.36 cm. The gradations of coarse and fine aggregate were determined immediately using a concrete sieve in accordance with ASTM C136 [26]. CBA was used as a concrete alternative.

After 28, 60, and 90 days, the compressive strengths of 100 mm dimension cube specimens were measured. A concrete compression machine is then used to assess the specimen's compressive strength. The compressive strength provided for each test age is the average of three concrete samples tested in the specified period for each mix design. Concrete compressive strength is measured using the ASTM C39 standard [27].

The water absorption of all samples was evaluated using a concrete cube soaked in water for 24 h and weighed in accordance with ASTM C1585, when the cube was placed in the kiln and weighed after 24 h to dry, and the wet cube was weighed. CBA in concrete was also employed in this study to minimize the quantity of natural aggregate needed in concrete.

3 Result and Discussion

3.1 *Sieve Analysis*

The sieve of crushed limestone and coarse coal bottom ash with sieve widths of 37.5, 20, 14, 10, 5, and 2.36 mm resulted in the particle size distribution and identification as presented in Fig. 2. This figure depicts the particle size distribution of coarse aggregates (crushed limestone) and coarse CBA. The curve is essentially smooth, indicating that the particle size distribution of CBA is in the same range as natural coarse aggregate, implying that there is an excellent match and implying that CBA has ideal properties, therefore enhancing the concrete quality, particularly in strength terms [13]. During the sieve process, CBA was below the lower limit, but it can still be considered because the gap is close to the range [28]. Most researchers simply sieve the fine CBA with sand and cement, resulting in a paucity of information for the coarse CBA.

The particle size distribution identification test of river sand with fine CBA was carried out in Fig. 3 with 7 difference size of sieve diameters from 150 μm to 10 mm. Fine aggregate (river sand) and fine CBA particle size distribution are illustrated in Fig. 3. The slope was smooth, showing that bottom ash particle sizes would be in the same range as natural river sand, indicating a strong link and implying that bottom ash possessed excellent characteristics [29]; as a result, the concrete quality is improved in terms of strength. Moreover, bottom ash is a grayish substance with grain size with coarse and fine size, and some of it was sand-like material. Fine aggregate that does not meet these grading standards shall meet the requirements of this section if

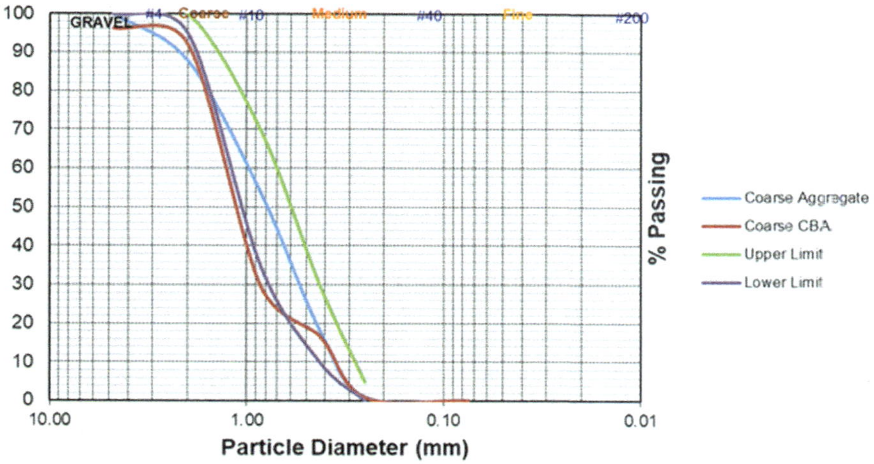

Fig. 2 Sieve analysis of coarse aggregate and coarse CBA

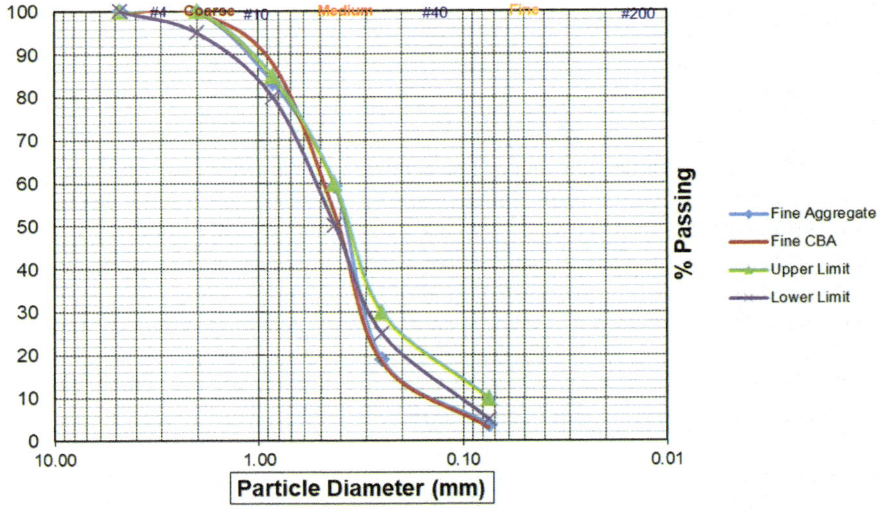

Fig. 3 Sieve analysis of fine aggregate and fine CBA

the supplier can show or indicate to the customer that the concrete of the specified class made using the fine aggregate in question will have the specified qualities [17].

According to a previous study, the particle size of CBA is about 50% smaller than 250 mm, making a higher amount of fine elements accessible in concrete [30]. The requirements of BS 3797:1990 and ASTM C 330 for the light fine aggregate requirements were satisfied, although they were consistently at the lower limit, showing that the Tanjung Bin CBA has a higher concentration of coarse particles [31].

3.2 Relationship Between Strength Development and Water Absorption

The findings of compressive strength compared with the water absorption experiments were obtained after 28, 60, and 90 days. Figure 4 illustrates a graph showing the relationship of compressive strength with the water absorption data.

This graph illustrates that the percentage of water absorption and compressive strength has a general negative relationship. The link between total water absorption and compressive strength is quite strong, as shown in Fig. 4, and the reduction in water absorption as compressive strength increases shows that the material is durable. After 28 and 60 days, the full replacement shows the highest water absorption between 1.5 and 2.7% and has a higher strength than the control concrete, but at 90 days it slightly decreases than the normal concrete. With partial replacement, the replacement of fine shows a smooth decrease from 28 to 90 days, while the coarse size shows a drastic fall from 28 to 60 days of water absorption. The compressive strength that both

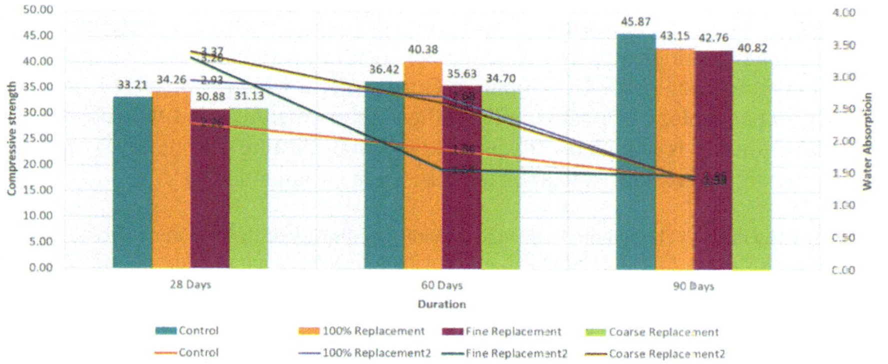

Fig. 4 Strength development correlation water absorption

partial replacements have is in increasing pattern and the fine replacement is higher than coarse replacement at 60 and 90 days. As curing time leads to less hydration of cement and lime base and sand particles, the water absorption at 90 days is the lowest for the normal concrete and another replacement due to the concrete dense along with the period time.

As previously stated [32], the amount of water a specimen can hold is used to determine its total liquid limit. In addition, the high pore count of the concrete specimen results in a low percentage of water absorption, which leads to high compressive strength. This is because the addition of CBA to the mix increases the pore size, which allows the specimen to absorb a lot of water [30]. Furthermore, the data showed an inverse relationship between the compressive strength of the sample and water absorption [33]. Material with good engineering properties, such as natural aggregate or control concrete has high strength but low water absorption, whereas CBA may have high water absorption, but moderate strength and because of the surface and interior structure of concrete directed to its total strength. Consequently, water absorption cannot be used to determine the strength of concrete [34]. Better building materials should have low water absorption and high strength. Therefore, it is advisable to research for a better type of material to replace the control concrete with coal ash.

4 Conclusion

In terms of strength, the greater the proportion of CBA replacement concrete can be almost parallel with control strength along with the period time. The concrete water absorption rises when CBA is substituted as proven that CBA affects the water absorption ratio. The greater replacement of CBA results in water absorption has low influence on the strength of the material. The particle size of aggregate replacement affects to the porosity area in concrete that can give impact to the percentage of water absorption. In the manufacturing of concrete, the use of coal waste material aids in the creation of ecologically friendly concrete. This innovation helped to improve the global goal or Sustainable Development Goals criteria which is in act 9.1 (develop quality, dependable, sustainable, and resilient infrastructure), and 9.4 (develop an infrastructure that is of high quality, dependable, long-lasting, and flexible) (upgrade infrastructure and retrofit industries to make them sustainable).

Acknowledgements The authors would like to thank the Ministry of Higher Education and Universiti Malaysia Pahang Al-Sultan Abdullah for providing financial support under RDU230301 and Universiti Malaysia Pahang for laboratory facilities.

References

1. Pepin, R.: The History of Concrete. Giatec Scientific Inc. Giatecscientific.Com (2017)
2. Berodier, E., Gibson, L.R., Burns, E., Roberts, L., Cheung, J.: Robust production of sustainable concrete through the use of admixtures and in-transit concrete management systems. Cem. Concr. Compos. **101**, 52–66 (2019). https://doi.org/10.1016/j.cemconcomp.2018.01.008
3. VDT, V.T.D.: Chapter 1: Components of concrete. **37**(176) 2 (2015)
4. Kirthika, S.K., Surya, M., Singh, S.K.: Effect of clay in alternative fine aggregates on performance of concrete. Constr. Build. Mater. **228**, 116811 (2019). https://doi.org/10.1016/j.conbuildmat.2019.116811
5. Graves, K.J.: McGraw-Hill Encyclopedia of Science & Technology. JAMA J. Am. Med. Assoc. **258**(21), 3174 (2003). https://doi.org/10.1001/jama.1987.03400210118038
6. Rahman, M.A.: Natural Resources: Extractions, Depletions, Protections and Sustainable Managements, pp. 53–58 (2017). https://doi.org/10.15242/heaig.er0917407
7. Mittal, I., Gupta, R.K.: Natural Resources Depletion and Economic Growth in Present Era, Sept 2015 (2017)
8. Salman, A.M., Akinpelu, M.A., Ahmed, G.A., Raheem, A.: Evaluation of variations of coarse aggregate types on hardened properties of concrete. J. Mater. Eng. Struct. **8**(February), 301–311 (2021)
9. Pitchaiah, P.S., Pradesh, A.: Impacts of sand mining on environment—a review **4**(1), 1–5 (2017)
10. Schreiber, B.A.: Concrete. In: The Editors of Encyclopaedia Britannica. Encyclopaedia Britannica, Inc. (2020)
11. Dai, S., et al.: Recognition of peat depositional environments in coal: a review. Int. J. Coal Geol. **219**(January), 103383 (2020). https://doi.org/10.1016/j.coal.2019.103383
12. Muthusamy, K., Rasid, M.H., Jokhio, G.A., Mokhtar Albshir Budiea, A., Hussin, M.W., Mirza, J.: Coal bottom ash as sand replacement in concrete: a review. Constr. Build. Mater. **236**, 117507 (2020). https://doi.org/10.1016/j.conbuildmat.2019.117507
13. Singh, N., Bhardwaj, A.: Reviewing the role of coal bottom ash as an alternative of cement. Constr. Build. Mater. **233**, 117276 (2020). https://doi.org/10.1016/j.conbuildmat.2019.117276
14. Rodríguez-álvaro, R., González-fonteboa, B., Seara-paz, S., Rey-bouzón, E.J.: Masonry mortars, precast concrete and masonry units using coal bottom ash as a partial replacement for conventional aggregates. Constr. Build. Mater. 283 (2021). https://doi.org/10.1016/j.conbuildmat.2021.122737
15. Singh, M., Siddique, R.: Effect of coal bottom ash as partial replacement of sand on workability and strength properties of concrete. J. Clean. Prod. **112**, 620–630 (2016). https://doi.org/10.1016/j.jclepro.2015.08.001
16. Bajare, D., Bumanis, G., Upeniece, L.: Coal combustion bottom ash as microfiller with pozzolanic properties for traditional concrete. Procedia Eng. **57**, 149–158 (2013). https://doi.org/10.1016/j.proeng.2013.04.022

17. ASTM C33: ASTM C33 (2013). https://doi.org/10.1520/C0033
18. Singh, M., Siddique, R.: Properties of concrete containing high volumes of coal bottom ash as fi ne aggregate **91** (2015). https://doi.org/10.1016/j.jclepro.2014.12.026
19. ASTM C 618: ASTM C618 :Standard Specification for Coal Fly Ash and Raw or Calcined Natural Pozzolan for Use in Concrete, ASTM International, West Conshohocken, PA, 2012. www.astm.org, ASTM Int., pp. 1–5 (2014). https://doi.org/10.1520/C0618
20. Athirah, N.: Effect of coal bottom ash as a partial replacement of sand on properties of cement mortar (2014)
21. Kurama, H., Kaya, M.: Usage of coal combustion bottom ash in concrete mixture. Constr. Build. Mater. **22**(9), 1922–1928 (2008). https://doi.org/10.1016/j.conbuildmat.2007.07.008
22. Mangi, S.A., Wan Ibrahim, M.H., Jamaluddin, N., Arshad, M.F., Putra Jaya, R.: Short-term effects of sulphate and chloride on the concrete containing coal bottom ash as supplementary cementitious material. Eng. Sci. Technol. an Int. J. **22**(2), 515–522 (2019). https://doi.org/10.1016/j.jestch.2018.09.001
23. Haldar, S.K.: Basic mineralogy (2020)
24. Elahi, M.A., Hossain, M., Karim, R., Fauzi, M., Zain, M., Shearer, C.: A review on alkali-activated binders: materials composition and fresh properties of concrete. Constr. Build. Mater. **260**, 119788 (2020). https://doi.org/10.1016/j.conbuildmat.2020.119788
25. Sadon, S.N., Beddu, S., Naganathan, S., Mohd Kamal, N.L., Hassan, H.: Coal bottom ash as sustainable material in concrete. A review. Indian J. Sci. Technol. **10**(36), 1–10 (2017). https://doi.org/10.17485/ijst/2017/v10i36/114595
26. ASTMC136: Standard Test Method for Sieve Analysis of Fine and Coarse Aggregates I (200), pp. 1–5 (2014). https://doi.org/10.1520/C0136
27. ASTM C39/C39M-14: ASTM C 39/C 39M—14. Standard Test Method for Compressive Strength of Cylindrical Concrete Specimens, pp. 3–9 (2014). https://doi.org/10.1520/C0039.
28. Singh, N., Kumar, P., Goyal, P.: Reviewing the behaviour of high volume fly ash based self compacting concrete. J. Build. Eng. **26**(July), 100882 (2019). https://doi.org/10.1016/j.jobe.2019.100882
29. Ramzi, N.I.R., Shahiron Shahidan, M.Z., Maarof Ali, N.: Physical and chemical properties of coal bottom ash (CBA) from Tanjung Bin Power Plant. IOP Conf. Ser. Mater. Sci. Eng. (2016). https://doi.org/10.1088/1757-899X/160/1/012056
30. Ernida, N., Abidin, Z., Haziman, M., Ibrahim, W.: The effect of bottom ash on fresh characteristic, compressive strength and water absorption of self-compacting concrete the effect of bottom ash on fresh characteristic, compressive strength and water absorption of self-compacting concrete, July 2015 (2014). https://doi.org/10.4028/www.scientific.net/AMM.660.145
31. Umar Abubakar, A., Baharudin, K.S.: Properties of concrete using Tanjung Bin Power Plant Coal Bottom Ash and Fly Ash. Int. J. Sustain. Constr. Eng. Technol. **3**(2), 2180–3242 (2012) [Online]. Available: http://penerbit.uthm.edu.my/ojs/index.php/IJSCET
32. Liang, K., Hou, Y., Sun, J., Li, X., Bai, J., Tian, W.: Theoretical analysis of water absorption kinetics of recycled aggregates immersed in water. Constr. Build. Mater. 302(April), 124156 (2021). https://doi.org/10.1016/j.conbuildmat.2021.124156
33. Jung, C.T., Siang, T.C., Kwong, T.H., Boon, K.: Compressive strength and water absorption of mortar incorporating silica fume. Int. J. Civ. Eng. **6**(8), 39–43 (2019). https://doi.org/10.14445/23488352/ijce-v6i8p106
34. Lawane, A., Minane, J.R., Vinai, R., Pantet, A.: Mechanical and physical properties of stabilised compressed coal bottom ash blocks with inclusion of lateritic soils in Niger. Sci. Afr. **6** (2019). https://doi.org/10.1016/j.sciaf.2019.e00198

Experimental Study on Column Strengthening Using Pre-tensioned Steel Straps Confinement

Jia-Yang Tan, Chau-Khun Ma, Chee-Loong Chin, Chin-Boon Ong, Abdullah Zawawi Awang, and Wahid Omar

Abstract Earthquake excitation can exert combined flexural, axial, shear, and torsional loading on reinforced concrete (RC) columns. Although behaviour of confined columns under cyclic pure bending load has been extensively investigated, the research on interaction between bending and torsion in confined columns is still limited. The studies of active confinement on its effectiveness in enhancing the behaviour of RC column subjected to combined loading are also scarce. This paper presents experimental studies on two similar concrete columns. One of the columns served as control specimen, while the other one confined with steel straps. Both columns were tested under constant axial and cyclic lateral load including torsion. This paper aims to study the effectiveness of the steel straps in upgrading the performance of the columns. Load deformation curves, energy dissipation, column cross-section curvature, rebar and steel straps strain value are being compared and presented in this paper. From the result, steel straps tensioning technique (SSTT) managed to improve the performance of column and achieve the purpose of strengthening the column under cyclic lateral load.

Keywords Cyclic load · Steel straps · Confinement · Column strengthening

1 Introduction

Columns are structural members that are used as support for the roof of building or upper storey. It transfers the weight of the structure to the footing. The column will sustain axial load and under compression state normally. However, columns may need to resist lateral forces to serve as the purposes of wind and earthquake design.

J.-Y. Tan · C.-L. Chin · C.-B. Ong · A. Z. Awang · W. Omar
Faculty of Civil Engineering, Universiti Teknologi Malaysia, 81310 Skudai, Malaysia

C.-K. Ma (✉)
Forensic Engineering Centre, Institute for Smart Infrastructure and Innovative Construction, University Teknologi Malaysia, 81310 Skudai, Malaysia
e-mail: machaukhun@utm.my

N. Sabtu (ed.), *Proceedings of AWAM International Conference on Civil Engineering 2022—Volume 2*, Lecture Notes in Civil Engineering 385,
https://doi.org/10.1007/978-981-99-6018-7_20

Columns are normally designed to always remain fit in real life. It is to ensure that the structure does not collapse and cause death. Unfortunately, some unforeseeable events such as earthquake may cause extra loading on columns leading to collapse of structure. Columns are subjected to torsional moments in addition to flexure, axial, and shear forces during earthquake excitation [1]. The addition of torsion is more likely in corner columns, skewed superstructure, different adjacent column heights, rigid decking, or abutment restraints [2]. The columns that are not able to withstand the earthquake load need to be demolished and rebuilt. Confinement becomes the solution to upgrade the capacity of the columns in line with sustainability and resiliency.

Many studies were conducted to improve the performance of columns using confinement. Behaviour of confined columns under cyclic pure bending with and without axial compression has been extensively investigated [3–7]. The effect of confinement on strengthening the column under cyclic pure bending was encouraging. However, the interaction between bending and torsion in confined columns is limited [8–12]. Until now, fibre reinforced polymer (FRP)-confined columns received the most attention. The research on using of active confinement in strengthening column is still lacking.

In this paper, the effect of steel straps (active confinement) on retrofitting column under lateral cyclic load including torsion is being investigated. Two columns (one is control specimen, and another is confined specimen) were tested by using quasi-static testing. The details of laboratory work and the performance of columns after being strengthening were discussed in the following topics.

2 Methodology

2.1 Specimens Detail

The laboratory study in this paper focused on two column specimens. Both column specimens were in the same dimension and the schematic drawing of specimen was as shown in Fig. 1. The specimens consist of a large base footing with dimension of 1150 mm × 600 mm × 300 mm to fix the specimens during laboratory work. A circular column with a diameter of 150 and 1000 mm clear height was casted on top of the footing. There was a column head with the dimension of 500 mm × 200 mm × 200 mm attached on top of the circular column to apply the lateral load. 16 mm diameter reinforcement bars were used in base footing and column head. For the main column, 6 numbers of 10 mm diameter reinforcement bars were used. Both column specimens were cast with high-strength concrete mix (60 MPa) due to its brittleness when the concrete failed [13].

Table 1 gives the differences between the two column specimens. Column A served as control specimen and was tested in bare condition. On the other hand, column B was confined by steel straps in two layers with 50 mm spacing as shown

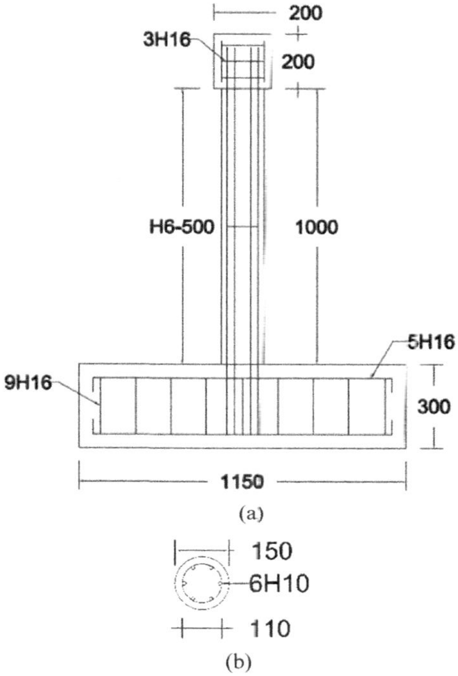

Fig. 1 **a** Column specimens, **b** cross section of column specimens (in mm)

Table 1 Specimens configuration

Column	T/M	Layers	Spacing (mm)
A	0.2	–	–
B	0.2	2	50

in Fig. 2. Both specimens were subjected to the same loading condition. The cyclic lateral load was applied at 200 mm from the centre of the column head to achieve 0.2 torque-to-moment ratio (T/M). Both columns were tested and analysed to know how far the steel straps can enhance the performance of columns under cyclic loading including torsion.

2.2 Laboratory Set-Up

Column A and B were tested in Structure Laboratory of School of Civil Engineering, Universiti Teknologi Malaysia. The steel frame was set as shown in Fig. 3 to test the specimens. The footing was fixed onto the strong floor by using UKB 178 × 102 × 19 during laboratory work. It prohibited both vertical and horizontal movements.

Fig. 2 **a** Unconfined specimen (column A), **b** confined specimen (column B)

The axial load in bridge columns typically falls between 5 and 10% [14]. So, the axial load was taken as 7% of the capacity of the columns to treat as the superstructure dead weight. Hydraulic jack was set on top of the column and implemented 120 kN of axial force. Two high strength thread rods with their specific nuts were utilised to maintain the axial load. It was ensured not to disturb the free horizontal displacement of the specimens during lateral cyclic event at the same time. Two hydraulic jack was fixed horizontally at the steel frame to create lateral force to the specimens. The cyclic loading point was situated 200 mm from the centre of the column to induce 0.2 T/M.

Linear variable displacement transformers (LVDTs) were used to measure the lateral displacement at corresponding lateral force applied. LVDT were placed at 300, 600, and 1000 mm (both side) from the footing surface during laboratory work. It is placed at both sides of the column to capture the reading when the cyclic load applied from both left and right direction. There were two LVDTs with 100-mm displacement range placed at the column head (one at the centre line and another one located at 75 mm from the centre) to record the twist angle of the specimens during testing.

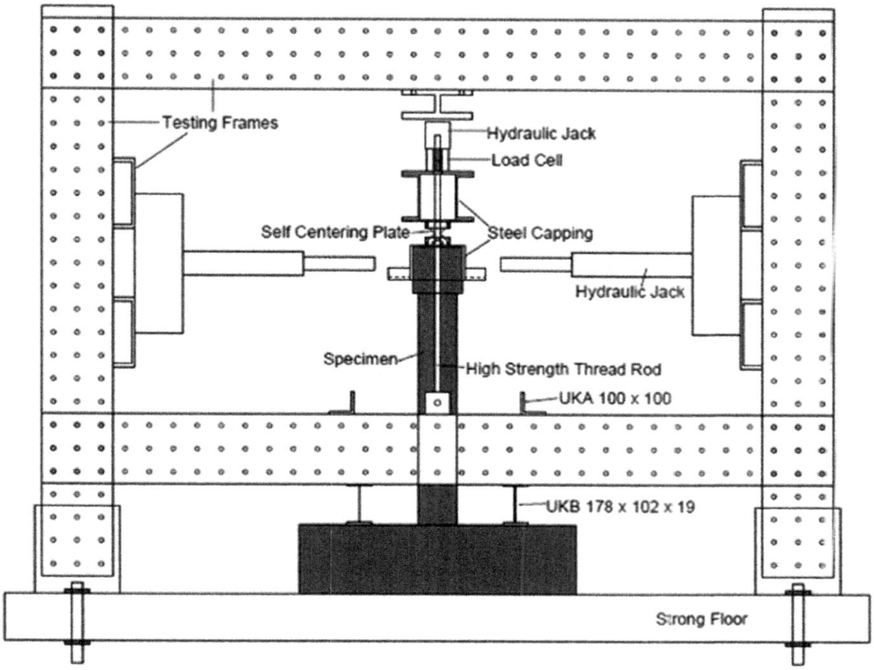

Fig. 3 Laboratory set-up in schematic drawing

2.3 Loading Protocol

Both specimens were subjected to the same loading condition. First, the column was axially loaded to 120 kN (7% of the ultimate load of column) and kept constant throughout the cyclic quasi-static test.

Lateral load was then started to apply to the column following the loading protocol in Fig. 4. The cyclic load was initiated with force control and then continued in displacement control. Each cycle was repeated twice as recommended by FEMA 461. 25% of its estimated flexural load was first applied laterally to the column and continued in the opposite direction. Then, it continued with 50 and 75% of its estimated flexural load. Next, the test was carried out under displacement control mode until the strength deteriorated to more than 80% of the peak capacity. Each cycle was increased by ductility amplitude of 1 from previous cycle ($\mu\Delta = 2, 3, 4, 5\ldots$). Both columns were tested under the same loading protocol which was similar to some other research studies [9, 11, 15–18].

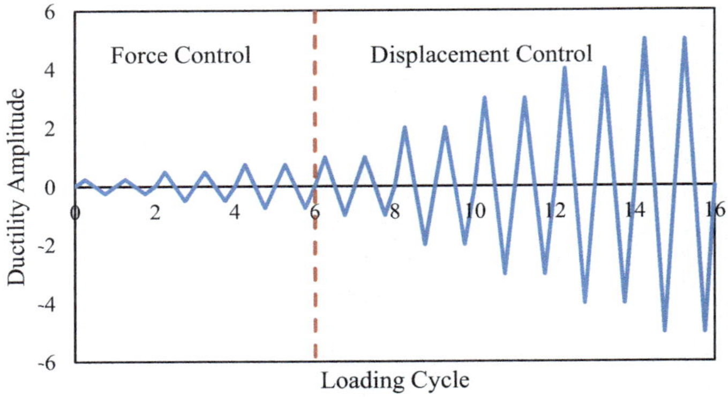

Fig. 4 Loading protocol

2.4 *Steel Strapping Method*

Steel straps tensioning technique (SSTT) was the confining method used to strengthen the column performance in this study. SSTT used the steel straps shown in Fig. 5 as confining method material. Steel strapping pneumatic tensioner as shown in Fig. 6 was used to tighten the steel straps with constant pressure of 0.2 MPa around the column. It was to ensure an effective lateral stress to be exerted to the column before loading. SSTT had emerged as a new effective strengthening and repairing technique for structural members due to its low cost and ease of installation [19–21].

3 Result and Analysis

All the column specimens were tested and analysed in terms of load deformation curve, energy dissipation, equivalent viscous damping, column cross-section curvature, longitudinal reinforcing bar, and steel straps strains values.

Fig. 5 Steel straps

Fig. 6 Steel strapping pneumatic tensioner

3.1 Load Deformation Curve

The lateral displacement of the column specimens with its corresponding lateral force was recorded and plotted in load deformation curve. Overall, the graphs showed hysteretic relationship. Figure 7 shows the load deformation curve of column A (control specimen), while Fig. 8 shows the load deformation curve of column B (confined with steel straps). The skeleton curve can be obtained from the envelope of the hysteresis curve. The skeleton curve of column A was compared with column B as shown in Fig. 9. This was carried out to know the contribution of steel straps towards the upgrading of performance of column specimen under cyclic load including torsion.

From Fig. 9, it can be seen clearly that the skeleton curve at first stage was almost similar for both specimens. The load deformation curve was in linear relationship until the lateral displacement of approximately 10 mm. Then, the curve changed to

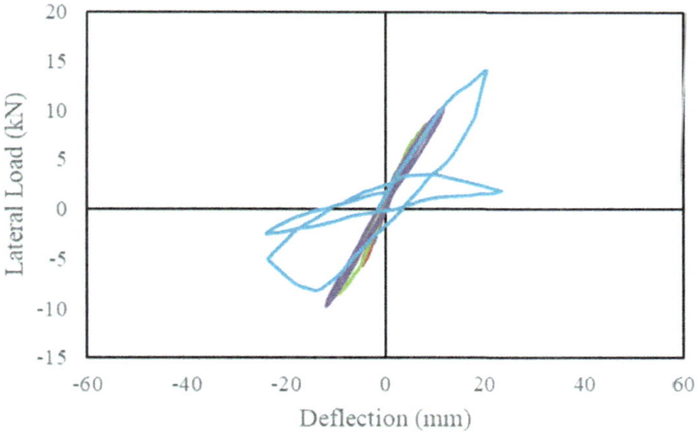

Fig. 7 Load deflection curve of column A (unconfined column)

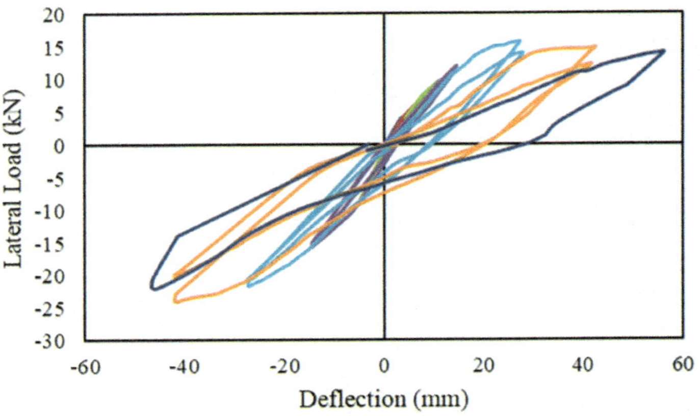

Fig. 8 Load deflection curve of column B (confined column)

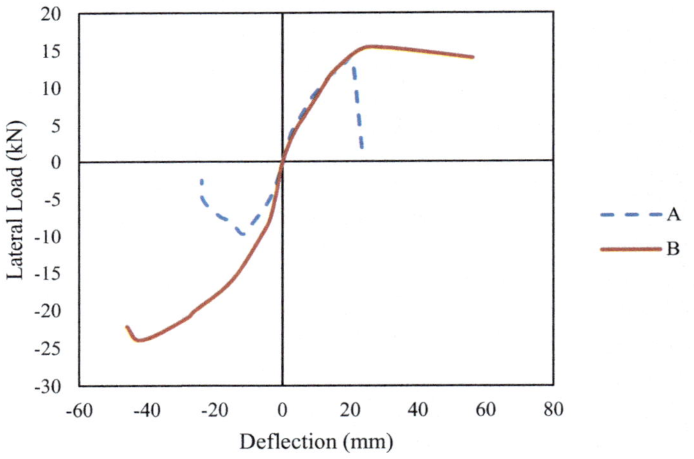

Fig. 9 Comparison between the skeleton curve of column A and B

nonlinear hysteretic loops until the end of the test for all specimens. At this stage, the displacement increased significantly, while the lateral load only increased slightly. This may be contributed by concrete cracking and spalling or even longitudinal reinforcement bar yielding.

When lateral load started to decrease, the load deformation curve of column A (control) showed a sharp decrease indicating a sudden failure. On the other hand, the smooth decreasing of lateral load with increasing displacement of the skeleton curves of confined column (column B) demonstrates that SSTT was beneficial to the ductility of the retrofitted column.

Table 2 Summary of performance of column

Column	Ultimate force, *F* (kN)	Ultimate displacement, Δ (mm)	Enhanced strength	Enhanced displacement
A	14.2	23.94		
B	23.9	56.21	68.3%	134.8%

The ultimate lateral force of column A was recorded as 14.2 kN. The ultimate lateral force of column B increased to 23.9 kN (increased by 68.3%) after being confined by steel straps. In terms of ultimate displacement, column B increased from 23.94 to 56.21 mm (increased by 134.8%) when compared with column A. It can clearly see that steel straps contributed to the strengthening in terms of strength and ductility. The summary of the result was tabulated in Table 2. The percentage of enhanced strength and displacement was calculated using Eqs. 1 and 2.

$$\text{Enhanced force}\ (\%) = \frac{F_R - F_O}{F_O} \times 100 \quad (1)$$

where F_R is the ultimate force of retrofitted column and F_O is the ultimate force of control column.

$$\text{Enhanced deformation}\ (\%) = \frac{\Delta_R - \Delta_O}{\Delta_O} \times 100 \quad (2)$$

where Δ_R is the ultimate deformation of retrofitted columns and Δ_O is the ultimate deformation of control column.

3.2 *Energy Dissipation*

Columns with high energy dissipation were desirable when subjected to earthquake [22]. Area under the load displacement curve can be calculated to know the amount of energy dissipated. Figure 10 shows the amount of energy dissipated at each ductility amplitude for the specimens. The average energy dissipated at each ductility amplitude was plotted.

There was very little energy dissipation at the initial stage of the testing work as the columns were still in good condition. However, there was an exponential increase of energy dissipation beyond 1 ductility amplitude. The energy curve of column A stopped at 2 ductility amplitude because the column had failed at that time, while loading on column B can be continued until 4 ductility amplitude due to the confinement effect of steel straps. SSTT helped the same column to achieve 1.6 times of energy dissipation and higher ductility amplitude. Column B also showed a

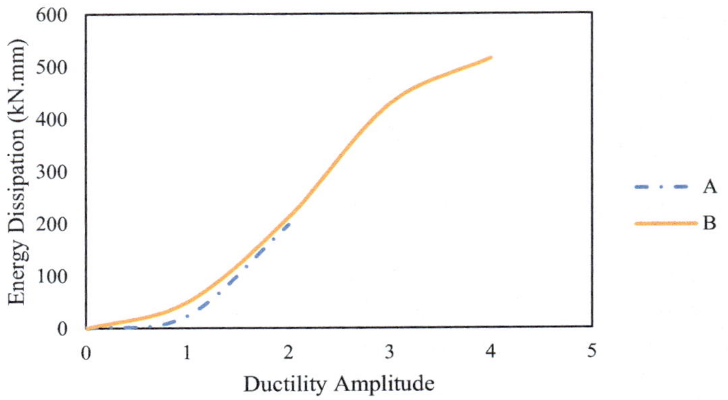

Fig. 10 Energy dissipation

slightly higher energy dissipation than column A, even at the same ductility amplitude. So, it can be concluded that steel straps had significant impact on columns energy dissipation.

3.3 Analysis of Column Cross-Section Curvature

Six LVDTs (three on each side) was set-up at 300, 600, and 1000 mm above the column footing to measure the lateral displacement. The average column cross-sections curvatures at three different heights of each cycle for column A and B were calculated. Figure 11 shows the curve of curvatures versus ductility amplitude for column A (control specimen) at three different levels above the column footing, including 0–300 mm, 300–600 mm, and 600–1000 mm; whereas the curve of curvatures versus ductility amplitude for column B (confined with steel straps) is shown in Fig. 12.

Based on the result, the curvature of the column from 0 to 300 mm was the highest compared with other heights. This is because this was the usual place for the column to undergo critical damage. The column cross-section curvature also increased with ductility amplitude. When comparing both graphs, the curvature reading for column B showed higher reading and approximately two times more than column A. This might be due to higher ductility of column B after confining with steel straps.

Figures 13 and 14 show the distribution of column curvatures along the column height above footing for column A and column B. Each ductility amplitude is represented by one line in the curve. From the curve, the specimens showed relatively high curvature concentration in the bottommost 300 mm along the column. The higher level of columns above the footings showed smaller curvature values.

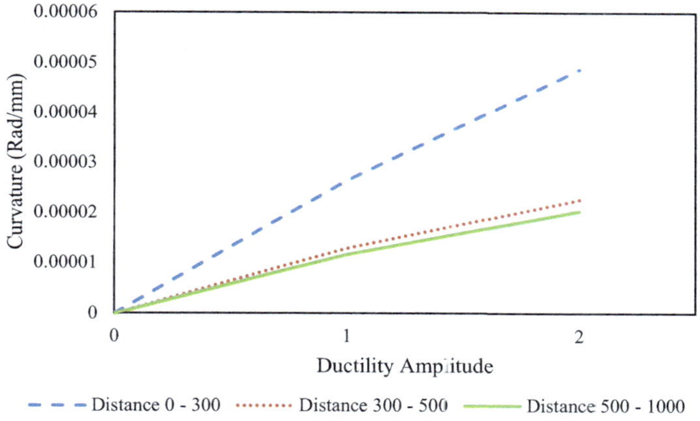

Fig. 11 Curvature versus ductility amplitude for column A

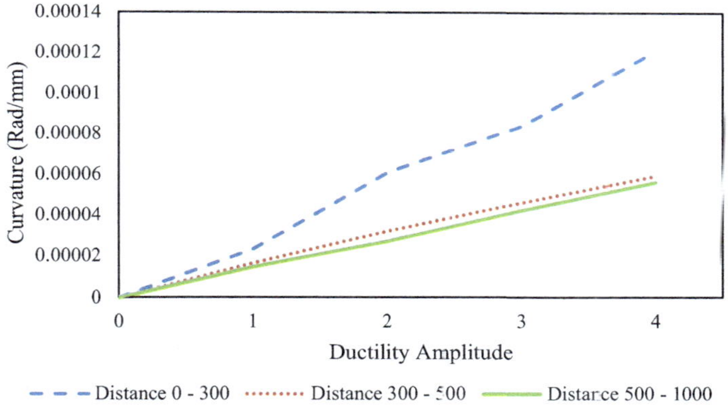

Fig. 12 Curvature versus ductility amplitude for column B

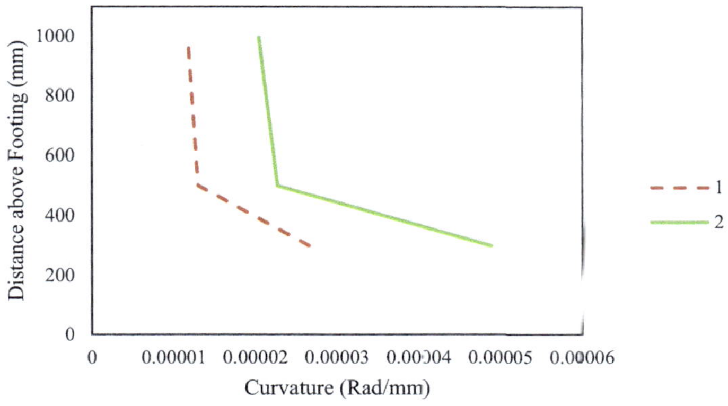

Fig. 13 Curvature distribution along column height of column A

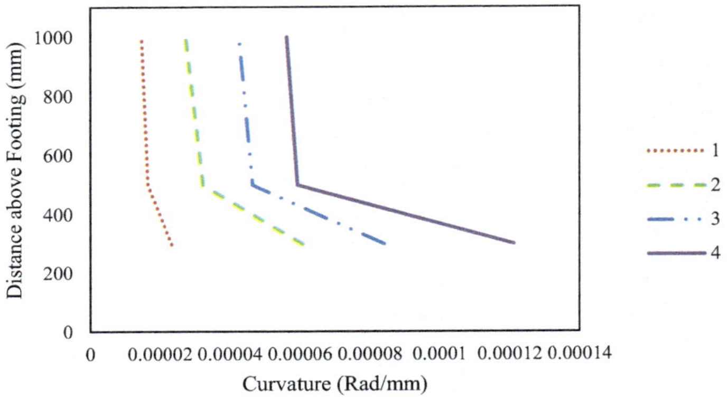

Fig. 14 Curvature distribution along column height of column B

3.4 *Analysis of Longitudinal Reinforcing Bar Strains*

Three strain gauges were installed on the most stress rebar in each column prior to casting. It measured the strain value at 150, 300, and 500 mm above the footing surface. Figures 15 and 16 show the rebar strain versus ductility amplitude for both columns measured at 150 mm as it showed the highest reading among all the readings. In this paper, positive strain refers to tensile strain while negative strain refers to compressive strain. When comparing column A and B, the maximum tensile and compressive rebar strains were seen to increase from 2889 micro strains to 12,578 micro strains. For the compressive rebar strains, it was increased from − 782 micro strains to − 11,482 micro strains.

From the result, the longitudinal bar for column A (unconfined) has not yielded when the column failed. The column had undergone critical damage of concrete spalling and failed even before the rebar managed to sustain the load and yielded. On the other hand, steel straps on the surface of column B managed to aid the rebar to reach higher strain value prior to failure. The confined column (column B) had relative higher lateral displacement that increased the chances for the concrete cracks around the rebar. Unlike column A, concrete was able to hold in place by steel straps. The unbonded region between rebar and concrete made the rebar elongation easier. Hence, it produced higher strain value and increased the ductility and energy absorption capacity in return [23].

3.5 *Analysis of Steel Straps Strains*

There were no steel straps attached at column A. Therefore, the analysis of steel straps was focused on column B. Three strain gauges were attached on the external surface of the steel straps at various locations. It is located at 124, 174, and 224 mm

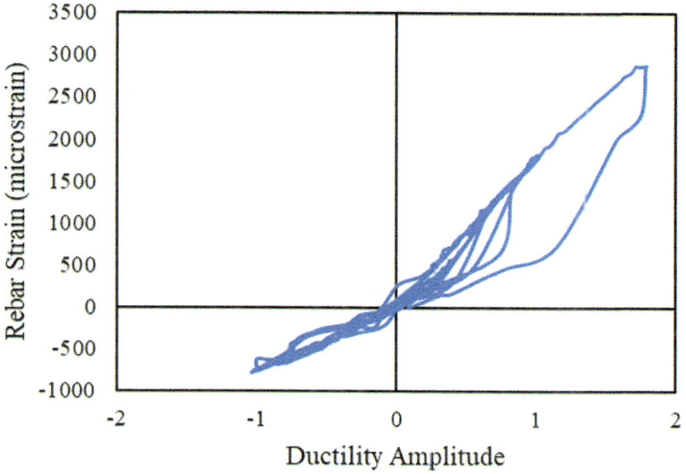

Fig. 15 Longitudinal rebar strain versus ductility amplitude for column A

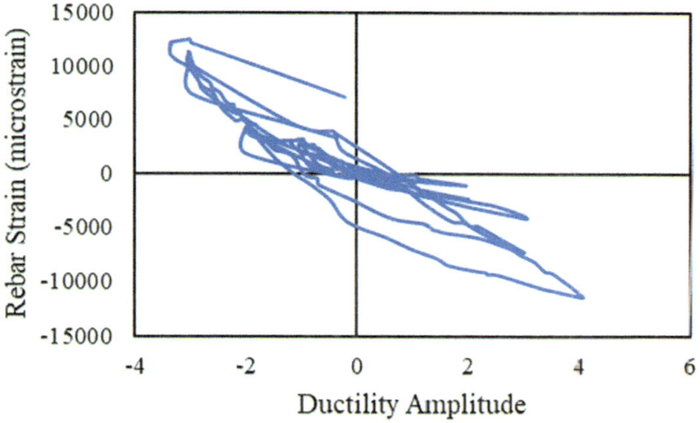

Fig. 16 Longitudinal rebar strain versus ductility amplitude for column B

above the footing surface. Figure 17 shows the steel straps strain hysteresis curve for column B. Steel straps strains increased significantly as the applied lateral drift increased. It recorded the highest strain value of 2655 micro strain. Figure 18 shows steel straps strains distribution along column height at different ductility amplitude for column B. It showed that the highest steel straps strain value focused on the lowest part of the column (124 mm) when compared with other parts of the column.

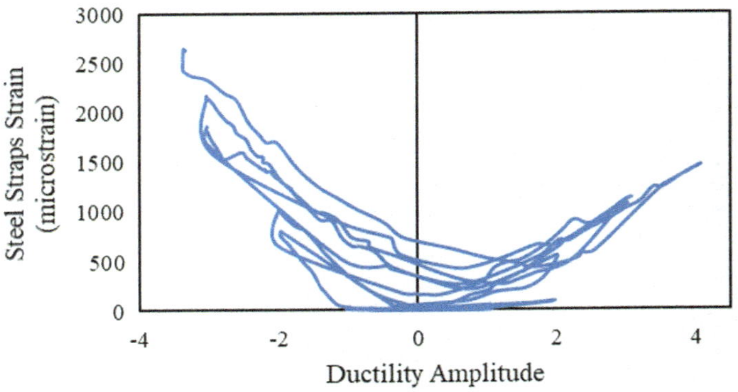

Fig. 17 Steel straps strains versus ductility amplitude for column B

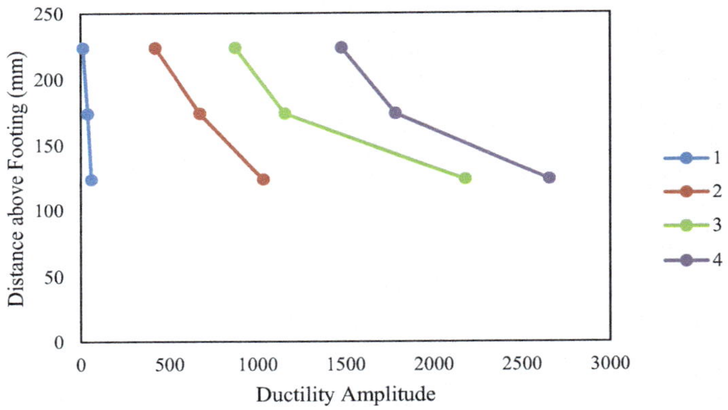

Fig. 18 Steel straps strains distribution along column height at different ductility amplitude for column B

4 Conclusion

This paper focused on the effect of steel straps in retrofitting columns under lateral cyclic load including torsion. From the result obtained, column B confined using steel straps showed good performance when compared with column A (control specimen). The results were summarised as follows:

- After the column was confined, the ultimate strength of column increased from 14.2 to 23.9 kN (68.3%), while ultimate displacement increased from 23.94 to 56.21 mm (134.8%).
- SSTT aided the column to dissipate 1.6 times more energy and reach higher ductility amplitude.

- The column cross-section curvature for column B showed higher reading and approximately two times more than column A.
- The curvature and the rebar strain gauges showed highest reading at the bottom part of the column as most of the damage occurred at this region. The rebar strain gauges value increased significantly after being confined.

Therefore, SSTT can be considered as an excellent alternative to strengthen the inadequately designed column.

Acknowledgements The authors gratefully acknowledge the support for this research from the Ministry of Higher Education, Malaysia under the Fundamental Research Grant Scheme (FRGS/1/2019/TK01/UTM/02/14) for the financial support. Financial supports were also received from the Universiti Teknologi, Malaysia (RMC) Grant No: [Q.J130000.2651.16J31] and [Q.J130000.2451.09G87], School of Civil Engineering, Faculty of Engineering, Universiti Teknologi, Malaysia.

References

1. Belarbi, A., Ayoub, A., Greene, G., Prakash, S., Mullapudi, R.: Seismic performance of reinforced concrete bridge columns subjected to combined loading including torsion. Struct. Eng. Res. Front., pp. 1–16 (2007)
2. He, R., Sneed, L.H., Belarbi, A.: Torsional repair of severely damaged column using carbon fiber-reinforced polymer. ACI Struct. J. **111**(3), (2014)
3. Ghobarah, A., Biddah, A., Mahgoub, M.: Rehabilitation of reinforced concrete columns using corrugated steel jacketing. J. Earthquake Eng. **1**(04), 651–673 (1997)
4. Lehman, D.E., Gookin, S.E., Nacamuli, A.M., Moehle, J.P.: Repair of earthquake-damaged bridge columns. Struct. J. **98**(2), 233–242 (2001)
5. Ye, L., Yue, Q., Zhao, S., Li, Q.: Shear strength of reinforced concrete columns strengthened with carbon-fiber-reinforced plastic sheet. J. Struct. Eng. **128**(12), 1527–1534 (2002)
6. Zhou, X., Liu, J.: Seismic behavior and shear strength of tubed RC short columns. J. Constr. Steel Res. **66**(3), 385–397 (2010)
7. Truong, G.T., Kim, J.C., Choi, K.K.: Seismic performance of reinforced concrete columns retrofitted by various methods. Eng. Struct. **134**, 217–235 (2017)
8. Tirasit, P., Kawashima, K.: Seismic performance of square reinforced concrete columns under combined cyclic flexural and torsional loadings. J. Earthquake Eng. **11**(3), 425–452 (2007)
9. Belarbi, A., Prakash, S., You, Y.M.: Effect of spiral reinforcement on flexural-shear-torsional seismic behavior of reinforced concrete circular bridge columns. Struct. Eng. Mech **33**(2), 137–158 (2009)
10. Nie, J.G., Wang, Y.H., Fan, J.S.: Experimental research on concrete filled steel tube columns under combined compression-bending-torsion cyclic load. Thin-Walled Struct. **67**, 1–14 (2013)
11. He, R., Sneed, L.H., Belarbi, A.: Rapid repair of severely damaged RC columns with different damage conditions: an experimental study. Int. J. Concr. Struct. Mater. **7**(1), 35–50 (2013)
12. Wang, Y.H., Wang, Y.Y., Hou, C., Zhou, X.H., Deng, R., Lan, Y.S., Luo, W., Kong, W.B.: Combined compression-bending-torsion behaviour of CFST columns confined by CFRP for marine structures. Compos. Struct. **242**, 112181 (2020)
13. Taerwe, L.: Brittleness versus ductility of high strength concrete. Struct. Eng. Int. **1**(4), 40–45 (1991)
14. Prakash, S.S., Belarbi, A., Ayoub, A.: Behavior of RC bridge columns under cyclic flexural-torsional loadings with moderate shear. In: 14th World Conference on Earthquake Engineering (WCEE) (2008)

15. Bayrak, O., Sheikh, S.A.: High-strength concrete columns under simulated earthquake loading. ACI Struct. J. **94**(6), 708–717 (1997)
16. Paultre, P., Eid, R., Robles, H.I., Bouaanani, N.: Seismic performance of circular high-strength concrete columns. ACI Struct. J. **106**(4), 395–404 (2009)
17. Prakash, S., Belarbi, A., You, Y.M.: Seismic performance of circular RC columns subjected to axial force, bending, and torsion with low and moderate shear. Eng. Struct. **32**(1), 46–59 (2010)
18. Liu, J., Zhou, X., Zhang, S.: Seismic behaviour of square CFT beam–columns under biaxial bending moment. J. Constr. Steel Res. **64**(12), 1473–1482 (2008)
19. Ma, C.K., Apandi, N.M., Sofrie, C.S.Y., Ng, J.H., Lo, W.H., Awang, A.Z., Omar, W.: Repair and rehabilitation of concrete structures using confinement: a review. Constr. Build. Mater. **133**, 502–515 (2017)
20. Ma, C.K., Awang, A.Z., Omar, W., Liang, M., Jaw, S.W., Azimi, M.: Flexural capacity enhancement of rectangular high-strength concrete columns confined with post-tensioned steel straps: experimental investigation and analytical modelling. Struct. Concr. **17**(4), 668–676 (2016)
21. Moghaddam, H., Samadi, M., Pilakoutas, K., Mohebbi, S.: Axial compressive behavior of concrete actively confined by metal strips; part A: experimental study. Mater. Struct. **43**(10), 1369–1381 (2010)
22. Youssf, O., ElGawady, M.A., Mills, J.E.: Static cyclic behaviour of FRP-confined crumb rubber concrete columns. Eng. Struct. **113**, 371–387 (2016)
23. Saadatmanesh, H., Ehsani, M.R., Jin, L.: Seismic retrofitting of rectangular bridge columns with composite straps. Earthq. Spectra **13**(2), 281–304 (1997)

Experimental Report on Compression Loading Test of Tempered Glass Panels

Saddam Hussain, Pei-Shan Chen, Yuta Matsuno, and Baoxin Liu

Abstract There is indeed a significant growth of glass panels in the modern contemporary building and construction industry. Primary load-bearing members used in glass structures, such as columns consisting of glass panels, are frequently subjected to in-plane loading. Its allowable stress for safety design appears to be a significant subject. In general, fragile and thin members are more susceptible to compression. The present paper focuses on the safety design of tempered glass panels under in-plane compression loading. The authors explored the mechanical features through in-plane compressive loading tests, including ultimate stress and Euler-Johnson buckling. Glass panels with high slenderness ratios satisfy the Euler buckling formula, while a lower slenderness ratio is not. Secondly, for the design of compression-resistant glass components, developing the limit for the under Euler and Johnson buckling curves based on a low slenderness ratio is a practical option, and so does the direct secondary stress analysis. The experimental results show that the ultimate loads of specimens with high slenderness ratios satisfy Euler's critical load formula.

Keywords Compression loading · In-plane loading · Tempered glass · Allowable stress · Slenderness ratio · Glass structures

S. Hussain · P.-S. Chen (✉) · Y. Matsuno · B. Liu
Department of Civil Engineering and Architecture, Kyushu Institute of Technology, Kitakyushu 804-0015, Japan
e-mail: chen@civil.kyutech.ac.jp

S. Hussain
e-mail: saddam.hussain372@mail.kyutech.jp

Y. Matsuno
e-mail: matuno.yuta569@mail.kyutech.jp

N. Sabtu (ed.), *Proceedings of AWAM International Conference on Civil Engineering 2022—Volume 2*, Lecture Notes in Civil Engineering 385,
https://doi.org/10.1007/978-981-99-6018-7_21

1 Introduction

Glass, especially tempered glass, can resist high compression stresses and has a unique architectural attraction due to its transparency. As a result, glass panels are increasingly used as (in-plane and out-plane) load-bearing elements such as beams, columns, and wall panels. However, it is a fundamental requirement to determine structural safety such that the stress of each part of the structure should be less than the allowable stress. Therefore, the theoretical and experimental research on the loading capacity of glass panels, especially tempered glass panels, has aroused wide concern recently [1–5].

Previous research on the buckling behavior of structural glass panels under boundary or loading conditions has recently become the theme of several articles. Analytical approaches for conservative verification have been proposed using the limit state design approach. Chen [1] in buckling and compression tests performed on toughened glass columns are presented. Luible [4], e.g., focused on the load-carrying characteristics of panels and conducted experiments to assess their reasonable response. The most critical parameters influencing buckling strength are these research-related in-plane compressive loading with two pinned column ends. Belis [5] focused on laminated and monolithic glass in-plane axial loading with pin end. Numerous peculiarities (production tolerances, initial defects, interlayer characteristics, and so on) made applying well-known design methodologies challenging. Blaauwendraad [6] analyzed three different expressions of the critical load available in the literature and ascertained that only the corresponding, respectively, to the sum of the buckling strength of the glass sheets and the buckling strength of a laminated panel with an infinite stiff foil. Proposed an analytical check of the maximum design axial loading of compression simply supported end and the corresponding deflection/tensile stress in the mid-span cross section of the column/panel. In-plane compressive loads introduced through boreholes were studied [7]. According to Danijel et al. [8], the compressive stresses became constant over the glass panel width at a certain distance from the bolt hole in the plane loading introduction point. The main reason for this is that stability issues typically govern the mechanical behavior of pressed glass components: several authors have demonstrated in recent years that buckling problems (flexural buckling, torsional buckling, lateral-torsional buckling, plate buckling, and shear buckling) are critical in determining the overall load-bearing capacity of structural elements. References [9, 10] and of glass compression members in particular [11, 12]. A sinusoidal wave pattern was found to determine the initial imperfection in glass beam-like elements. To enable the strength checking of LG beams columns strength [13, 14], Analytical processes and consolidated authentication methods are available in standards for the load-carrying components composed of conventional materials; however, this is not the case for the glass [15, 16].

However, in the case of glass structures, few experimental results were reported on the compression loading of simple supported single-layer tempered glass panels. The calculation method for allowable tress of glass panels subjected to in-plane loading is still important for further research. Therefore, this paper concentrates on the tempered glass panels subjected to compression. Tempered glass under the condition of supporting two sides glass is performed, so there is presumption regarding the buckling of the line element and its behavior. Because surface pre-stress is introduced in tempered glass panels during manufacture, mechanical behavior is investigated. This experimental investigation also shows if the surface pre-stresses alter the glass panel's compression capacity regarding slenderness ratio. As a result, this study will propose the allowable stress for safe design based on experimental results in the future.

2 The Experimental Setup and Specimens

This session explains the experimental setup and specimens of the tempered glass panels, which will be the foundation of research for the next step.

2.1 *Preliminary Test Setup*

Compression tests were performed on tempered glass panels with four different types of thicknesses (4, 6, 8, and 10 mm). The test setup arrangement represents the ideal column buckling with simple supporting at the upper and lower edges. And the steel supports frames were fabricated to curb the load that can be acted downwards vertically to achieve this goal. Specimens were clamped at both end-edges using special steel supports such as the adapter in Fig. 1a. The adapter was mounted on the top and bottom edges of the specimens. Aluminum connectors have dimensions of 300 × 300 mm (length and width) to hold the load on top and bottom. For the slit dimensions for the adapter shown in Fig. 1b, the outside width is 20 mm, the inside width is 10 mm, and the depth is 6 mm. In Fig. 1b, the adapter penetrates itself into the connector as top/bottom of the adapter. Figure 1c shows the inserted rubber material as a cushioning material between the glass and the adapter. To prevent the horizontal movement of glass panels, the frame is designed with bolt holes and welded into the external support frame.

The methods for supporting the structural glass edge are shown in Fig. 1d. This design technique involves connecting the glass sides edges to a support frame and sandwiching a sealant material between the support frame and the glass to avoid out-of-plane or in-plane movement. The Amsler testing machine loads the test panels, which is supported as shown in Fig. 1. With a strain gauge, measure the vertical strains of specimens, and the gauge is attached to both sides of the panels in the center. The displacement gauge measures the displacement and is attached at the center of the glass panel on both sides. In Fig. 1c, the loading below the adapter was measured using a load cell.

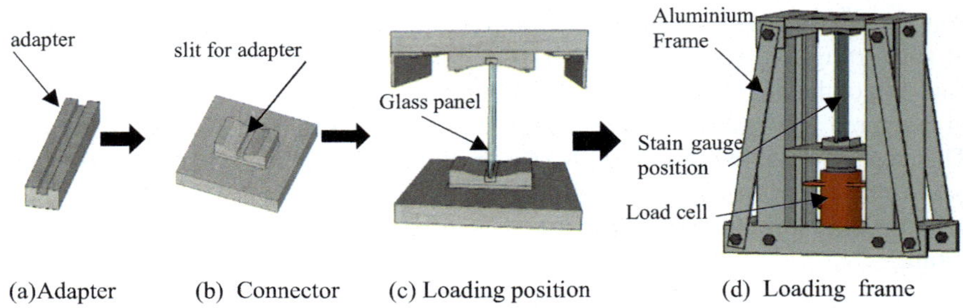

Fig. 1 Experimental setup

2.2 Specimens Description

The number, dimensions, and size of the tempered glass specimens are selected by considering the slenderness ratio. When the experimental values are graphed, they are evenly distributed along the theoretical buckling line. The details are given in Table 1. Static loading studies are done on 30 test pieces of 10 types of slenderness as TG4 (1, 2, 3), TG6 (1, 2, 3), TG8 (1, 2, 3), and TG10 (1, 2, 3), and the frame is designed to limit deviations other than the loading direction.

The symbols in Table 1, t: thickness (mm), b: width (mm), h: height (mm), I_x: moments of inertia of area (mm^4), A: the cross-sectional area (mm^2), i: cross secondary section radius(mm), l_k: the buckling coefficient member length to consider (mm), λ: the slenderness ratio.

Table 1 Specimens of the glass panel with different slenderness ratios

Name	t	b	h	$I_x = bt^3/12$	$A = bt$	$i = \sqrt{I_x/A}$	$l_k = h$	$\lambda = l_k/i$
TG4 ①	4	100	100	533	400	1.154	100	87
TG4 ③	4	100	350	533	400	1.154	350	303
TG6 ⑤	6	100	100	1800	600	1.732	100	58
TG6 ⑥	6	100	375	1800	600	1.732	375	217
TG8 ⑦	8	100	250	4267	800	2.309	250	108
TG8 ⑧	8	100	400	4267	800	2.309	400	173
TG1 ⑨	10	100	100	8333	1000	2.887	100	35
TG1 ⑩	10	100	400	8333	1000	2.887	400	139

3 Experimental Results

This session explains the stress–strain relationship and Euler buckling theory justification of tempered glass panels under in-plane loads. The followings are experimental tests, as shown in Fig. 2.

3.1 Stress–Strain and Displacement Relationship

A compression experiment is carried out over the specimens shown in Fig. 2. Young's modulus for tempered glass is 71,600 N/m^2. In addition, the influence occurs when the specimens are loaded, and the aluminum alloy plate of the upper plate is significantly tilted from the graph of the state at loading and various experimental values.

As the load increased, the glass flexed significantly and then immediately before it cracked, the broken sound in the glass was observed the moment before it cracked. Particularly excellent results are shown below of the stress–strain relationship and the stress-horizontal displacement relationship until this experiment breaks the specimen glass. These are shown in Fig. 1b order, in which the investigation was completed. Moreover, the experimental values divided by the compressive stress are located on these two curves.

For Fig. 3. TG4, the stress–strain relationship and stress-horizontal displacement relationship obtained from the experiment are maximum loads: 46,933.6 N, maximum stress: 124.24 N/mm^2 maximum strain: 6.14 × 10^{-3}, and maximum horizontal displacement: 2.34 mm.

As shown in Fig. 4. for TG6, the stress–strain relationship and the stress-horizontal displacement relationship obtained from the experiment are maximum loads: 4,118,667 N, maximum stress: 244.24 N/mm^2, maximum strain: 8.20 × 10^{-3}, and maximum horizontal displacement: 5.54 mm.

Nevertheless, as shown in Fig. 5 TG8, the dimensional values and installation of the glass test panel and the stress–strain relationship and the stress-horizontal displacement relationship obtained from the experiment are shown as a maximum

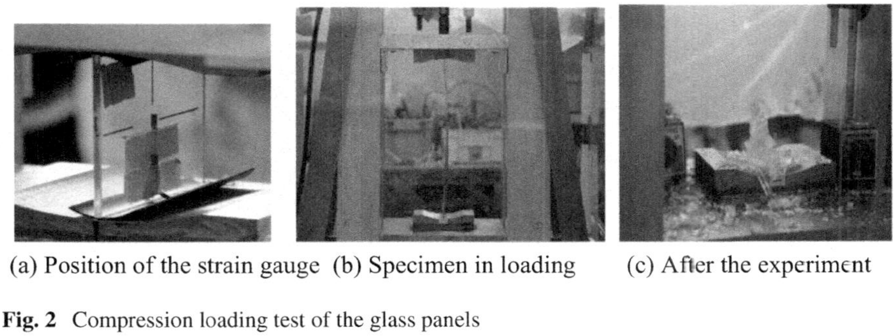

(a) Position of the strain gauge (b) Specimen in loading (c) After the experiment

Fig. 2 Compression loading test of the glass panels

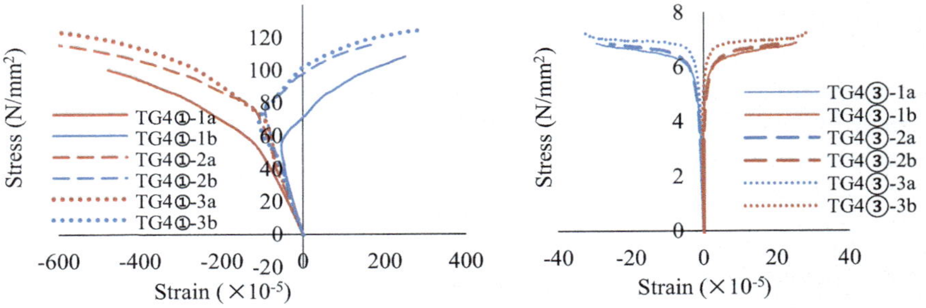

(a) Stress-Strain curves TG4 (b) Stress-Strain curves TG4

Fig. 3 Stress–strain curves of TG4

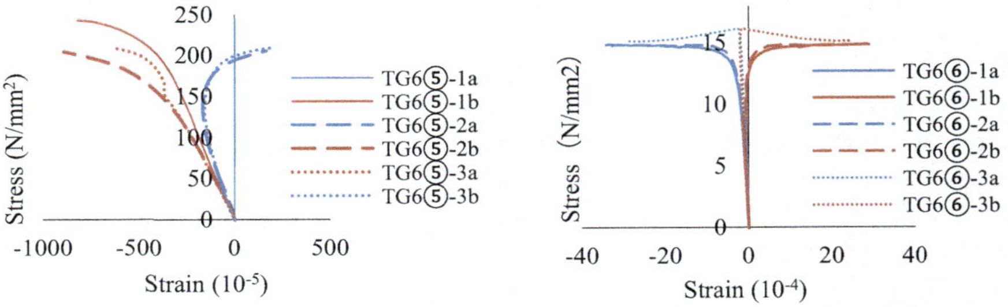

(a) Stress-Strain curves TG6⑤ (b) Stress-Strain curves TG6⑥

Fig. 4 Stress–strain curves of TG6

load: 37,600.17 N, maximum stress: 24.44 N/mm^2, maximum strain: 3.50 × 10^{-4}, and maximum horizontal displacement: 5.46 mm.

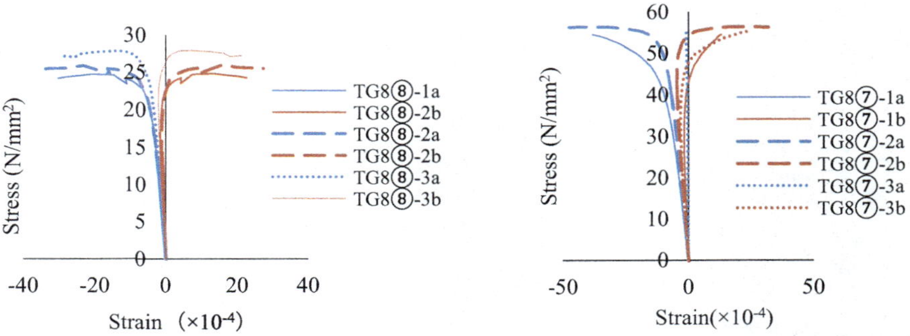

(a) Stress-Strain curves TG8⑧ (b) Stress-Strain curves TG8⑦

Fig. 5 Stress–strain curves of TG8

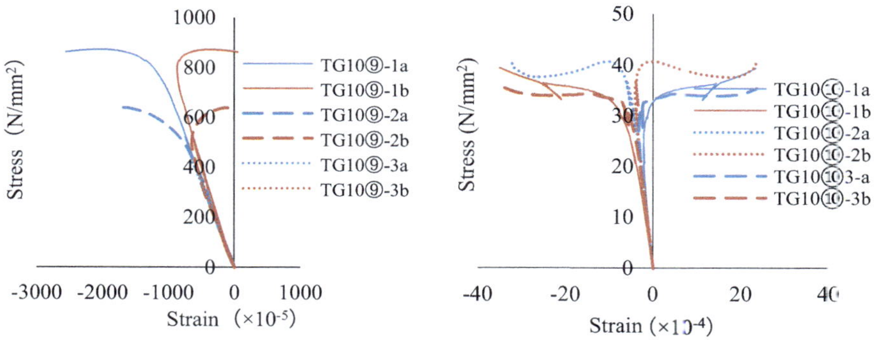

(a) Stress-Strain curves TG10⑨

(b) Stress-Strain curves TG10⑩

Fig. 6 Stress–strain curves of TG10

However, in Fig. 6. for samples TG.6, the dimensional values and installation of the glass test panel and the stress–strain relationship and the stress-horizontal displacement relationship obtained from the experiment are shown as a maximum load: 38,600.17 N, maximum stress: 39.44 N/mm^2, maximum strain: 3.50×10^{-3}, and maximum horizontal displacement: 9.26 mm in Fig. 7.

Besides determining the ultimate strength curve of the glass material in Fig. 8, the high scope of application of glass panel of dimensions motivates interest in investigating their compression. The investigation of the ultimate stress of the panels is to confirm whether the tempered glass is helpful to the theory of compression described in this section, as given in Table 2. In the future, it helps to evaluate the allowable stress. Furthermore, the state of loading and after loading justifies the tempered glass characteristics. Concerning the selection methodology of the test, body glass panels, the number of panels and directions, and the slenderness ratio were considered. During graphing the experimental values, the tempered glass panels were chosen so that the slenderness ratio would be constant to divide it equally on the buckling line.

3.2 The Buckling Theory Justification

In the experiments, the glass panels are performed with a simple (pin) supporting at upper and lower edges, so it is assumed that the line element (column) buckling, and Euler's Formula can predict its behavior. Inelastic buckling theory of line elements, the slenderness ratio of the member does not reach the limit slenderness ratio λ; the member causes inelastic buckling according to the Johnson buckling theory. σ_y is yield stress but such as glass has no yield stress then used $\overline{\sigma}$ average of the ultimate stress.

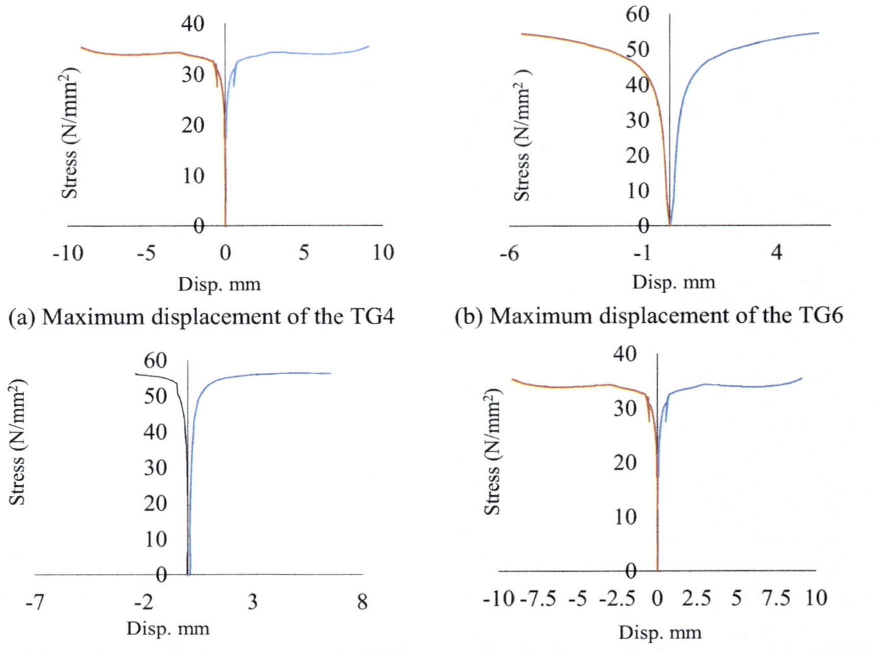

(a) Maximum displacement of the TG4

(b) Maximum displacement of the TG6

(c) Maximum displacement of the TG8

(d) Maximum displacement of the TG10

Fig. 7 Maximum stress-displacement curves

Fig. 8 Experimental results of the ultimate strength for slenderness ratio

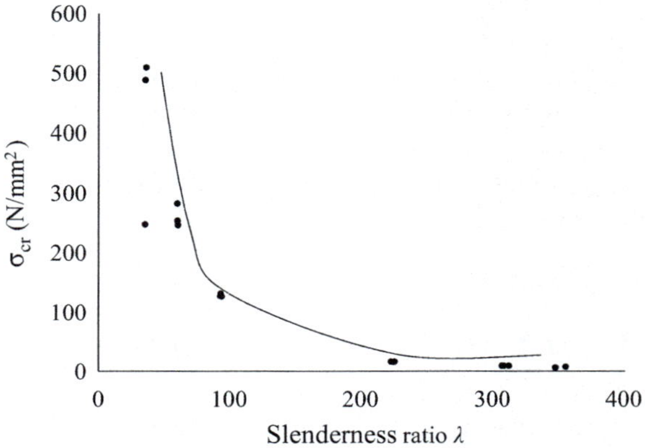

$$\sigma_{cr} = \sigma_y - \frac{\sigma_y^2}{4\pi^2 E}\lambda^2 \rightarrow \sigma_{cr} = \overline{\sigma} - \frac{\overline{\sigma}^2}{4\pi^2 E}\lambda^2 \tag{1}$$

Elastic buckling theory of wire elements (Euler's Formula), the slenderness ratio of the member exceeds the critical slenderness ratio λ; the member causes elastic

Table 2 Maximum values of the glass panels with different slenderness ratio

Samples	TG4	TG6	TG8	TG10
Max. load (N)	46,933.6	141,867	37,600.17	38,600.17
Max. stress (N/mm^2)	124.24	244.24	24.44	39.44
Max. strain	6.14×10^{-3}	8.20×10^{-3}	3.50×10^{-4}	3.50×10^{-3}
Max. disp. (mm)	2.34	5.54	5.46	9.26

buckling according to Euler's buckling theory.

$$\sigma_{cr} = \pi^2 \frac{EI}{\lambda^2} \tag{2}$$

Figure 9 shows the results of comparing the theoretical curves and experimental values of Euler and Johnson's buckling theory, which are explained in detail. The high slenderness ratio satisfied the Euler buckling theory, and the low slenderness ratio did not. For maximum results are mentioned above in Table 2. In addition, based on this experiment, tempered glass will be calculated. In the case of glass with a small slenderness ratio, the strength varies widely as a loading experiment. Euler's formula is only appropriate for long slenderness satisfied the Euler buckling theory, in which the buckling process takes precedence over direct compression, and hence compressive stress is neglected. Considerable variance when buckling failure can be predicted when the slenderness ratio is small of a thin solid approach 100. At lower slenderness ratios, panels made of more rigid and brittle materials will buckle. Euler's critical load formula clearly illustrates that a panel failure is determined by arranging the panel's ends.

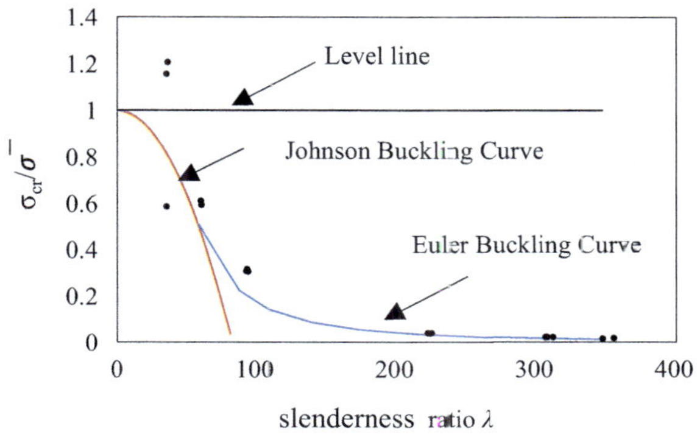

Fig. 9 Buckling results from slenderness ratio, according to Johnson and Euler

4 Discussion

In the case of glass with a small slenderness ratio, the strength varies widely as in the static loading experiment, and it is necessary to conduct many experiments in the future and make a statistical judgment. In addition, the reason for the fracture stress of 10 mm thick glass with the smallest slenderness ratio but the fracture stress increases due to repeated loading. It is difficult to think that it was done. Overall, in the current experiments, the deformation behavior and stress values of the line elements supported on the upper and lower sides of tempered glass have stress in the cross section according to Euler's buckling theory and Johnson's buckling theory. It turned out that tempered glass could be applied. The results of this experiment suggest that Euler's buckling theory can be applied to glass with a large slenderness ratio, but glass with a small slenderness ratio has Euler and Johnson's buckling theory cannot be applied.

Therefore, in this paper, the slenderness ratios 35, 57, and 87 are not satisfied the Euler buckling theory and found the allowable stress for that specimen with the help of the normal distribution method. However, the deformation behavior and stress vary widely.

5 Conclusion

The present paper focuses on the safety design of single-layer tempered glass panels under in-plane compression loading. In-plane compressive loading tests were performed on various slenderness mentioned in Table 1. The buckling theory is challenged for three low slenderness: slenderness ratio 35, 58, and 87 of tempered glass, in the future compression of area for the 35, 58, and 87 initiations of the safe design. A high slenderness ratio justifies the Euler buckling, but lower slenderness ratios do not. Developing the limit line curve under Euler and Johnson buckling curves or doing direct secondary stress analysis is alternatives for designing compression-resistant glass components. This paper also provides the foundation for future research on the allowable stress of a tempered glass panel with various slenderness.

Acknowledgements I would like to express sincere gratitude to the research advisor, and this paper was developed with the Laboratory of Building Structure at Kyushu Institute of Technology.

References

1. Chen, P.-S., Tsai, M.-T.: On configuration and structural design of frameless glass structures. In: Structures and Architecture: Bridging the Gap and Crossing Borders, pp. 628–637. CRC Press (2019)
2. Bedon, C., Amadio, C.: Design buckling curves for glass columns and beams. Proc. Inst. Civ. Eng.-Struct. Build. **168**(7), 514–526 (2015)
3. Amadio, C., Bedon, C.: Buckling verification of laminated glass elements in compression. J Civ Eng Sci **1**(3), 90–101 (2012)
4. Hussain, Future importance and demand of frameless glass structure. World J. Adv. Sci. Res. **4**(2), 1–19 (2021)
5. Luible, A.: Stabilität von Tragelementen aus Glas, Dissertation Thèse N0 3014, EPFL Lausanne, Switzerland (2004)
6. Luible A.: Stabilität von tragelementen aus glas. Doctoral thesis EPFL 3014. Lausanne: Ecole Polytechnique Fédérale de Lausanne (2004)
7. Luible, A., Crisinel, M.: Auf Biegen und Brechen. tec21, No., pp. 29–36. Zürich: SIA (2002)
8. Belis, J., Mocibob, D., Luible, A., Vandebroek, M.: On the size and shape of initial out-of-plane curvatures in structural glass components. Constr. Build. Mater. **25**, 2700–12 12/2002
9. Blaauwendraad, J.: Buckling of laminated glass columns. Heron T.U. Delft, 52(1–2) (2007)
10. Amadio, C., Bedon, Buckling of laminated glass elements in compression. J. Struct. Eng. **137**(8), 803–810 (2011)
11. Luible, A., Crisinel, M.: Buckling strength of glass elements in compression. Struct. Eng. Int. (2), 120–125 (2004)
12. Huang, X., Cui, M., Liu, Q., Nie, J.: An experimental study on the structural behavior of laminated glass members under combined axial compression and in-plane bending. Mater. Struct. **53**, 1–14 (2020)
13. Pešek, O., Horáček, M., Melcher, J.: Experimental verification of the buckling strength of structural glass columns. Procedia Eng. **161**, 556–562 (2016)
14. Aiello, S., Campione, G., Minafò, G., Scibilia, N.: Compressive behavior of laminated structural glass members. Eng. Struct. **33**(12), 3402–3408 (2011)
15. Sonck, D., Belis, J.: Elastic lateral-torsional buckling of glass beams with continuous lateral restraints. Glass Struct. Eng. **1**(1), 173–194 (2016)
16. Huang, X., Cui, M., Liu, Q., Nie, J.: An experimental study on the structural behaviour of laminated glass members under combined axial compression and in-plane bending. Mater. Struct. **53**, 39 (2020)
17. Liess, J.: Bemessung druckbelasteter Bauteile aus Glas, Dissertation, Universität Kassel, Germany (2001)
18. Luible, A., Crisinel, M.: Plate buckling of glass panels. In: Glass Proceedings Days (2005)

Experimental Study of Reinforced Concrete Beam-Column Joint Using Pre-tensioned Stiffened Steel Angles

Chin-Boon Ong, Chee-Loong Chin, Chau-Khun Ma, Jia-Yang Tan, Abdullah Zawawi Awang, and Wahid Omar

Abstract Retrofit is one of the effective and sustainable ways to upgrade seismically deficient structures. The beam-column joint is an important structural member that should be prevented from damaging during a seismic event. An innovative retrofit technique named pre-tensioned stiffened steel angles had been proposed to retrofit the joint shear deficient beam-column joints which aimed to achieve the "strong column-weak beam" design philosophy. Two half-scale exterior joints without joint transverse reinforcement had been constructed and tested under cyclic load. The results showed that this proposed technique successfully eliminated the joint shear failure and resulted in the beam flexural failure as well as enhanced its seismic properties. Hence, this proposed retrofit technique is effective to retrofit the seismically deficient beam-column joints.

Keywords Retrofit · Pre-tensioned stiffened steel sngles · Beam-column joint · Plastic hinge relocation · Cyclic load

1 Introduction

Past seismic events [1–6] not only caused damage to the structures but a great impact on the economy and life losses as well [7, 8]. It is not always sustainable and economically efficient to demolish and rebuild those constructed non-seismically designed structures in the seismic active zones. For this reason, retrofitting the structure or structural members becomes a sustainable solution. In the field investigation of the damaged structures after the seismic event, it was evident that the non-seismically detailed beam-column joints caused major damage to the structures [9]. Hence, the

C.-B. Ong · C.-L. Chin · J.-Y. Tan · A. Z. Awang · W. Omar
Faculty of Civil Engineering, Universiti Teknologi Malaysia, 81310 Skudai, Malaysia

C.-K. Ma (✉)
Forensic Engineering Centre, Institute for Smart Infrastructure and Innovative Construction, Universiti Teknologi Malaysia, 81310 Skudai, Malaysia
e-mail: machaukhun@utm.my

N. Sabtu (ed.), *Proceedings of AWAM International Conference on Civil Engineering 2022—Volume 2*, Lecture Notes in Civil Engineering 385,
https://doi.org/10.1007/978-981-99-6018-7_22

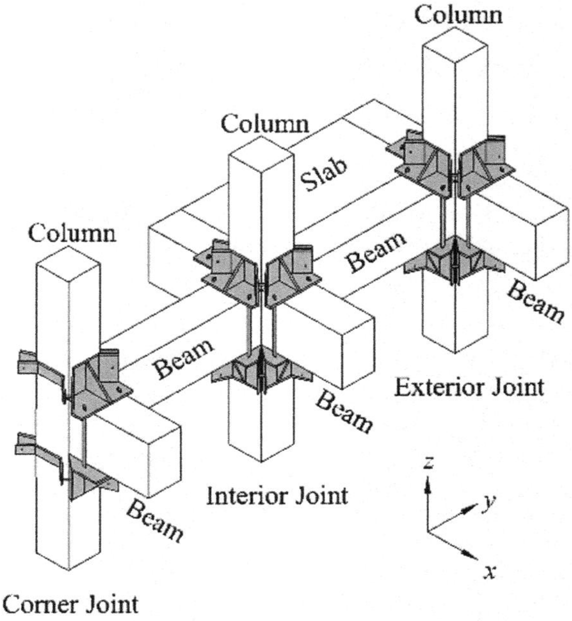

Fig. 1 3D view of the proposed technique (highlighted) in a structure

retrofit of the beam-column joint is vital to prevent the structure from collapsing during a seismic event.

The seismic design codes of practice [10–12] have highlighted the "strong column-weak beam" concept in seismically designed structures. This concept is important to ensure the column together with the joint region that beams connecting to it remained intact during and after the seismic events. To this end, a retrofit strategy, namely pre-tensioned stiffened steel angles was proposed to retrofit the joint shear deficient beam-column joints. This proposed technique is aimed to encourage beam failure instead of joint shear failure as well as enhance the seismic properties of the deficient beam-column joint. Therefore, a "strong column-weak beam" concept can be achieved in the structure. Besides, this proposed technique is able to retrofit various types of beam-column joints, such as exterior joint, interior joint, and corner joint as shown in Fig. 1. This proposed technique was specially designed to enable retrofit around the columns. In other words, the *x*-direction and *y*-direction of the structure can be retrofitted using this proposed technique as illustrated in Fig. 1.

2 Literature Review

Pampanin et al. [13] proposed the double diagonal haunch elements to retrofit the plain bar constructed beam-column joints. The diagonal haunch elements were placed on the top and bottom of the beam. This technique was able to transform the joint shear failure into the beam failure. Dang, Dinh [14], and Truong et al. [15] used a single diagonal haunch element as the retrofit strategy on deficient beam-column joints. The single diagonal haunch element proved able to enhance the seismic properties of the retrofitted beam-column joints. It was found that this single haunch element encouraged the beam failure, although hairline cracks appeared in the joint region. A similar failure mode can be observed in the recent experimental study done by Kanchanadevi, Ramanjaneyulu [16]. Moreover, the fully fastened haunch elements were proposed by Marchisella et al. [17] for retrofitting the deficient beam-column joints. The results confirmed that these different types of haunch elements were able to transform the failure mode from the joint into the beam, as well as enhance the seismic properties of retrofitted beam-column joints.

Torabi, Maheri [18] repaired the seismically deficient beam-column joint and transformed the joint shear failure into the beam failure using the stiffened steel plate. Shafaei et al. [19] introduced a technique, namely joint enlargement using pre-stressed steel angles to retrofit the beam-column joints. This technique was adopted by Adibi et al. [20] to retrofit beam-column joints constructed with plain reinforcement bars. Recently, Maddah et al. [21] and Khodaei et al. [22] applied the retrofit technique to the beam-column joints with a slab. These results showed that the haunch elements, stiffened steel plate, and steel angles were effective to achieved beam failure mode instead of joint shear failure. Besides, the seismic properties of the retrofitted beam-column joints were enhanced as well. It is noted that the beam failure occurred far away from the column face to the cut-off point of the technique used.

3 Methodology

Two half-scaled exterior beam-column joints, namely the control specimen and specimen J-FST were constructed under the same reinforcement details and dimensions in this experimental work. These specimens represent the non-seismically designed beam-column joints without joint transverse reinforcement in accordance with BS 8110 [23]. Figure 2 showed the details of the sizes and reinforcement of the specimens.

3.1 Materials Properties

The specimens were cast in a single batch of ready-mix concrete with characteristic concrete cylindrical strength of 30 MPa. Six 150 mm sides of concrete cubes were prepared for determining the concrete strength. The average compressive strength of concrete at the age of 28th days was 56.01 MPa.

The main longitudinal reinforcements (T10) used were 10 mm diameter high tensile steel bar, while the transverse reinforcements (R6) used were 6 mm diameter

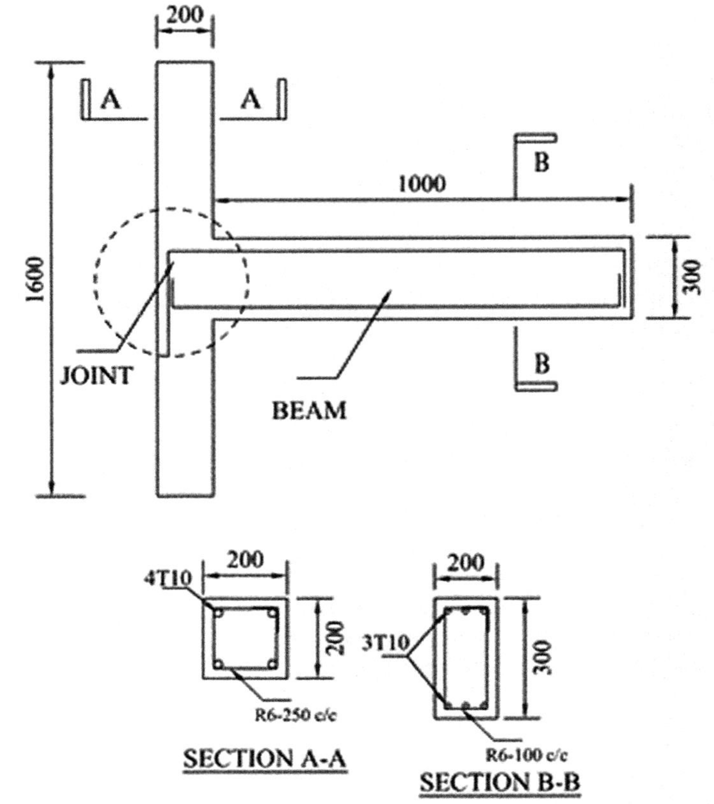

Fig. 2 Detailing and geometric of exterior beam-column joints

mild steel bar. Six tensile tests were conducted on each of the reinforcement bars and structural steel to determine the tensile properties. The yield stress and strain for T10 were 656 MPa and 0.315%, the R6 was 413 MPa and 0.254%, and the structural steel was 339 MPa and 0.142%, respectively.

3.2 Retrofitting Scheme

One of the specimens (J-FST) was retrofitted using the proposed retrofit technique, namely pre-tensioned stiffened steel angles. This technique was fabricated using structural steel (S275). In this experimental work, the size of the stiffened steel angles' used was 140 mm. The column axial load (225 kN) was applied on the tip of the column before the retrofitting works. Figure 3 depicted the pre-tensioned stiffened steel angles assembled using bolts, studs, and nuts. The pre-tensioning force was set

Fig. 3 Retrofitted exterior beam-column joint (J-FST)

to 60% of the bolt yield strength (936 MPa) which was measured by the strain gauges attached to the studs as shown in Fig. 3.

3.3 *Experimental Test*

The retrofitted beam-column joint was tested under quasi-static cyclic load at the beam tip. The column axial load was maintained throughout the experimental test. Figure 4 showed the quasi-static cyclic loading protocol was in accordance with FEMA 461 [24]. The drift ratio of the beam-column joints was calculated by dividing the beam tip deflection by the distance of the loading point to the column centre.

All the loads applied were measured using load cells. A total of four linear variable displacement transducers (LVDTs) were located at the loading point of the beam tip and joint region of the beam-column joint to measure the deflection and joint distortion, respectively, throughout the test. The strains of longitudinal reinforcement were measured by bonding the strain gauges where the significant strain would be expected as illustrated in Fig. 5.

A test rig was used to hold the beam-column joints in a stable vertical position as shown in Fig. 5. It is noted that at the column ends, the supports were placed to replicate hinge conditions which allow rotation but not translation.

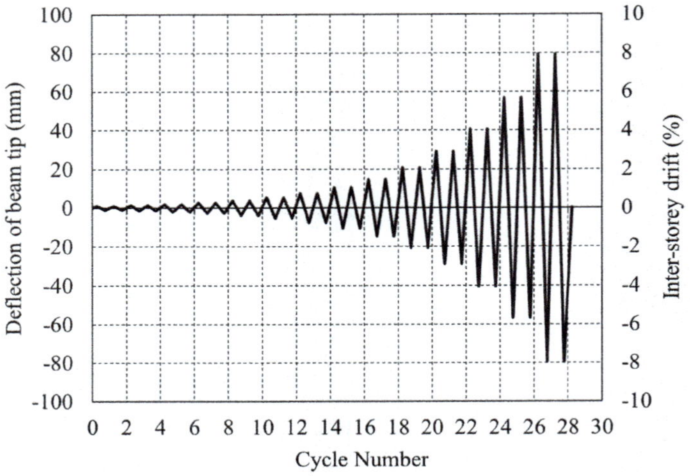

Fig. 4 Quasi-static cyclic loading protocol

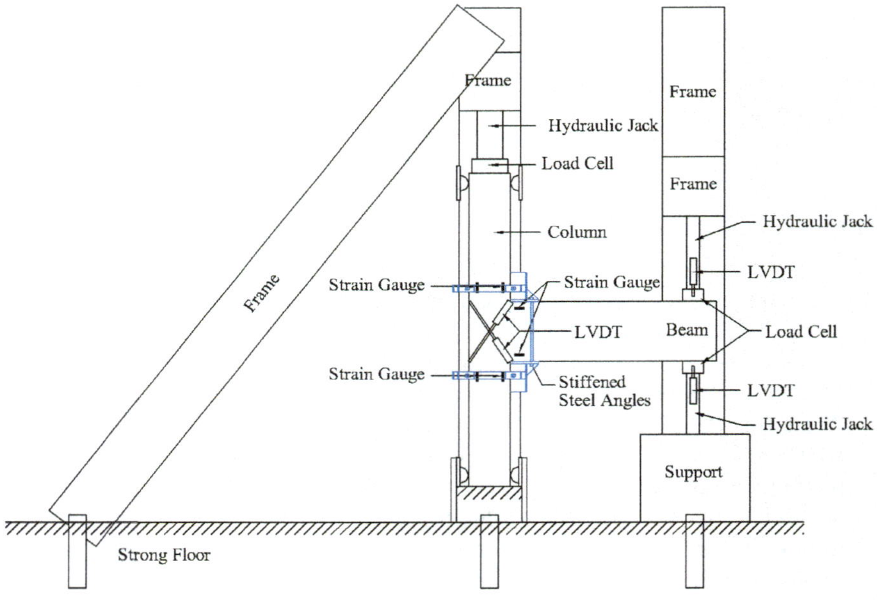

Fig. 5 Schematic diagram of the test rig and setup

4 Result and Discussion

4.1 General Observations

Generally, different failure modes, i.e. the joint shear failure and beam flexural failure can be observed on the control specimen and specimen J-FST, respectively, as shown in Fig. 6. The first hairline crack appeared at the beam-joint interface at a displacement of 2.75 mm (drift ratio 0.27%) for the control specimen. As the loading continue, the beam cracks appeared and stabilised at a displacement of 10.0 mm (drift ratio of 1.0%) until the specimen failed. A joint diagonal crack appeared at a displacement of 14.70 mm (drift ratio of 1.47%) with a value of 39.6 kN. Another joint diagonal crack appeared at a displacement of 29.00 mm (drift ratio of 2.9%) forming an X-pattern crack in the joint region. The diagonal cracks propagated towards the back of the column face as loading continue. Finally, severe damage and concrete spalling at the joint region were observed with buckling of column longitudinal reinforcements. The results showed that the inadequate joint shear reinforcement lead to joint shear failure in the control specimen.

For specimen J-FST, the first flexural cracks were observed at the beam-joint interface and stiffened steel angle's cut-off point which is 140 mm measured from the column face at a displacement of 5.53 mm (drift ratio of 0.55%). As the cyclic load increases, several cracks gradually formed and propagated over the whole length of the beam. At a displacement of 40.69 mm (drift ratio of 4.06%), the concrete cover at the bottom of the beam spalled. Finally, the beam flexural failure was achieved

(a) Control specimen (b) Specimen J-FST

Fig. 6 Failure mode of tested specimens, **a** control specimen and **b** specimen J-FST

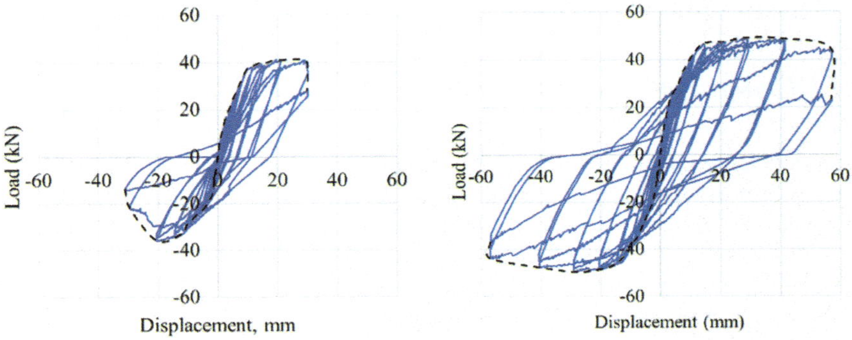

(a) Control specimen (b) Specimen J-FST

Fig. 7 Load–displacement hysteresis response of tested specimens **a** control specimen and **b** specimen J-FST

with buckled beam longitudinal reinforcement as well as the anchorage of the beam transverse reinforcement opened at a displacement of 56.95 mm (drift ratio of 5.69%).

4.2 Load–Displacement Hysteretic Response and Envelope Curve

The load–displacement hysteretic and envelope responses are shown in Fig. 7. It was observed that the specimen J-FST has larger hysteresis responses. The control specimen and specimen J-FST exhibited an initial linear load–displacement response up to a displacement of 7.50 mm (drift ratio of 0.75%) and 10.00 mm (drift ratio of 1.0%), respectively. The maximum load-carrying capacities for the control specimen and specimen J-FST are 41.0 kN (displacement of 20.00 mm) and 49.8 kN (displacement of 29.00 mm), respectively. Subsequently, both the load-carrying capacities decreased after attaining the maximum load-carrying capacity due to the failure of the joint and beam for the control specimen and specimen J-FST, respectively. As compared with the control specimen, the load-carrying capacity of specimen J-FST was enhanced by 21.4%.

4.3 Ductility

Ductility is an important seismic property of a structure or structural member. This is because high ductility will lead to a larger inelastic deformation with an acceptable stiffness and strength reduction. Thus, the structure or structural member will be able to absorb large seismic energy via a hysteresis behaviour [25]. Ductility, μ_Δ, is the ratio of ultimate displacement, Δ_u, to yield displacement, Δ_y, as shown in Eq. (1):

$$\mu_\Delta = \frac{\Delta_u}{\Delta_y} \quad (1)$$

In order to include the positive and negative hysteresis curves under cyclic load, Eq. (2) was adopted which is termed as the average ductility displacement, $\mu_{\Delta\,avg}$.

$$\mu_{\Delta avg} = \frac{|\Delta_u^+| + |\Delta_u^-|}{|\Delta_y^+| + |\Delta_y^-|} \quad (2)$$

where $\Delta_u{}^+$ and $\Delta_u{}^-$ are the positive and negative ultimate displacement, respectively; $\Delta_y{}^+$ and $\Delta_y{}^-$ are the positive and negative yield displacement, respectively.

The yield and ultimate displacements were determined using an equivalent bilinear curve as shown in Fig. 8. The area under both the hysteresis envelope curve and the equivalent bilinear curve are the same by using this method, which has the same energy absorption capacities among these two curves [18, 26–32]. Table 1 summarised the ductility of the tested beam-column joints. It was clearly shown that the ductility of the specimen J-FST significantly enhanced as compared with the control specimen by 103 and 125% for positive and negative directions using Eq. (1), respectively, while 117.2% using Eq. (2). This result indicated that the seismic properties of deficient beam-column joints were effectively enhanced by using the pre-tensioned stiffened steel angles.

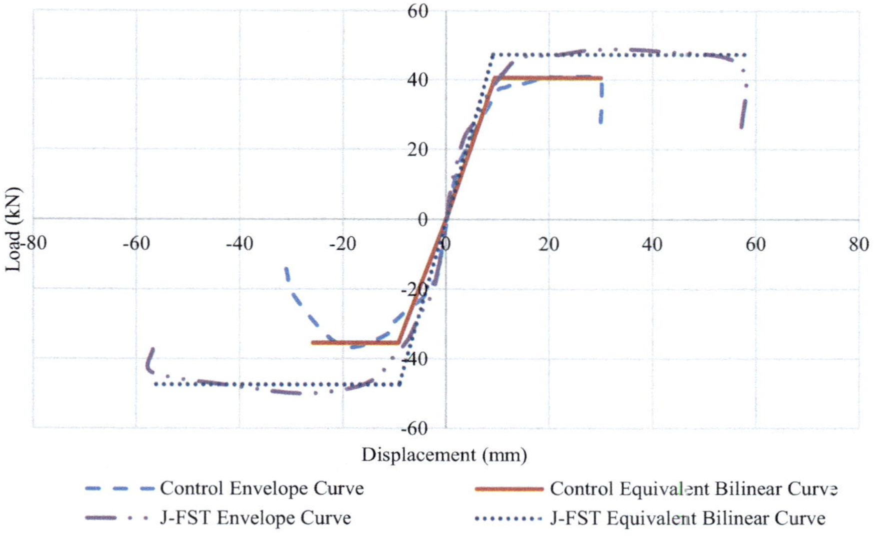

Fig. 8 Equivalent bilinear curve converted from experimental hysteresis envelope curves

Table 1 Displacement ductility of tested specimens

Specimen	Displacement ductility, μ_{Δ}				Average displacement ductility, $\mu_{\Delta\ avg}$	
	Positive (+)	Difference (%)	Negative (−)	Difference (%)	Equation (2)	Difference (%)
Control	3.1	–	2.8	–	2.9	–
J-FST	6.3	103	6.3	125	6.3	117.2

4.4 Energy Dissipation Capacity

Energy dissipation capacity is another important seismic property of a structure and structural member subjected to seismic load. Higher energy dissipation capacity means the structure or structural members able to dissipate more induced seismic energy through the hysteresis behaviour of reinforced concrete prior to the collapse of a structure. The energy dissipation capacity is determined from the enclosed area of hysteresis curves at each cycle in the load–displacement hysteresis curves. The cumulative energy dissipation capacity is the total energy dissipation capacity of all cycles in the load–displacement hysteresis curves.

Figure 9 showed the comparison of energy per cycle and the cumulative energy dissipation for the control specimen and specimen J-FST. The energy dissipation capacity was almost similar up to the displacement of 10.50 mm (drift ratio of 1.05%) for both specimens. As the cyclic displacement increases, the energy dissipation capacity of specimen J-FST continues increasing. Hence, a higher cumulative energy dissipation capacity could be observed for specimen J-FST as compared with the control specimen which was significantly enhanced by 296%.

4.5 Stiffness Degradation

Stiffness is the ability of a structure or structural member to resist deformation under load. In other words, the higher the stiffness, the higher the load-carrying capacity of a structure or structural member can be achieved under a small displacement. The degradation of stiffness will progressively occur under cyclic load. The stiffness of a structure or structural member can be obtained by measuring the peak-to-peak line slope of the positive and negative hysteresis curves. Stiffness degradation was determined by normalising the stiffness of each hysteresis cycle to the first cycle stiffness as shown in Fig. 10. The initial stiffness of specimen J-FST was increased by 7% and a stable stiffness degradation could be observed for both specimens during testing.

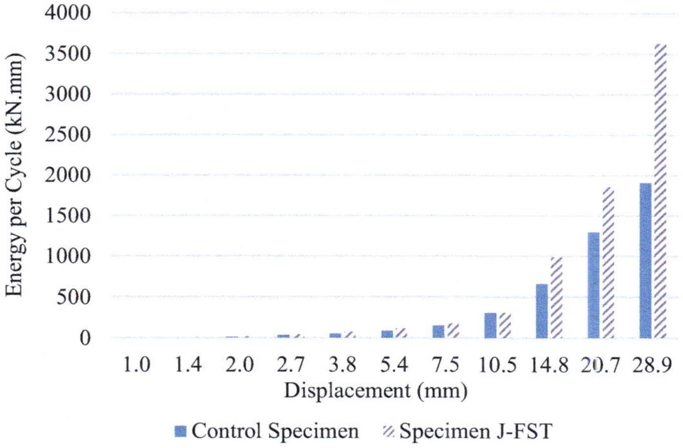

(a) Energy dissipation per cycle

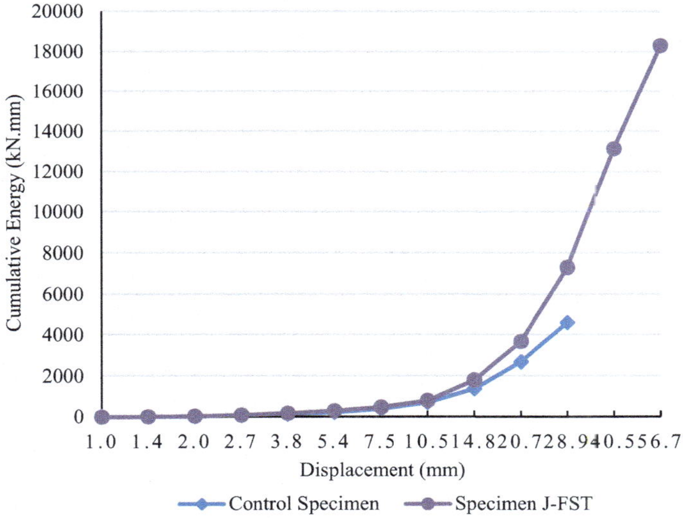

(b) Cumulative energy dissipation

Fig. 9 Comparison of **a** energy dissipation per cycle and **b** cumulative energy dissipation for control specimen and specimen J-FST

4.6 *Strain Measurement*

The recorded beam longitudinal reinforcement strains were presented in Fig. 11. The yield reinforcement stain (3150 $\mu\varepsilon$) was indicated by dashed lines. The strain measured for the control specimen and specimen J-FST were approximately 48,295

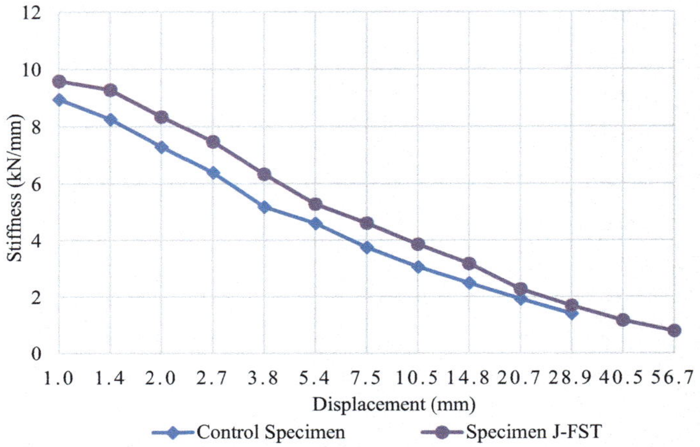

Fig. 10 Stiffness degradation of control specimen and specimen J-FST

$\mu\varepsilon$ and 27,865 $\mu\varepsilon$, respectively. The pre-tensioned stiffened steel angles is effective in reducing the plastic strains of beam longitudinal reinforcements as observed in Fig. 11. This is due to the plastic hinge had been transformed from the joint into the beam-joint interface and the pre-tensioned stiffened steel angle's cut-off point which can be observed in the crack patterns of specimen J-FST.

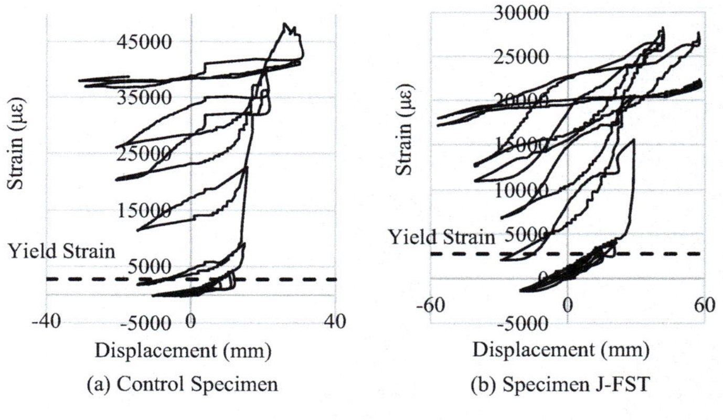

Fig. 11 Strain measurement of beam longitudinal reinforcement for **a** control specimen and **b** specimen J-FST

4.7 Joint Shear Distortion

The beam-column joint will undergo distortion under load which is known as joint shear distortion as illustrated in Fig. 12. The higher the imposed loading, the higher the joint shear distortion indicating an increase in joint distress. Two diagonal LVDTs had been placed on the joint surface to measure the average joint distortion during cyclic load testing. The joint shear distortion, γj, is measured according to Eq. (3) using the diagonal LVDTs [14, 33, 34].

$$\gamma_j = \gamma_1 + \gamma_2 = \frac{\sqrt{l_h^2 + l_v^2}}{2 l_h l_v}(\delta_1 - \delta_2) \quad (3)$$

where γ_1 and γ_2 are the vertical and horizontal joint distortion, respectively; δ_1 and δ_2 are the changes of diagonal LVDTs (+ for lengthening and − for shortening); l_h and l_v are the joint length in horizontal and vertical, respectively.

The joint shear distortions of the control specimen and specimen J-FST were shown in Fig. 13. It was shown that the initial joint distortion of specimen J-FST was lower than the control specimen indicating the improvement of joint stiffness. As the cyclic load increases, the joint shear distortion of specimen J-FST was higher than the control specimen up to 9.88×10^{-3} radians recorded at the end of the test. Although the joint distortion of specimen J-FST is higher than the control specimen, no crack was observed in the joint region.

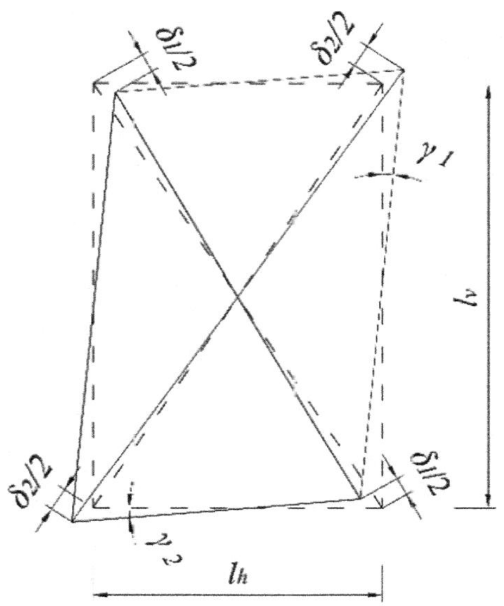

Fig. 12 Joint shear distortion definition

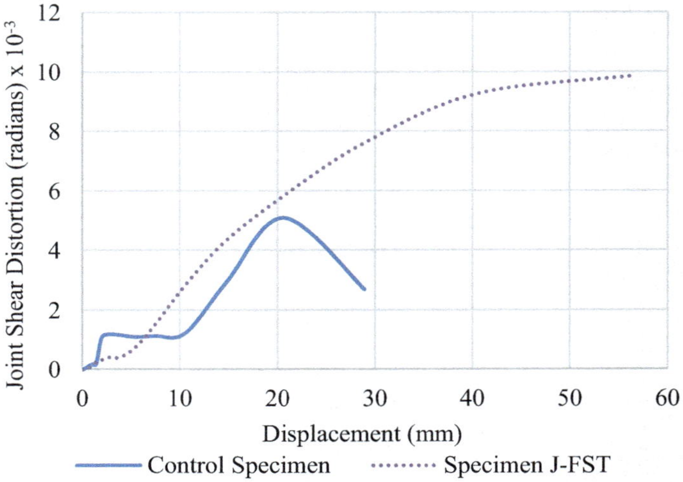

Fig. 13 Maximum joint shear distortion of control specimen and specimen J-FST

5 Conclusion

The proposed retrofit technique, namely "pre-tensioned stiffened steel angles" successfully enhanced the seismic properties of the deficient beam-column joints in terms of load-carrying capacity, ductility, energy dissipation, and initial stiffness, as well as reduced stiffness degradation and initial joint shear distortion. Besides, the results showed that this technique was able to transform the failure mode from joint shear into the beam failure. Hence, this technique can be adapted to retrofit the deficient beam-column joints as a sustainable solution. The summary of the experimental results is as follows:

- The proposed technique not only successfully transformed the plastic hinge from the joint into the beam region but a distance away from the column face.
- The concrete of the column and the joint region remained intact in accordance with the "strong column-weak beam" concept of the seismic designs' codes of practice.
- The control specimen showed the typical failure mode of joint shear failure, while the retrofitted specimen J-FST failed in beam flexural failure.
- The seismic properties of the specimen J-FST enhanced the load-carrying capacity, average displacement ductility, energy dissipation capacity, and initial stiffness for 21.4%, 117.2%, 296%, and 7%, respectively, as compared with the control specimen.

References

1. Zhao, B., Taucer, F., Rossetto, T.: Field investigation on the performance of building structures during the 12 May 2008 Wenchuan earthquake in China. Eng. Struct. **31**(8), 1707–1723 (2009). https://doi.org/10.1016/j.engstruct.2009.02.039
2. Mulas, M.G., Perotti, F., Coronelli, D., Martinelli, L., Paolucci, R.: The partial collapse of "Casa dello Studente" during L'Aquila 2009 earthquake. Eng. Fail. Anal. **34**, 566–584 (2013). https://doi.org/10.1016/j.engfailanal.2013.02.031
3. Ates, S., Kahya, V., Yurdakul, M., Adanur, S.: Damages on reinforced concrete buildings due to consecutive earthquakes in Van. Soil Dyn. Earthq. Eng. **53**, 109–118 (2013). https://doi.org/10.1016/j.soildyn.2013.06.006
4. Ruiz-Pinilla, J.G., Adam, J.M., Pérez-Cárcel, R., Yuste, J., Moragues, J.J.: Learning from RC building structures damaged by the earthquake in Lorca, Spain, in 2011. Eng. Fail. Anal. **68**, 76–86 (2016). https://doi.org/10.1016/j.engfailanal.2016.05.013
5. Kam, W.Y., Pampanin, S., Elwood, K.: Seismic performance of reinforced concrete buildings in the 22 February Christchurch (Lyttleton) earthquake. Bull. N. Z. Soc. Earthq. Eng. **44**(4), 239–278 (2011)
6. Sharma, K., Deng, L., Noguez, C.C.: Field investigation on the performance of building structures during the April 25, 2015, Gorkha earthquake in Nepal. Eng. Struct. **121**, 61–74 (2016). https://doi.org/10.1016/j.engstruct.2016.04.043
7. Gokhale, V.A., Joshi, D.R., Abhayankar, A.: The Psychological and Socio Economic Aspects of Earthquake Occurrence (2004)
8. Sinha, A., Kumar, S.: Economic consequences of earthquakes. Rehabilitation **8**(28), 1 (2017)
9. Alih, S.C., Vafaei, M.: Performance of reinforced concrete buildings and wooden structures during the 2015 Mw 6.0 Sabah earthquake in Malaysia. Eng. Fail. Anal. **102**, 351–368 (2019). https://doi.org/10.1016/j.engfailanal.2019.04.056
10. EN 1998-1: Design of Structures for Earthquake Resistance-Part 1: General Rules, Seismic Actions and Rules for Buildings, p. 229. EN 1998-1: 2004. CEN (Comité Européen de Normalisation), Brussels (2004)
11. ACI Committee 318: ACI Committee 318. Building Code Requirements for Structural Concrete (ACI 318-14) and Commentary (ACI 318R-14). American Concrete Institute: ACI Committee 318, Farmington Hills, MI (2014)
12. NZS 3101:Part 1: Concrete Structures Standards, Part 1—The Design of Concrete Structures. New Zealand Standard, Wellington, New Zealand (2006)
13. Pampanin, S., Christopoulos, C., Chen, T.-H.: Development and validation of a metallic haunch seismic retrofit solution for existing under-designed RC frame buildings. Earthquake Eng. Struct. Dynam. **35**(14), 1739–1766 (2006). https://doi.org/10.1002/eqe.600
14. Dang, C.-T., Dinh, N.-H.: Experimental study on structural performance of rc exterior beam-column joints Retrofitted by Steel Jacketing and Haunch element under cyclic loading simulating earthquake excitation. Adv. Civ. Eng. **2017**, 1–11 (2017). https://doi.org/10.1155/2017/9263460
15. Truong, G.T., Dinh, N.H., Kim, J.-C., Choi, K.-K.: Seismic performance of exterior RC beam-column joints retrofitted using various retrofit solutions. Int. J. Concr. Struct. Mater. **11**(3), 415–433 (2017). https://doi.org/10.1007/s40069-017-0203-x
16. Kanchanadevi, A., Ramanjaneyulu, K.: Non-invasive hybrid retrofit for seismic damage mitigation of gravity load designed exterior beam-column sub-assemblage. J. Earthquake Eng. **25**(8), 1590–1615 (2021). https://doi.org/10.1080/13632469.2019.1592790
17. Marchisella, A., Muciaccia, G., Sharma, A., Eligehausen, R.: Experimental investigation of 3d RC exterior joint retrofitted with fully-fastened-haunch-retrofit-solution. Eng. Struct. **239**, 112206 (2021). https://doi.org/10.1016/j.engstruct.2021.112206
18. Torabi, A., Maheri, M.R.: Seismic repair and retrofit of RC beam-column joints using stiffened steel plates. Iran. J. Sci. Technol., Trans. Civ. Eng. **41**(1), 13–26 (2017)

19. Shafaei, J., Hosseini, A., Marefat, M.S.: Seismic retrofit of external RC beam–column joints by joint enlargement using prestressed steel angles. Eng. Struct. **81**, 265–288 (2014). https://doi.org/10.1016/j.engstruct.2014.10.006
20. Adibi, M., Marefat, M.S., Esmaeily, A., Arani, K.K., Esmaeily, A.: Seismic retrofit of external concrete beam-column joints reinforced by plain bars using steel angles prestressed by cross ties. Eng. Struct. **148**, 813–828 (2017). https://doi.org/10.1016/j.engstruct.2017.07.014
21. Maddah, A., Golafshar, A., Saghafi, M.H.: 3D RC beam–column joints retrofitted by joint enlargement using steel angles and post-tensioned bolts. Eng. Struct. **220**, 110975 (2020). https://doi.org/10.1016/j.engstruct.2020.110975
22. Khodaei, M., Saghafi, M.H., Golafshar, A.: Seismic retrofit of exterior beam-column joints using steel angles connected by PT bars. Eng. Struct. **236**, 112111 (2021). https://doi.org/10.1016/j.engstruct.2021.112111
23. BS 8110: Structural Use of Concrete–Code of Practice for Design and Construction, Part 1, BS 8110. British Standards Institute, London UK (1985)
24. FEMA 461: Interim Protocols for Determining Seismic Performance Characteristics of Structural and Nonstructural Components Through Laboratory Testing. Federal Emergency Management Agency (FEMA) (2007)
25. Elnashai, A.S., Di Sarno, L.: Fundamentals of Earthquake Engineering. Wiley Online Library (2008)
26. Esmaeeli, E., Barros, J.A.O., Sena-Cruz, J., Fasan, L., Li Prizzi, F.R., Melo, J., et al.: Retrofitting of interior RC beam–column joints using CFRP strengthened SHCC: cast-in-place solution. Compos. Struct. **122**, 456–467 (2015). https://doi.org/10.1016/j.compstruct.2014.12.012
27. Esmaeeli, E., Barros, J.A.O., Sena-Cruz, J., Varum, H., Melo, J.: Assessment of the efficiency of prefabricated hybrid composite plates (HCPs) for retrofitting of damaged interior RC beam–column joints. Compos. Struct. **119**, 24–37 (2015). https://doi.org/10.1016/j.compstruct.2014.08.024
28. Esmaeeli, E., Danesh, F., Tee, K.F., Eshghi, S.: A combination of GFRP sheets and steel cage for seismic strengthening of shear-deficient corner RC beam-column joints. Compos. Struct. **159**, 206–219 (2017). https://doi.org/10.1016/j.compstruct.2016.09.064
29. Mostofinejad, D., Hajrasouliha, M.: 3D beam–column corner joints retrofitted with X-shaped FRP sheets attached via the EBROG technique. Eng. Struct. **183**, 987–998 (2019). https://doi.org/10.1016/j.engstruct.2019.01.038
30. Ilia, E., Mostofinejad, D.: Seismic retrofit of reinforced concrete strong beam–weak column joints using EBROG method combined with CFRP anchorage system. Eng. Struct. **194**, 300–319 (2019). https://doi.org/10.1016/j.engstruct.2019.05.070
31. Mostofinejad, D., Akhlaghi, A.: Experimental investigation of the efficacy of EBROG method in seismic rehabilitation of deficient reinforced concrete beam-column joints using CFRP sheets. J. Compos. Constr. **21**(4), 04016116 (2017). https://doi.org/10.1061/(ASCE)CC.1943-5614.0000781
32. Yurdakul, Ö., Tunaboyu, O., Avşar, Ö.: Retrofit of non-seismically designed beam-column joints by post-tensioned superelastic shape memory alloy bars. Bull. Earthq. Eng. (2018). https://doi.org/10.1007/s10518-018-0323-y
33. Le-Trung, K., Lee, K., Lee, J., Lee, D.H., Woo, S.: Experimental study of RC beam–column joints strengthened using CFRP composites. Compos. B Eng. **41**(1), 76–85 (2010). https://doi.org/10.1016/j.compositesb.2009.06.005
34. Li, B., Lam, E.S., Wu, B., Wang, Y.: Seismic behavior of reinforced concrete exterior beam-column joints strengthened by ferrocement composites. Earthq. Struct. **9**(1), 233–356 (2015)

Compressive Strength of Light-Transmitting Concrete (LTC) Incorporated with Optical Fibre

Shing Mei Chiew, Izni Syahrizal Ibrahim, and Mohd Azreen Mohd Ariffin

Abstract This paper aims to investigate the effect of incorporating optical fibre on the concrete compressive strength by varying the fibre diameter and fibre grid (spacing) in light-transmitting concrete (LTC). Optical fibres made of *poly-methymethacrylate* (PMMA) are used as translucent material for the production of LTC. The fibre diameters considered are 0.75 mm, 1.0 mm, 1.5 mm, and 2.0 mm, while the fibre grids are 4 × 4 grids (25 mm), 5 × 5 grids (18.75 mm) and 6 × 6 grids (15 mm). Self-compacting concrete (SCC) was used for casting. The specimens were tested for UPV test and compressive strength test on the 7th and 28th day. SEM analysis was also carried out to observe the fibre-matrix interface of the LTC specimens. The addition of optical fibre did not necessarily deteriorate the strength of the concrete. It was found that LTC with a fibre diameter of 1.0 mm had the highest compressive strength, while thereafter the strength started to decrease with increasing fibre diameter. It was found that the smaller fibre spacing reduced the concrete compressive strength. All specimens reached the design strength of 40 MPa. The UPV results showed that all specimens were of good quality regardless of the incorporation of the optical fibre. The microstructure analysis showed that there were no visible gaps within the fibre-matrix interface of all LTC specimens except for LTC with a fibre diameter of 1.5 mm. Overall, the effect of the use of optical fibre on the concrete compressive strength is insignificant

Keywords Light-transmitting concrete · Compressive strength · Microstructure analysis

---

S. M. Chiew (✉) · I. S. Ibrahim · M. A. Mohd Ariffin
Forensic Engineering Centre, Institute for Smart Infrastructure and Innovative Construction, Faculty of Civil Engineering, Universiti Teknologi Malaysia, 81310 Johor Bahru, Johor, Malaysia
e-mail: shing_mei@hotmail.com

I. S. Ibrahim
e-mail: iznisyahrizal@utm.my

M. A. Mohd Ariffin
e-mail: mohdazreen@utm.my

N. Sabtu (ed.), *Proceedings of AWAM International Conference on Civil Engineering 2022—Volume 2*, Lecture Notes in Civil Engineering 385,
https://doi.org/10.1007/978-981-99-6018-7_23

1 Introduction

Light-transmitting concrete (LTC) is an emerging concrete that has been an innovative and creative construction material in the past decade. It attracts attention from engineers and researchers due to its ability to transmit light which promotes the conservation of light energy. LTC was first introduced by a Hungarian architect named Aron Losonczi in 2001 [1–5] using optical fibre.

It has been reported that artificial lighting accounts for about 19% of global energy consumption [6]. This phenomenon is mainly observed in developed countries where there are many high-rise buildings, and people are highly dependent on artificial lighting. Shen and Zhou [7] conducted a simulation of the light transmittance of LTC and concluded that about 20% of light energy consumption can be reduced by using LTC. The reduction of light energy consumption eventually reduces the production of greenhouse gas and minimises the impact of global warming. LTC allows light to pass through the opaque concrete and reduces the light energy demand in a building, which, in addition to increase the aesthetic value of the concrete [8].

There are several applications of LTC around the world, for instance, the Italian Pavilion at Shanghai Expo 2010 [9, 10], the Al Aziz Mosque in Abu Dhabi [11], the translucent facades of Aachen University, Germany [12], the walls and partitions in Bank of Georgia [12] and the walkway constructed by Lucem Lichbeton, Germany [13]. There are various types of translucent materials used for the fabrication of LTC, mainly waste glass aggregates, glass fibres, plastic optical fibres, glass rods and epoxy resin. However, among all these translucent materials, plastic optical fibres are most commonly used for LTC fabrication due to their excellent light transmittance properties [11] and chemical resistance [14]. Therefore, only LTC incorporated with plastic optical fibres is considered and discussed in this study.

Optical fibres basically composed of three parts, namely the core, the cladding and the coating or also known as jacket [15]. Since light passes through the optical fibre by means of total internal reflection, it is important that the core and cladding have a suitable refractive index so that total internal reflection can take place. The refractive index is a parameter that expresses the speed of light in a medium. The refractive index of the core is always greater than that of the cladding. When light passes through an optical fibre, the light beam in the core exceeds the critical angle and total internal reflection occurs instead of refracton. Due to the continuous total internal reflection, light is transmitted through the optical fibre.

2 Background of Problem

In addition to the aesthetic value brought by LTC, it is also important to study the mechanical properties of LTC as it has the potential to be used as a construction material. A study by Chiew et al. [16] stated that the most considered parameters in the LTC investigation incorporated with optical fibre in previous studies are fibre volumetric fraction, fibre diameter and fibre spacing. This is because these three parameters have a significant direct impact on the light transmittance of LTC. Among these three parameters, the fibre volumetric fraction has the highest percentage considered in either light transmittance or mechanical properties of LTC. However, most of the experimental studies on LTC focused on the light transmittance properties rather than the mechanical properties. Hence, this study is aimed to investigate the effect of incorporating optical fibre on the concrete compressive strength by varying the fibre diameter and fibre spacing.

3 Literature Review

Li et al. [17] investigated the effect of filamentous optical fibre and bunchy optical fibre towards the light transmittance and compressive strength of LTC. They concluded that LTC with filamentous optical fibre possessed higher compressive strength than LTC with bunchy optical fibre form. This is because the filamentous fibres are more evenly distributed in the concrete matrix than the bunchy fibre, resulting in a stronger bond between the fibre-matrix interface. It is also reported that the compressive strength of LTC only decreased when the fibre volumetric fraction exceeds 2% [18]. However, a different conclusion has been reported by Kumar and Ahlawat [19] that the inclusion of optical fibre has no significant effect on the compressive strength of concrete if the fibre volumetric fraction does not exceed 4%. Another study also agreed with this statement [20].

Li et al. [21] found that the inclusion of optical fibre possessed lower compressive strength than normal concrete. In contrast, the increment of fibre volumetric fraction was reported to increase the concrete compressive strength [12, 22–24]. Tahwia et al. [25] reported that the concrete compressive strength increased with the increment of fibre volumetric fraction and fibre diameter in LTC. Several studies came to the opposite conclusion, namely that the concrete compressive strength decreased with the increment of fibre volumetric fraction [26, 27]. Henriques et al. [28] indicated that the compressive strength of LTC reduced by 20% compared to the control sample when the concrete was incorporated with 5% of fibre volumetric fraction.

In relation to this contradiction, Tahwia et al. [25] provided the explanation that the lower compressive strength obtained in the previous studies was due to the use of mortar instead of self-compacting concrete (SCC). Although mortar does not contain aggregates, the flowability of the mortar is the main concern in placing the concrete, because the mortar must be able to flow into every corner surrounding the congested optical fibre. In this issue, self-compacting concrete (SCC) or self-compacting mortar (SCM) is preferable because of its high flowability, as it directly minimises the voids and gaps, especially between the fibre-matrix interface. Another reason is that the arrangement of optical fibre in the specimens by Tahwia et al. [25] is in bunchy form which the holes and spacing were fixed in the specimens. The increment of optical fibre volumetric fraction or fibre diameter in the fixed holes and spacings increased the contact area between the fibre and the concrete matrix, resulting in a stronger concrete structure.

On the other hand, it has been found that the inclusion of optical fibre into the concrete decreased the flexural strength of the concrete drastically compared to the normal concrete [26, 28, 29]. Henriques et al. [28] and Li et al. [26] performed scanning electron microscopic test (SEM) on the LTC specimens to observe the fibre-matrix interface. They noted the presence of gaps between the optical fibre and concrete matrix and concluded that this was the main reason for the low compressive strength and flexural strength of the concrete when incorporated with optical fibre. To obtain further evidence of the presence of gaps between the fibre-matrix interface, Henriques et al. [30] also performed capillary water absorption test on both control specimen and LTC specimen. They found that the capillary water absorption of the LTC specimens increased with the inclusion of optical fibre as well as with the increment of fibre volumetric fraction. In addition, the surface of the optical fibre is extremely smooth and water resistant which also resulted in weaker bonding between the fibre-matrix interface. Furthermore, Navabi et al. [31] suggested that the lower strength of concrete could also be due to the large volume of optical fibre in concrete which act like pores in concrete even though the modulus of elasticity and optical fibres are different.

It has been reported that the incorporation of larger fibre diameter provides higher compressive strength when the fibre volumetric fraction was kept constant [32, 33]. Robles et al. [33] suggested that the bond between fibre-matrix interface is mechanically weak due to the low surface roughness of the optical fibre as failure began at the fibre-matrix interface during compression. However, the study found that the strength increased with the increment of the fibre diameter due to the lower interface area within the transition zone. Besides that, Altlomate et al. [12] found that although the increment of fibre diameter increased the compressive strength, the strength decreased as the distance between fibre decreased. This could be due to the fact that the smaller interconnecting distance between the fibre for the macrocracks to propagate when the LTC was subjected to compressive loading. Nevertheless, the same authors also conducted the ultrasonic pulse velocity (UPV) tests on LTC and concluded that the LTC is homogeneous and of good to excellent quality (UPV > 3.5 m/s) with the incorporation of optical fibre.

In summary, the incorporation of optical fibre generally deteriorates the compressive strength of LTC because the optical fibres adhere poorly to the concrete matrix and gaps exist between fibre-matrix interface. However, several studies reported otherwise. This may be due to different specimen dimensions, compaction of the concrete, type of concrete used, fibre diameter, fibre spacing and mechanical properties of the aggregates used in concrete casting [34]. In this study, different fibre diameters and fibre spacing of optical fibres were incorporated into LTC. In addition to investigating the compressive strength of LTC with different parameters, UPV tests and microstructure analysis (SEM) were also conducted to understand the mechanical properties of LTC.

4 Materials and Methods

4.1 Construction Materials

In this study, SCC is used for the construction of LTC. Construction materials included Ordinary Portland Cement (OPC), Class F fly ash, sand, coarse aggregate, water, polycarboxylate ether-based superplasticiser and polymethamethylcrylate (PMMA) optical fibres. The design strength of SCC is 40 MPa and the mix proportion was determined from the trial mixes and the reference from the previous study [35]. The maximum size of the coarse aggregate used in this study was 10 mm. The mix design of the constituents is given in Table 1.

In this study, fly ash was used as a cement substitute at 30% by weight, as recommended in previous studies [36, 37] which is the optimum dosage of fly ash. Fly ash was used in SCC to increase the flowability of SCC, reduce the shrinkage cracking of concrete after casting and enhance the adhesion of fibres within the concrete matrix. Besides that, the smaller size of fly ash particles also acted as a filler to fill the voids in the concrete matrix. The composition of fly ash used is presented in Table 2.

On the other hand, PMMA optical fibres were used as the translucent materials for the fabrication of LTC due to its mechanical flexibility [38] and excellent light transmittance properties. Four different fibre diameters were used in this study, namely 0.75 mm, 1.0 mm, 1.75 mm, and 2.0 mm. The optical fibres were arranged by means

Table 1 Mix proportions of SCC

Constituent	Mix proportions (kg/m^3)
Cement	385
Fly ash	165
Fine aggregate	910
Coarse aggregate	590
Water	253
Superplasticiser	10.6

Table 2 Composition of class F fly ash

Composition	Content (%)
Al_2O_3	21.3
SiO_2	59.5
SO_3	0.6
Fe_2O_3	7.1
CaO	7.18
MgO	1.42
K_2O	1.53
Na_2O	1.42

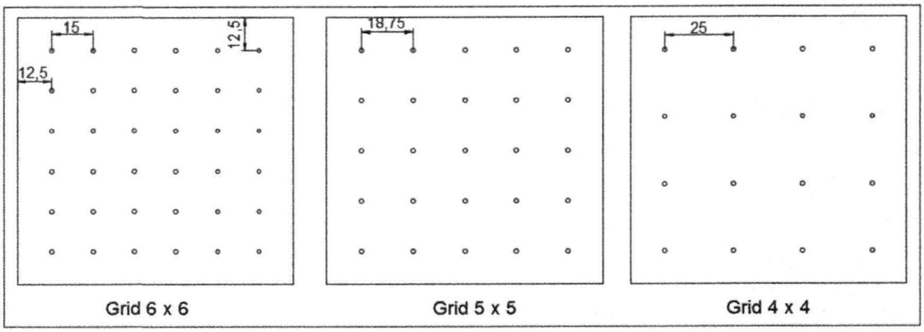

Fig. 1 Design of fibre spacings by means of grid

of grid with fibre spacing of 25 mm for the 4 × 4 grid, 18.75 mm for the 5 × 5 grid and 15 mm for the 6 × 6 grid, as shown in Fig. 1. The minimum spacing of fibres in this study is 15 mm, allowing 10 mm of coarse aggregates passes without any hindrance. The designation of specimens in this study was "LTC D", where D represented the fibre diameter.

4.2 Fabrication of LTC

The size of the LTC specimen in this study is 100 mm long × 100 mm wide × 100 mm high. Plywood was used to construct the formwork of LTC. Holes were drilled in the plywood according to the fibre spacing. The optical fibres were manually inserted into the formwork before casting. The ends of the optical fibres were fixed with a hot glue gun to prevent any mess in the arrangement during concrete casting. The formwork of LTC specimens is shown in Fig. 2. The casting procedures were in accordance with BS 1881:125:2013 [39]. During casting, fine and coarse aggregates were first poured into the mixer to mix well with a small amount of water, followed by cement and fly ash. Water was poured consistently into the mixture during casting. To achieve the desired flowability, superplasticiser was added at the end of the mixture. Before placing the concrete in the formwork, fresh concrete properties tests such as slump test, V-funnel test and L-box test were carried out to ensure that the fresh concrete properties requirements stated in EFNARC [40] were met. A total of three specimens were cast with the same parameters (both fibre diameter and fibre spacing). A total of 78 specimens were cast, including the control specimens prior to testing. The cast specimens were left for 24 h before the formwork was opened. The specimens were then cured in the curing tank for 28 days prior to testing.

Fig. 2 Formwork for LTC specimens

4.3 Compression Test, UPV and Microstructure Analysis

LTC specimens were tested for the compressive strength on 7th day and 28th day. The specimens were removed from the curing tank one day before the testing for drying. Three samples with the same parameters were tested for UPV analysis and compressive strength tests. For the compressive strength tests on the 7th and 28th day, the average strength of the three samples were recorded. The compression machine used in this study is NL Scientific compression machine with capacity of 3000 kN. The LTC specimens were placed in the compression machine with the loading direction perpendicular to the fibre arrangement of LTC. The LTC specimens were compressed at a loading rate of 0.6 ± 0.2 MPa/s as specified in BS 12,390-3:2009 [41]. Prior to compressive strength testing of LTC specimens on the 28th day, the specimens were tested for UPV on all three sides of the specimens in accordance with BS EN 12,504-4:2004 [42]. The results measured for three sides of the sample were not averaged due to the different orientations of optical fibre in relation to the concrete sides. Additional specimens were also cast and small pieces of specimens with different fibre diameters were ground for use in microstructure analysis.

Table 3 Fresh properties of SCC

Test	Unit	Result	Criteria in EFNARC [40]
Slump flow	mm	600–700	550–850
V-funnel	s	2–4	≤ 8
L-box	–	0.9–1.0	0.8–1.0

5 Results and Discussions

5.1 *Fresh Properties of SCC*

Table 3 gives the fresh properties of SCC in this study, where the slump, V-funnel test and L-box test were conducted to determine its flowability, viscosity and passing ability of SCC, respectively. The results of the fresh concrete properties were within the requirements specified in EFNARC. In slump flow test, it was found that no segregation occurred and the concrete mix flowed well into the formwork. To minimise air voids within the concrete mix, the formwork was lightly struck from the side after placing the concrete. It was also observed that the surface of LTC specimens was smooth and no voids were visible on the concrete surface after opening the formwork.

5.2 *Ultrasonic Pulse Velocity (UPV)*

Ultrasonic pulse velocity (UPV) test is a very useful test for determining concrete quality and strength [12]. The velocity of the ultrasonic pulse is measured. If the concrete is homogeneous and of good quality, the measured velocity is high. Otherwise, if there are voids or cracks in the concrete, the velocity of the pulse is relatively slow due to a different medium, *t*. The recommendation of UPV depending on the concrete quality is given in Table 4. There are different methods to measure UPV, including direct transmission, semi-direct transmission and indirect transmission. In this study, direct transmission was used to measure UPV of LTC specimen with two transducers placed on opposite sides of the specimen.

From Fig. 3, it can be seen that almost all UPV results of the LTC specimens as well as the normal concrete were above 4000 m/s, despite the incorporation of optical fibre and also the increase in fibre diameter, which is in agreement with the study by

Table 4 Concrete quality in in relation to UPV

UPV (km/s)	Concrete quality (Grading)
Above 4.5	Excellent
3.5–4.5	Good
3.0–3.5	Medium
Below 3.0	Doubtful

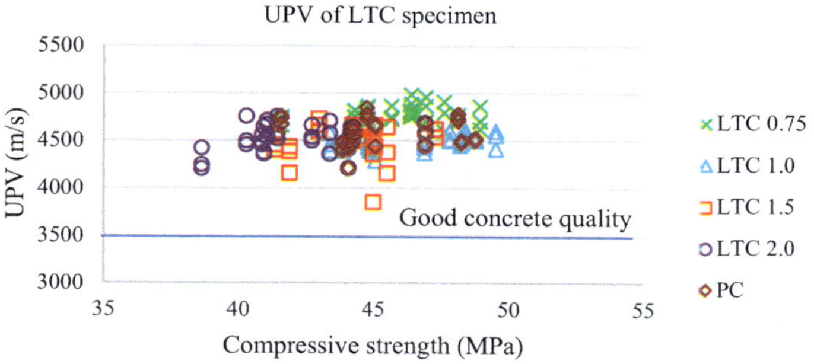

Fig. 3 UPV results for LTC specimens with different diameter

Altlomate et al. [12]. This indicates that all the specimens were of good quality. It is worth noting that the specimens for LTC with fibre diameter of 0.75 mm (LTC 0.75 in Fig. 3) had the highest UPV values of all samples with excellent concrete. This shows that the incorporation of optical fibres does not always deteriorate the quality of concrete, but improves the homogeneity of the concrete specimens. Furthermore, the difference in UPV value between the normal concrete and LTC is insignificant despite the different fibre diameters.

5.3 Compressive Strength of LTC

Compressive strength is one of the most important mechanical properties of concrete. It indicates how much compressive stress a concrete can withstand when subjected to compressive loads. The light transmittance of LTC depends on the fibre volumetric fraction, the diameter of the fibres and also the distances between the fibres incorporated in the concrete. It is important to investigate the compressive strength of LTC in addition to its light transmittance properties, as it has the potential to be used as a building material.

Figures 4 and 5 illustrate the compressive strength of LTC with different fibre diameters and fibre spacings on 7th day and 28th day, respectively. On 7th day, the compressive strength of all the LTC specimens was lower than the normal concrete with compressive strength of 32.49 MPa. Nonetheless, the strength achieved two-thirds of the design strength of 40 MPa for all specimens, which is desirable. The compressive strength of concrete decreased with bigger fibre diameter. The effect of fibre spacings did not show any significant trend on the concrete compressive strength. On the other hand, it was noteworthy that LTC 0.75 and LTC 1.0 had slightly higher compressive strength on the 28th day compared to normal concrete.

This is contrary to the findings in most of the literature that incorporation of optical fibre deteriorates the concrete compressive strength [21, 23, 26, 43, 44]. This could be due to the presence of coarse aggregates and the use of SCC. In fabricating

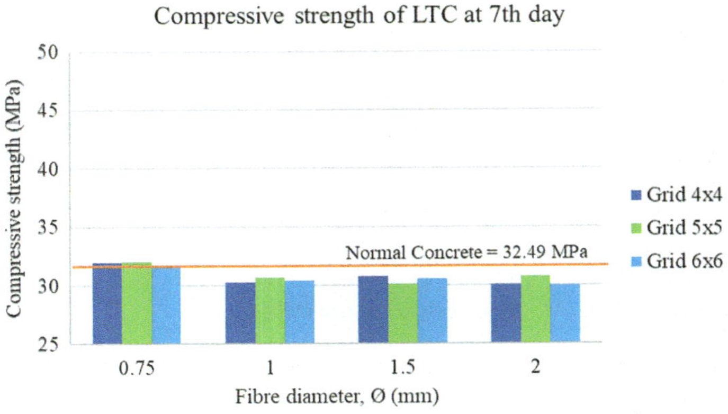

Fig. 4 Compressive strength of LTC on 7th day

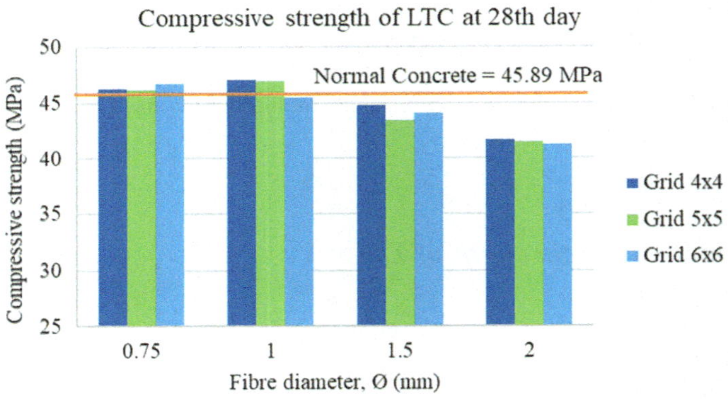

Fig. 5 Compressive strength of LTC on 28th day

LTC, most studies preferred the use of mortar instead of concrete, omitting the coarse aggregates to maximise the arrangement of optical fibre. Tahwia et al. [25] reported that the strength of concrete containing either fine or coarse aggregate was lower than that of concrete containing both fine and coarse aggregate which resulted in a more desirable distribution in concrete. Besides that, SCC is highly workable that it filled all voids and corners in the formwork more efficiently compared to non-self-compacting concrete. Therefore, the compressive strength of concrete with the inclusion of optical fibre was in line with Tahwia et al. [25].

It is worth noting that the compressive strength of LTC of grid 6 × 6 with 15 mm spacing decreased linearly with the increment of fibre diameter. This could be due to the smooth and low roughness of surface of optical fibre which led to a weaker bond between fibre-matrix interface. LTC with the 6 × 6 grid had a larger surface area of optical fibre in contact with the concrete matrix than LTC with the 4 × 4 grid (25 mm spacing) and the 5 × 5 grid (18.75 mm spacing). The same reason applied to

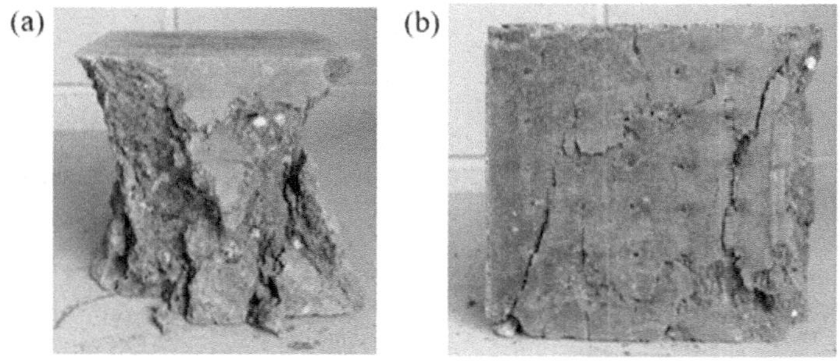

Fig. 6 Failure mode of **a** normal concrete (control specimen) and **b** LTC with grid 6 × 6

the decreased in compressive strength when the fibre diameter increased to 1.5 and 2.0 mm. Besides that, when the fibre spacing was kept constant, the increment of fibre diameter also decreased the interconnecting distance between the macrocracks to propagate when subjected to compression.

Figure 6 shows the failure mode of the control specimen and LTC specimen. It is worth noting that cracks initiated at the transition zone for almost all the LTC specimen as shown in Fig. 6b. However, the inclusion of optical fibre enhanced the failure mode of the concrete from being crushed severely as shown in Fig. 6. This shows that the optical fibre has the potential to alter the failure pattern of normal concrete from brittle to ductile mode. This phenomenon requires further investigation in the future research.

Figure 7 shows the comparison of the compressive strength of the current study with the previous studies [12, 25]. These studies were selected for comparison because due to similar fibre spacing was used (10–16 mm). It is worth noting that the trend of LTC compressive strength in this study was different from the other two studies where larger fibre diameter resulted in higher compressive strength. This could be due to the variation in fibre diameter, type of filler, mineral admixtures used to produce SCC and fresh concrete properties of SCC. Despite the study by Altlomate et al. [12], the difference in the normalised compressive strength of LTC compared to normal concrete in both the study by Tahwia et al. [25] and the current study is insignificant, which is ~0.1.

5.4 Scanning Electron Microscope Analysis (SEM)

Figures 8 and 9 show the images of the fibre-matrix interface of LTC 0.75 and LTC 2.0, respectively, under scanning electron microscope analysis (SEM). No visible voids or gaps were found around the fibre-matrix interface, which is in contrast to previous studies [26, 30]. A similar image is found for LTC 1.0. This could be due to the type of concrete used in casting, where SCC was not used in concrete castings in

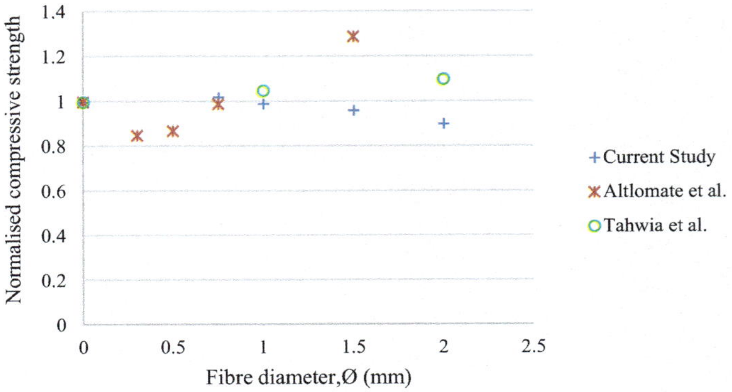

Fig. 7 Comparison of LTC compressive strength with the previous studies [12, 25]

fabricating LTC in these studies, as stated by Tahwia et al. [25]. However, it is worth noting that gaps were found in LTC 1.5 in this study, which were probably caused by the smooth surface roughness of optical fibre of LTC. However, the approximate width of the gap is 7.85 μm which is relatively small. Li et al. [26] and Henriques et al. [30] concluded that the presence of gaps between the fibre-matrix interface is the main contribution for the deterioration of the compressive strength. Comparing the compressive strength and the SEM images of LTC 1.5 with LTC 2.0 in Figs. 8 and 9, the absence of gaps between the fibre-matrix interface does not guarantee higher compressive strength because LTC 2.0 has lower compressive strength than LTC 1.5 as shown in Fig. 5. There are other factors that may contribute to lower strength of LTC, such as the difference in elastic modulus of LTC and optical fibre and the larger surface area of interaction of optical fibre within the transition zone at larger fibre diameter.

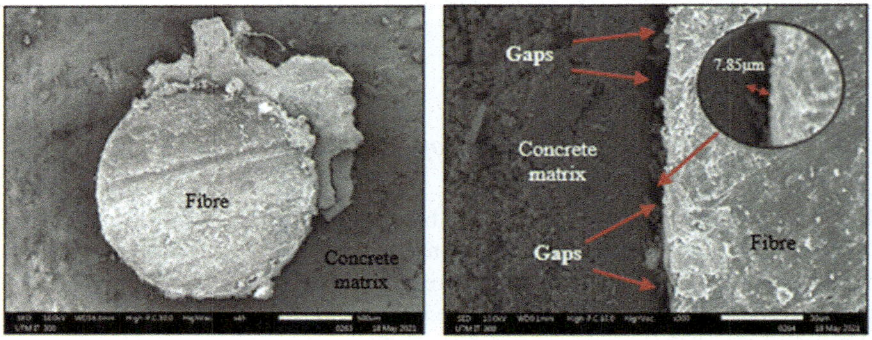

Fig. 8 SEM images of LTC 0.75 with the magnification of ×100 (left) and ×500 (right)

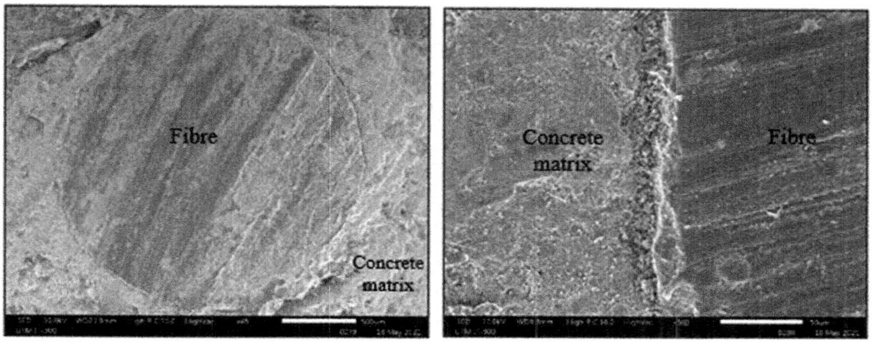

Fig. 9 SEM images of LTC 2.0 with the magnification of ×45 (left) and ×500 (right)

6 Conclusion

The inclusion of optical fibre does not deteriorate the concrete compressive strength if the fibre dimeter incorporated in concrete does not exceed 1.0 mm. If the fibre diameter exceeds 1.0 mm, the compressive strength decreased. If a larger fibre diameter is used, a high flowability of SCC with mineral admixtures and fillers is recommended to fabricate LTC in order to minimise the gap between fibre-matrix interface and also increase the grip and bond within the transition zone.

Besides that, the smaller the fibre spacing, the lower the concrete compressive strength because the interconnection distance between the macrocracks becomes smaller when the concrete is subjected to compression load. However, the effect of fibre spacing on concrete compressive strength is insignificant compared to the effect of fibre diameter. It is worth noting that LTC incorporated with optical fibres showed a ductile failure mode compared to the control specimens which fails in brittle when subjected to compression. This required further investigation in the future research.

On the other hand, it is noted that the inclusion of optical fibre did not deteriorate the quality and homogeneity of the concrete. The results showed that the LTC incorporated with different fibre diameter and spacing had good or excellent concrete quality. As for SEM analysis, it was found that there were no visible voids or gaps between the fibre-matrix interface except for specimen LTC 1.5. This could be due to the smooth surface of optical fibre which find little grip in the concrete matrix. Nonetheless, the gaps present between the fibre-matrix interface may not be the main reason for the lower compressive strength, as no visible gaps were observed in the transition zone of specimen LTC 2.0 which possessed a lower compressive strength than LTC 1.5.

In summary, although the inclusion of optical fibre has a slight effect on the compressive strength, it is insignificant. It is recommended to consider the optimum of fibre diameter and fibre spacing in both mechanical strength and light transmittance properties in the fabrication of LTC, instead of focusing only on the light transmittance properties of LTC. For the application of LTC in the construction industry, it is recommended to improve the manufacturing method of LTC to minimise the labour and manpower, which are the main causes of the high cost of LTC. LTC is more expensive than conventional concrete, but considering the long-term energy savings in light consumption, it can be considered economical and sustainable.

Acknowledgements The authors would like to send their warmest appreciation for the technical support from the Structural and Material Laboratory at Universiti Teknologi Malaysia (UTM) and the financial research support by the Ministry of Higher Education (MoHE) of Malaysia under the Fundamental Research Grant Scheme (FRGS) No. FRGS/1/2020/TK0/UTM/02/72 (R.J130000.7809.5F287).

References

1. Ugale, A.B., Badnakhe, R.R., Nanhe, P.P.: Smart technologies for energy. Environ. Sustain. Develop. Springer (2019). https://doi.org/10.1007/978-981-13-6148-7_25
2. Timina, A., Yanova, R., Popov, A., Sorokoumova, T.: Modern translucent materials and their impact on architectural forming. In: E3S Web of Conferences, p. 01035, EDP Sciences (2019). https://doi.org/10.1051/e3sconf/20199701035
3. Zielińska, M., Ciesielski, A.: Analysis of transparent concrete as an innovative material used in civil engineering. In: IOP Conference Series: Materials Science and Engineering, vol. 2, p. 022071. IOP Publishing (2017). https://doi.org/10.1088/1757-899X/245/2/022071
4. Said, S.H.: State-of-the-art developments in light transmitting concrete. Mater. Today Proceed. (2020)
5. Covaleov, G.A.: Translucent concrete. In: IOP Conference Series: Earth and Environmental Science, vol. 1, p. 012144. IOP Publishing (2019)
6. Shitote, S., Tuaum, A., Oyawa, W.O.: Experimental evaluation on light transmittance performance of translucent concrete. Int. J. Appl. Eng. Res. **13**, 1209–1218 (2018)
7. Shen, J., Zhou, Z.: Light transmitting performance and energy-saving of plastic optical fibre transparent concrete products. Indoor Built Environ. 1420326X20903368 (2020). https://doi.org/10.1177/1420326X20903368
8. George, E.H., Rajul, A.A.: An intensive review on using translucent concrete as energy efficient source on green buildings. Sustainability Agric. Food Environ. Res. **10**(1) (2022)
9. Cement, H.: Transparent Cement for the Italian Pavilion at Expo 2010 in Shanghai. In: Halyps Building Materials SA, Greece. https://www.heidelbergcement.com/en/italian-pavilion-shanghai (2021). Accessed 5/12/2021
10. Juan, S., Zhi, Z.: Some progress on smart transparent concrete. Pacific Sci. Rev. **15**(1), 51–55 (2013)
11. Han, B., Zhang, L., Ou, J.: Smart and Multifunctional Concrete Toward Sustainable Infrastructures. Springer (2017). https://doi.org/10.1007/978-981-10-4349-9
12. Altlomate, A., Alatshan, F., Mashiri, F., Jadan, M.: Experimental study of light-transmitting concrete. Int. J. Sustain. Build. Technol. Urban Dev. **7**(3–4), 133–139 (2016). https://doi.org/10.1080/2093761X.2016.1237396
13. Nachrichten. Award-winning translucent concrete technology for smart cities. https://www.bft-international.com/en/artikel/bft_Award-winning_translucent_concrete_technology_for_smart_cities_3068721.html (2021). Accessed 5/12/2021.
14. Merchant, D., Scully, P., Schmitt, N.: Chemical tapering of polymer optical fibre. Sens. Actuators A **76**(1–3), 365–371 (1999). https://doi.org/10.1016/S0924-4247(99)00008-4
15. Luhar, I., Luhar, S., Savva, P., Theodosiou, A., Petrou, M.F., Nicolaides, D.: Light transmitting concrete: a review. Buildings **11**(10), 480 (2021)

16. Chiew, S.M., Ibrahim, I.S., Ariffin, M.A.M., Lee, H.-S., Singh, J.K.: Development and properties of light-transmitting concrete (LTC)–a review. J. Clean. Prod. **284**, 124780 (2021). https://doi.org/10.1016/j.jclepro.2020.124780
17. Li, Y., Xu, Z.Y., Gu, Z.W., Bao, Z.Z.: Research on the light transmitting cement mortar. In: Advanced Materials Research, pp. 397–401, Trans Tech Publ (2012). https://doi.org/10.4028/www.scientific.net/AMR.450-451.397
18. Jiménez-Muñoz, E., Fernández-Martínez, F.: Construction and Building Research. Springer (2014). https://doi.org/10.1007/978-94-007-7790-3_15
19. Kumar, A., Ahlawat, R.: Experimental study on light transmitting concrete. Int. J. Innov. Sci. Eng. Technol. 4 (6):201–210 (2017)
20. Taneja, E.K., Joshi, T.M., Dave, U.: Evaluation of mechanical properties and light transmission of light-transmitting concrete. In: Technology Drivers: Engine for Growth: Proceedings of the 6th Nirma University International Conference on Engineering (NUiCONE 2017), November 23–25, 2017, Ahmedabad, India, p. 53, CRC Press (2018).
21. Li, Y., Li, J., Wan, Y., Xu, Z.: Experimental study of light transmitting cement-based material (LTCM). Constr. Build. Mater. **96**, 319–325 (2015). https://doi.org/10.1016/j.conbuildmat.2015.08.055
22. Huong, O.W., Kassim, U.: Translucent concrete by plastics fibre optics as a sustainable material that benefit to residential building (2019)
23. Tuaum, A., Shitote, S., Oyawa, W., Biedebrhan, M.: Structural performance of translucent concrete façade panels. Adv. Civil Eng. (2019)
24. Shahmir, N.G., Bhat, M.: Structural and luminance properties of light transmitting concrete. In: Annales de Chimie-Science des Matériaux, vol. 3, pp. 185–190 (2020)
25. Tahwia, A.M., Abdel-Raheem, A., Abdel-Aziz, N., Amin, M.: Light transmittance performance of sustainable translucent self-compacting concrete. J. Build. Eng. 102178 (2021). https://doi.org/10.1016/j.jobe.2021.102178
26. Li, Y., Li, J., Guo, H.: Preparation and study of light transmitting properties of sulfoaluminate cement-based materials. Mater. Des. **83**, 185–192 (2015). https://doi.org/10.1016/j.matdes.2015.06.021
27. Sawant, A., Jugdar, R., Chougule, V.: Experimental work on light transmitting concrete by using optical fiber. Int. J. Adv. Technol. Eng. Sci. **2**(12), 636–645 (2014)
28. Henriques, Td.S.: Dal Molin DC, Masuero ÂB: Optical fibers in cementitious composites (LTCM): Analysis and discussion of their influence when randomly arranged. Constr. Build. Mater. **244**, 118406 (2020). https://doi.org/10.1016/j.conbuildmat.2020.118406
29. Salih, S.A., Joni, H.H., Mohamed, S.A.: Effect of plastic optical fiber on some properties of translucent concrete. Eng. Technol. J. **32** (12 Part (A) Engineering):2846–2861 (2014)
30. Henriques, Td.S., Dal Molin, D.C., HMasuero, A.B.: Study of the influence of sorted polymeric optical fibers (POFs) in samples of a light-transmitting cement-based material (LTCM). Constr. Build. Mater. **161**, 305–315 (2018). https://doi.org/10.1016/j.conbuildmat.2017.11.137
31. Navabi, D., Javidruzi, M., Hafezi, M.R., Mosavi, A.: The high-performance light transmitting concrete and experimental analysis of using polymethylmethacrylate optical fibers in it. J. Build. Eng. **38**, 102076 (2021). https://doi.org/10.1016/j.jobe.2020.102076
32. Bashbash, B.F., Hajrus, R.M., Wafi, D.F., Alqedra, M.: Basics of light transmitting concrete. Basics Light Transmitt. Concr. **2**(3), 76–83 (2013)
33. Robles, A., Arenas, G.F., Stefani, P.M.: Light transmitting cement-based material (LTCM) as a green material for building. J. Appl. Res. Technol. Eng. **1**(1), 9–14 (2020). https://doi.org/10.4995/jarte.2020.13832
34. Huang, B., Wang, Y., Lu, W., Cheng, M.: Fabrication and energy efficiency of translucent concrete panel for building envelope. Energy **248**, 123635 (2022)
35. Siddique, R.: Properties of self-compacting concrete containing class F fly ash. Mater. Des. **32**(3), 1501–1507 (2011)
36. Dinakar, P., Reddy, M.K., Sharma, M.: Behaviour of self compacting concrete using Portland pozzolana cement with different levels of fly ash. Mater. Des. **46**, 609–616 (2013). https://doi.org/10.1016/j.matdes.2012.11.015

37. Kim, J., Han, S., Park, Y., Noh, J., Park, C., Kwon, Y.: Experimental research on the material properties of super flowing concrete. Product Methods Workabil. Concr. **32**, 271 (2004). https://doi.org/10.1201/9781482271782
38. Van Eijkelenborg, M.A., Large, M.C., Argyros, A., Zagari, J., Manos, S., Issa, N.A., Bassett, I., Fleming, S., McPhedran, R.C., de Sterke, C.M.: Microstructured polymer optical fibre. Opt. Express **9**(7), 319–327 (2001). https://doi.org/10.1364/OE.9.000319
39. Institution, B.S.: Testing concrete—methods for mixing and samplingfresh concrete in the laboratory (2013)
40. (EFNARC) TEFoSCCaCS.: The European guidelines for self-compacting concrete: Specification, production and use (2005)
41. Institution, B.S.: Testing Hardened Concrete: Compressive Strength of Test Specimens (2009)
42. Institution, B.S.: Testing Concrete—Part 4: Determination of ultrasonic pulse velocity (2004)
43. Sawant, A., Jugdar, R., Sawant, S.: Light transmitting concrete by using optical fiber. Int. J. Inventive Eng. Sci. **3**(1), 23–28 (2014)
44. Li, Y., Xu, Z.Y., Gu, Z.W., Bao, Z.Z.: Preparation of light transmitting cement-based material with optical fiber embedded by the means of parallel arrange. In: Advanced Materials Research, pp. 677–682 (2012)

Development of Adaptive Neuro-Fuzzy Inference System to Predict Concrete Compressive Strength

S.J.S Hakim, N. Jamaluddin, K.H. Boon, S.N. Mokhatar, A. Nasradeen Khalifa, and Z. Jamellodin

Abstract Predicting the compressive strength of concrete is one of the complex problems in civil engineering because different parameters and factors must be considered. There is several research that have predicted the compressive strength of normal concrete using neuro-fuzzy systems. However, little research has been done to predict the strength of high strength concrete. Recently, machine learning techniques such as artificial neural networks (ANNs), fuzzy logic, and adaptive neuro-fuzzy inference system (ANFIS) are becoming extensively established in predicting complex problems. ANFIS has the advantages of both ANNs and fuzzy systems and is most suitable in engineering complicated applications. This study focuses on the development of ANFIS in predicting the compressive strength of high strength concrete. A total of 550 experimental datasets of concrete were used in this research. Each dataset was consisting of six input variables that were water, cement, fine and coarse aggregates, silica fume, and superplasticizer. The compressive strength of high strength concrete was considered as the output of the ANFIS model. In this study, 440 datasets were assigned as training datasets and 110 datasets were considered as testing sets to verify the ANFIS model. The mean square error (MSE) for the training set was 0.00573, and 0.00647 for the testing datasets. The ANFIS model was able to quickly predict the concrete compressive strength with high accuracy. Also, in this research, a sensitivity analysis was applied to study the contribution of input parameters to predict the compressive strength of concrete.

Keywords Adaptive neuro-fuzzy inference system (ANFIS) · Compressive strength · Artificial neural networks (ANNs) · Mean square error (MSE)

S. Hakim (✉) · N. Jamaluddin · K.H. Boon · S.N. Mokhatar · A. N. Khalifa · Z. Jamellodin
Department of Civil Engineering, Faculty of Civil Engineering and Built Environment, University Tun Hussein Onn Malaysia (UTHM), Johor, Malaysia
e-mail: seyedhakim@uthm.edu.my

N. Sabtu (ed.), *Proceedings of AWAM International Conference on Civil Engineering 2022—Volume 2*, Lecture Notes in Civil Engineering 385,
https://doi.org/10.1007/978-981-99-6018-7_24

1 Introduction

High strength concrete is a complex mixture, inducing aggregate, water, cement, mineral, and chemical admixture and is used where it is very significant to reduce the weight of the structure or where small support of structural elements is required. High strength concrete is more efficient than normal concrete in terms of carrying loads and it can reduce the total amount of materials, consequently decreasing the overall cost of the concrete structures. Fly ash and silica fume as two pozzolans are regularly used in high strength concrete. Also, chemical admixtures like superplasticizers are usually used to produce high strength concrete mixtures. The superplasticizer makes the concrete workable and increases the compressive strength by reducing the ratio of cement to water in the concrete. However, the prediction of the compressive strength of high strength concrete became more complex because different factors were considered.

Recently, artificial intelligence (AI) techniques are becoming extensively established in predicting complex problems [1–8]. Artificial neural networks (ANNs), single and multiple linear regression, fuzzy logic, and adaptive neuro-fuzzy inference system (ANFIS) are examples of artificial intelligence systems, and they are most consistent when used to predict the compressive strength of concrete [9–14]. Artificial intelligence techniques can learn complicated relationships from datasets and offer more precise predictions. Using these techniques eliminates the need for costly laboratory tests.

According to previous studies, ANN and ANFIS make more accurate predictions for the compressive strength of concrete [15–19]. ANFIS is the combination of both ANNs and fuzzy systems that incorporates the learning abilities of ANNs with fuzzy inference system (FIS) for complex problems such as compressive strength prediction of concrete and has both ANN and fuzzy logic advantages in an individual framework [20–22]. Several research on the application and development of ANFIS have been conducted and both their pros and cons have been discussed. For example, Vakhshouri and Nejadi [23] studied on self-compacting concrete and applied ANFIS to predict the compressive strength in this kind of concrete. The authors used the basic information from 55 experimental studies as input parameters. The results ascertained that the ANFIS technique could predict the strength of concrete with a high level of accuracy. In this study, the influence of the aggregate volume and size have not been considered.

In another study, Golafshani et al. [24] proposed a novel ANFIS model in combination with ANN for predicting concrete compressive strength. In their study, an optimizer, namely Gray Wolf (GW) is applied in training procedure to eliminate the weakness of overtraining. Input parameters were the number of various materials such as coarse and fine aggregates, water, cement, blast furnace slag, fly ash, and the age of the concrete specimens. The authors concluded that the ANFIS-based model could predict the compressive strength better than the ANN-based model.

In another recent study, Pei and Wei [25] predicted the bond strength of fiber reinforced polymer (FRP) using the ANFIS technique. Eight variables including the compressive strength, tensile strength, thickness, and modulus of elasticity of FRP, aggregate size, adhesive tensile strength, length, and width of FRP are used as input parameters, and the bond strength is considered as the output of ANFIS. The authors applied Ant Colony Optimization (ACO) algorithm in the combination with ANFIS to improve the predicting accuracy of the proposed model. The results exhibited excellent accuracy of ANFIS to predict the bond strength of FRP under direct tension.

ANFIS was also used in a further study to identify the compressive strength of concrete combined with fly ash as a geopolymer by Ahmad et al. [26]. Experimental and analytical studies were conducted in this research. The results concluded that the ANFIS model could provide a novel method for the design of geopolymers to predict compressive strength. This approach was able to provide a good prediction for an insufficient or incomplete dataset of geopolymers. In a similar order, the ANFIS technique was implemented to predict the compressive strength of concrete which was contained fly ash as geopolymer and furnace slag by Dao et al. [27]. The results ascertained that ANFIS had a robust prediction capability when compared to the ANN technique.

Suffice to say, in recent years, the interest in applying ANFIS to predict concrete compressive strength has increased, and several attempts have been made [28–32]. Despite the valuable efforts by researchers, there is still inadequate information and outstanding need for the application and development of ANFIS to predict the compressive strength of high strength concrete, and there is a need to accomplish more study in this area. Given that, the main highlight of this research was to develop an ANFIS system to predict the compressive strength of high strength concrete using experimental datasets.

2 Adaptive Neuro-Fuzzy Inference System (ANFIS)

ANFIS is developed by Jang [33] and is the combination of both ANNs and fuzzy systems that incorporates the learning abilities of ANNs with fuzzy inference system (FIS) for complex problems such as compressive strength of concrete. ANFIS merges both ANN and fuzzy logic concepts and has the advantages of both methods in an individual structure. Therefore, ANFIS is very strong in nonlinear function approximation and has a powerful generalization capability.

ANFIS works very well with adaptive techniques and linear approaches. Therefore, this approach has a superior processing time for a weighted average defuzzification and is well-suited for mathematical analysis. The architecture of the ANFIS has five layers and is demonstrated in Fig. 1. These layers consist of a fuzzification layer, product layer, normalized layer, defuzzification layer, and a total output layer. In the fuzzification layer, inputs are connected to the membership functions (MFs) and {ai, bi, ci} is the parameter set of the adaptive node. The parameters in the fuzzification layer are named premise parameters. In this study, the ANFIS model is one of the applications of a Sugeno fuzzy inference system.

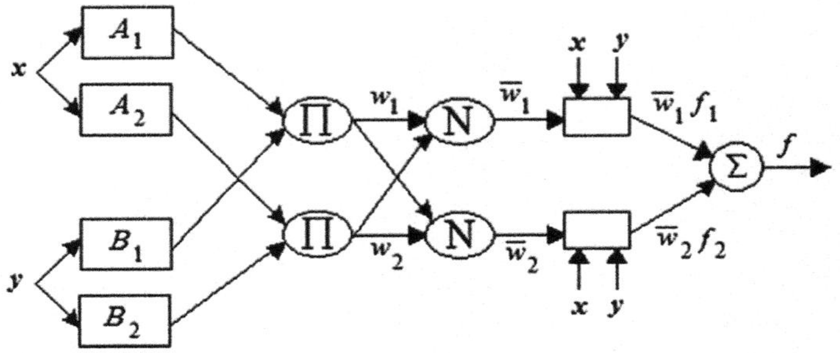

Fig. 1 Architecture of ANFIS for two-input Sugeno method [34]

Each node in the second layer is fixed and the AND operator is used to obtain the strength of the rule. The output from the second layer is the product of all the signals. In the third layer, each node is also fixed and labeled as Norm. In layer 4, namely defuzzification, the outputs are normalized firing strengths. Each node in this layer accepts all nodes from the rule. Also, in layer 4, each node is an adaptive node with a node function and {pi, qi, ri} is the parameter set of this node. The parameters in layer 4 are referred consequent parameters. Finally, the single node in layer 5 calculates the entire output as the total of all incoming signals.

Backpropagation and hybrid algorithm are two ANFIS training algorithms. ANFIS creates a FIS, whose MFs can be adjusted using a backpropagation algorithm or in combination with a least square approach. This modification allows FIS to acquire information from the datasets. In general, ANFIS uses a hybrid algorithm to train the inference system, which is the combination of least mean squares (LMS) and the gradient descent method.

The gradient descent approach is applied to adjust premise factors ({ai, bi, ci}), while the least squares method is applied to identify consequent factors ({pi, qi, ri}). In the training process, the LMS approach is first applied to find the consequent factors, while the premise factors such as membership functions (MFs) are fixed for the recent cycle during the training procedure. Then, the gradient descent is applied to update the premise factors, while the consequent factors remain fixed. In this present study, ANFIS used a hybrid method for membership function parameter estimation.

3 Development of ANFIS Model

At the first step, training of the ANFIS using a hybrid learning algorithm is started to fix all the adjustable factors such as the number of membership functions (MFs) and fuzzy rules to make the ANFIS output match the training datasets. The hybrid algorithm is a combination of a backpropagation algorithm and the least mean square approach, and it converges faster than the original backpropagation method applied in ANNs. In this algorithm, premise and consequent parameters will be updated after each dataset is presented in the algorithm. The hybrid algorithm has two steps consisting of forward and backward passes. In forward phase of the hybrid algorithm, the input membership functions are fixed, and the output membership functions are trained and identified using the least mean square approach. In the backward phase, the output membership functions are fixed, and the premise parameters which are the input membership function parameters will be adjusted using the gradient descent approach.

In order to construct the ANFIS models for predicting the strength of concrete, 550 high strength concrete mix design datasets were gathered from the research laboratory of concrete technology. These data were containing water, coarse and fine aggregate, cement, silica fume, and superplasticizer. The strength of concrete was from 30 to 140 MPa. Out of the 550 datasets, 440 (80%) were considered as training sets, and 110 were used as testing datasets. The dividing of the datasets was performed randomly between the training and testing sets, and each dataset has been statistically checked to make sure that it covers the range of input variables.

In this study, ANFIS was trained using MATLAB 2015b and various ANFIS architectures using input datasets with different types and number of membership functions (MFs), and fuzzy rules were trained to obtain the output with high precision. The bell membership functions were applied in this research and the number of iterations in the training phase was set to 5000. After these number of iterations, the mean square error reached a value of 0.00573, and the ANFIS model reaches an acceptable level after training. According to the training results, the correlation coefficient (R^2) was obtained to 0.988 for training datasets. The comparison between the predicted compressive strength of concrete by the ANFIS model and laboratory results is illustrated in Fig. 2.

The architecture of the ANFIS to predict compressive strength had 31 membership functions and 31 rules for layers 2 and 3. In the next step, the ANFIS model was

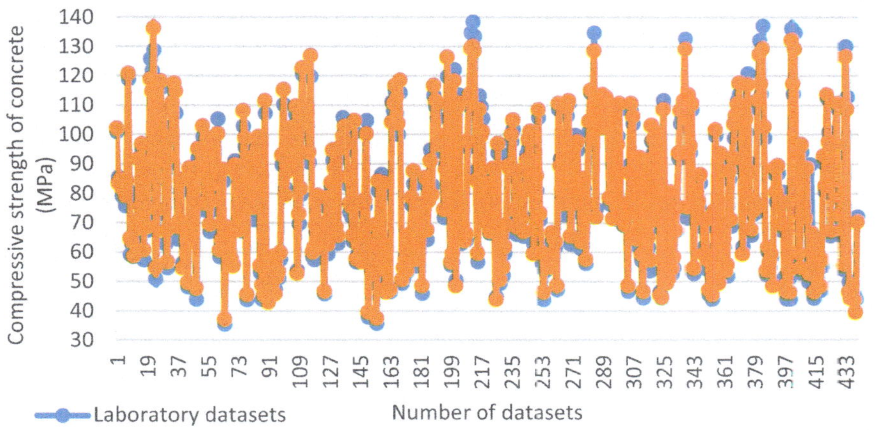

Fig. 2 Comparison of the predicted results by ANFIS with laboratory training datasets

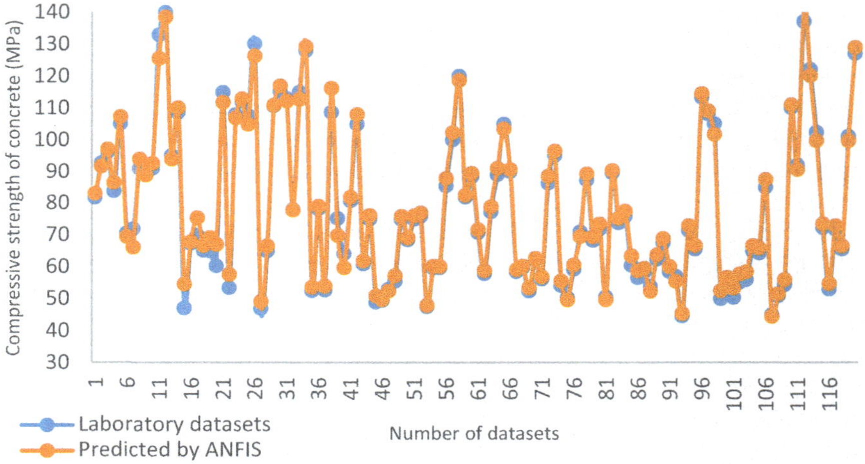

Fig. 3 Comparison of the predicted results by ANFIS with laboratory testing datasets

tested to check the performance of the architecture to predict compressive strength of concrete and to evaluate how well the network is trained. The predicted results from testing datasets have been compared with laboratory results and are shown in Fig. 3.

According to the testing results, the ANFIS could predict the concrete compressive strength with the mean square error of 0.00647. The correlation coefficient (R^2) value was 0.976 for testing datasets. The results from ANFIS were very close to the laboratory results and indicated that the developed ANFIS could be applied as an efficient technique for predicting the concrete compressive strength.

Also, in this research, a sensitivity analysis is carried out to explore the sensitivity of the ANFIS to the input parameters to predict the compressive strength of concrete. The sensitivity analysis on six input parameters was carried out and the outcomes have demonstrated in Fig. 4. It is obvious from Fig. 4 that the cement has the maximum impact on the compressive strength of concrete. If the contribution of cement is 1, the contribution of water, silica fume, superplasticizer, fine aggregate, and coarse aggregate will be 0.39, 0.29, 0.14, 0.32 and, 0.30 respectively. Figure 4 shows that superplasticizer has only minor contributions to the variation in compressive strength of concrete.

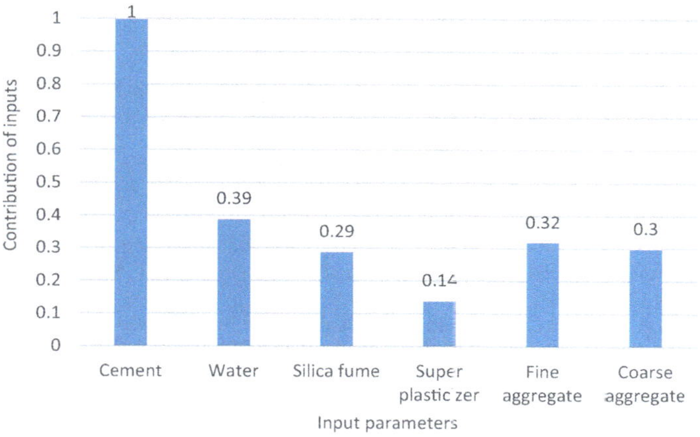

Fig. 4 Sensitivity analysis of ANFIS input variables

4 Conclusion

In this present study, an ANFIS model for the prediction of compressive strength of high strength concrete based on laboratory datasets is developed. Six parameters consisting of water, cement, fine and coarse aggregates, silica fume, and superplasticizer were identified as input variables, and the compressive strength was considered as the output of the ANFIS model.

Different architectures of ANFIS models were proposed and implemented. From the results, it was determined that ANFIS predicted the compressive strength of high strength concrete with a very small mean square error (MSE) and a high coefficient of correlation. The results demonstrated that the ANFIS with hybrid algorithm has the potential to give a very encouraging level of performance for the problem of concrete strength prediction. According to this study, using ANFIS as a computational intelligent method can provide a quick means of predicting the compressive strength of concrete based on some of its influence factors.

Also, in this research, a sensitivity analysis was implemented to study the contribution of input parameters to predict the compressive strength of concrete. The results of sensitivity analysis ascertained that cement had the maximum impact on the compressive strength of concrete and superplasticizer had only minor contributions to the variation in compressive strength of concrete. The application of ANFIS to extract relational knowledge from datasets to concrete strength prediction is practical and has a good future.

Acknowledgements Communication of this research is made possible through monetary assistance by Universiti Tun Hussein Onn Malaysia and the UTHM Publisher's Office via Publication Fund E15216.

References

1. Ahmad, A., Ahmad, W., Aslam, F.: Compressive strength prediction of fly ash-based geopolymer concrete via advanced machine learning techniques. Case Stud Constr Mater **16**, e00840 (2022)
2. Guçluer, K., Ozbeyaz, A., Goymen, S.: A comparative investigation using machine learning methods for concrete compressive strength estimation. Mater Today Commun **27**, 102278 (2021)
3. Hakim, S.J.S., Abdul Razak, H: Frequency response function-based structural damage identification using artificial neural networks-a review. Res. J. Appl. Sci. Eng. Technol. 7(9), 1750–1764 (2014)
4. Akrami, S.A., Nourani, V.: Development of nonlinear model based on Wavelet-ANFIS for rainfall forecasting at Klang Gates Dam. Water Resour. Manage. **28**(10), 2999–3018 (2014)
5. Akrami, S.A., Nourani, V., Hakim, S.J.S., Mohammadhassani, M: Using modified adaptive neural fuzzy and ANFIS-Wavelet for prediction of rainfall data. Environ. Eng. Manage. J. 13(9), 2201–2211 (2014)
6. Hakim, S.J.S., Abdul Razak, H: Application of artificial neural network on vibration test data for damage identification in bridge girder. Int. J. Physic. Sci. 6(35), 7991–8001 (2011)
7. Hakim, S.J.S., Irwan, J.M., Ibrahim, M.H.W., Ayop, S.S.: Structural damage identification employing hybrid Intelligence using artificial neural networks and vibration-based methods. J. Appl. Res. Technol. 20(2), 221–236 (2022)
8. Hakim, S.J.S., Ibrahim, M.H.W., Mohammadhassani, M., Yeoh, D., Jaini, Z.M., Chik, T.N.T.: Implementing fuzzy-based artificial intelligence approach for location of damage in structures. Civil Eng. Architect. 10(4), 1564–1573 (2022)
9. Kar, S., Pandit, A.R., Biswal, K.C: Prediction of FRP shear contribution for wrapped shear deficient RC beams using adaptive neuro-fuzzy inference system (ANFIS). Structures 23, 702–717 (2020)
10. Khademi, F., Akbari, M., Jamal, S.M: Prediction of compressive strength of concrete by data-driven models, i-manager's. J. Civil Eng. 5(2), 16–23 (2015).
11. Khademi, F., Jamal, S.M., Deshpande, N.: Londhe, S: Predicting strength of recycled aggregate concrete using artificial neural network, adaptive neuro-fuzzy inference system and multiple linear regression. Int. J. Sustain. Built Environ. **5**(2), 355–369 (2016)
12. Noorzaei, J., Hakim, SJS., Jaafar, M.S., Abang Ali, A.A., Thanoon, W.A.M: An optimal architecture of artificial neural network for predicting compressive strength of concrete. Indian Concrete J. 81(8), 17–24 (2007)
13. Hakim, SJS., Noorzaei, J., Jaafar, M.S., Jameel, M., Mohammadhassani, M: Application of artificial neural networks to predict compressive strength of high strength concrete. Int. J. Physic. Sci. 6(5), 975–981 (2011)
14. Noorzaei, J., Hakim, S.J.S., Jaafar, M.S., Thanoon, W.A.M.: Development of artificial neural networks for predicting concrete compressive strength. Int. J. Eng. Technol. 4(2), 141–153 (2007)
15. Mansouri, I., Ozbakkaloglu, T., Kisi, O.: Xie, T: Predicting behavior of FRP-confined concrete using neuro fuzzy, neural network, multivariate adaptive regression splines and M5 model tree techniques. Mater. Struct. **49**(10), 4319–4334 (2016)
16. Noorzaei, J., Jaafar, M.S., Thanoon, W.A.M.: Hakim, SJS: Development of an artificial neural network model for prediction of ultimate soil bearing capacity. Civil-Comp Proceed. **82**, 102646 (2005)
17. Hakim, S.J.S., Razak, H.A: Damage identification using experimental modal analysis and adaptive neuro-fuzzy interface system (ANFIS). In: Conference Proceedings of the Society for Experimental Mechanics Series, 5, pp. 399–405 (2012)
18. Mohammadhassani, M., Nezamabadi-Pour, H., Jumaat, M., Jameel, M., Hakim, S.J.S.: Zargar, M: Application of the ANFIS model in deflection prediction of concrete deep beam. Struct. Eng. Mech. **45**(3), 323–336 (2013)

19. Hakim, S.J.S.: Abdul Razak, H: Adaptive neuro fuzzy inference system (ANFIS) and artificial neural networks (ANNs) for structural damage identification. Struct. Eng. Mech. **45**(6), 779–802 (2013)
20. Hakim, S.J.S., Abdul Razak, H., Ravanfar, S.A.: Ensemble neural networks for structural damage identification using modal data. Int. J. Damage Mechan. 25(3), 400–430 (2016).
21. Hakim, S.J.S., Razak, H.A., Ravanfar, S.A., Mohammadhassani, M.: Structural damage detection using soft computing method. Struct. Health Monitor. 5, 143–151 (2014).
22. Ozel, C.: Prediction of compressive strength of concrete from volume ratio and Bingham parameters using adaptive neuro-fuzzy inference system (ANFIS) and data mining. Int. J. Physic. Sci. **6**(31), 7078–7094 (2011)
23. Vakhshouri, B.: Prediction of compressive strength of self-compacting concrete by ANFIS models. Neurocomputing **280**, 13–22 (2018)
24. Golafshani, E.M., Behnood, A., Arashpour, M.: Predicting the compressive strength of normal and High-Performance Concretes using ANN and ANFIS hybridized with Grey Wolf Optimizer. Constr. Build. Mater. 232, 117266 (2020)
25. Pei, Z.: Prediction of the bond strength of FRP-to-concrete under direct tension by ACO-based ANFIS approach. Compos. Struct. **282**, 115070 (2022)
26. Ahmad, M., Rashid, K., Tariq, Z.: Utilization of a novel artificial intelligence technique (ANFIS) to predict the compressive strength of fly ash-based geopolymer. Constr. Build. Mater. **301**, 124251 (2021)
27. Dao, D.V., Ly, H.B., Trinh, S.H., Le, T.T., Pham, B.T.: Artificial intelligence approaches for prediction of compressive strength of geopolymer concrete. Materials 12 (2019)
28. Song, H., Ahmad, A., Farooq, F., Ostrowski, K.A., Maslak, M., Czarnecki, S.: Predicting the compressive strength of concrete with fly ash admixture using machine learning algorithms. Constr. Build. Mater. **308**, 125021 (2021)
29. Nunez, I., Marani, A., Flah, M., Nehdi, M.L.: Estimating compressive strength of modern concrete mixtures using computational intelligence: a systematic review. Constr. Build. Mater. **310**, 125279 (2021)
30. Chaabene, W.B., Flah, M., Nehdi, M.L.: Machine learning prediction of mechanical properties of concrete: critical review. Constr. Build. Mater. **260**, 119889 (2020)
31. Nedushan, B.A.: Prediction of elastic modulus of normal and high strength concrete using ANFIS and optimal nonlinear regression models. Constr. Build. Mater. **36**, 665–673 (2012)
32. Yusoff, A.R.M., Mokhatar, S.N., Budiea, A.M.A., Hakim, S.J.S.: A review: study on spent garnet as construction material. Int. J. Integrat. Eng. 14(5), 73–80 (2022).
33. Jang, J.S.R.: ANFIS: adaptive network-based fuzzy inference systems. IEEE Trans. Syst. Man Cybern. **23**(3), 665–685 (1993)
34. Boga, A.R., Ozturk, M., Topcu, I.B.: Using ANN and ANFIS to predict the mechanical and chloride permeability properties of concrete containing GGBFS and CNI. Composites 45, 688–696 (2013)

Investigation on Mechanical Properties of High Strength Ductile Engineered Geopolymer Composites (EGC)

Vishal Avinash Ramesh, Ehsan Nikbakht Jarghouyeh, Nurizzati Amirzali, and Dayana Eddie Hirawan

Abstract In this paper, high strength ductile engineered geopolymer composite (EGC) is investigated. EGC is an environmentally friendly and more importantly it is bendable geopolymer with tensile strain-hardening behaviour which possesses enhanced mechanical properties compared to normal geopolymer concretes and ordinary Portland cement (OPC) concrete. Bond performance or the bond strength between the rebar and the EGC is the main study in this research as it is an important factor as it influences the structural load bearing capacity and failure modes of elements. The idea of casting and creating the best mix design is required in order to generate higher strength EGC cured in ambient temperature. On top of that, the mechanical property is also a part that was discussed in this research which includes the compressive strength. Hence, in this study, the compressive strength of ductile fibrous EGC cured in ambient temperature is investigated. Moreover, bond strength between EGC and rebar with various diameters was used as variables in experiments to examine the results leading to bonding strength using the pull-out test method and from the analysis of bond strength it was observed that the bond strength of EGC increases with the increase in rebar diameter. Other than that, the slip behaviour of the EGC specimens decreases with the increase of rebar diameter. Critical analysis was made and calculated to conclude this investigation.

Keywords Bond behaviour · Reinforced concrete · Geopolymer · Engineered geopolymer composite (EGC) · Pull-out test

---

V. A. Ramesh · E. N. Jarghouyeh (✉) · N. Amirzali · D. Eddie Hirawan
Civil Engineering Department, Universiti Teknologi PETRONAS, Seri Iskandar, Perak, Malaysia
e-mail: ehsan.nikbakht@utp.edu.my

V. A. Ramesh
e-mail: vishal_20001985@utp.edu.my

N. Amirzali
e-mail: nurizzati_16005033@utp.edu.my

D. Eddie Hirawan
e-mail: dayana_16003735@utp.edu.my

N. Sabtu (ed.), *Proceedings of AWAM International Conference on Civil Engineering 2022—Volume 2*, Lecture Notes in Civil Engineering 385,
https://doi.org/10.1007/978-981-99-6018-7_25

1 Introduction

The production of the OPC concrete leaves a huge impact to the environment. The ordinary concrete is liable for around 10% of worldwide anthropogenic ozone harming substance emanations, e.g., carbon dioxide releases [1]. As a solution to these issues, geopolymer concrete is being introduced and implemented as it is an environmentally friendly product which indirectly decreases the environmental impact and gives many advantages [2]. However, recently due to the brittle behaviour of geopolymer concrete, engineered geopolymer composite (EGC) is being developed and attracted more attention from both researches and engineers due to its ductile, bendable properties. The main difference between (EGC) is that there will be no coarse aggregate included in the EGC matrix. Other than that, the important part of EGC is using fibre reinforcement bar as it improves the ductility of the geopolymers and allows certain characteristics functioning as high tensile and strain-hardening behaviour [3].

Even though, several studies have been conducted on the bonding behaviour between steel rebar and geopolymer concrete, however there is limited research on EGC mechanical properties and structural behaviour. Therefore, due to the huge margin in the matrix formation between the engineered geopolymer composite and conventional concrete, it is necessary to thoroughly study the bonding performance of EGC with the rebars [4].

Other than that, there are other certain elements that can influence the mechanical strength of the EGC. One of the main factors is the quality of the source materials. Source materials that have a high content of aluminium and silicate with high reactivity are able to enhance the strength of the geopolymer concrete [5]. Finally, since geopolymer needs heat for curing is one of the drawbacks geopolymer concrete faces as it is not suitable for on-site pouring and casting [6]. One of the research gaps that was identified in the previous literatures includes the curing method where most of the researchers focused their studies on curing the normal geopolymer composite cured at elevated temperature. Hence, in this study, the EGC with enhanced mechanical properties and high strength is developed and investigated in the ambient condition due to its suitability. The scope of this research emphasizes on the mechanical behaviour and performance of the engineered geopolymer composite. The main mechanical properties that was tested include the compressive strength. Furthermore, this research focuses on the parameter which includes the rebar diameter of the engineered geopolymer composite which will be investigated on how they will influence the bond behaviour of the EGC. Moreover, the pull-out test has been conducted to study the bond behaviour of EGC and various rebar diameters. The main objective in this research comprises investigating the influence of rebar diameter in affecting the bond strength and bonding slip behaviour of the engineered geopolymer composite using the pull-out test method.

2 Experimental Programme

2.1 Materials and Processing

In this study, materials used as ingredients to produce EGC are fine aggregate (silica sand), fly ash (FA), ground granulated blast furnace slag (GGBS), water, sodium silicate (Na_2SiO_3), sodium hydroxide (NaOH), superplasticizer and polyvinyl alcohol fibre (PVA). Locally sourced low-calcium fly ash (Class F) (ASTM C618) was used in this study where it was obtained from a power plant in the state of Perak, Malaysia. The alkaline activators that were utilized to bind together with the dry ingredients to produce the EGC were sodium silicate and sodium hydroxide. Furthermore, the molarity that was chosen to be used in this study is constant which was 12 M. PVA fibre is utilized for the reinforcement in the EGC mixture with 2% of volume fraction.

Few trial mixes were examined to ensure the quantity needed for the production of EGC and the variables that was involved in preparing the mix design involves firstly the alkaline solution to fly ash ratio that was used in this mix was 0.4 and sodium silicate to sodium hydroxide was fixed at a constant value of 2.5, and the molarity of sodium hydroxide was fixed at 12 M. Table 1 gives the mixture proportions that was utilized in this study.

In this research, the pull-out test is used to study the bond behaviour of EGC and steel rebar with the parameter of various rebar diameters. The diameter of rebars that was used are 8, 10, 14 mm, respectively. Cylinder samples of 100 × 200 mm (diameter × height) with varying rebar diameter was used. The mechanical property of the rebar utilized for this study is provided in Table 2.

Table 1 Mix design proportions

Mix proportion	Kg/m^3
GGBS	220
Silica fume	220
Silica sand	230
Fly ash	880
NaOH	112
Na_2Sio_3	283
PVA	26
Water	88
Curing temperature	27

Table 2 Mechanical property of deformed bars

Type of rebar	Rebar diameter (mm)	Yield strength (MPajl)	Ultimate strength (MPa)	Modulus of elasticity (GPa)
Deformed bar	8	376	521	207
Deformed bar	10	436	587	204
Deformed bar	14	454	581	203

2.2 *Testing Procedures*

Compressive test is conducted on 100 mm × 100 mm × 100 mm EGC cubes. The cubes are tested on 7, 14 and 28 days. For each mixture that was produced, 3 cubes were tested. The pull-out test for the cylinder specimens was conducted following a procedure ASTM C234 (ASTM 1988).

3 Results and Discussion

3.1 *Compressive Strength*

The compressive strength that was obtained from the 100 mm EGC cubes is given in Table 3. From the two-mix design produced, it can be observed that the usage of fine aggregate and PVA fibre.

The compressive strengths of two mixture designs with same proportions of constituents but with a different type of fly ashes (mix one) and (mix two) are summarized in Tables 3 and 4. Both fly ash was type class *F* but from different

Table 3 Compressive strength of mix 1

Marking date cast	Age days	Dimension (mm)	Density (kg/m^3)	Compressive strength (MPa)	Average strength (MPa)
1	1	100 × 100 × 100	1954	12.58	12.58
2	7		1945	46.36	46.42
3	7		1940	47.02	
4	7		1975	45.88	
5	14		1840	46.88	47.45
6	14		1924	46	
7	14		1958	49.46	
8	28		1965	58.77	59.7
9	28		1960	60.92	
10	28		1954	59.42	

Table 4 Compressive strength of mix 2

Marking date cast	Age days	Dimension (mm)	Density (kg/m^3)	Compressive strength (Mpa)	Average strength (Mpa)
1	1	100 × 100 × 100	1953	9.22	9.22
2	3		2002	21.83	22.18
3	3		1953	21.14	
4	3		1957	23.57	
5	7		1949	35.35	35.28
6	7		2019	35.25	
7	7		1858	35.25	
8	14		1949	35.02	40.29
9	14		1822	39.54	
10	14		1863	46.3	
11	28		1965	50.52	53.3
12	28		1988	53.30	
13	28		1878	56.06	

sources were provided for this study. The samples of both mix designs after the compressive strength test are displayed in Fig. 1. As the results show, mix design one has slightly higher strength compared to mix design two because of different chemical compositions of fly ashes used in their mix design. Moreover, according to previous research papers, the dosage and reactivity of the activator solution with the binders have an impact on the strength of geopolymer concrete formed [7]. Thus, the geopolymerization process that took place in mix two was lesser which affected the compressive strength and durability of mix two compared to mix one.

3.2 *Bond Behaviour*

The bond stress of the cylinder specimens is calculated as below:

$$\tau = \frac{P_{max}}{\pi d_b L_d}$$

where P_{max} = maximum pull-out load
d_b = rebar diameter.
L_d = embedded bar length.

The results of bond strength and bond failure of specimens with rebar diameters of 8, 10 and 14 are compared in Table 5. Based on Table 5, it shows that the maximum ultimate bond strength occurred at the 14 mm bar diameter which is 8.672 MPa. The lowest bond strength occurred at the 8 mm bar diameter which is 8.025 Mpa.

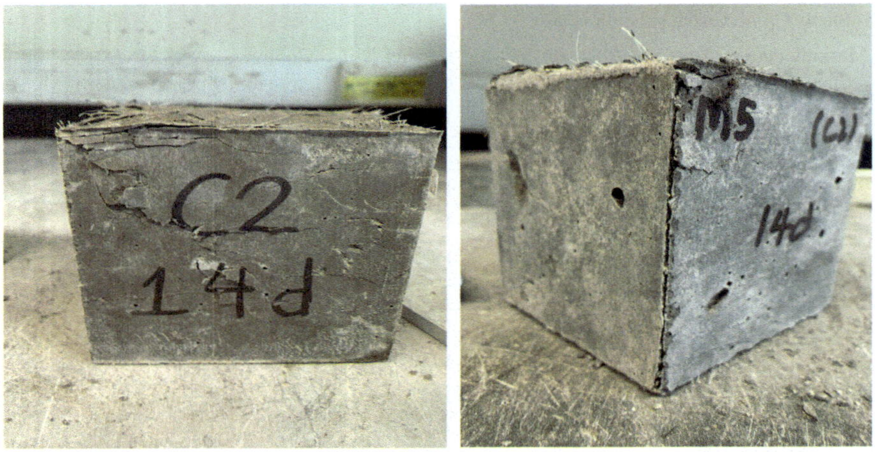

(a) Mix 1 (b) Mix 2

Fig. 1 Compressive failures of samples mix 1 and mix 2 with different type of fly ashes

Table 5 Pull-out test results of EGC with varying rebar diameter

No.	Bar diameter (mm)	Embedded length (mm)	Maximum pull-out load (KN)	Ultimate bond strength (MPa)	Mode of failure
1	8	100	21.352	8.025	P-S
2	10	100	25.211	8.496	P-S
3	14	100	32.671	8.672	P-S

Based on Fig. 2, it is observed that bond strength increases with the increase in rebar diameter. One of the main reasons that can be observed on the high bond strength in 14 mm rebar is due to the greater bond area of 14 mm bar diameter with the EGC compared to the other rebar diameter. Hence, higher load was needed in order to pull out the rebar from the surrounding of EGC.

3.2.1 Bond Failure Modes and Slip Behaviour

As the results show in this study, the main type of failure in the EGC specimens with varying rebar diameter is pull-out splitting failure (P-S), From Table 5, P-S failure consists of pulling out and splitting colonical cracks. The failure modes of the cylinder specimens after the pull-out test are shown in Fig. 3. Ribbed bars were used as the rebars in this study, and from the researchers view, pull-out splitting failure normally occurs in ribbed rebars [8].

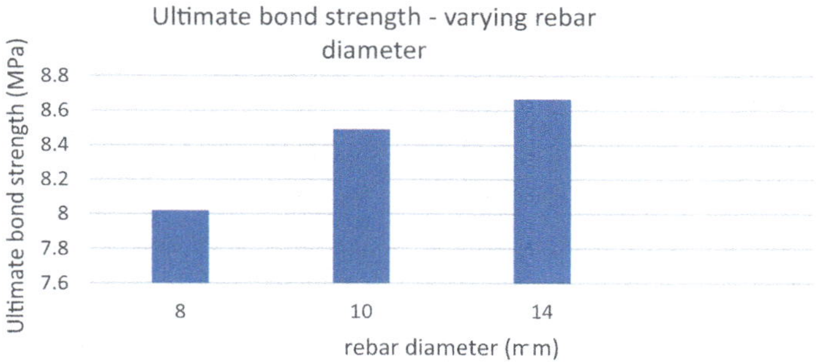

Fig. 2 Ultimate bond strength of EGC with varying rebar diameter

Moreover, the slip behaviour of EGC with varying parameter which includes the rebar diameter is shown in Fig. 4. During the pull-out test, the maximum bond strength shifts progressively from the loaded to the unloaded part of the rebar, while the bond strength value at the loaded end drops significantly, resulting in a nonlinear stress distribution along the rebar. The bond strength dependency on rebar diameter may be described by this migration and nonlinear strength distribution [9]. As shown in Fig. 4, among all specimens, the EGC specimen with 14 mm bar diameter exhibited less slip compared to other specimens with 8 and 10 mm, where the higher the load of the EGC the lower the slip bond that corresponds to higher tensile strength. One of the reasons that correspond to the lower bond slip in the 14 mm bar diameter can be due to the good chemical adhesion between the rebar and the concrete which have been also discussed by Mathew et al. [10]. The failure load for 8, 10, 14 mm rebar of 100 mm embedded length is 21.352, 25.211, 32.692 kN, respectively. Whereas the slip of 8, 10, 14 mm of rebar was 0.47, 0.57, 0.17 mm, respectively. The bond-slip value overall observed was <1 mm. As the bar diameter increase, the pull-out load capacity of the bond-slip specimen also increases.

4 Conclusions

This paper investigated and provided the impact of behaviour mechanical properties of EGC including compressive strength and bond-slip behaviour of EGC with different bar diameters. A few cylindrical specimens was conducted with different size of deformed rebar.

For the mechanical property of the EGC, the compressive strength is analysed and it was shown that EGC achieves early high strength concrete, and the usage of silica sand and PVA fibre used in this mix design has lower the void specimen because the components has dispersed in early strength accordingly.

Furthermore, the bond-slip tests in this study show there is excellent bonding between rebar and EGC. Mode of failure of all specimens with varying rebar diameter

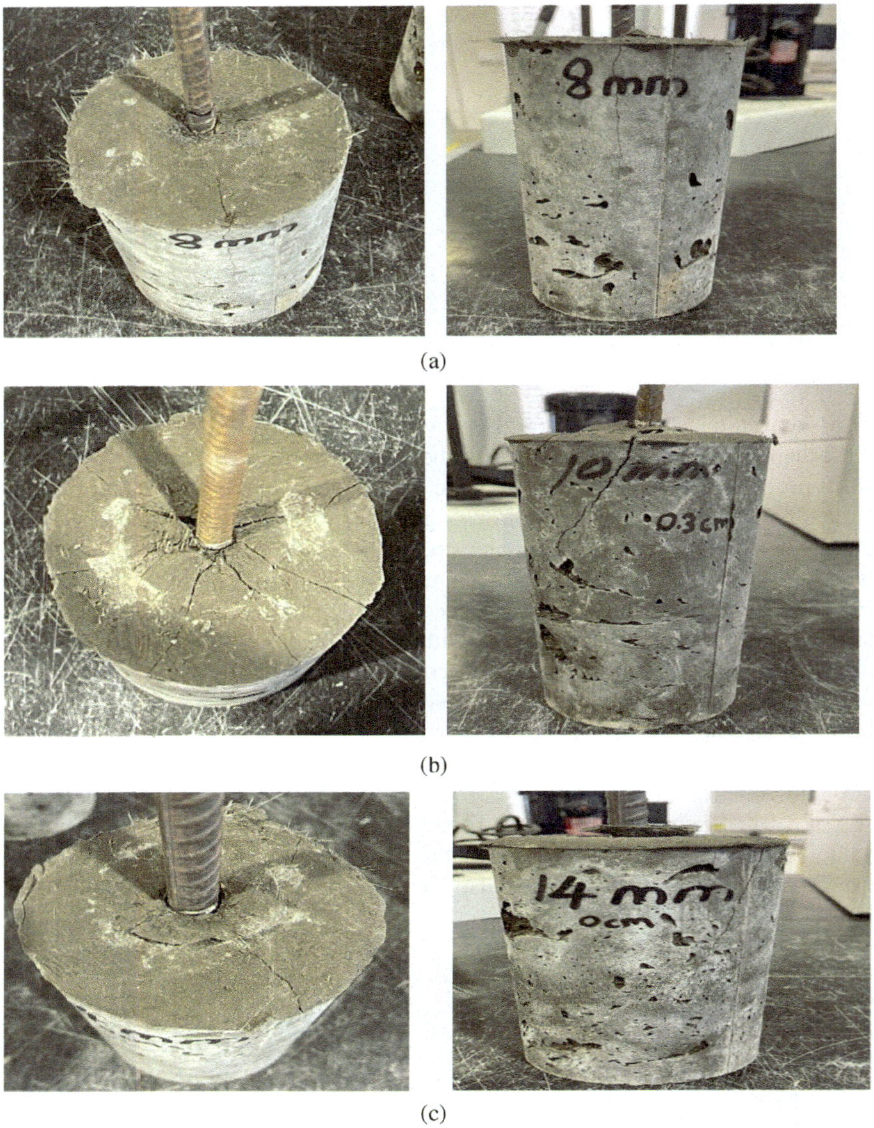

Fig. 3 Typical failure modes of cylinder specimens: **a** 8 mm, **b** 10 mm, **c** 14 mm

was P-S failure. As the results showed, the rebar slip in all specimens with various diameters overall was small. Moreover, as the bar diameter increased, the pull-out load capacity of specimens also increased proportionally.

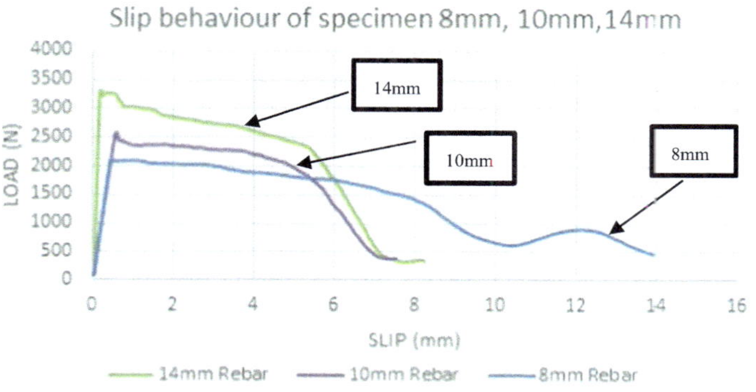

Fig. 4 Slip behaviour of EGC with varying rebar diameter

Acknowledgements The authors would like to express their gratitude for the financial assistance they have received from the Universiti Teknologi PETRONAS Malaysia under grant number YUTP 015LC0-177.

References

1. Hassan, A., Arif, M., Shariq, M.: A review of properties and behaviour of reinforced geopolymer concrete structural elements-a clean technology option for sustainable development. J. Clean. Prod. **245**, 1–15 (2020)
2. Nikbakht, E., Lok, J.W.W., Teo, W.: Structural behaviour of novel composite beams consisting of geopolymer concrete and high-performance concrete. Structures **32**, 106–115 (2021)
3. Zahid, M., Shafiq, N., Razak, S.N.A., Tufail, R.F.: Investigating the effects of NaOH molarity and the geometry of PVA fibers on the post-cracking and the fracture behaviour of engineered geopolymer composite. Construct. Build. Mater. **265**, 1–13 (2020)
4. Sumajouw, D.M., Hardjito, D., Wallah, S.E., Rangan, B.V.: Fly ash based geopolymer concrete: study of slender reinforced columns. Adv. Geopolym. Sci. Technol. **42**, 3124–3130 (2007)
5. Trindade, A.C.C., Borges, P.H.R., Silva, F.D.A.: Mechanical behavior of strain-hardening geopolymer compositesreinforced with natural and PVA fibers. Mater. Today Proceed. **8**, 753–759 (2019)
6. Gholampour, A., Ozbakkaloglu, T., Ng, C.-T.: Ambient- and oven-cured geopolymer concretes under active confinemen. Constr. Build. Mater. **228**, 1–8 (2019)
7. Herwani, I., Pane, I., Budiono, B.: Compressive strength of fly ash-based geopolymer concrete with a variable of sodium hydroxide (NaOH) solution molarity. In: MATEC Web of Conferences (2018)
8. Xiao, J., Long, X., Ye, M., JIang, H., Liu, L., Zhai, K.: Experimental study of bond behavior between rebar and PVA-engineered cementitious composite (ECC) using pull-out tests. Front. Mater. **8**, 1–15 (2021)
9. Zhao, R., Sanjayan, J.G.: Geopolymer and Portland cement concrete in simulated fire. Mag. Concr. Res. **63**(3), 163–173 (2011)
10. Mathew, G., Sureshbabu, N.: Bond-slip behavior of geopolymer concrete after exposure to elevated temperature. Jordan J. Civil Eng. **15**, 570–585 (2021)

Effect of Chevron, Single and Cross Bracing in Optimized Planar Steel Braced Frames on the Mitigating Progressive Collapse

Vahid Jahangiri, Mehdi Maghfouri, and Seyed Mohammadreza Ghadiri

Abstract Improving the behavior of structures can be a vital requirement in order to the enhancement of the building strength against progressive collapse. To investigate this, structural optimization methods are used to cost-effectively design three types of steel moment frames, namely chevron-, single- and cross-braced frames. The frames were assessed by using an alternative path method (APM), based on the General Services Administration (GSA) code, as well as the linear static analysis. One meta-heuristic algorithm was also used to optimize different layouts of braced frames. In this regard, eighteen different steel frames in five and ten-story buildings with four-bay steel dual systems, which are combined with chevron-, single- and cross-braced frames, were assumed to investigate the vertical displacements and weight of the structures through OpenSees, MATLAB and "TEKLA structures" software. However, the design requirements were on the basis of "LRFD360-10" and "GSA2003" code provisions. The studied models were subjected to different scenarios of one-column sudden removal on the ground floor. The obtained results showed that the chevron-braced frame has the desired structural behavior for the progressive collapse due to less value of vertical displacements in the location of column removal, and can be more economical in comparison with single- and cross-braced frame optimized weight of the structures.

Keywords Progressive collapse · Bracing layout · Dual system · LRFD · Meta-heuristic algorithm · Optimization

---

V. Jahangiri (✉)
Department of Civil Engineering, Faculty of Engineering, Kharazmi University, Tehran, Iran
e-mail: Vahid.jahangiri.eng@gmail.com

M. Maghfouri
Department of Civil Engineering, Faculty of Engineering, University of Malaya, 50603 Kuala Lumpur, Malaysia

S. M. Ghadiri
School of T&L, Malaysia University of Science and Technology, Petaling Jaya, Malaysia

N. Sabtu (ed.), *Proceedings of AWAM International Conference on Civil Engineering 2022—Volume 2*, Lecture Notes in Civil Engineering 385,
https://doi.org/10.1007/978-981-99-6018-7_26

1 Introduction

In the construction industry, the safety of structures has been always the main concern for design engineers, as they are in a prime position to influence the safety aspects of the structures and to prevent any failure [1]. Structural failures can happen for many reasons, but the results are very often devastating. It could be life-threatening or cause severe physical injury. In some particular circumstances, an initial local failure of a building structure could be developed and ultimately leads to partial or even collapse of an entire building, progressively [2]. This phenomenon, which is technically called progressive collapse, caused by the removal of one or more load-bearing element—due to accidental actions—is followed by a chain of failures through the structure. This dynamic failure is also called disproportionate collapse—the notion of "disproportionality" is common to all definitions of progressive collapse but is ambiguous—for the reason that the collapse is out of proportion to the triggering event [3, 4]. The General Services Administration [5] describes the progressive collapse as a specific situation where local failure of a primary structural member causes the collapse of adjoining members which, in turn, leads to additional collapse. Progressive collapse can result from a type of abnormal loads not considered routinely in the design such as internal or external explosions, severe fires, vehicular collisions and severe unexpected environmental loads such as earthquakes, hurricanes, floods and tornadoes that stress the building system beyond the design assumptions and criteria [6]. The importance of preventing buildings from the progressive collapse was not considered until the late 1960s, after the partially failure of a 22-story precast concrete building—a gas explosion at eighteenth story led to the collapse of load-bearing precast concrete panels near the corner of the recently opened Ronan Point council estate in London (Fig. 1)—was occurred [7, 8].

Furthermore, recent findings in high-rise concrete and steel buildings include advanced framing systems, the efficient usage of concrete with high strength and

Fig. 1 Progressive collapse of 22-story Ronan Point apartment building (May 16, 1968) [8]

increased level of precision in the analysis approaches. These result in building designs with a considerable smaller safety margin, leading to greater vulnerability during progressive collapse [9, 10]. Considering severe pressures to obtain economy and efficiency in designing or versatility in occupancy may also lead to systems having a little capacity of inherent energy absorbing or built-in preventing the progressive collapse, which may be vulnerable to load conditions outside the design envelope. In the late of twentieth century (1980s), design codes and standards in the U.S. started incorporating requirements for "general structural integrity" in order to consider nominal resistance to progressive collapse [11]. Structural integrity was to be achieved by providing ductility, redundancy and continuity. These days, however, U.S. model building provisions do not include specific standards to provide resistance under progressive collapse. Therefore, the progressive collapse of buildings attracted the attention of structural engineers toward the collapse of buildings in a progressive manner.

Several studies are available to solve the progressive collapse behavior in structures. The strategies including infill steel panels [12], viscous dampers [13], rotational friction dampers [14], appropriate column arrangements, more redundancy in the structure, use of beam-to-column connections have been considered. These are the possibility of axial force transformations that result from the destructed column removal and a higher capacity of energy absorption in the structure. In the last decades, there have been design requirements and guidelines for retrofitting a structure to withstand progressive collapse [15, 16]. National Building Code of Canada (NBC) [17] provides some general statements regarding the need for structural integrity and includes a recommendation for reinforcement continuity, structural mechanisms and good layout to resist progressive collapse after local loss of support. Some design provisions including Precast Concrete Institute (PCI) guide and ACI 318 for precast concrete bearing wall buildings (PCI 1976) provide provisions for all structural integrity minimum levels [2]. Other corresponding design methods, such as the tie force method, alternative load path (ALP) method, key element design approach and risk-based method have been also implemented. Previous researchers reported that the ALP method, which is also known as the notional member removal approach, is a simple yet effective method widely employed in many research programs. In this method, one or more supporting components such as columns and walls are selectively removed, and the structural robustness can be estimated to see whether there are sufficient ALPs to sustain the load originally carried by the missing columns or walls [18].

Kim and An [19] studied the resisting capacity of progressive collapse in steel moment frames with the use of the alternate path method (APM) and the results were compared with the ones obtained from linear static and nonlinear dynamic analyses. It is concluded that nonlinear dynamic analysis caused to larger structural responses depending on the column removal position, number of stories and applied loads. In another investigation, they also analyzed the progressive collapse behavior of a steel moment frame using dynamic and nonlinear static analysis. The obtained results confirmed that ductility and resisting capacity of progressive collapse in the frames are increased with increasing the number of stories [20]. Fu [21] studied the response of a multi-story steel braced building considering consecutive column removal scenarios using 3-dimensional finite element models. It was observed that different column removal sequences produce different plasticity formations. Accordingly, the author recommended several measures to mitigate progressive collapse in future designs. Tavakoli and Hasani [22] evaluated the seismic progressive collapse capacity of steel special moment resisting frames. Analyses showed that the potential of progressive collapse depends on the removed column, number of stories and earthquake characteristics. Previous researchers concluded that during nonlinear static and nonlinear dynamic analyses, the upper column and the columns within the range of failure do not have the potential for progressive collapse. According to their research, the nodes' displacements in linear static analysis is approximately 40% higher than that of nonlinear dynamic analysis, and the displacement of nodes in the nonlinear static analysis is almost 30% as much as that of nonlinear dynamic analysis [23].

Braced frames are widely used in constructions because they are not only easily analyzable but also economic. The concentrically braced frame (CBF), moment resisting frame (MRF) and eccentrically braced frame (EBF) are the most common structural systems which are able to resist the lateral loading [24]. Although the CBF system has considerable stiffness, it provides some weaknesses such as low deformation capacity compared with the EBF system. Eccentrically connection of braces to the beam-column joints in the EBF system has positive advantages in comparison with CBF and MRF systems. Besides, it dissipates more energy by developing moment and shear hinges in the link beams [25, 26]. Khandelwal et al. [27] analyzed the progressive collapse resistance of CBF and EBF systems and found out that the latter ones are less vulnerable to progressive collapse compared to the CBF. In addition, the simulation results indicated that in spite of their ductility, shear-tab connections performed in gravity fail to have enough strength to resist progressive collapse when a gravity column is lost [3]. Liu et al. [28] used a structural optimization method in order to cost-effectively design of steel moment frames with resistance against progressive collapse. They optimized each structure by using seismic design without progressive collapse consideration (MWD), Progressive Consideration by Linear Static Analysis (PCLS), Progressive Damage by Nonlinear Static Analysis (PCNS) and Progressive Collapse by Nonlinear Dynamic Analysis (PCND), and finally the weights of the structures were compared. The results showed that the weight of the optimized structure in PCLS, PCNS and PCND configurations increased by 38.8, 13.2 and 8.2%, respectively. This means that linear static analysis gives more conservative results in comparison with others. Moreover, the genetic algorithm (GA) was used not only for weight minimization of steel frames, but also for solving the member-sizing problem among different combination of section types.

According to the literature review, it can be concluded that although there has been much research on the collapse performance of steel braced frames, the studies on 2 and 3D structures are still quite limited, as is specially emphasized in Ren et al. [18]. To make an effort to fill the knowledge gaps, the present paper investigates the resistance capacity of progressive collapse in 2-dimensional of a 5 and 10-story building CBF frames with various types of braces and braced bays, namely chevron-, single diagonal- and cross-braced frames, under gravity loads. To calculate the vertical displacements and weight of the structures, some of the columns in different locations on the first floor and different types of braced frames were removed. The progressive collapse behavior of the structure was then investigated using alternate path method according to the General Services Administration (GSA) code, as well as the linear static analysis.

2 Investigated Structures

In the present study, a 5-story and 10-story 4-bay planar building frame equipped with single diagonal, chevron and cross braces were used. The height of the stories and the length of the spans are assumed 3 and 5 m, respectively. The 5-story and 10-story steel buildings considered to investigate the structural response under column removal scenarios are presented in plan, elevation and perspective view in Figs. 2 and 3.

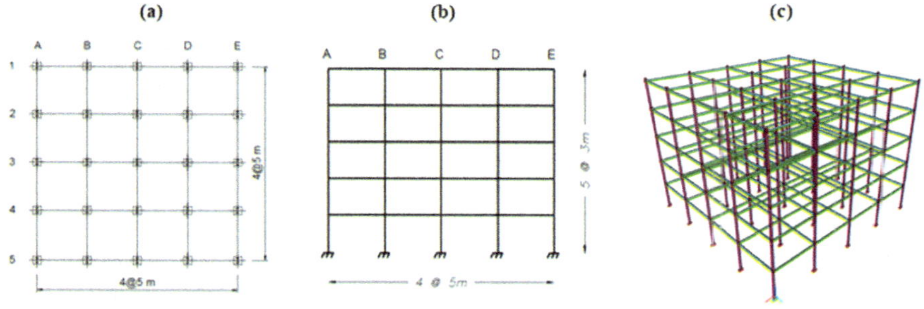

Fig. 2 5-story 4-bay frames in **a** plan, **b** elevation and **c** perspective view

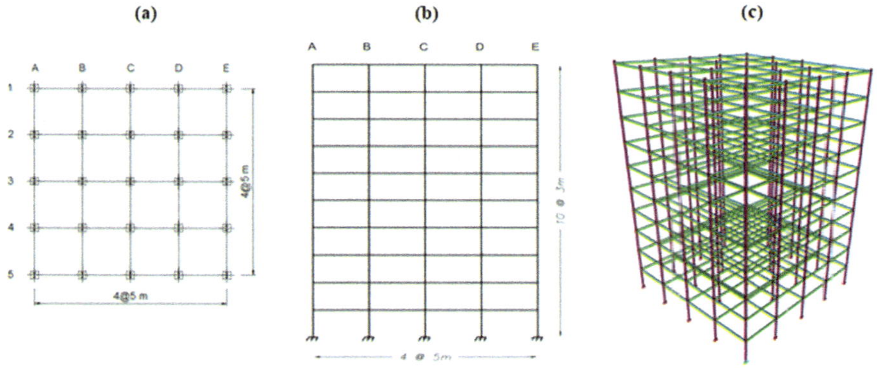

Fig. 3 10-story and 4-bay frames in **a** plan, **b** elevation and **c** perspective view

Table 1 APM analysis case (scenarios)

APM scenario	Frame name	Removed column	Optimized weight (kg)	Number of iterations
1	5S-V-A1	A1	5138	71
2	5S-V-B1	B1	6217	63
3	5S-V-C1	C1	5480	57
4	10S-V-A1	A1	13,624	78
5	10S-V-B1	B1	16,533	88
6	10S-V-C1	C1	13,774	85
7	5S-single-A1	A1	5781	82
8	5S-single-B1	B1	6735	90
9	5S-single-C1	C1	5336	65
10	10S-single-A1	A1	14,292	97
11	10S-single-B1	B1	18,338	120
12	10S-single-C1	C1	14,536	63
13	5S-X-A1	A1	6209	67
14	5S-X-B1	B1	6976	67
15	5S-X-C1	C1	5464	70
16	10S-X-A1	A1	15,441	69
17	10S-X-B1	B1	19,900	75
18	10S-X-C1	C1	14,684	90

In total, there are 18 dual-frame models with different scenarios of column removal. All scenarios are introduced in Table 1. The frame names have formed of 3 parts. The first part including 5S or 10S indicates the number of stories. The second part shows the type of bracing system for each model, namely V (chevron), X (cross-braced) or single. The third part refers to the location of the removed column, which A1, B1 and C1 are representative for the removal of the first, second and third columns, respectively.

3 Progressive Collapse Analysis Methodology

Linear static procedure (LSP) coupled with optimization methods based on LRFD 360-10 approach and Alternative Path Method (APM) according to the 2003 GSA guideline were performed to analyze dual frames.

3.1 Column Removal Scenarios

One important and critical parameter in the strength of various structures against progressive collapse risk is the type of lateral load resisting systems. By changing the load path on the effect of removing a critical member such as a column, the forces of the removed member as a result of this event must be transmitted to other parts of the structure by neighboring members. Considering the geometric symmetry and the loading symmetry of the structure, the first three columns were progressively eliminated one by one. In the first model, the corner column (A1) was removed, and in the second and third models, the second column (B1) and third column (C1) were removed and then analyzed, respectively.

3.2 Modeling

To assess the resistance capacity of progressive collapse among the different types of braces, a two-dimensional five-story and ten-story steel building frame has been considered and the details are reported in Table 2. The plan of the building consists of 4×5 and 4×10 frames has been analyzed and the obtained results were compared. Yield stress (Fy) and module of elasticity (E) were assumed 2400 and 2,000,000 kg/cm^2, respectively.

3.3 Linear Analysis Method

Linear static analysis was considered to minimize the long-lasting of the structure optimization. To do structural analysis, a link between open source software of OpenSees and MATLAB was developed. The results were then compared and validated with Common Structural Analysis Software. An alternative load path approach was also applied for loading the structures under progressive collapse. The load combination was calculated according to the Eqs. 1–3, respectively.

$$\mathrm{DL} = 500 \times 2.5 = 1250\,\frac{\mathrm{kg}}{\mathrm{m}};\ \mathrm{LL} = 300 \times 2.5 = 750\,\frac{\mathrm{kg}}{\mathrm{m}} \tag{1}$$

$$2 \times (\mathrm{DL} + 0.25\,\mathrm{LL}) = 3000\,\frac{\mathrm{kg}}{\mathrm{m}} \tag{2}$$

$$\mathrm{DL} + 0.25\,\mathrm{LL} = 1500\,\frac{\mathrm{kg}}{\mathrm{m}} \tag{3}$$

where DL and LL are dead load and live load, respectively.

Table 2 Sections of the beams, columns and bracings

Number of stories	The label of beams and columns	Beams and columns section	Label of bracings	Bracings section
5-story	1	IPE160	10	BOX 70 * 70 * 8
	2	IPE180	11	BOX 80 * 80 * 10
	3	IPE200	12	BOX 80 * 80 * 12.5
	4	IPE220	13	BOX 90 * 90 * 10
	5	IPE240	14	BOX 90 * 90 * 16
	6	IPE270	15	BOX 90 * 90 * 12.5
	7	IPE300	16	BOX 100 * 100 * 16
	8	IPE330		
	9	IPE360		
10-story	1	IPE180	13	BOX 70 * 70 * 8
	2	IPE200	14	BOX 80 * 80 * 10
	3	IPE240	15	BOX 80 * 80 * 12.5
	4	IPE270	16	BOX 90 * 90 * 10
	5	IPE300	17	BOX 90 * 90 * 16
	6	IPE330	18	BOX 90 * 90 * 12.5
	7	IPE360	19	BOX 100 * 100 * 16
	8	HEB260		
	9	HEB280		
	10	HEB300		
	11	HEB320		
	12	HEB340		

Figure 4 shows the vertical loads in the frame systems loaded through the uniform forces method. The first, second and third columns were removed when the structures had been under vertical loads.

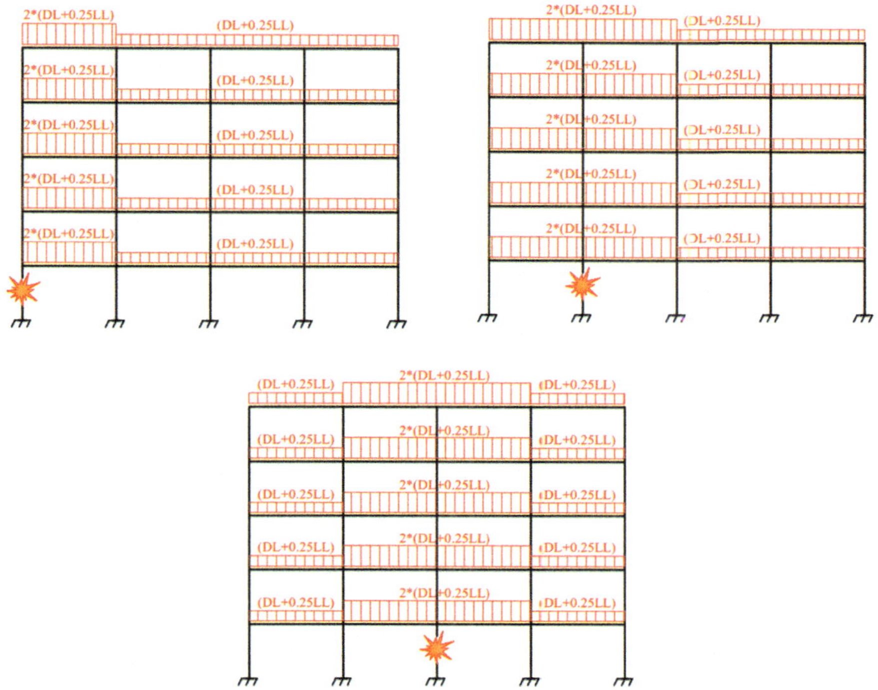

Fig. 4 Loading the frames under progressive collapse analysis by the 2003 GSA approach after eliminating different columns [5]

4 Optimization

4.1 The Symbiotic Organism Search Algorithm (SOS)

The proposed SOS algorithm simulates the interactive behavior seen among organisms in nature. Organisms rarely live in isolation due to reliance on other species for sustenance and even survival. This reliance-based relationship is known as symbiosis. Previous researchers used the term symbiosis for the first time in 1878 to explain the symbiotic behavior of the anomalous creatures. Today, the term "symbiosis" is used to explain the relationship between the two known species. Symbiotic relationships may be compulsory, meaning that two organisms are dependent on each other for survival or facultative, meaning the two organisms choose to cohabitate in a mutually beneficial but nonessential relationship. The most common symbiotic relationships found in nature are mutualism, commensalism and parasitism. Mutualism denotes a symbiotic relationship between two different species in which both benefit. Commensalism is a symbiotic relationship between two different species in which one benefits and the other is unaffected or neutral. Parasitism is a symbiotic relationship between two different species in which one benefits and the other is actively harmed.

In other words, the two organisms decide to coexist in a mutually beneficial but unnecessary way. Similar to other population-based algorithms, the proposed SOS iteratively uses a population of candidate solutions to promising areas in the search space in the process of seeking the optimal global solution. SOS begins with an initial population called the ecosystem. In the initial ecosystem, a group of organisms is generated randomly to the search space. Each organism represents one candidate solution to the corresponding problem. Each organism in the ecosystem is associated with a certain fitness value, which reflects the degree of adaptation to the desired objective [29].

4.2 Particle Swarm Optimization Algorithm (PSO)

The Particle Swarm Optimization algorithm (PSO) is a group stochastic optimization algorithm, inspired by the social behavior of a group of birds. In this algorithm, each member of the community changes its position according to personal experiences and the experiences of the entire community. Previous researches have been based on birds' efforts to maintain a shorter path between themselves and their neighbors. In a collection of creatures, each creature can benefit from the experiences of its predecessors in finding food. This can be crucial in the competition for food. The social sharing of information between members of a community has a number of evolutionary advantages, and this hypothesis is the basis of the bird algorithm [30].

4.3 Selecting the Proper Algorithm

In this study, two different kinds of algorithms including PSO and the symbiotic organism search algorithm (SOS) have been used to compare the convergence speed and fitness results. For a comparison, five-story frames with an X-braced system (Fig. 5) were used in both PSO and SOS optimization methods. Due to the same results, selecting just a type of bracement was sufficient.

According to Table 3, the SOS algorithm has more speed of convergence. Although it was needed 70 iterations for SOS to reach the best response, the number of iterations for PSO was 150, indicating that SOS is approximately two times faster than PSO. The weight of the optimized structure measured from the SOS algorithm has been calculated around 5.5 tons, which is almost 2 tons less than that of one estimated with PSO. As can be seen from Fig. 6, the PSO algorithm is not able to achieve a better result with a sufficient number of bracements. However, the SOS algorithm has better results with only two couple of bracements. With this statistical finding, it was decided to use the SOS algorithm because of obtaining lighter optimized frames in a shorter time.

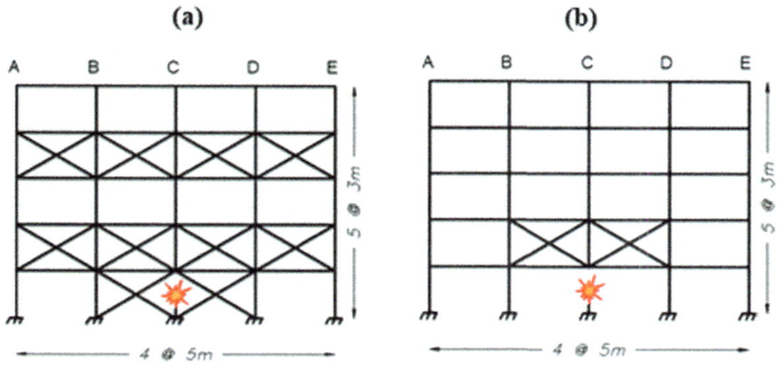

Fig. 5 Comparison of PSO and SOS algorithms. **a** Optimized layout by PSO and **b** Optimized layout by SOS

Table 3 Weight and number of iteration values for PSO and SOS algorithm

	PSO	SOS
Weight (kg)	7527	5464
Number of iterations	150	70

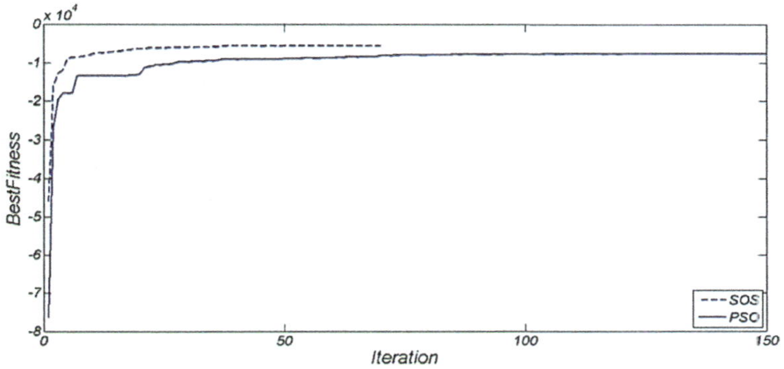

Fig. 6 Comparison of the speed of convergence using SOS and PSO algorithms

4.4 Optimization Problem Formulation

The goal is to find vector X consisting of structural sections that have a member as much as element groups of the structure. Groups are considered in such a way that symmetric members and some floors are placed in one group. The constraints used in this study include the constraint of stress ratio for beams, columns and bracings based on LRFD 365-10. In the analyzed structures, it is assumed that the entire bays have been covered from up to down with the bracing, and there is the possibility of

Table 4 Constraint of the beam, column, beam-column and bracing stress ratio regarding LRFD 360-10 [32]

$\frac{P_u}{2\phi_c P_n} + \frac{M_u}{\phi_c M_n} - 1 \le 0$	The constraint of the beams stress ratio
$\left\{\frac{P_u}{\phi_c P_n} + \frac{M_u}{\phi_c M_n}\right\} - 1 \le 0$	The constraint of the beams and beam-columns stress ratio $\frac{P_u}{0.9\times P_n} < 0.2$ if
$\left\{\frac{P_u}{\phi_c P_n} + \frac{8}{9}\left(\frac{M_u}{\phi_c M_n}\right)\right\} - 1 \le 0$	The constraint of the beams and beam-columns stress ratio $\frac{P_u}{0.9\times P_n} \ge 0.2$ if
$\frac{P_u}{\phi_c P_n} - 1 \le 0$	The constraint of the bracings stress ratio

$\phi_c = 0.9$
where M_{ux}, M_{nx}, P_u and P_n are the final flexural load of the section, nominal flexural capacity of the section, final axial load of the section and nominal axial force capacity of the section, respectively [32].

choosing each one by the desired algorithm being able to select each bracing group. It has been also determined a "zero-section" for specifying the presence or absence of bracing in a bay, which helps the bracing element that does not place into the desired bay. Since the braced frames optimizing is a constrained problem, penalty weight strategy has been assumed, and the fitness function of the penalty weight was utilized, which is equal with the penalty weight in terms of quantity and symmetry sign. The less worthy results are dropped, and the algorithm then moves toward the category of the results with more fitness, leading to obtaining the maximum fitness function. The algorithm is also planned if the convergence diagram shows the same values for 20 steps, the optimization will be stopped. The intended parameters are described below:

$$X = \{X_1, X_2, \ldots, X_m\} \tag{4}$$

$$\text{Fitness}(X) = -\left(\sum_{i=1}^{n} W_i\right) \times (1 + \text{Kp} \times \text{Infeasibility}) \tag{5}$$

where n, m, W, Kp and Infeasibility are the number of members, the number of groups, the membership weight, violation factor of the specified constraints and sum of violations of the specified constraints, respectively [31]. The constraints, based on the criteria of the method of Limit Sate, are given in the Table 4.

4.5 Grouping

The grouping of the members is in such a way that the symmetrical columns and beams are categorized into a group. According to the number of floors, each two or three floors are placed in one group. But the grouping of bracings is in such a way that each two symmetric bracings on each floor are placed in one group, leading to having more groups than beams and columns.

5 Results and Discussion

5.1 Structural Analytical Models

In this study, a total of 18 different scenarios have been modeled and analyzed under progressive collapse to assess the potential of failure in variable sorts of scenarios. Nine of these models are 5-story and others are 10-story steel dual (bracing-moment) frames. There are four five-meter bays for all the frames and also three-meter height stories. Frames are segregated from each other with types of bracing configurations, and the position of the removed column. Models are shown in three groups (Figs. 7, 8 and 9), which identified with bracing layout system. In each group, six scenarios have been accommodated and the numbers beside the members are presentative of section number which refers to member sections in Table 1.

5.2 Conventional Frames in Comparison with Optimized Frames

The following six types of optimized frames obtained from the SOS algorithm compare with some of the conventional layouts. These are subjected to loads due to progressive collapse, in terms of the period of the first mode and the vertical displacement of the column removal location. In addition, these compared frames are the same in terms of weight and sections used and just the configuration of braces is different.

The results show that the distribution and arrangement of braces are very effective to obtain a suitable load path. In addition, different types of braces have various effects to neutralize progressive failure. For example, conventional chevron bracing systems have outstanding performance in reducing the potential of progressive failure and the percentage of element failure compared with single and cross bracing systems. However, the stiffness of the frames has less effect on progressive failure.

Single diagonal-braced frame. Table 5 shows that the first mode period in optimized and conventional single diagonal-braced frames are 1.12 and 0.49 s, respectively. The vertical displacement of the removed column in optimized and conventional frames are 11 and 22 mm, respectively. These results indicate that the structural stiffness is higher in the conventional frame, while the collapse potential in the corner of the conventional frame is higher than that of the optimized frame. This confirms

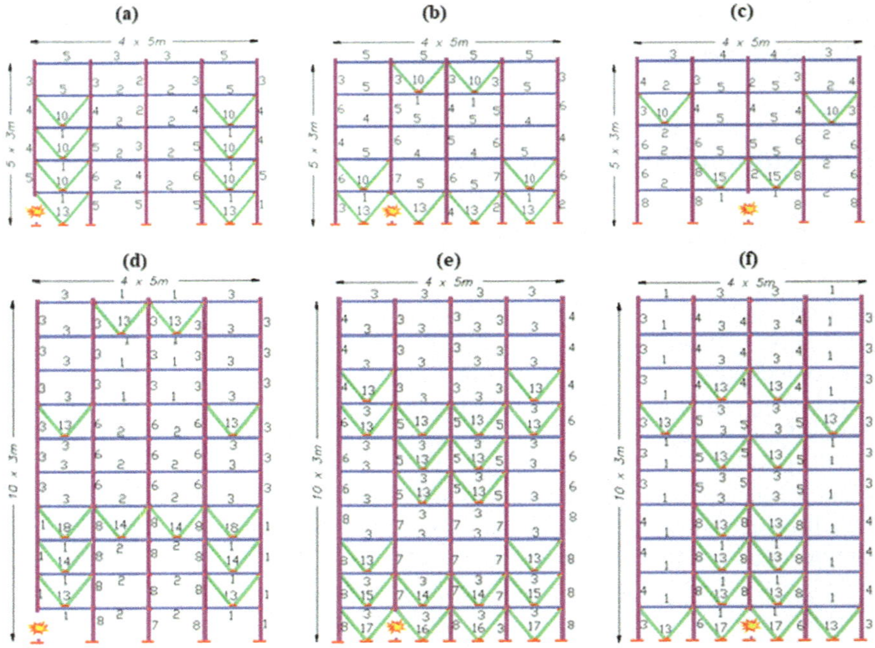

Fig. 7 Optimum layout of 5-story chevron-braced frames after removing the **a** first column (5S-V-A1), **b** second column (5S-V-B1) and **c** third column (5S-V-C1), and optimized layout of 10-story chevron-braced frames when removing the **d** first column (10S-V-A1), **e** second column (10S-V-B1) and **f** third column (10S-V-C1)

the importance of bracing topology and its effect on mitigating damage from the structural collapse.

Although the stiffness of the conventional single diagonal-braced frame is more than the optimized one, the potential of the progressive collapse in the conventional frame is more than the optimized frame. Furthermore, 10–20% of the members in single diagonal-braced frames failed to satisfy the designing constraints.

Figure 10 compares the conventional layout of a five-story single diagonal-braced frame with the optimized one, which is analyzed under progressive collapse. Ten elements (the red ones) are stressed and exceeded over the specified constraint, which means that 18% of the members are not satisfied with the designing qualifications. Figure 10a shows that the chosen conventional frame is one of the most used typical single bracing layouts in some countries, which is configured by using 10 single bracing members on both sides of the frame. Besides, the elements in the conventional frame cannot find the proper alternative path to stability the collapsed element and are not able to carry out the redistributed forces. Moreover, by removing the corner column due to an explosion (sudden removal), it can be deduced that elements on both sides of a conventional frame should be able to carry out the redistributed load

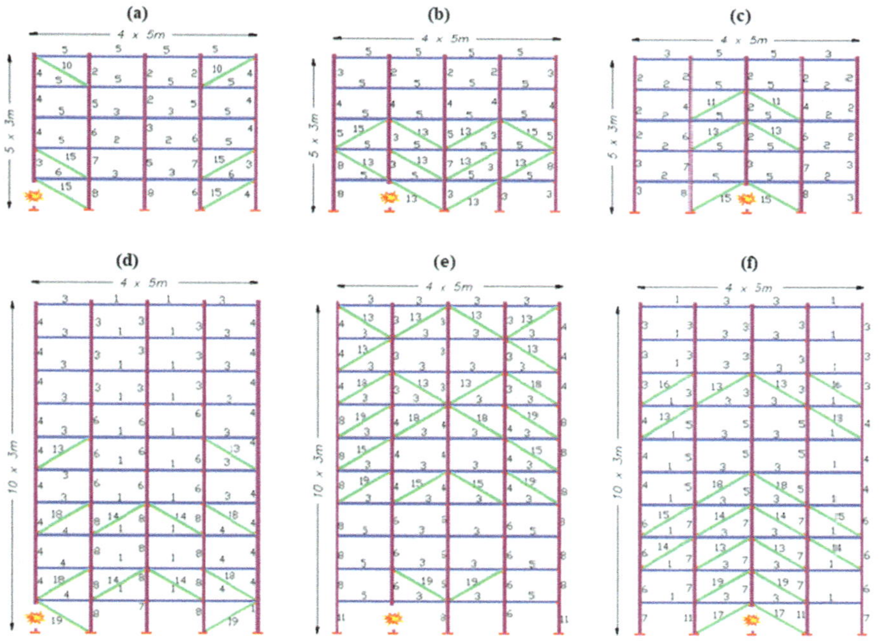

Fig. 8 Optimum layout of 5-story single-braced frames after removing the **a** first column (5S-S-A1), **b** second column (5S-S-B1) and **c** third column (5S-S-C1), and optimized layout of 10-story single-braced frames when removing the **d** first column (10S-S-A1), **e** second column (10S-S-B1) and **f** third column (10S-S-C1)

in order to find a suitable load path. Furthermore, an appropriate configuration of bracings helps to mitigate the damage of the progressive collapse.

For a ten-story single diagonal-braced frame, all the results are the same as those of a five-story frame. Figure 11 represents the conventional frame which is configured by using 20 bracing elements. 11 out of 110 elements (red ones) overstressed which means that 10% of the members could not pass the designing qualifications. Moreover, after the removal of the column in the vicinity of the middle one, the middle elements need to have a proper configuration to connect the collapsed member. Therefore, the configuration of bracings in some bays cannot be an effective solution to prevent progressive collapse in conventional single diagonal-braced frames.

Chevron-Braced Frame

The reported results in Table 6 show that the optimized chevron-braced frame in all members have satisfied the specified constraints, while about 5% of the elements in the conventional chevron-braced frame are overstressed due to lack of suitable load path for the redistributed load. The first mode period in the optimized and conventional frame is calculated at 0.55 and 0.88 s, respectively. Besides, vertical displacement of the removed column in the optimized and conventional frame is

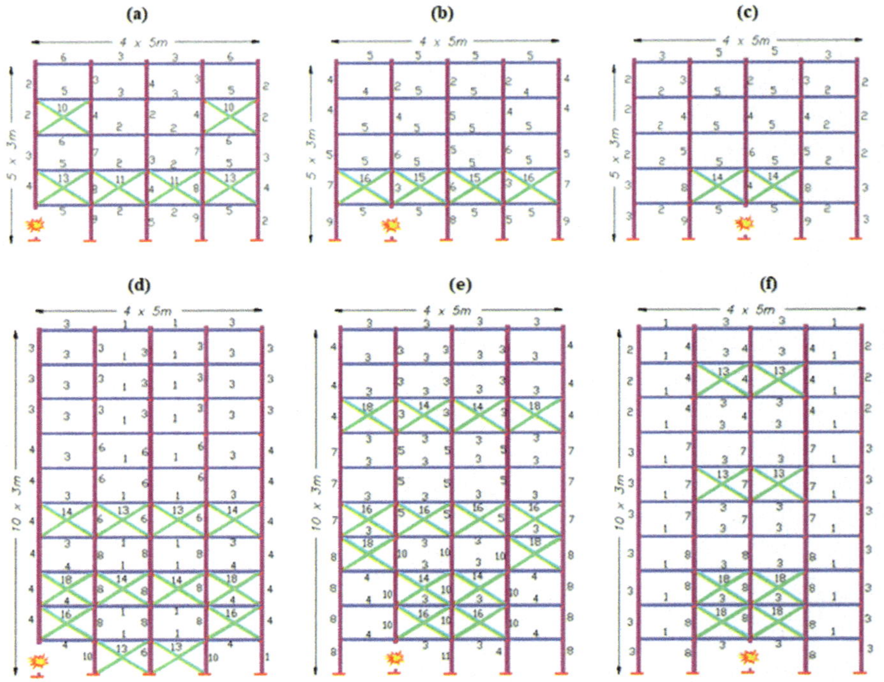

Fig. 9 Optimum layout of 5-story cross-braced frames after removing the **a** first column (5S-X-A1), **b** second column (5S-X-B1) and **c** third column (5S-X-C1), and optimized layout of 10-story cross-braced frames when removing the **d** first column (10S-X-A1), **e** second column (10S-X-B1) and **f** third column (10S-X-C1)

Table 5 Comparison of some characteristics between optimized and conventional single diagonal-braced frames

Single diagonal-braced frame	5-story frames		10-story frames	
	Optimized	Conventional	Optimized	Conventional
Weight (ton)	5.84	5.84	18.3	18.3
Displacement (mm)	11	22	13.9	14
Period (s)	1.12	0.49	1.49	0.97
Over-stressed members (%)	–	18%	–	10%

obtained at 3.6 and 3.8 mm, respectively. The results also indicate that just 5% of members in conventional frames are unable to satisfy the designing constraints, meaning that other elements are appropriately able to carry out the redistributed load.

Figure 12a, b shows that the potential for collapse in the vicinity of the middle column of the conventional frame is nearly equal to the optimized one that just a few members fail under the progressive collapse loading.

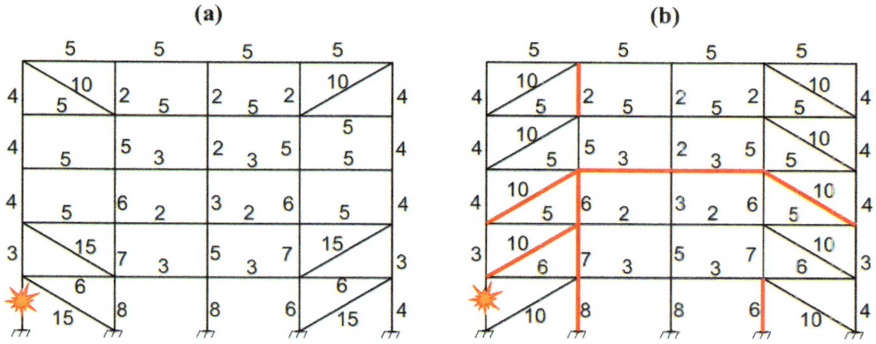

Fig. 10 **a** Optimized layout of a 5-story single diagonal-braced frame after removing the column A1 and **b** Conventional layout of a 5-story single diagonal-braced frame after removing column A1 with over-stressed elements (red sections)

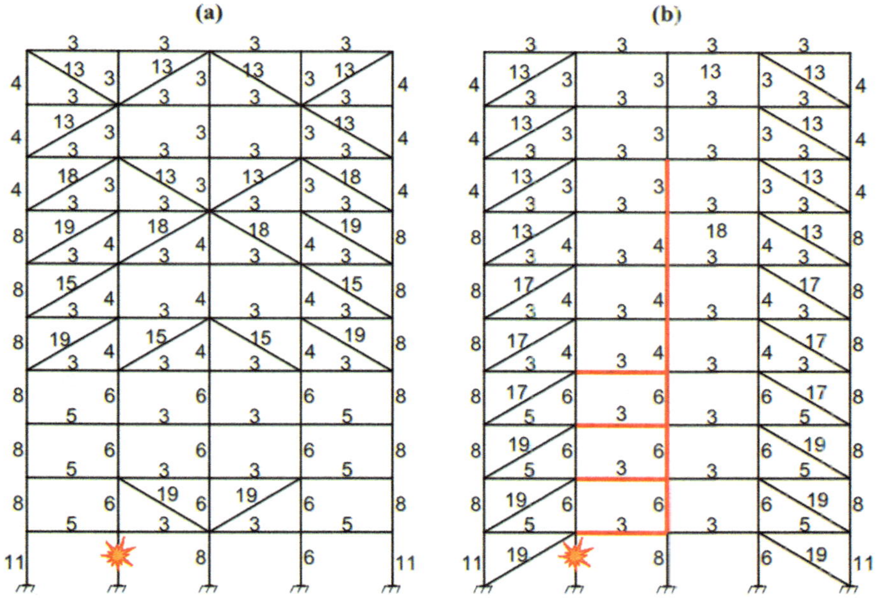

Fig. 11 **a** Optimized layout of a 10-story single diagonal-braced frame after removing column B1 and **b** conventional layout of 10-story single diagonal-braced frame after removing column B1 with over-stressed elements (red sections)

For a ten-story chevron-braced frame, the results are a little different. In the conventional frame which is configured by using 32 bracing elements, as shown in Fig. 13, just 6 out of 122 elements (red ones) overstressed meaning that 5% of the members could not pass the designing qualifications. Moreover, Table 6 compares the first mode period in conventional and optimized frames, and the results show

Table 6 Comparison of some characteristics between optimized and conventional chevron-braced frames

Chevron-braced frame	5-story frames		10-story frames	
	Optimized	Conventional	Optimized	Conventional
Weight (ton)	6.2	6.2	13.8	13.8
Displacement (mm)	3.6	3.8	3.7	6.3
Period (sec)	0.88	0.55	0.97	1.04
Over-stressed members (%)	–	5%	–	5%

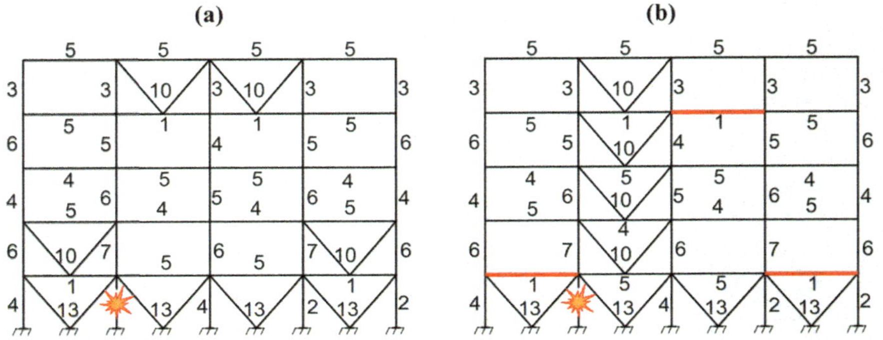

Fig. 12 **a** Optimized layout of 5-story chevron-braced frame after removing column B1 and **b** conventional layout of 5-story chevron-braced frame after removing column B1 with over-stressed elements (red sections)

that both of them have almost equal stiffness. However, the vertical displacement of the removed column in the conventional frame is two times greater than that of the optimized frame. This indicates that stiffness cannot be the one important parameter to evaluate progressive collapse and other factors such as type and configuration of bracing systems can be more effective to mitigate progressive collapse.

Cross-Braced Frame

Table 7 shows that the progressive collapse potential in the conventional cross-braced frame is nearly 2–5 times greater than that of the optimized cross-braced frame. Furthermore, 10–45% of the members in the conventional frame exceed the allowable designing limits. In addition, although the conventional frame is relatively stiffer than the optimized frame, the vertical displacement at the column removal location is approximately 5 times greater than that of the optimized frame. These findings show that the arrangement of braces in conventional cross-braced frames needs to be improved to have a better performance against progressive failure. This means that the correct distribution of cross braces in these frames is crucial to find a proper load path against progressive collapse.

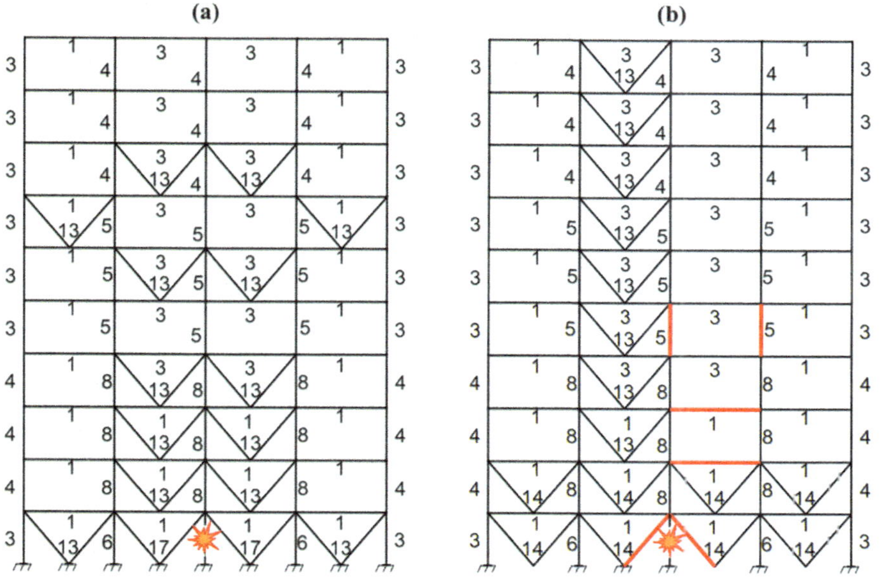

Fig. 13 **a** Optimized layout of a 10-story chevron-braced frame after removing column C1 and **b** conventional layout of a 10-story chevron-braced frame after removing column C1 with over-stressed elements (red sections)

Table 7 Comparison of some characteristics between optimized and conventional cross-braced frames

Cross-braced frame	5-story frames		10-story frames	
	Optimized	Conventional	Optimized	Conventional
Weight (ton)	5.5	5.5	15.4	15.4
Displacement (mm)	6.7	31	14	22
Period (s)	1.52	0.94	0.97	1.11
Over-stressed members (%)	–	45%	–	10.3%

As can be seen from Fig. 14a, by removing the middle column in the optimized frame, only 2 pairs of braces are needed, meaning that the frame shows a better performance against column removal with less stiffness. However, the conventional 5-story cross-braced frame has a poor performance against loading resulted from progressive failure. This is because almost half of the elements (red sections) failed to meet the design constraints.

The results are slightly different for a 10-story cross-braced frame. To deal with the removal of the first floor column in the 10-story optimized frame (Fig. 15a), a proper horizontal configuration of braces needs to be considered in all bays, which is crucial to create a continuous chain transmitting forces. According to Fig. 15b, just 12 of the 116 available elements (red sections) could not withstand the redistributed

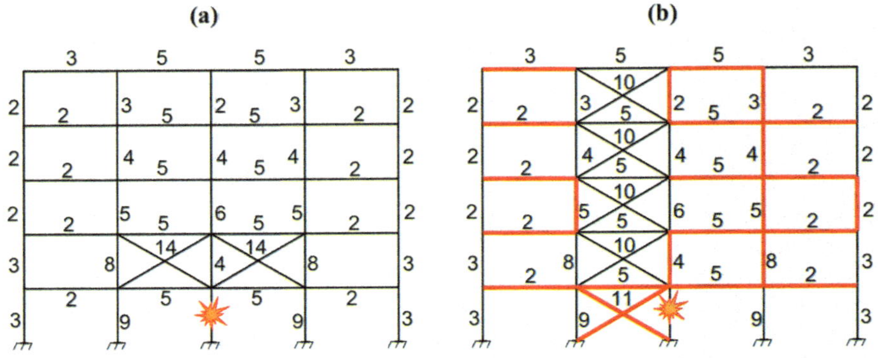

Fig. 14 **a** Optimized layout of 5-story cross-braced frame after removing column C1 and **b** conventional layout of 5-story cross-braced frame after removing column C1 with over-stressed elements (red sections)

load due to the removal of the corner column. This is more acceptable compared to the 5-story cross-braced frame. Comparing the vertical displacement at the location of the removed column also shows that the potential of progressive failure in the conventional cross-braced frame is about 1.5 times higher than that of the optimized frame.

5.3 Comparison of Weight and Displacement

To evaluate the effect of column removal on the strength of structures against the progressive collapse, the graph of weight and displacement resulted from the removal of the three columns is separately represented in Figs. 16 and 17. As can be shown from Fig. 16, the peak vertical displacement of the node corresponding to the top of the removed first column (A1) is generally more than that of other columns. This indicates that the failure potential in the corner of the structure is more compared with other locations. The results are also shown that displacements in chevron-braced frames are generally less than those of single- and cross-braced frames, confirming that the chevron bracing system has a higher resistance to progressive collapse. In addition, Fig. 17 shows that the weight of chevron-braced frames is lower than that of single- and cross-braced frames, respectively. The obtained results in this study suggest that the chevron-braced frames are less vulnerable to progressive collapse than the other types of braced frames, which may be due to the fewer bracings.

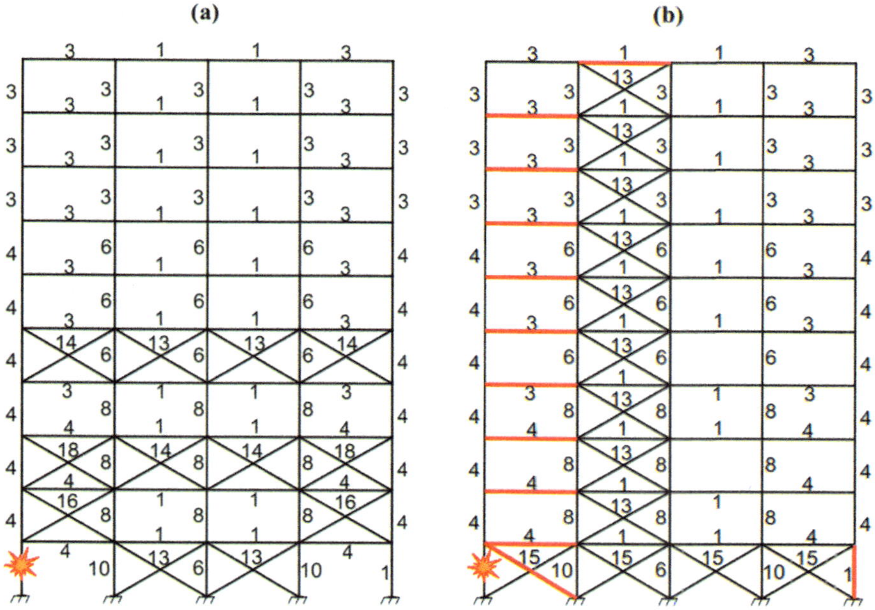

Fig. 15 **a** Optimized layout of 10-story cross-braced frame after removing column A1 and **b** conventional layout of 10-story cross-braced frame after removing column A1 with over-stressed elements (red sections)

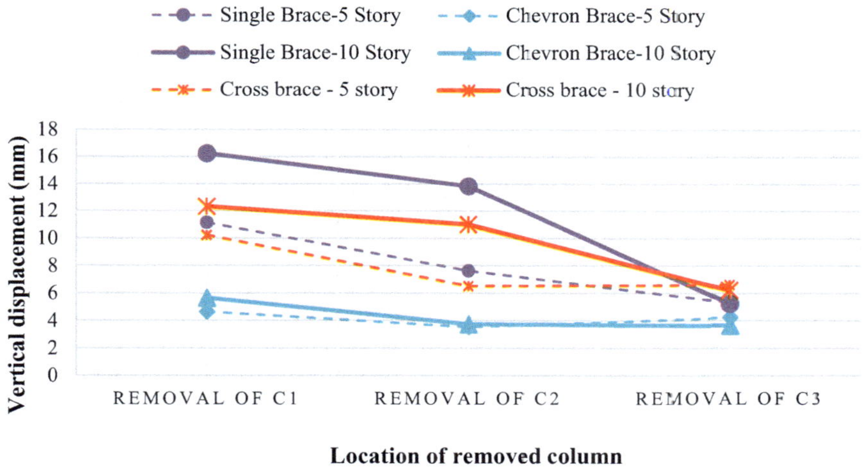

Fig. 16 Vertical displacement at the column removal location of the frames

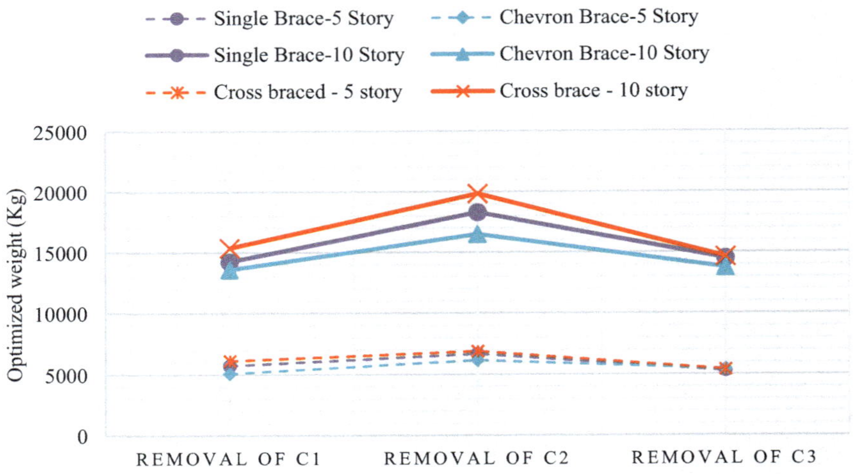

Fig. 17 The optimized weight of different frame configurations

6 Conclusion

In the present study, macro-models for three popular types of braced frames, namely single-, chevron- and cross-braced frames were investigated. The developed models were then utilized within the context of AP method to compare the progressive collapse resistance of the braced frames in 5 and 10-story frames equipped with various braces, according to contemporary seismic design specifications and practices. All the models were also optimized with the SOS meta-heuristic algorithm extracted from MATLAB software to find out the best-optimized model, having the lowest weight and displacements. The obtained results are listed below:

- The bracing layout has a significant impact on the progressive collapse analysis and counteracts the collapse risk of the structure.
- In all the structures, removing corner columns causes the concentration of the bracings at the side bays.
- In frames which the middle column is removed, the concentration of bracings is measured at middle bays.
- The weight of structures installed with a single diagonal and cross bracing system are almost 6 and 12% more than that of structures equipped with the chevron bracing.
- It was found that the structures with second column removal (B1) have the highest weight indicating these structures need a larger bracing network to deal with the impact of a progressive collapse. However, the structures with corner column removal (A1) have the lowest weight, meaning that it is more economical.

- The results showed that the vertical displacement due to the removal of corner column (A1) in single diagonal-, chevron- and cross-braced layouts was the highest amount, as opposed to the displacement at the points of middle column removal (C1) having the lowest value. Therefore, progressive collapse potential in the corner of the structure is more likely to occur and the middle column has the least potential for progressive collapse.
- Vertical displacement in single diagonal- and cross-braced structures is averagely twice as much as that of chevron-braced frames. This means that the potential of progressive collapse in a single diagonal and cross bracing layout is more than the chevron bracing.
- Some elements in the conventional frames cannot find the proper ALP to bridge over the collapsed element in order to carry out the redistributed forces, proving the necessity of paying attention to catenary action.
- Frames with chevron braces have a better performance against column removal compared to single diagonal and cross braces.
- The rigidity of the frames does not have a direct effect on reducing or increasing the potential for progressive failure, while the proper distribution and arrangement of braces as well as the type of braces have a greater effect on creating a suitable load path, resulted in reducing the effect of progressive failure.

References

1. Asadi, I., Shafigh, P., Hashemi, M., Akhiani, A., Maghfouri, M., Sajadi, B., Mahyuddin, N., Esfandiari, M., Rezaei Talebi, H., Simon CornelisMetselaar, H.: Thermophysical properties of sustainable cement mortar containing oil palm boiler clinker (OPBC) as a fine aggregate. In: Construction and Building Materials, p. 121091 (2020)
2. Ellingwood, B.R., Smilowitz, R., Dusenberry, D.O., Duthinh, D., Lew, H.S., Carino, N.J.: Best Practices for Reducing the Potential for Progressive Collapse in Buildings (2007)
3. Khandelwal, K., El-Tawil, S., Sadek, F.: Progressive collapse analysis of seismically designed steel braced frames. J. Constr. Steel Res. **65**(3), 699–708 (2009)
4. Wang, H., Zhang, A., Li, Y., Yan, W.: A review on progressive collapse of building structures. Open Civil Eng. J. **8**(1) (2014)
5. Gsa, U.: Progressive Collapse Analysis and Design Guidelines for New Federal Office Buildings and Major Modernization Projects. Washington, DC (2003)
6. Mahmoud, Y.M., Hasan, M.M., Mourad, S.A., Sayed, H.S.: Assessment of progressive collapse of steel structures under seismic loads. Alex. Eng. J. **57**(4), 3825–3839 (2018)
7. Nair, R.S.: Progressive collapse basics. Mod. Steel Constr. **44**(3), 37–44 (2004)
8. Ślaga, Ł.: Catenary Action in Precast Skeleton Structures–Protection Against Progressive Collapse. Łukasz Ślaga, Brno (2013)
9. Al-Salloum, Y., Abbas, H., Almusallam, T.H., Ngo, T., Mendis, P.: Progressive collapse analysis of a typical RC high-rise tower. J. King Saud Univ.-Eng. Sci. **29**(4), 313–320 (2017)
10. Maghfouri, M.: Drying shrinkage strain development of agro-waste oil palm shell lightweight aggregate concrete by using the experimental result, ACI and Eurocode prediction models. Int. J. Integr. Eng. **11**(9), 255–263 (2019)
11. Καραγιάννης, Ν.Σ.: Progressive Collapse Analysis: A Case of a Furnance Steel Tower in Dubhai (2016)

12. Kim, J., Kong, J.: Progressive collapse behavior of rotor-type diagrid buildings. Struct. Design Tall Spec. Build. **22**(16), 1199–1214 (2013)
13. Kim, J., Lee, S., Choi, H.: Progressive collapse resisting capacity of moment frames with viscous dampers. Struct. Design Tall Spec. Build. **22**(5), 399–414 (2013)
14. Kim, J., Choi, H., Min, K.W.: Use of rotational friction dampers to enhance seismic and progressive collapse resisting capacity of structures. Struct. Design Tall Spec. Build. **20**(4), 515–537 (2011)
15. Qian, K., Weng, Y.-H., Li, B.: improving behavior of reinforced concrete frames to resist progressive collapse through steel bracings. J. Struct. Eng. **145**(2), 04018248 (2019)
16. DoD: Design of buildings to resist progressive collapse. In: Unified Facilities Criteria (UFC) 4-023-03 (2009)
17. Code, N.R.C.o.C.A.C.o.t.N.B.: National Building Code of Canada. Associate Committee on the National Building Code. National Research Council (1990)
18. Ren, L.-M., Yang, B., Chen, K., Sun, Y.: Progressive collapse of 3D composite floor systems with rigid connections under external column removal scenarios. J. Struct. Eng. **146**(11), 04020244 (2020)
19. Kim, J., An, D.: Evaluation of progressive collapse potential of steel moment frames considering catenary action. Struct. Design Tall Spec. Build. **18**(4), 455–465 (2009)
20. Kim, H.-S., Kim, J., An, D.-W.: Development of integrated system for progressive collapse analysis of building structures considering dynamic effects. Adv. Eng. Softw. **40**(1), 1–8 (2009)
21. Fu, F.: Progressive collapse analysis of high-rise building with 3-D finite element modeling method. J. Constr. Steel Res. **65**(6), 1269–1278 (2009)
22. Tavakoli, H.R., Hasani, A.H.: Effect of Earthquake characteristics on seismic progressive collapse potential in steel moment resisting frame. Earthq. Struct **12**, 529–541 (2017)
23. Mahmoudi, M., Koozani, H., Teimoori, T.: Stability Assessment of Steel Moment Frames Against Progressive Collapse (2016)
24. Mirjalali, H., Ghasemi, M., Labbafzadeh, M.: Effect of bracing type and topology on progressive collapse resistance of eccentrically braced frames. Int. J. Steel Struct. **19**(5), 1497–1510 (2019)
25. Roeder, C.W., Popov, E.P.: Inelastic behavior of eccentrically braced steel frames under cyclic loadings. STIN **78**, 20375 (1977)
26. Roeder, C.W., Popov, E.P.: Eccentrically braced steel frames for earthquakes. J. Struct. Div. **104**(3), 391–412 (1978)
27. Khandelwal, K., El-Tawil, S.: Pushdown resistance as a measure of robustness in progressive collapse analysis. Eng. Struct. **33**(9), 2653–2661 (2011)
28. Liu, M.: Progressive collapse design of seismic steel frames using structural optimization. J. Constr. Steel Res. **67**(3), 322–332 (2011)
29. Cheng, M.-Y., Prayogo, D.: Symbiotic organisms search: a new metaheuristic optimization algorithm. Comput. Struct. **139**, 98–112 (2014)
30. Eberhart, R., Kennedy, J.: A new optimizer using particle swarm theory. In: MHS'95. Proceedings of the Sixth International Symposium on Micro Machine and Human Science. IEEE (1995)
31. Arora, J.S.: Introduction to Optimum Design. Elsevier (2004)
32. American Institute of Steel Construction: Specification for Structural Steel Buildings (ANSI/AISC 360-10). American Institute of Steel Construction, Chicago-Illinois (2010)

Natural Localized Corrosion of Steel Bar in 44-Years Old Cracked RC Beam Structures Exposed to Marine Tidal Environment

Pinta Astuti, Rahmita Sari Rafdinal, Daisuke Yamamoto, and Hidenori Hamada

Abstract The chloride-induced corrosion processes of exposed structures are very complex. Locally isolated anodically and cathodically active steel surface areas may form on reinforcing bars of RC structures. Corrosion processes may only be examined indirectly using electrochemical methods due to the obscurity of the cathodically active steel surface areas. In this study, an experiment was undertaken to identify the development of natural corrosion on broken, 44-year-old RC structures exposed to the actual marine environment. Observation of the inadequate appearance of the concrete and electrochemical measurements were used to predict the probability of the deterioration phase. The greatest corrosion risk was recognized in the middle tensile section, and a check of the actual state of the steel bars was conducted there. Natural localized corrosions were obviously generated on the middle tensile steel bar of the pre-cracked RC beam as a result of the accelerated development of macro-cell corrosion current brought on by the first cracks in the constructions.

Keywords Deterioration · Chloride-induced corrosion · Localized corrosion · Macro-cell corrosion · Marine environment

P. Astuti (✉)
Department of Civil Engineering, Universitas Muhammadiyah Yogyakarta, Jalan Brawijaya, Kasihan, Bantul, Daerah Istimewa Yogyakarta, Indonesia
e-mail: pinta.astuti@ft.umy.ac.id

R. S. Rafdinal
Maintenance Engineering Group, PS Mitsubishi Construction Co. Ltd., Harumi Center Building 3F, 2-5-24, Harumi, Chuo-ku, Tokyo, Japan
e-mail: s-ramita@psmic.co.jp

D. Yamamoto
National Institute of Technology, Oita College, Maki 1666, Oita-Shi, Oita, Japan
e-mail: d-yamamoto@oita-ct.ac.jp

H. Hamada
Department of Civil and Structural Engineering, Kyushu University, 744, Motooka, Nishi-ku, Fukuoka, Japan
e-mail: h-hamada@doc.kyushu-u.ac.jp

N. Sabtu (ed.), *Proceedings of AWAM International Conference on Civil Engineering 2022—Volume 2*, Lecture Notes in Civil Engineering 385,
https://doi.org/10.1007/978-981-99-6018-7_27

1 Introduction

The endurance of reinforced concrete (RC) buildings exposed to maritime environments is a global engineering and scientific problem. The chloride intrusion was the primary cause of corrosion damage, which might lead to the most damaging impact of deficient structures and the performance decline. The presence of a fissure on the concrete's surface will increase chloride penetration to the steel bar's surface, therefore destroying the passivity coating. The progression of degradation during this phenomenon is in the "acceleration" stage. As seen in Fig. 1, there is a considerable decline in performance at this stage [1].

As shown in Fig. 2, there are two types of corrosion generation in electrochemistry: micro-cell corrosion and macro-cell corrosion. The term "micro-cell corrosion" refers to the circumstance in which active corrosion and the accompanying cathodic half-cell reaction occur at the adjacent section of a steel bar. Macro-cell corrosion may develop when an actively corroding portion is connected to a passive part, either due to differing material qualities or distinct environmental conditions [2].

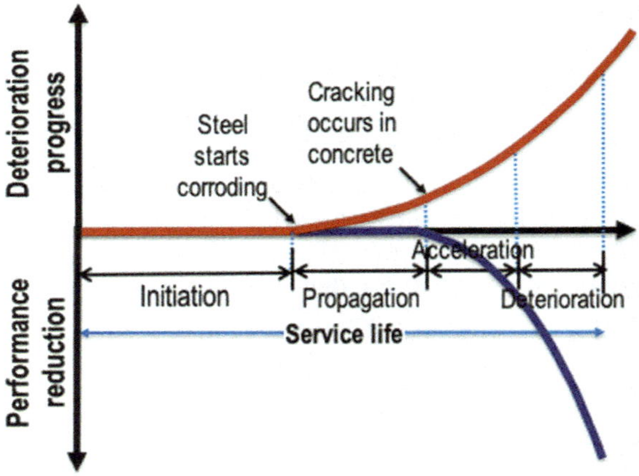

Fig. 1 Service life model of RC structures

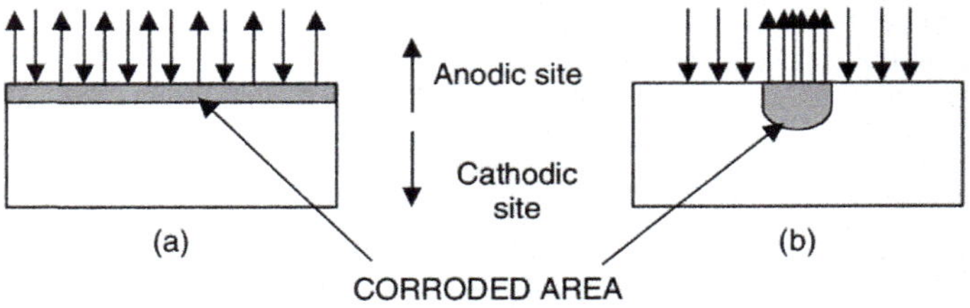

Fig. 2 Schematic illustration of **a** uniform or micro-cell corrosion and **b** localized or macro-cell corrosion [12]

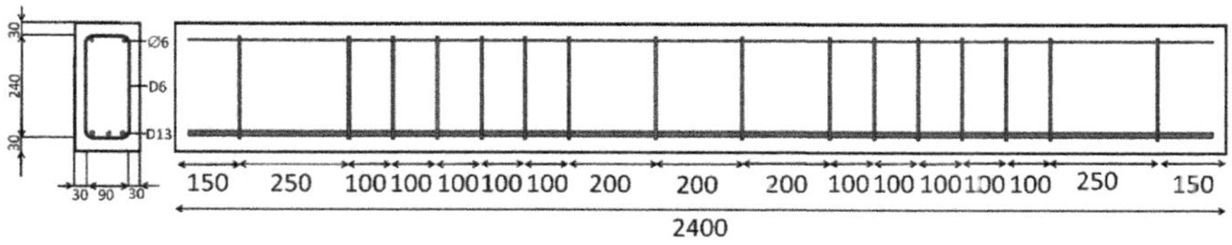

Fig. 3 Detail of steel bar position in RC specimen

Many preventative strategies are available to address both micro-cell and macro-cell chloride-induced corrosion damage. Cathodic protection is a well-known and efficient approach. The research of real structures is essential in this approach to establish the degradation level, so that the cathodic protection design may be simply constructed. Other study has revealed that certain factors should be prepared before using the maintenance system, such as electrochemical examination [3–5] and fundamental characteristics of repair technique [6–9]. Furthermore, the use of corrosion prevention methods might suppress corrosion on the proper target [10, 11].

This investigation identified the onset of natural corrosion on 44-year-old, deteriorated RC structures exposed to a real maritime environment. The purpose of this research is to investigate the association between the production of natural corrosion and electrochemical parameters so that effective countermeasures to safeguard RC structures over their service life may be developed.

2 Detail of Structure

In this experiment, two identical RC beam constructions (RC-A and RC-B) were cast in 1974 and examined at 44 years of age. From 1975 to 1995, the constructions were exposed to the natural tidal maritime environment at Sakata Port in Japan's northwest. Prior to their transfer to the Kyushu University facility in Fukuoka, Japan, the beams were housed and protected from the rain at the Port and Airport Research Institute (PARI) laboratory in Yokosuka, Japan [13, 14], from 1995 to 2010. Figure 3 depicts the dimensions of beams, arrangement of reinforcing bars, and cross-sectional details. The beam marked RC-B was subjected to an initial bending pre-crack using 0 25 of its ultimate moment capacity.

3 Investigation Method

Several observation techniques used in this investigation are described below.

Table 1 Evaluation criteria of deterioration degree [14]

Evaluation items	Deterioration degree					
	0	1	2	3	4	5
Corrosion of reinforcing steel (X_1)	None	Rust spots found on concrete surface	Partial rust stains found on concrete surface	Significant rust staining	Significant floating rust	Dramatically increased amount of floating rust
Cracking (X_2)	None	Partial cracks found on concrete surface	Some cracks	Many cracks, including some of several millimeters or more in width	Many cracks of several millimeters in width	–
Spalling of concrete cover (X_3)	None	None	Partial floating concrete found	Partial spalling found	Significant spalling	Drastic spalling

Total deterioration progress value = $X_1 + X_2 + X_3$

3.1 Visual Observation of Defective Appearance

As shown in Table 1, the degree of degradation of buildings may be determined using corrosion of reinforcing steel (X_1), cracking condition (X_2), and concrete cover spalling (X_3). Total assessment of degradation value is defined as the sum of all elements' values.

3.2 Half-Cell Potential

The half-cell potential test of rebar in accordance with ASTM C876-15 [15] was performed using a saturated calomel electrode (SCE) and high impedance multimeter (voltmeter setting) on every 50 mm square grid after an hour of pre-wetting. The reference electrode is linked to the negative terminal of the multimeter [16], while the reinforcing bar is attached to the positive terminal (Table 2).

Table 2 Deterioration degree of specimens

Evaluation items	Specimen	
	RC-A	RC-B
Corrosion of reinforcing steel (X_1)	3	3
Cracking (X_2)	3	2
Spalling of concrete cover (X_3)	1	1
Total	7	6

3.3 Corrosion Current Density

On tensile and compressive steel bars located 15, 40, 80, and 120 cm from the edge, AC impedance measurement was used to determine the corrosion current density.

3.4 Actual Steel bar Corrosion

The concrete was crushed with dimensions of 80 cm × 15 cm × 7 cm, and steel bars were found to be in the intermediate tensile region. Table 1 describes the visual observation assessment criteria.

4 Results

4.1 Visual Observation of Defective Appearance

The cracks in the concrete surface were found on both sides, as shown in Figs. 4 and 5. Cracks were presented to rebar along the beam, with the majority of cracking occurring in the central tensile region. Many longitudinal and transversal fractures in both the tensile and compressive areas are detected in RC-A, while cracks occur only on both sides and the tensile surface in other beams. When compared to other beams, RC-A also causes the most severe visual damage. The longitudinal fractures are parallel to the tensile and compressive rebar positions. The transverse fractures centered in the tensile area's middle span were caused by pre-cracking or stirrup corrosion [4]. The maximum crack width for RC-A is 2.2 mm, whereas it is 1.5 mm for RC-B. On the surface of each beam, air bubbles 117 mm in diameter developed. The visual degradation degree of RC-A is greater than that of RC-B. Both examples had the same extent of corrosion and spalling of the concrete cover. Furthermore, the cracking condition of RC-A seems to be more severe than that of RC-B. Table 3 shows the estimation of degradation level based on visual inspection.

4.2 Half-Cell Potential

At 25 °C, Fig. 6 shows the half-cell potential value of steel bar in front surface versus CSE norm. It said that in RC-A, 20% of the area is classed as a corrosion zone and 73% as an uncertainty region, while 21% of the area in RC-B is classified as a corrosion region and 11% as an uncertainty region. The images further revealed consistent corrosion mapping of RC-A. The corrosion probability of the tensile steel bar position at RC-B, on the other hand, is greater than that of the compressive steel

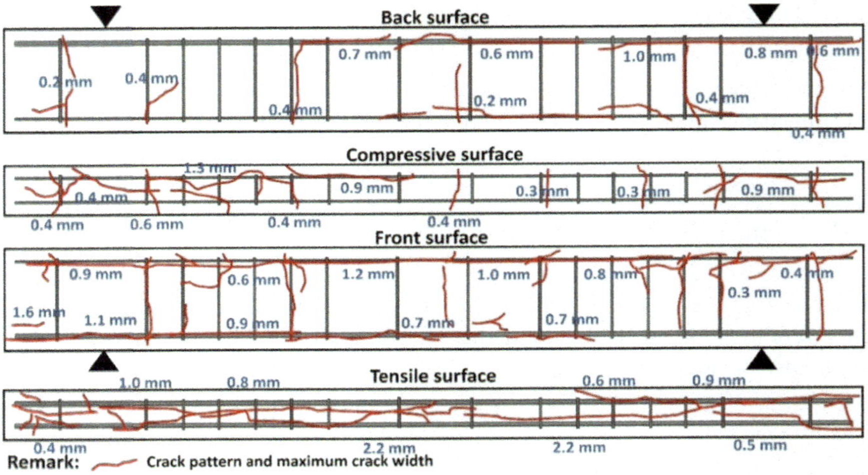

Fig. 4 Crack width and crack pattern on the concrete surface of RC-A

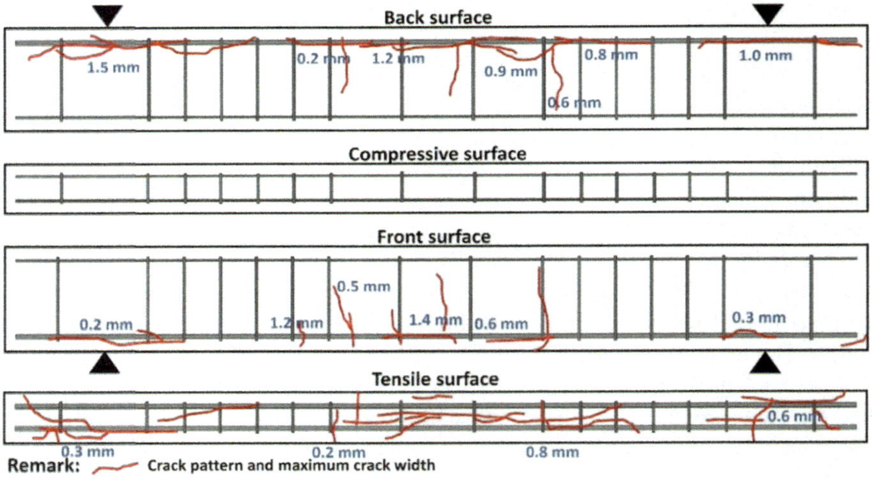

Fig. 5 Crack width and crack pattern on the concrete surface of RC-B

Table 3 Corrosion current density

Measurement point	Corrosion current density (μA/m^2)	
	RC-A	RC-B
1	1.50	0.44
2	1.48	0.89
3	0.52	0.15
4	0.34	0.16

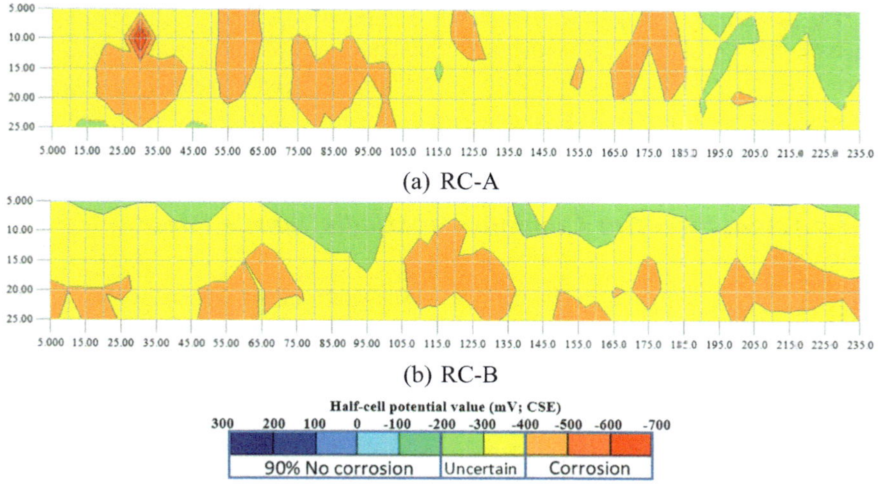

Fig. 6 Half-cell potential test value of **a** RC-A and **b** RC-B

bar. This half-cell potential result follows the fracture pattern in that the cracks are localized in the central tensile section of RC-B. As a result, the next observation concentrated on the central tensile portion of each specimen.

4.3 Corrosion Rate

Corrosion risk was quantified using the corrosion rate or corrosion current density parameter. In this study, the AC impedance approach was employed as an experimental program to assess current density. Figure 7 depicts the measurement spots: two sites at tensile rebar (1 and 2) and two points at compressive rebar (3 and 4). Table 3 summarizes the corrosion current density value data. It should be noted that the corrosion risk for tensile rebar is larger than for compressive rebar. According to the CEB standard [17], current densities ranging from 0.5 to 1.0 A/m^2 are classified as medium–high corrosion risk, while values more than 1.0 A/m^2 are classified as extremely high risk.

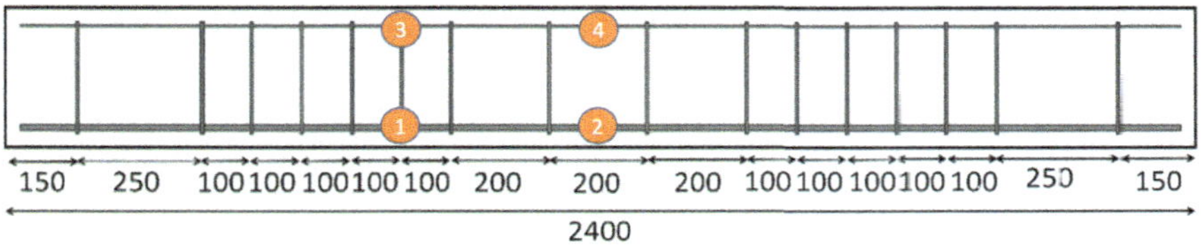

Fig. 7 Assessment point of corrosion rate by using AC impedance method

The corrosion rate findings agreed well with the observation of defective appearance on the concrete surface, indicating that the central section of the tensile rebars was in the most severe cracking state. As a result, real steel condition inspection is required, as described in Sect. 4.4.

4.4 Actual Steel Bar Condition in Middle Tensile Part

The concrete was broken in the middle tensile portion with dimensions of 80 cm × 15 cm × 7 cm in order to evaluate the real state of steel bar corrosion based on the non-destructive test result indicating the greatest chance of corrosion. Figure 8 depicts the actual steel bar conditions of both specimens.

Even though higher fracture widths were seen at points A and B in RC-A specimens, rust formed evenly along the tensile rebars in the crushed region, and a distinct corrosion profile was observed in RC-B specimens. The comparative characteristics of points A, B, and C for both specimens are recorded in Tables 4 and 5.

Observations indicate that homogeneous corrosion developed at RC-A. In this instance, uniform corrosion is produced by uniformly extremely high chloride concentrations in the region of the steel owing to the specimen's initial twenty-year

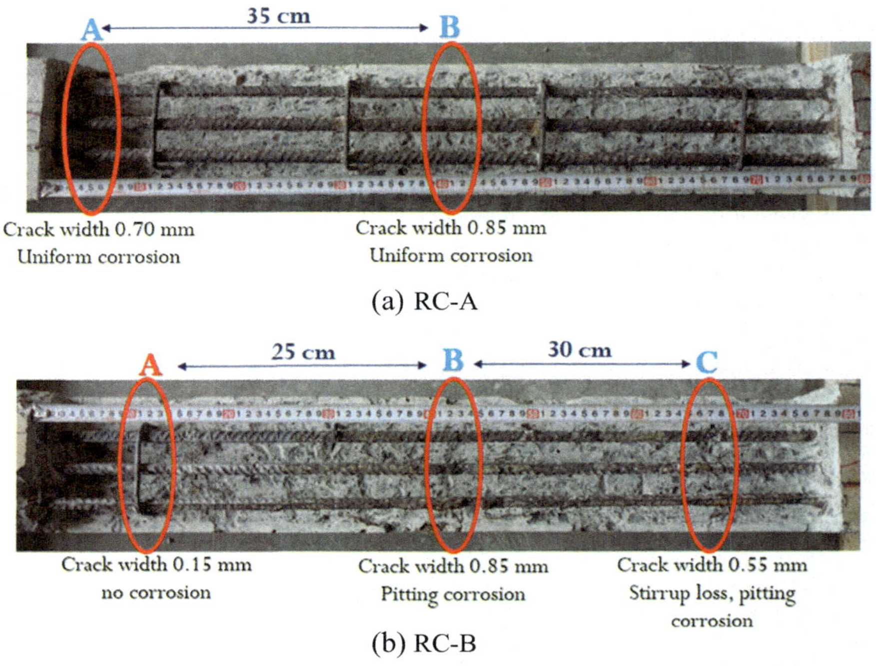

Fig. 8 Actual steel bar condition at the middle tensile part of **a** RC-A and **b** RC-B

Table 4 Evaluation of crack condition at the middle tensile part of RC-A

Crack location	Half-cell potential (mV; CSE)	Crack width (mm)	Total cross section area (mm^2)	Visual condition
A	– 423	0.70	350.2	Uniform corrosion
B	– 332	0.85	347.4	Uniform corrosion
–	–	–	–	–

Table 5 Evaluation of crack condition at the middle tensile part of RC-B

Crack location	Half-cell potential (mV; CSE)	Crack width (mm)	Total cross section area (mm^2)	Visual condition
A	– 262	0.15	352.8	No corrosion
B	– 377	0.85	248.1	Pitting corrosion
C	– 356	0.55	339.3	Pitting corrosion

exposure to the marine tidal environment. This results in the production of so-called micro-cells, which consist of pairs of anodes and cathodes that are directly nearby. These are tiny in size, thus from the outside they seem to generate uniform steel removal [18], as shown in Fig. 8a. In chloride-induced corrosion, macro-cells are often seen. They consist of anodically acting zones, often where the critical chloride concentration has been attained, and big cathodes that are next to the anodes or, on occasion, a few meters away [18].

In this experiment, the steel surface of RC-B exhibited non-uniform corrosion, as shown in Fig. 8b. Cross-sectional steel loss demonstrated the presence of pitting corrosion at locations (B) and (C). Even though the fracture occurred, the surface quality of the steel bar at point (A) is fine, and there is no corrosion. In the example of macro-cell corrosion, it is projected that the rebar in position A acted cathodically, whereas positions B and C acted anodically. Typically, the section loss of a steel bar in the anodically acting surface region is substantially greater in macro-cell corrosion because large cathodically acting areas encourage local iron removal [18]. Although the distance between anodes and cathodes may be many meters, in this experiment, the anodes and cathodes were just 25 and 55 cm apart.

Macro-cell corrosion was typically caused by high driving voltages because the potential of the anodes is very low due to the high chloride content, the depassivation of the steel, and the high water content, while the potential of the cathode is very positive due to the high oxygen content and the surface passivity of the steel. In these instances, the driving forces vary from several hundred millivolts to almost half a volt [18]. Due to inhomogeneity in oxygen delivery, a fracture in the concrete, or localized chloride ion concentration, macro-cell corrosion was developed [19]. Additionally, in this instance, the macro-cell is expedited by $0.25M_u$ of initial pre-cracking administered to the RC-B specimen prior to its exposure to the marine environment.

5 Conclusions

On the basis of the experimental program, it was determined that the formation of uniform (micro-cell) and non-uniform (macro-cell) corrosion may be investigated by the use of visual inspection of faulty concrete surface and electrochemical testing. Natural localized (macro-cell) corrosion was clearly formed on the most severe deficient appearance of 44-year-old pre-cracked RC beams exposed to the sea environment, as determined by cross-sectional loss. Initial pre-cracks placed to the specimen before to its exposure to the maritime environment enhance the development of macro-cell corrosion. In the macro-cell phenomena, the distance between anodes and cathodes may be several meters; however, in this experiment, anodes and cathodes were only separated by 25 and 55 cm.

Acknowledgements The authors would like to express their gratitude to the Port and Airport Research Institute (PARI) for providing the RC beam specimens seen in this work, as well as to all laboratory members who supported this research.

References

1. JSCE: Standard Specification for Concrete Structures (Part: Maintenance). Japan Society of Civil Engineers (2007)
2. Hansson, C.M., Poursaee, A., Laurent, A.: Macrocell and microcell corrosion of steel in ordinary Portland cement and high performance concretes. Cement Concr. Res. **36**, 2098–2102 (2006)
3. Dasar, A., Hamada, H., Sagawa, Y., Yamamoto, D.: Deterioration progress and performance reduction of 40-year-old reinforced concrete beams in natural corrosion environment. Constr. Build. Mater. **147**, 690–704 (2017)
4. Astuti, P., Rafdinal, R.S., Mahasiripan, A., Hamada, H., Sagawa, Y., Yamamoto, D.: Potential development of sacrificial anode cathodic protection applied for severely damaged RC beams aged 44-years. J. Thailand Concr. Assoc. **6**(2), 24–31 (2018)
5. Rafdinal, R. S.: *Life Extension of RC Structure by Cathodic Protection Using Zinc Sacrificial Anode Embedded in Concrete*. Thesis (PhD) Department of Civil and Structural Engineering, Kyushu University Fukuoka Japan (2016)
6. Astuti, P., Rafdinal, R.S., Hamada, H., Sagawa, Y., Yamamoto, D.: Time dependency of corrosion rate on rusted and non-rusted rebar protected by sacrificial zinc anode in patch repair section. In: Japan Concrete Institute (JCI) Symposium, September 2018, Tokyo Japan (2018)
7. Astuti, P., Rafdinal, R.S., Hamada, H., Sagawa, Y., Yamamoto, D., Kamarulzaman, K.: Effectiveness of rusted and non-rusted reinforcing bar rotected by sacrificial anode cathodic protection in repaired patch concrete. In: IOP Conference Series: Earth and Environmental Science, vol. 366 (012013) (2018)
8. Astuti, P., Kamarulzaman, K., Rafdinal, R. S., Hamada, H., Sagawa, Y., Yamamoto, D.: Influence of rust removal process on the effectiveness of sacrificial anode cathodic protection in repair concrete. In: IOP Conference Series: Material Science and Engineering, vol. 819 (012088) (2020)
9. Mahasiripan, A., Astuti, P., Rafdinal, R.S., Hamada, H., Sagawa, Y., Yamamoto, D.: Experimental study on the performance of sacrificial anode cathodic protection under low temperature around—17 C. In: Proceedings of the 3rd International Symposium on Concrete and Structures 2018, September 2018, Bangkok, Thailand (2018)

10. Astuti, P., Rafdinal, R.S., Hamada, H., Sagawa, Y., Yamamoto, D.: Application of sacrificial anode cathodic protection for partially repaired RC beams damaged by corrosion. In: Proceedings of 4th International Symposium on Concrete and Structures (CSN2019) Kanazawa, Japan, 17–19 June 2019, pp 284–291 (2019)
11. Astuti, P., Rafdinal, R.S., Kamarulzaman, K., Hamada, H., Sagawa, Y., Yamamoto, D.: Repair method of deteriorated RC beams by sacrificial anode cathodic protection and corrosion inhibitor. In: Proceedings of 3rd ACF Symposium on 2019 Assessment and Intervention of Existing Structures, Hokkaido, Japan, September 2019 (2019)
12. Akid, R.: Chapter 11—Corrosion of engineering materials. In: Wessel, J.K. (eds.) Handbook of Advanced Materials. Wiley, New York (2004)
13. Hamada, H., Otsuki, N., Haramo, M.: Durability of concrete beams under marine environments exposed in port of Sakata and Kagoshima (after 10 year's exposure). Technical Note of the Port and Airport Research Institute, vol. 614, pp. 3–43 (1988)
14. Yokota, H., Akiyama, T., Hamada, H., Mikami, A., Fukute, T.: Effect of degradation of concrete on mechanical properties of reinforced concrete beams exposed to marine environment (for 20 years in Sakata). In: Report of the Port and Airport Research Institute, vol. 38, no. 2 (1999)
15. ASTM C876-15: Standard test method for half-cell potentials of uncoated reinforcing steel in concrete. ASTM Int (2015)
16. Elsener, B.: Half-cell potential measurements—potential mapping on reinforced concrete structures RILEM TC 154-EMC: 'electrochemical techniques for measuring metallic corrosion.' Mater. Struct. **36**, 461–471 (2003)
17. CEB: Comité Euro International du Béton 1989 Durable concrete structures—CEB Design Guide. Bull. D' Inform. **182**, 310 (1989)
18. Raupach, M.: Chloride-induced macrocell corrosion of steel in concrete–theoretical background and practical consequences. Constr. Build. Mater. **10**(5), 329–338 (1996)
19. Qian, S., Zhang, J., Qu, D.: Theoretical and experimental study of microcell and macrocell corrosion in patch repairs of concrete structures. Cement Concr. Compos. **28**(8), 685–695 (2006)

Influence of Waste Crumb Rubber as a Partial Replacement for Fine Aggregates on Concrete Properties

Che Azrul Asyraaf Che Ani, Mohammed Yahya Mohammed Al-Fasih, Izni Syahrizal Ibrahim, Noor Nabilah Sarbini, and Khairul Hazman Padil

Abstract This study investigated the effect of using waste crumb rubber as a partial replacement for the fine aggregates on the fresh and hardened properties of normal mix concrete. Waste crumb rubber with particles sizes of 2 mm–4 mm was supplanted in the mix design of the concrete as part of the fine aggregates. Three different percentages of waste crumb rubber were considered to replace the fine aggregate which are 5%, 10%, and 15% by volume. The properties of fresh concrete were evaluated through standard slump test, while the hardened properties were examined through compressive, splitting tensile, and flexural strength tests after curing at 7 days and 28 days. Furthermore, ultrasonic pulse velocity (UPV) tests were also carried out to evaluate the integrity and quality of the crumb rubber concrete. The test results revealed that the partial replacement of the fine aggregate with waste crumb rubber significantly decreased the workability and mechanical properties of the concrete. In addition, the concrete workability decreased with the increasing substitution of the waste crumb rubber replacement to that of the fine aggregate. Similarly, by increasing substitution of the waste crumb content, the compressive strength, splitting tensile strength, and flexural strength showed a decreasing trend from the control mixture at 7 days and 28 days.

C. A. A. Che Ani · M. Y. Mohammed Al-Fasih · I. S. Ibrahim (✉) · N. N. Sarbini · K. H. Padil
Forensic Engineering Centre, Institute for Smart Infrastructure and Innovative Construction, Faculty of Civil Engineering, Universiti Teknologi Malaysia, Johor Bahru, 81310 Johor, Malaysia
e-mail: iznisyahrizal@utm.my

M. Y. Mohammed Al-Fasih
e-mail: mymalfasih@gmail.com

N. N. Sarbini
e-mail: noornabilah@utm.my

K. H. Padil
e-mail: khairulhazman@utm.my

M. Y. Mohammed Al-Fasih
Department of Civil Engineering, College of Engineering and Information Technology, Onaizah College, Onaizah 56447, Saudi Arabia

N. Sabtu (ed.), *Proceedings of AWAM International Conference on Civil Engineering 2022—Volume 2*, Lecture Notes in Civil Engineering 385,
https://doi.org/10.1007/978-981-99-6018-7_28

Keywords Crumb rubber concrete · Modified mixture · Strength performance · Sustainable concrete

1 Introduction

Environmental concerns are becoming one of the most pressing challenges affecting the production of natural aggregates. Several industrial and activity developments are essential to realizing this objective. Environmental saving primarily comes from the dodged effects of concrete manufacturing, removal of waste tires and solid trash in landfills, and decreased in transportation distance. The use of natural aggregates has a significant impact on the environment when depending on natural resources of using fine aggregates in the concrete. Waste-stream materials such as crumb tires, fly ash, palm oil fuel ash (POFA), and others can provide verifiable natural life-cycle effects and immediate cost savings when compared to that of the standard concrete production. The economic advantages of fly ash are obvious; either the material is discarded at a cost or sold for a minimal monetary price. Meanwhile, recycled aggregates (waste crumb tires and demolition wastes) are not found to be economically recycled, and can be reduced in transportation costs to lesser the distance from the source to its destination [1].

Crumb rubber is recycled or re-processing (shredding) rubber produced from the disposed of automotive tires [2]. In recent decades, recycled or re-used of disposed of tires had been made in many ways around the world. However, the number of stockpiles increased more than the utilization of waste tires. It is more important to find new ways to make use of these discarded tires. There are five applications for shredding rubber in landfill construction and operation such as a drainage layer for cap closures, as permeable backfill in gas venting systems, as a material for daily cover, permeable aggregate for leachate collection systems, and in operational layers. These new uses of waste tires have been through various refinements in a variety of applications. For some reason, waste tire refinements are a confined concern that is difficult to obtain for industrial usage. However, Civil Engineering products have been making a significant contribution and corporation to this effort by making this waste tire into various sizes also known as crumb rubber [3]. In the early 1990s, the usage of crumb rubber increases to make an invention known as rubberized concrete [4]. According to the Rubber Manufacturer Association, in 2007, Civil Engineering projects that used waste tires as crumb rubber was about 12% for their respective fields which includes modified asphalt concrete pavements, road base for highway projects, speed bumps, and filtration systems for water treatment [5]. Crumb rubber is also applied to other fields such as sports surfaces for kindergarten playgrounds and recreation areas, school sports areas, and athletic tracks [6].

Crumb rubber is particles of fine granular or in various refinements. During the shredding process, steel and fibers were removed separately, leaving only the tire rubber. The process was then continued in the cracker mill and with aid of mechanical means, its further reduced the size of particles to generate crumb rubber. There are two technologies for producing crumb rubber, which are mechanical grinding and cryogenic grinding [6]. The cryogenic grinding is more expensive, but they produced smoother and smaller size rubber tire. According to Mushunje et al. [7], a typical scrap tire contains 70% of recoverable rubber, 15% of steel, 3% of fiber, and 12% of extraneous material (e.g., inert fillers), and tire rubber or crumb rubber is the production by the most common process of either ambient and cryogenic. These two processes give different size particles of the crumb rubber. The ambient grinding process can be described in a granular and cracker mill. In the ambient system, the rubber of the tire will be at room temperature when it enters the granulator and cracker mill. The rubber particles produced through the granulation process have a cut at the shape and rough surface texture. Meanwhile, rubber particles or crumb rubber produced by cracker mill are generally long and narrow in shape, and they have high surface area [7]. The cryogenic process is basically using liquid nitrogen or other materials to freeze the rubber particle before making the size reduction. In this stage, most of the rubber physically become solid glass at temperature below − 80 °C. Size reduction of the rubber tires can be applied to the cryogenic temperature at any stage. The materials produced through this process will appear shiny, clean, fractured surfaces and have low steel and fiber content. The size of particles production range from finer 40 mesh to 60 mesh, and for very fine crumb rubber (below 60 mesh) it required a secondary grinding system [7]. Eldin and Senouci [3] stated that waste tire is classified into three sizes, namely shredded or chipped rubber (13 mm – 73 mm), crumb rubber (0.425 mm – 4.75 mm), and ground rubber (0.075 mm – 0.425 mm).

According to Pierce [8], the feature and of the concrete mixture are influenced by the components that are consolidated to produce the final product. Rubberized concrete cement had been tested using materials commonly used in traditional concrete compositions, where coarse and fine aggregate parts were replaced with scrap tire particles of various sizes. Some of the replacement ranges from 1% to 100% by volume of crumb rubber to that of the natural aggregates. Crumb rubber size of 76 mm long is commonly used in the replacement of fine aggregates in concrete. The particle size of the rubber and the percentage of replacement directly affected the fresh properties of the concrete mixture. The recycled or scrap tire market for a specific district typically indicates the particle size of the rubber used in the past research projects. Crumb rubber size of 30 mesh has become a typical material in rubberized concrete. According to Khatib and Bayomy [9], for concrete with 80% of rubber content (or about 40% by total aggregates), the concrete slump was close to zero and was not well mixed under hand mixing. Many studies had been conducted to achieve the ability of hardened concrete properties when incorporating crumb rubber in the concrete mixture. Most of the experimental tests focused on compressive, splitting tensile, and flexural strength with different percentages of aggregate substituted with recycled rubber. According to Eldin and Senouci [3], the strength of rubberized concrete depended on various sizes of the crumb rubber proportions and its surface textures that can affect the compressive strength of the rubberized concrete. Crumb rubber concrete resulted to lower compressive and splitting tensile strengths as compared with normal ordinary Protland cement (OPC). This reduction or losses in the rubberized concrete is due to poor adhesion between the particle of crumb rubber and cement paste [10]. Sukontasukkul [11] found that replacing 10%–30% of fine aggregate with crumb rubber reduced the unit weight of concrete from 14% to 28%. In addition to being lighter, it had better sound absorption and heat transfer properties than the plain concrete panel. The crumb rubber has the potential to be used as a freeze or thaw resistant compound in concrete in which 6% of crumb rubber content gives very good freeze or thaw resistance [12].

This study is aimed to investigate the potential of utilizing waste crumb rubber in the concrete mixture. The concrete mixture is designed by incorporating waste crumb rubber as partial replacement to fine aggregates at certain percentages. The mechanical properties of rubberized concrete at fresh and hardened states were investigated, and the optimum percentage replacement of fine aggregates with waste crumb rubber was determined.

2 Methodology

2.1 Raw Concrete Materials Characteristics

The concrete mixture consists of cement, coarse aggregates, fine aggregates, and water. The OPC used in the concrete mixture met the requirement of ASTM C150-04 Type I [13]. In addition, to produce rubberized concrete, crumb rubber (CR) particle is added to the concrete mixture. CR is sourced from Malaysia's rubber industry which is an engineering artifact generated by shredders and shearing machines converting discarded scrap tires into its required particle size. These CRs were graded based on ASTM C 136 [14] approach. Two different ranges of CR were obtained from the sieve testing in accordance with ASTM C33 [15], with grain sizes ranging from 1 mm to 4 mm and from 5 mm to 8 mm. From here, the selected grain size of between 2 mm and 4 mm of CR was partially replaced with fine aggregates in the concrete mix design. Natural aggregates of river sand quartz are used as fine aggregates according to the requirements of ASTM C33 [15]. Prior to the mixing process, the natural fine aggregates (sand) were maintained at saturated-surface dry (SSD) to obtain opposite amounts of water and cement. In addition, the fine aggregates were also graded using sieve analysis according to ASTM C136 [14]. The bulk of the particles in the fine aggregates are natural sand passing through a 3/8″ filter, while coarse aggregates are particles greater than 0.19″, generally ranging between 3/8″ and 1.5″ in diameter. To properly balance a concrete mixture that includes CR, it is important to calculate an exact specific gravity. Based on the specific gravity testing, the size of CR used in this study passed through a 2 mm passing sieve. To compute the specific gravity employed in the concrete mixture design, the final mass is obtained. The specific gravity of natural fine aggregates (sand) is 2.67 while CR is 1.11. Sieve analysis was performed on the components used in the concrete mix design to compare the gradation of the aggregates and to provide a complete picture of the particle size distribution in the mix. The goal of this experiment was to replace fine aggregates

Table 1 Concrete mix design with various percentages replacement of CR

Mix	Weight per cubic meter (kg/m^3)				
Percentage of CR replacement	Cement	Water	Crumb rubber (CR)	Fine aggregate (Sand)	Coarse aggregate
Control 0%	18.7	8.21	–	26.40	53.46
CR 5%	18.7	8.21	0.54	25.08	53.46
CR 10%	18.7	8.21	1.07	23.76	53.46
CR 15%	18.7	8.21	1.61	22.44	53.46

with the equivalent percentage of CR. As a result, it was decided to substitute natural fine aggregates (sand) with CR in the percentages of 5%, 10%, and 15% of the concrete volume. For this study, the passing fine aggregates of 600 μm was 54% and used in the concrete mix design.

2.2 *Mix Design*

This study consisted of six concrete samples as the control and 24 CR concrete samples at 7 days and 28 days. The concrete mixture was designed to achieve the compressive strength of 30 N/mm^2 at 28 days. The water-to-cement ratio in all mixtures was kept constant at 0.45, and the CR was substituted with various percentages to that of natural fine aggregates (sand). Meanwhile, the coarse aggregates were designed to use size of 20 mm. Since the fine aggregates (sand) and CR had varying specific gravities, CR replacement was done by volume rather than weight. The percentage volume replacement of fine aggregates (sand) with CR at 5%, 10%, and 15% of the total volume is shown in Table 1.

2.3 *Testing Sample*

In order to determine how the fresh concrete behaved, a standard slump examination was carried out following ASTM C 143 [16]. Then, hardened concrete properties through the compressive strength, flexural strength, and splitting tensile strength tests were carried out according to ASTM C109, ASTM C78, and ASTM C 496 [17–19], respectively. The samples were tested for the compressive strength at 7 days and 28 days of the curing period. Meanwhile, splitting tensile strength and flexural strength tests were performed at 28 days of the curing period.

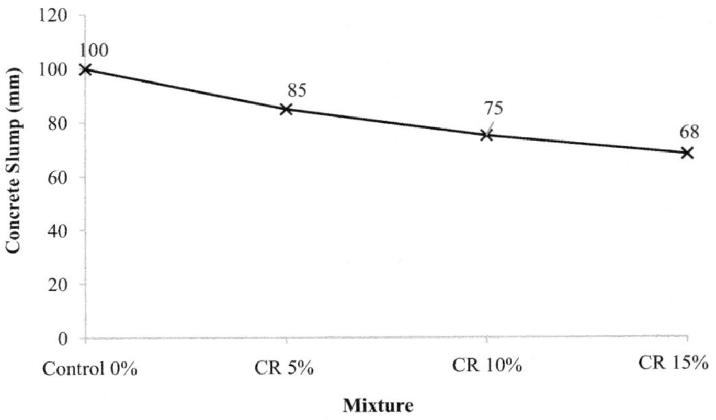

Fig. 1 Concrete slump test results between the control (0%) and CR at 5%, 10% and 15%

3 Result and Discussion

3.1 Fresh Concrete Properties

The fresh properties at the time of concrete mixing are known as concrete slump test. The slump of the concrete presented the workability of the mixture. Figure 1 shows the results of the concrete slump with the various percentages replacement of fine aggregates (sand) by CR. The concrete slump decreases with the increasing percentage of CR replacement in the mix design. Concrete with CR of 5%, 10%, and 15% led to the decreasing concrete slump from 100 mm (0% control) to 85 mm, 75 mm, and 68 mm, respectively. Compared with the control, this indicates that the workability of concrete with CR was constantly decreasing as the percentages of replacement increases. This is because, concrete mixes containing higher quantities of CR had greater specific surface areas and accompanying roughness, thus significantly reducing the workability [20].

3.2 Hardened Concrete Properties

Compressive Strength. The compressive strength tests were conducted for 24 cubes of 100 mm × 100 mm × 100 mm at the curing age of 7 days and 28 days. For each test day and concrete mixture, 3 cubes were tested to determine its average. The average compressive strength for each mixture at the curing age of 7 days and 28 days are shown in Fig. 2a. From Fig. 2a, the compressive strength of the control (0%) mixture increases from 34.57 N/mm^2 to 45.97 N/mm^2 at 7 days and 28 days, respectively. The same pattern was also observed for CR concrete, where the compressive strength increased from the 7 days to 28 days. However, the pattern shows a decreasing trend when compared with the control (0%) at both tested days. For instance, the compressive strength at 7 days decreased from 34.57 N/mm^2 (control 0%) to 30.97 N/mm^2, 31.48 N/mm^2, and 27.45 N/mm^2 when the percentage replacement of CR increased from 0% to 5%, 10%, and 15%, respectively. Meanwhile, the compressive strength at 28 days also dropped from 45.97 N/mm^2 (control 0%) to 42.22 N/mm^2, 41.38 N/mm^2, and 37.34 N/mm^2 when increasing the percentage of CR replacement from 0% to 5%, 10%, and 15%, respectively. It was observed that the compressive strength dropped by comparison of the control mixture with the CR concrete as the percentage of replacement increases. However, all concrete mixtures achieved higher compressive strength than the 28 days design concrete strength of 30 N/mm^2. The failure mode of each mixture is shown in Fig. 2b. The failure modes are consistent for all mixtures where straight parallel and crushing were observed. This also indicates that minimum time can be reached to its ultimate load for the CR concrete without any unexpected failure observed.

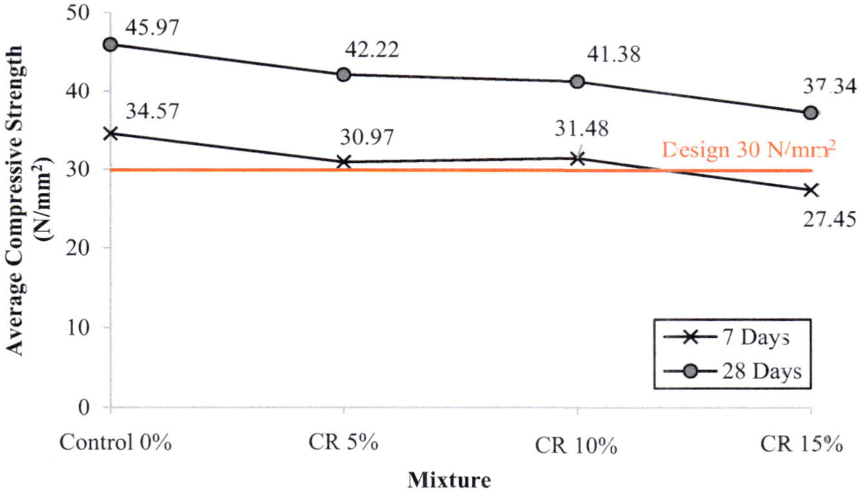

(a) Compressive strength relationships at 7 days and 28 days

Control (0%) CR 5% CR 10% CR 15%

(b) Failure mode of cube specimens

Fig. 2 Compressive strength relationship and failure mode of the control (0%), CR 5%, CR 10%, and CR 15%

Splitting Tensile Strength. The splitting tensile tests were conducted for 12 cylinders of 150 mm diameter × 300 mm high at 28 days. For each mixture, three cylinders were tested to determine their average splitting tensile strength. The relationship is shown in Fig. 3a for the control (0%) and CR concrete. The implying of CR from 0% to 5%, 10%, and 15% as a replacement to fine aggregates (sand) resulted in the decreased of the splitting tensile strength from 3.32 N/mm^2 to 2.94 N/mm^2, 2.96 N/mm^2, and 2.81 N/mm^2, respectively. The decreasing pattern maybe due to the poor bonding between the CR and cement paste as compared with the bonding between the natural fine aggregate (sand) and cement paste. CR which was also lacking in stiffness were further a contributing factor in the strength decreasing pattern [21, 22]. However, the results show that the splitting tensile strengths are still higher than the design strength of 2.60 N/mm^2, calculated based on the expression given in EN 1992-1-1: 2010. The failure mode from the splitting tensile test shown in Fig. 3b shows improvement of the cracking pattern. This can be seen from the CR concrete, where the failure for the CR concrete does not split the cylinder into two parts as compared with the control (0%).

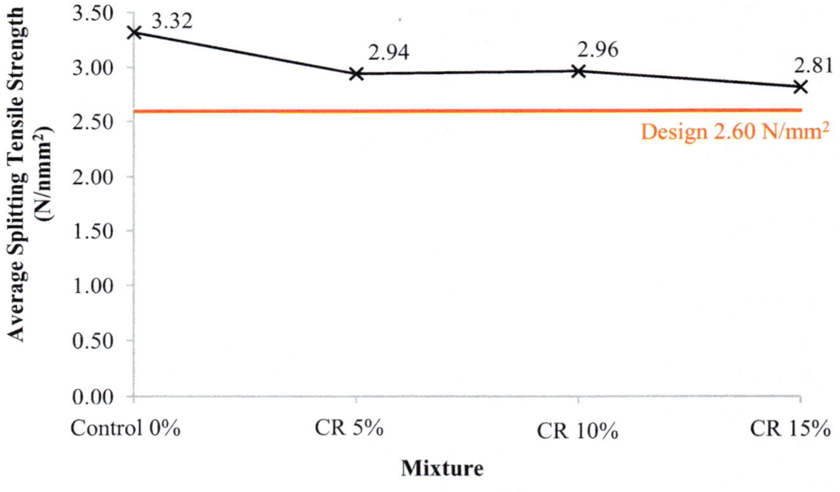

(a) Splitting tensile strength relationships at 28 days

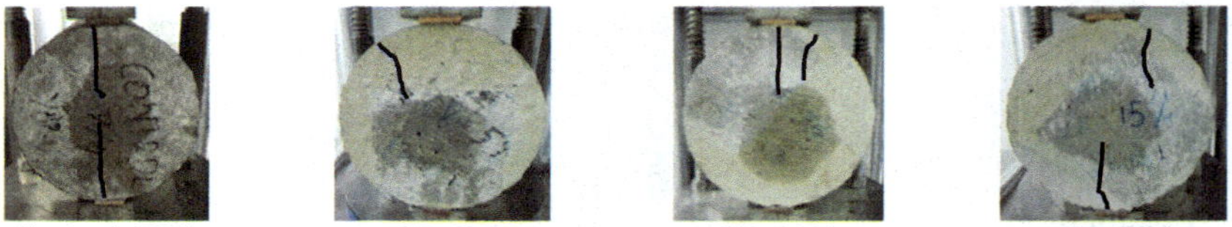

Control (0%) CR 5% CR 10% CR 15%

(b) Failure mode of cylinder specimens

Fig. 3 Splitting tensile strength relationship and failure mode of the control (0%), CR 5%, CR 10%, and CR 15%

Flexural Strength. The flexural strength tests were conducted for 12 prisms of 100 mm × 100 mm × 500 mm long at 28 days. Figure 4a shows the flexural strength relationship between the control (0%) and CR 5%, CR 10%, and CR 15% mixtures. The flexural strength in the figure was taken as the average of the three prisms for each concrete mixture. From Fig. 4a, it was found that the flexural strength reduced significantly when the percentage replacement of natural aggregates (sand) with CR was at 15%. The replacement of CR by 5% and 15% to that of the natural fine aggregate (sand) resulted to the decrease in the flexural strength from 4.62 N/mm^2 of the control (0%) to 4.24 N/mm^2 and 4.11 N/mm^2, respectively. However, the mixture with 10% of CR replacement increased to 4.74 N/mm^2 from 4.62 N/mm^2. In comparison, flexural strength is normally taken as 10%–20% of the compressive strength. Following the expression given in IS 456: 2000, the design flexural strength was used as a comparison with the current study. Generally, with the increase percentage of CR in the mixture, the flexural strength dropped significantly. Similar findings were also found by Eldin and Senouci [3]. However, the strength for all mixtures are still higher than the design flexural strength of 3.50 N/mm^2. The main reason for the lower flexural strength is likely due to exceedingly poor bonding between the CR and the cement pastes where the rubber filled up the gaps in the concrete matrix, resulting in a decrease in the concrete density [23, 24]. This can be further proven from the field emission scanning electron microscopy (FESEM) analysis as shown in Fig. 5, showing gaps between the CR and the cement paste. The failure mode of prisms is shown in Fig. 4b, showing consistencies for all mixtures. All specimens failed due to the separation of the prisms in two parts located at the mid-span region.

In order to improve the adhesion, the rubber surface can be roughened to increase the mechanical strength of the concrete [25, 26]. Previous study also shows that pre-coating the rubber aggregates with cement paste can increase the compressive strength by 30% and 50% [25] due to increase in adhesion between the rubber surface and cement paste.

Ultrasonic Pulse Velocity Testing (UPV). The UPV testing was used to determine the consistency of the concrete composition between the control (0%) and CR concrete. Apart from that, it can also be used on the structural element to indicate the quality of the concrete material. Slower velocities may suggest concrete with many fractures or voids, whereas higher velocities signify good quality and consistency of the material. Table 2 shows the results for all specimens and mixtures tested at 28 days. The results were taken from the average of three specimens for each mixture of the cube, cylinder, and prism. In general, the average velocity (in m/s) for the cube, cylinder, and prism specimens slightly decreased with the increase percentage of CR replacement in the concrete as compared with the control (0%). The cube and prism specimens show that the average velocity is above 4500 m/s, showing excellent concrete quality of the mixture. However, the average velocity of the cylinder specimens was slightly lower than 4500 m/s, but still considered as good quality concrete composition. This also shows that the mixing process was consistent throughout the experimental work.

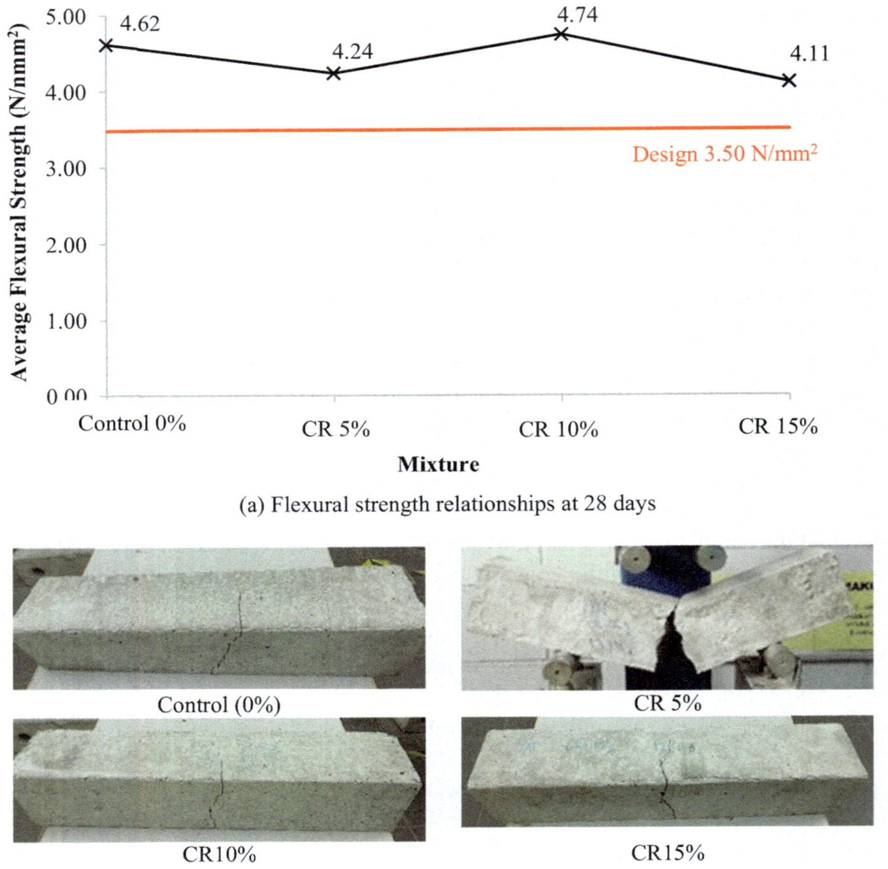

(b) Failure mode of prism specimens

Fig. 4 Flexural strength relationship and failure mode of the control (0%), CR 5%, CR 10%, and CR 15%

4 Conclusion

I. Adding CR into concrete as partial replacement to natural fine aggregates (sand) significantly decreased the properties of the concrete. The decreasing effect caused by CR in the mixture was higher than that produced by the fine aggregates (sand).

II. Fresh concrete properties displayed that the increased percentage of CR in the mixture led to lower concrete slump and poor workability.

III. Compressive strength, splitting tensile strength, and flexural strength decreased with the increased percentage of CR into the mixture. However, they achieved the design strength for concrete grade C30. These properties might be enhanced

Fig. 5 Poor bonding between CR and cement paste for specimen with CR15%

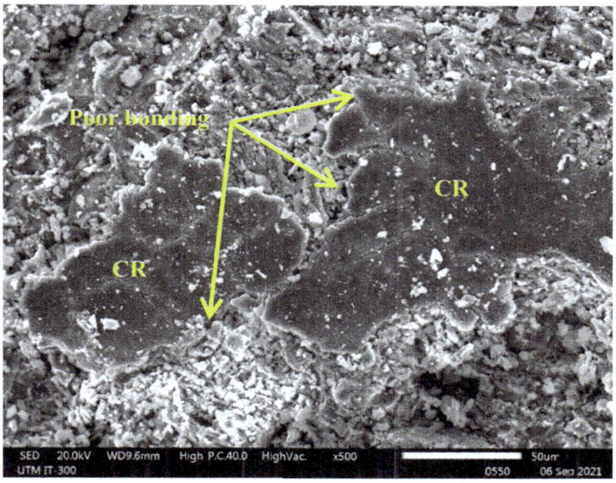

(a) Magnification 500×

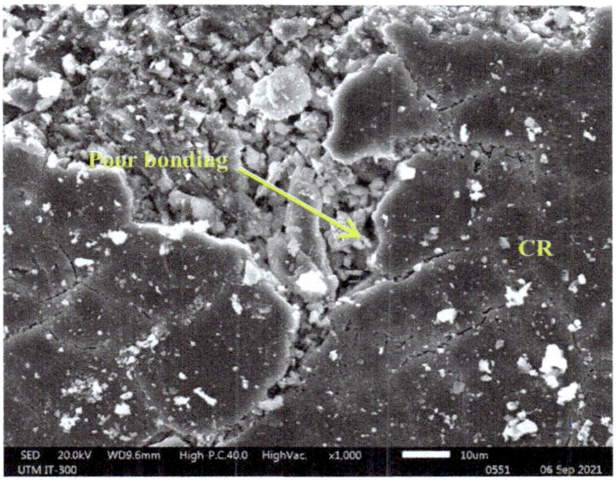

(b) Magnification 1000×

Table 2 UPV results for the cube, cylinder, and prisms specimens at 28 days

Percentage of CR replacement	Cube		Cylinder		Prism	
	Average transmission Time (μs)	Average velocity (m/s)	Average transmission time (μs)	Average velocity (m/s)	Average transmission time (μs)	Average velocity (m/s)
Control	19.40	5154.64	66.30	4524.89	105.90	4721.44
5%	20.20	4950.50	66.90	4484.30	106.87	4678.73
10%	20.97	4769.48	67.03	4475.39	109.00	4587.16
15%	21.77	4594.18	68.13	4403.13	110.17	4538.58

if the use of the rubber aggregates with a roughened surface and pre-coating rubber aggregates due to an increase in the adhesion between the rubber surface and cement paste.

IV. CR 5%, CR 10%, and CR 15% concrete met the design strength requirement in this study. However, an optimum percentage of CR replacement as fine aggregate in the mixture can be executed until 10%. For the replacement of CR as fine aggregates (sand) higher than 15%, it can cause misplacement of CR in the mixture and directly affected a high decrease of the concrete strength.

Acknowledgements The authors would like to send their warmest appreciation for the technical supports from the Structural & Material Laboratory at Universiti Teknologi Malaysia and the financial research support by Universiti Teknologi Malaysia and Hanyang University, South Korea, under the Matching Research Grant Vote No. R.J130000.7309.4B688.

References

1. Kardos, A.J.: Beneficial Use of Crumb Rubber in Concrete Mixtures. University of Colorado at Denver (2011)
2. Chesner, W.H., Collins, R.J., MacKay, M.H., Emery, J.: User Guidelines for Waste and By-product Materials in Pavement Construction. Federal Highway Administration (FHWA). FHWA-RD-97-148 (1998)
3. Eldin, N.N., Senouci, A.B.: Rubber-tire particles as concrete aggregate. J. Mater. Civ. Eng. **5**(4), 478–496 (1993)
4. Ellis, D., Gandhi, P.: Innovative use of Recycled Tires in Civil Engineering Applications. Swinburne University of Technology, Melbourne (2009)
5. Rubber Manufacturer's Association. USA (2014)
6. Gomes, F.O., Rocha, M.R., Alves, A., Ratola, N.: A review of potentially harmful chemicals in crumb rubber used in synthetic football pitches. J. Hazard. Mater. **409**, 124998 (2021)
7. Mushunje, K., Otieno, M., Ballim, Y.: Partial replacement of conventional fine aggregate with crumb tyre rubber in structural concrete–effect of particle size on compressive strength and time dependent deformations. In: MATEC Web of Conferences, vol. 199, p. 11002. EDP Sciences (2018)
8. Pierce, C., Blackwell, M.: Potential of scrap tire rubber as lightweight aggregate in flowable fill. Waste Manage. **23** (2002)
9. Khatib, Z.K., Bayomy, F.M.: Rubberized Portland cement concrete. J. Mater. Civ. Eng. **11**(3), 206–213 (1999)
10. Xi, Y., Li, Y., Xie, Z., Lee, J.S.: Utilization of solid wastes (waste glass and rubber particles) as aggregates in concrete. In: International Workshop on Sustainable Development and Concrete Technology, Beijing, China, pp. 45–52 (2004)
11. Sukontasukkul, P.: Use of crumb rubber to improve thermal and sound properties of pre-cast concrete panel. Constr. Build. Mater. **23**(2), 1084–1092 (2009)
12. Paine, K.A., Dhir, R.K., Moroney, R., Kopasakis, K.: Use of crumb rubber to achieve freeze thaw resisting concrete. In: Proceedings of the International Conference on Concrete for Extreme Conditions, University of Dundee, Scotland, UK, pp. 486–498 (2002)
13. ASTM C150-04: Standard Specification for Portland Cement, Annual Book of ASTM Standards, Cement; Lime; Gypsum, Section 4, Volume 4.01, pp. 150–155, West Conshohocken, PA, USA (2004)

14. ASTM C 33: Standard Specification for Concrete Aggregates. In: American Society for Testing and Materials, ASTM Specification, Philadelphia (1999)
15. ASTM C136: Standard test method for sieve analysis of fine and coarse aggregates. In: Annual Book of American Society for Testing Materials Standards, vol. C 04.02 (2006)
16. ASTM C 143: Standard test method for slump of hydraulic-cement concrete. In: Annual Book of ASTM Standards, vol. 04.02. American Society for Testing and Materials, Philadelphia (2004)
17. ASTM C109: Standard Test Method for Compressive Strength of Hydraulic Cement Mortars. ASTM International, West Conshohocken, PA (2008)
18. ASTM C78: Standard Test Method for Flexural Strength of Concrete (Using Simple Beam with Third-Point Loading). American Society for Testing and Materials, Philadelphia (2010)
19. ASTM-C496: Standard Test Method for Splitting Tensile Strength of Cylindrical Concrete. American Society for Testing and Materials West Conshohocken, PA (1996)
20. Al-Fasih, M.Y.M., Huseien, G.F., bin Ibrahim, I.S., Sam, A.R.M., Algaifi, H.A., & Alyousef, R.: Synthesis of rubberized alkali-activated concrete: experimental and numerical evaluation. Constr. Build. Mater. **303**, 124526 (2021)
21. Aslani, F., Ma, G., Wan, D.L.Y., Le, V.X.T.: Experimental investigation into rubber granules and their effects on the fresh and hardened properties of self-compacting concrete. J. Clean. Prod. **172**, 1835–1847 (2018)
22. Ganjian, E., Khorami, M., Maghsoudi, A.A.: Scrap-tyre-rubber replacement for aggregate and filler in concrete. Constr. Build. Mater. **23**(5), 1828–1836 (2009)
23. Thomas, B.S., Gupta, R.C.: Properties of high strength concrete containing scrap tire rubber. J. Clean. Prod. **113**, 86–92 (2016)
24. Thomas, B.S., Gupta, R.C., Panicker, V.J.: Recycling of waste tire rubber as aggregate in concrete: durability-related performance. J. Clean. Prod. **112**, 504–513 (2016)
25. Cairns, R., Kew, H., Kenny, M.J.: The Use of Recycled Rubber Tyres Inconcrete Construction. The University of Strathclyde, Glasgow (2004)
26. Tantala, M.W., Lepore, J.A., Zandi, I.: Quasi-elastic behavior of Rubber Included Concrete (RIC) using waste rubber tires. In: Proceedings of the 12th International Conference on Solid Waste Technology and Management, Philadelphia, USA (1996)

Numerical Analysis of Arch Reinforced Concrete Deep Beams with Various Cross Sections

Martyana Dwi Cahyati, Muhammad Rizki Fadillah, Risky Dwi Erlinda, Hakas Prayuda, Fanny Monika, and Muhammad Mirza Abdillah Pratama

Abstract Beam member is one of the important components in a building structure with the functions to transfer loads to column and foundation. The main purpose of this study is to determine the effect of cross-section arch deep beam structure with a different type of cross-section. The deep beam in this study consists of two types of longitudinal versions called non-arch and deep arch beam. Besides, and four different cross-sections are applied as a variation to investigate the effect of cross-sections on flexural behavior. This study only used finite element analysis to investigate the stress distribution, load, and displacement relationship, ductility, and stiffness of the model. The loading used in this analysis is in 4 points where the load is located at 1/3 and 2/3 of the span. The results show that the deep arch beam experience a decrease in flexural performance compared with the normal deep beam. Deep beam with rectangular section produced high performance compared with other cross-sections. However, the damage pattern indicates that shear failure is still dominant in all models.

M. D. Cahyati · M. R. Fadillah · R. D. Erlinda · H. Prayuda (✉) · F. Monika
Department of Civil Engineering, Faculty of Engineering, Universitas Muhammadiyah Yogyakarta, Jl. Brawijaya, Tamantirto, KasihanBantul, Yogyakarta 55182, Indonesia
e-mail: hakasprayuda@umy.ac.id

M. D. Cahyati
e-mail: martyana.dc@ft.umy.ac.id

M. R. Fadillah
e-mail: muhammad.rizki.2016@ft.umy.ac.id

R. D. Erlinda
e-mail: risky.dwi.2016@ft.umy.ac.id

F. Monika
e-mail: fanny.monika.2007@ft.umy.ac.id

M. M. A. Pratama
Department of Civil Engineering, Faculty of Engineering, Universitas Negeri Malang, Jl. Semarang No. 5, Kota Malang, East Java 65145, Indonesia
e-mail: mirza.abdillah.ft@um.ac.id

N. Sabtu (ed.), *Proceedings of AWAM International Conference on Civil Engineering 2022—Volume 2*, Lecture Notes in Civil Engineering 385,
https://doi.org/10.1007/978-981-99-6018-7_29

Keywords Deep beam · Arch beam · Flexural behavior · FE analysis

1 Introduction

Deep beam reinforced concrete is often used as a structural member in modern constructions such as high-rise buildings, bridge structures, foundation components, and underground structures [1–3]. In general, reinforced concrete beam structures are designed to withstand a sustained transverse load to keep the structure in balance condition. Based on the ratio of shear span to effective depth (*a*/*d*), reinforced concrete beams are divided into three (3) general categories, namely $a/d \leq 1.0$ (deep beam), $1.0 < a/d \leq 2.5$ (moderately deep beam), and $a/d \geq 2.5$ (ordinary shallow beams) [4, 5]. The damage patterns in deep beam structures are different from conventional beams. Due to the small *a*/*d* ratio, the damage in the deep beam is a shear failure, whereas, in ordinary beams, flexural failure often occurs when testing using bending loads.

Several previous researchers have proven the shear damage that occurs in the deep beam of reinforced concrete. Several studies have been found regarding repairing the damage that occurs in the deep beam with symmetrical geometry, including by using the strut and tie strengthening method [6, 7], carbon fiber polymer laminates [8–10], glass fiber reinforced polymer [11–13], and steel fiber [14, 15]. Many factors affect the stiffness of deep beam structures, including the young modulus (E), which depends on the type and quality of the material used, and inertia (I) which depends on the geometry, especially the cross-section dimensions used in deep beam structures.

The geometry of reinforced conventional concrete beams has been improved in recent years to meet the needs of the construction industry. Several studies with different beam sections have been found, including web opening [16–18], a non-prismatic beam [19, 20], and other variants of cross-sectional shapes. However, a different cross-section type has not been used in the analysis of deep beam characteristics due to flexural load. The construction advancements, particularly in reinforced concrete beam construction, necessitate an examination of the effects of non-symmetrical beam shapes in both the transverse and longitudinal directions.

This study investigated using numerical analysis on four types of beams with different cross-sections, namely rectangular section, trapezoidal section, T-section, and I-section. Each cross-section beam is divided into two types of longitudinal sections: symmetrical shape (non-arch) and arch shape. This study has eight beam models to investigate their flexural behavior and compare the results. The main objective of this research is to determine the effect of different cross-sections and the effect of different longitudinal sections on deep beam reinforced concrete. This study discusses the values of flexural strength, elastic stiffness, ductility, and damage pattern in each model.

2 Model Information

2.1 Geometry

Details of geometry and dimensions on deep reinforced concrete beams with symmetrical (non-arch) and longitudinal arch sections can be seen in Figs. 1 and 2. Figure 1 shows the length of the beam, which is 4 m, while the height varies according to the type of cross-section, but all models have the same concrete volume. Nowadays, the price of concrete still depends on its volume. This study can be one of the references to compare deep beam capacity with identical volume and quality but different geometries. The cross-section comparison consists of four types of cross-section ash shown in Fig. 2, where each cross-section has two types of longitudinal configuration called non-arch and arch deep beam.

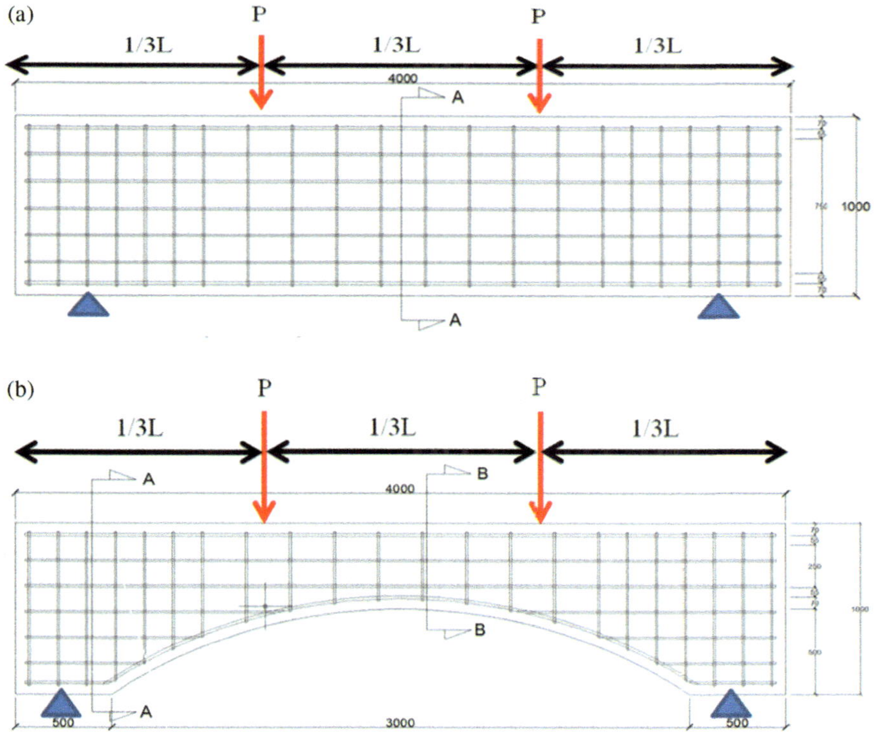

Fig. 1 Reinforced deep beam concrete. **a** Symmetrical longitudinal section (non-arch); **b** arch longitudinal section

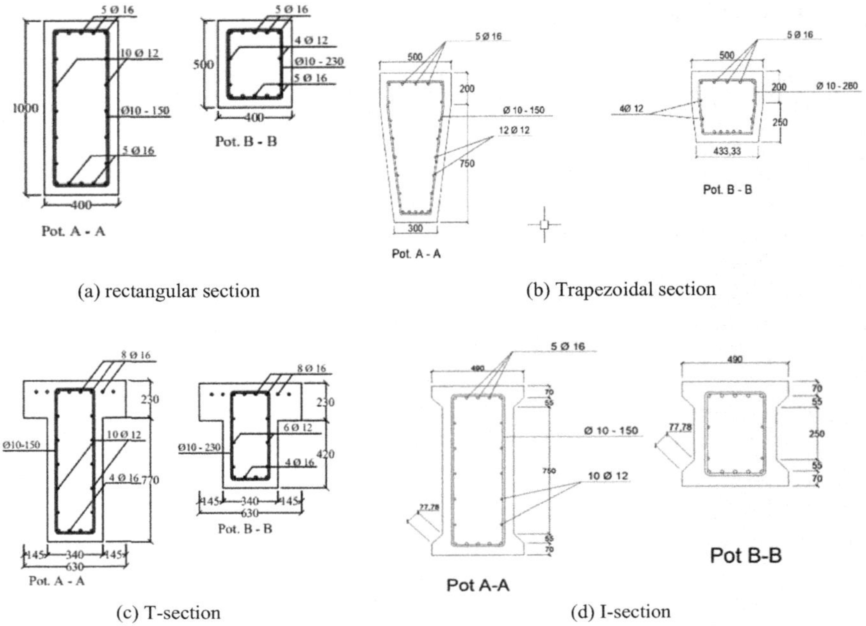

Fig. 2 Variations of cross-section

2.2 *Finite Element Analysis*

Finite element (FE) analysis in this study used Abaqus CAE software. Concrete elements are presented using 3D solid extrusion, while rebar reinforcement is presented using wire planar. Several studies using the same element type have also been found in several previous studies [21–23]. An eight-node linear brick element (C3D8R) idealized for concrete type element and type element used for transversal and longitudinal rebar is truss T3D2. The pinned joint was applied in the support as a boundary condition, as seen in Fig. 1. General static in nonlinear condition using full newton selected as a numerical method. The interaction between concrete and rebar is embedded interaction. The view of geometry for non-arch and arch models is shown in Fig. 3, respectively.

2.3 *Material Properties*

Two material conditions are required as input data in ABAQUS, such as elastic and plastic conditions. According to the compressive strength test, the concrete modulus elasticity of 24,870 MPa and poison ratio of 0.15 is applied as an elastic parameter. Besides, the damaged plasticity data was chosen to represent the plastic behavior of

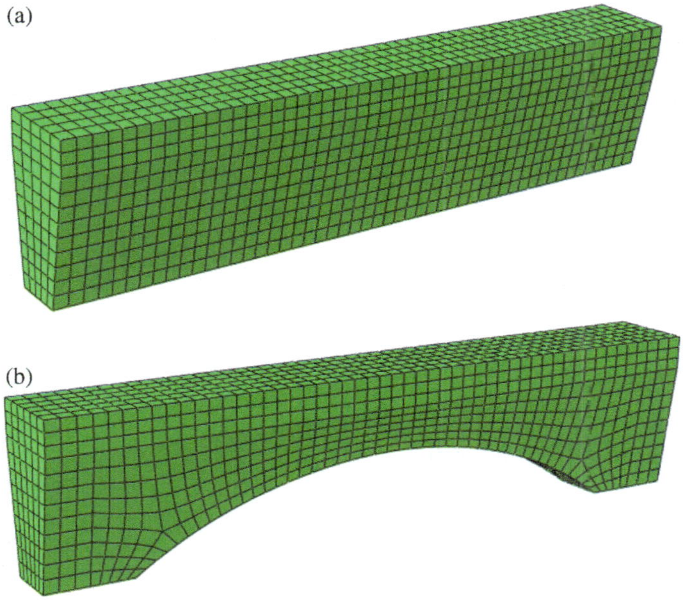

Fig. 3 Type element of concrete beam for trapezoidal section, **a** non-arch deep beam; **b** arch deep beam

concrete, as shown in Table 1. The stress–strain relationship behavior of concrete in this research can be seen in Fig. 4a, which includes compression and tension parts. The second material used in this study is steel rebar for assigning the flexural and shear reinforcement. This study uses three different dimensions are used in this study such as D10, D12, and D16. Young modulus of 200,000 MPa and poison ratio 0.3 are used for elastic behavior of steel rebar. However, the stress–strain relationship of rebar reinforcement can be seen in Fig. 4b, representing the inelastic behavior of steel properties.

Table 1 Concrete damaged plasticity [22]

Dilatation angle	Eccentricity	F_{b0}/f_{c0}	K	Viscosity
30	0.1	1.16	0.67	0

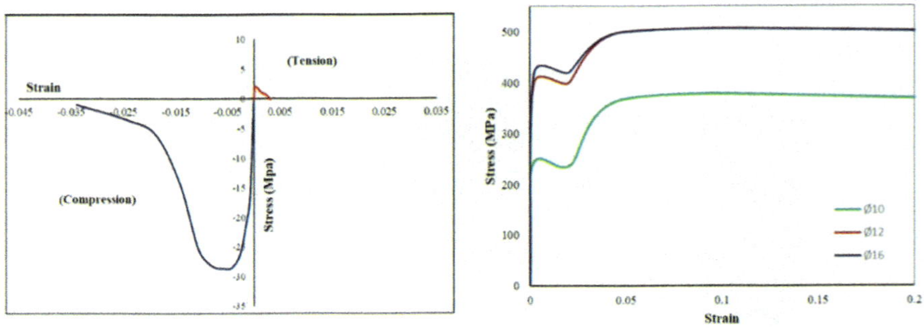

(a) (b)

Fig. 4 Stress and strain relationship of **a** concrete, **b** rebar reinforcement [21, 22]

3 Results and Discussion

3.1 Load and Displacement Relationship

The relationship between load and displacement is one of the behaviors of reinforced concrete beams that must always be considered. The relationship between load and displacement in each beam is shown in Fig. 5. The analysis results show that all reinforced concrete deep beams with arch sections have decreased capacity to withstand loads compared to beams with symmetrical (non-arch) sections. The inertia is decreased when the dimension of the deep beam decreases, which causes the reduction of capacity.

The load capacity investigation in this numerical simulation is carried out in two stages. The first stage is to look into the load when the deep beam concrete cracked for the first time, as shown in Fig. 6a. The results show that the deep beam with the I-section (non-arch) form has the first crack with the highest load, which is 1242.82 kN, while the I-section beam with the arch shape is cracked after experiencing a load of 1077.77 kN. The other cross-section type also shows a similar pattern to I-section, where the arch deep beam produces lower capacity in the first crack than the non-arch shape. The applied load when the first crack occurs in the rectangular section is 869.23 kN for the non-arch beam and 738.46 kN for the arch beam, the trapezoidal section for the non-arch and arch beam is 985.73 and 979.96 kN, while the T-section is 645.54 and 618.51 kN. A more in-depth and detailed study was required to reduce the error value. The applied load investigation at the time of the first crack resulted from an approach based on visual observations that occurred in each model.

The results of the peak load investigation in this study are shown in Fig. 6b. Beam with a non-arch rectangular section produces a load of 2646.04 kN, whereas the rectangular arch section of 2500.29 kN reduces the load capacity by 5.51%, as shown in Fig. 7. Non-arch rectangular section beam is used as a control to evaluate the performance of other beam types. The trapezoidal cross-section produces a peak load of 2221.61 kN for non-arch and 1933.67 kN for deep arch beams. Trapezoidal

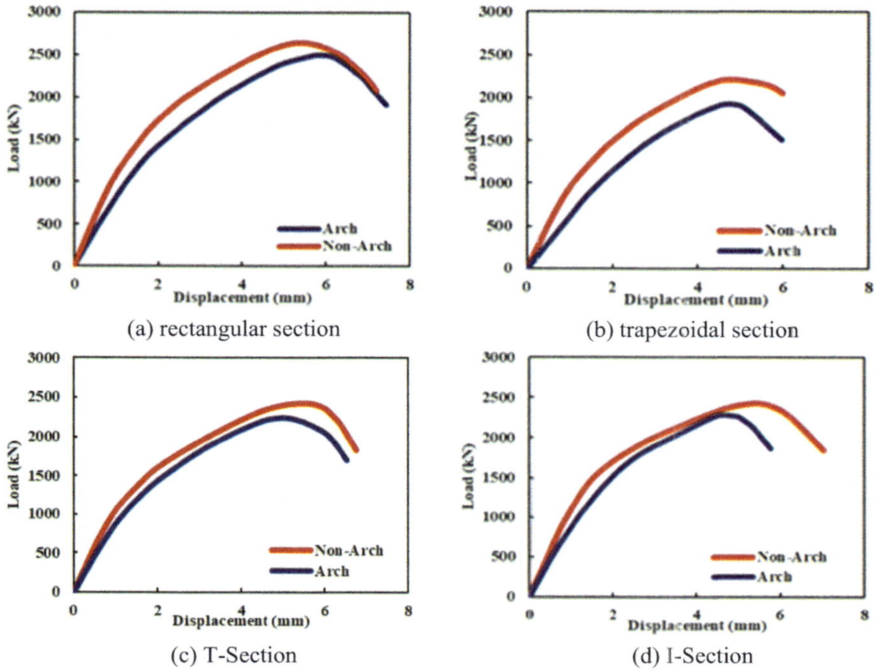

Fig. 5 Load–displacement relationship

cross-section reduces load capacity by 16.04% for non-arch and 26.92% for deep arch beams. The highest load reduction experienced on beams with trapezoidal cross-section shows that this trapezoidal cross-section is not effective enough to maintain the capacity compared to other types of cross-sections.

Meanwhile, the T-section deep beam can accept the load of 2430.57 kN for non-arch and 2248.91 kN with arch beam, and this peak point is reduced to 8.14% for non-arch 15.01% for deep arch beam compared to control beam (non-arch rectangular section deep beam). I-section non-arch deep beam can accept loads of 2440.70 kN while 2284.12 KN for deep arch beam. The peak load of the I-section is decreased by 7.76% for non-arch and 13.68% for deep arch beams. Based on the overall results of peak load, it can be concluded that a deep beam with a rectangular section is the most effective to accept high load both for non-arch and arch longitudinal sections compared to other types of cross-sections.

3.2 *Elastic Stiffness*

Another analysis is conducted to determine arch deep beams effectiveness and bending behavior by evaluating the elastic stiffness. The stiffness results of each

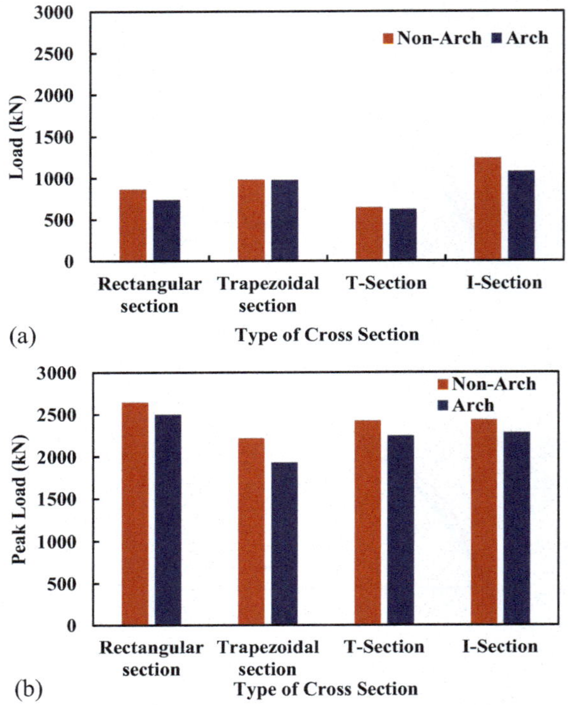

Fig. 6 Load comparison for each model. **a** First crack, **b** peak load

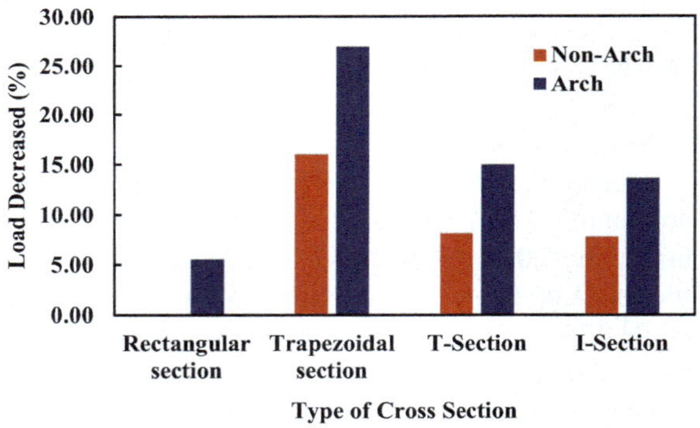

Fig. 7 Peak load percentage of decrease in load capacity

deep beam, both for non-arch and arch, are shown in Fig. 8a. Meanwhile, the stiffness comparison for each model can be seen in Fig. 8b. The highest stiffness is produced by a non-arch rectangular cross-section deep beam, which is 1337.37 kN/mm. Meanwhile, the rectangular arch section experienced a decrease in the stiffness of 37.96–839.73 kN/mm. Non-arch trapezoidal beam produces stiffness of 940.81 kN/mm, while arch trapezoidal section stiffness is 595.86 kN/mm. Compared to other beams, the trapezoidal cross-section had the highest reduction in stiffness,

29.65% for non-arch and 55.44% for deep arch beams. Because all beams with trapezoidal cross-sections have a high reduction in stiffness, it can be concluded that the beam with this cross-section is ineffective in terms of stiffness compared to other types of beams. Although the volume of concrete and reinforcement used for all beam types is the same.

The stiffness of T-section deep beams is 1152.71 kN/mm for non-arch deep beams and 909.57 kN/mm for the arch. The reduction in stiffness was 13.80% in non-arch and 31.98% in arch T-section beams. The stiffness decreased also occur on I-section deep beam, which was 19.88%, or the stiffness was 1071.39 kN/mm for non-arch, while decreased 37.04% or the stiffness was 843.01 kN/mm for deep arch beam. It can be concluded that the reduction in stiffness of non-arch deep beam for all cross-sections is highest than 10% compared to the control beam. However, stiffness decreased significantly for all types of arch beams, which is more than 30%.

Fig. 8 **a** Stiffness for each model; **b** decreased of stiffness

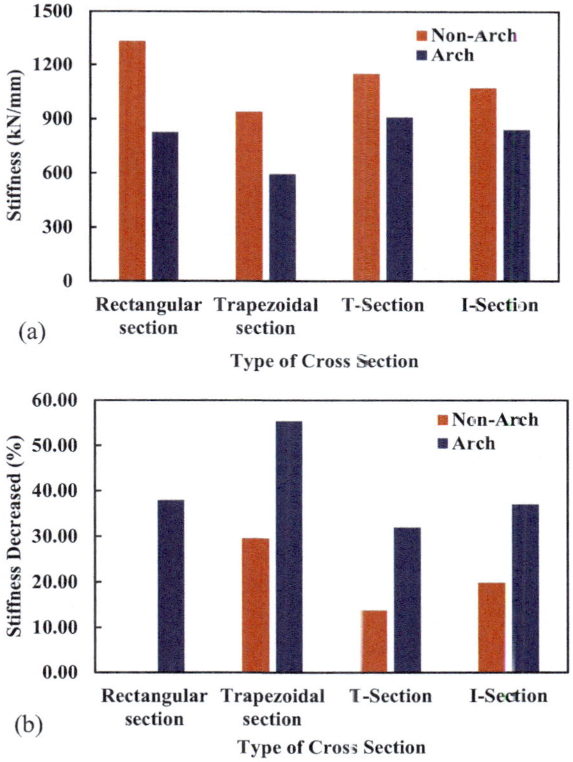

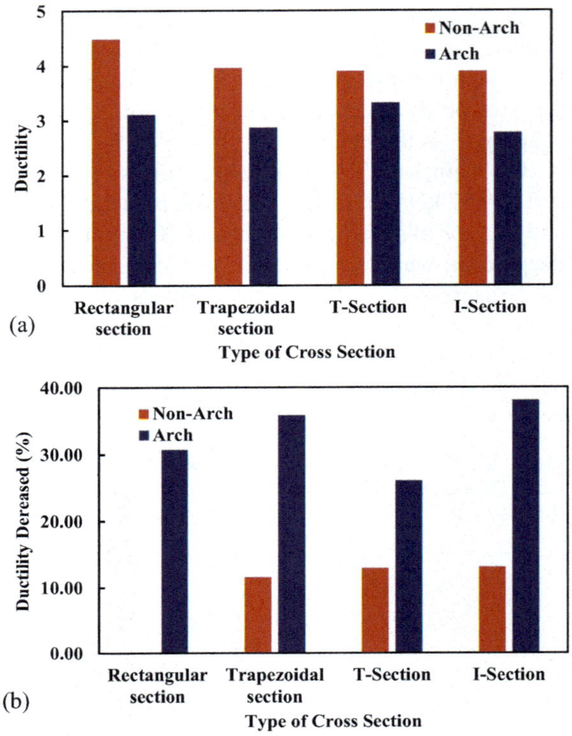

Fig. 9 **a** Ductility for each model; **b** decreased ductility

3.3 Ductility

The effect of the cross-section on the ductility of the beam was investigated in this study. The results of each beam are shown in Fig. 9a, while the percentage reduction in the ductility of each beam is shown in Fig. 9b. The ductility of the non-arch rectangular deep beam was 4.49 and 3.11 for the arch section or a decrease of 30.73% compared to the beam with the highest ductility. The decrease in ductility also occurs in the trapezoidal section, T-section, and I-section deep beam. The trapezoidal cross-section produces ductility of 3.97 for non-arc and 2.88 for deep arch beams. For T-section and I-section of non-arch produces ductility of 3.91 and 3.90, while the arch is 3.32 and 2.78, respectively.

3.4 Damage Pattern

Based on the simulation results, the distribution of cracks in the beam can be seen in the plastic strain output (PEMAG), as shown in Fig. 10. The illustration of cracks distribution along the span in non-arch and arch beams. The failure contour, distributed from blue to red, represents the gradation of plastic strain from the lowest to the highest value. The results show that all deep beams have similar cracks visualization in non-arch and deep arch beams in the failure condition. The damage starts to occur from the support system and develops diagonally to the loading point known as a shear failure mechanism.

The damage process with shear mechanism has been described in detail on the nonlinear fracture mechanic theory and softening models in concrete [24, 25]. Cracks on the deep beam are caused by plastic tensile stress exceeding the tensile stress capacity of concrete. When the load is increased, stress on the reinforcement also increases so that it becomes yield and the distribution of the crack be more intensive. Based on the damage pattern investigation, it can be concluded that reinforced deep beam with rectangular section produces the highest performance both in non-arch and longitudinal arch sections compared to other types of cross-sections. Meanwhile, the trapezoidal cross-section deep beam produces the lowest performance in all effects investigated in this study.

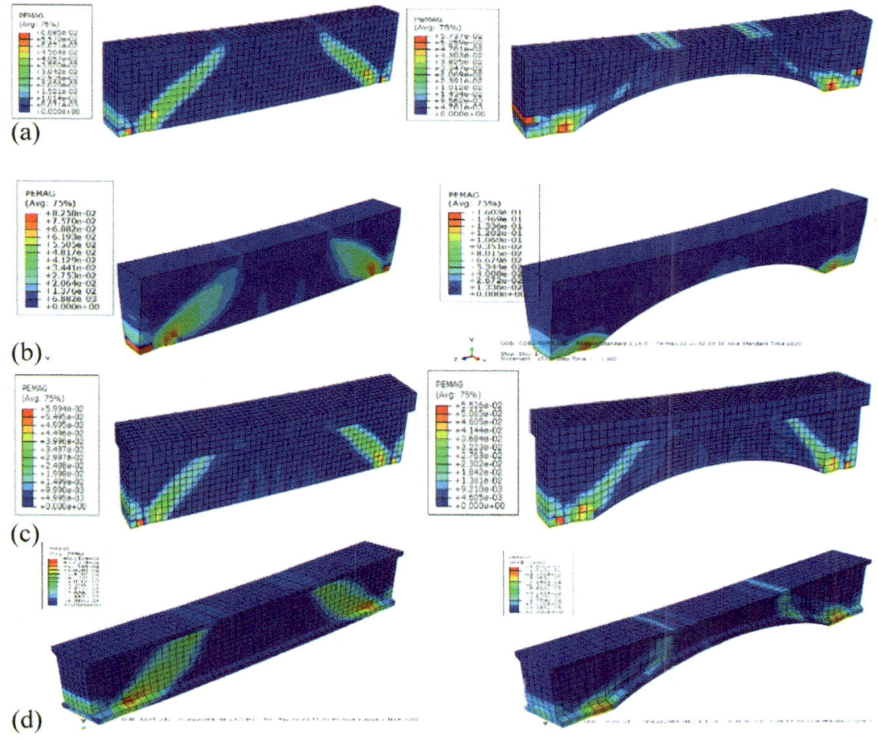

Fig. 10 Damage pattern **a** rectangular section; **b** trapezoidal section; **c** T-section; **d** I-section

4 Conclusion

Based on the simulation results on the deep beam with various cross-sections, it can be concluded that the arch beam always produces lower performance compared to the non-arch deep beam. Besides, the cross-section affected the flexural behavior of the beam, such as load–displacement relationship, elastic stiffness, and ductility. The investigation results show that rectangular deep beam cross-sections have the best performance in non-arch and arch forms. In contrast, other cross-section types have decreased peak load, stiffness, and ductility capacity.

Reinforced deep beam with trapezoidal cross-section produces the lowest performance compared to other types of cross-sections. Moreover, the stiffness of the trapezoidal arch deep beam decreased by 55.44% even though the volume of concrete and reinforcement used is the same. Thus, the same volume of concrete results in a different performance if the geometric of the beam are designed differently. The geometry of the deep beam plays an important role in maintaining performance.

References

1. Rahim, N.I., Mohammed, B.S., Al-Fakih, A., Wahab, M.M.A., Liew, M.S., Anwar, A., Amran, Y.H.M.: Strengthening the structural behavior of web opening in RC deep beam using CFRP. Materials **13**, 1–21 (2020)
2. Chen, H., Wang, L., Zhong, J.: Study on an optimal strut-and-tie model for concrete deep beams. Appl. Sci. **9**, 1–14 (2019)
3. Campione, G.: Flexural behavior of steel fibrous reinforced concrete deep beams. J. Struct. Eng. **138**, 235–246 (2012)
4. Al-Gasham, T.S., Mhalhal, J.M., Abid, S.R.: Flexural behavior of laced reinforced concrete moderately deep beams. Case Stud. Constr. Mater. **13**, 1–22 (2020)
5. Slowik, M.: Shear failure mechanism in concrete beams. Procedia Mater. Sci. **3**, 1977–1982 (2014)
6. Park, S., Aboutaha, R.S.: Strut and tie method for CFRP strengthened deep RC members. J. Struct. Eng. **135**, 632–643 (2009)
7. Tang, C.Y., Tan, K.H.: Interactive mechanical model for shear strength of deep beams. J. Struct. Eng. **130**, 1534–1544 (2004)
8. Zhang, Z., Hsu, C.T., Moren, J.: Shear strengthening of reinforced concrete deep beams using carbon fiber reinforced polymer laminates. J. Compos. Constr. **8**, 403–414 (2004)
9. Chalioris, C.E., Kosmidou, P.K., Papadopoulos, N.A.: Investigation of a new strengthening technique for RC deep beams using carbon FRP ropes as transverse reinforcements. Fibers **6**, 1–18 (2018)
10. Hanoon, A.N., Jaafar, M.S., Zaidee, S.R.A., Hejazi, F., Aziz, F.N.A.A.: Effectiveness factor of the strut and tie model for reinforced concrete deep beams strengthened with CFRP sheet. J. Build. Eng. **12**, 8–16 (2017)
11. Mohamed, K., Farghaly, A.S., Benmokrane, B., Neale, K.W.: Nonlinear finite element analysis for the behavior prediction and strut efficiency factor of GFRP reinforced concrete beams. Eng. Struct. **137**, 145–161 (2017)
12. Zinkaah, O.H., Ashour, A.: Load capacity predictions of continuous concrete deep beams reinforced with GFRP bars. Structures **19**, 449–462 (2019)

13. Zinkaah, O.H., Ashour, A., Sheehan, T.: Experimental test of two span continuous concrete deep beams reinforced with FGRP bars and strut and tie method evaluation. Compos. Struct. **216**, 112–126 (2019)
14. Moradi, M., Esfahani, M.R.: Application of the strut and tie method for steel fiber reinforced concrete deep beams. Constr. Build. Mater. **131**, 423–437 (2017)
15. Ma, K., Qi, T., Liu, H., Wang, H.: Shear behavior of hybrid fiber reinforced concrete deep beams. Materials **11**, 1–14 (2018)
16. El-Kareim, A., Arafa, A., Hassanin, A., Atef, M., Saber, A.: Behavior and strength of reinforced concrete flanged deep beams with web opening. Structures **27**, 506–524 (2020)
17. Smarzewski, P.: Analysis of failure mechanics in hybrid fiber reinforced high performance concrete deep beams with and without openings. Materials **12**, 1–24 (2019)
18. Jasim, W.A., Tahnat, Y.B.A., Halahla, A.M.: Behavior of reinforced concrete deep beam with web opening strengthened with CFRP sheer. Structures **20**, 785–800 (2020)
19. Prayuda, H., Syandy, R.R., Soebandono, B., Maulana, T.I.: Numerical study on beam column connection of cantilever precast concrete beam with asymmetric shape under static load. In: MATEC Web of Conferences, vol. 195 (2019)
20. Cahyati, M.D., Prayuda, H., Monika, F., Pratama, M.M. A., Siswantoro, E.P., Pradana, P.: Finite element analysis on flexural behavior of non-prismatic longitudinal section reinforced concrete deep beam. In: IOP Conference, Series: Materials Science and Engineering, vol. 1144 (2021)
21. Prayuda, H., Cahyati, M.D., Soebandono, B.: Analisis Tegangan Regangan dan Defleski pada Sambungan Balok-Kolom Beton Bertulang Menggunakan Beban Statik. Media Komunikasi Teknik Sipil **24**, 122–130 (2018). (in Indonesian)
22. Apriyatno, H., Utomo, K.S., Dewi, M.S.: Analysis of exterior beam-column connection on structures of Marrakech suites apartment by using Abaqus CAE V6.14 program on the earthquake area 4. Int. J. Latest Res. Sci. Technol. **8**, 12–16 (2019)
23. Chiriki, S.S., Harsha, G.S.: Finite element analysis of RC deep beams strengthened with I-section and truss reinforcement. Mater. Today 1–6 (2020)
24. Hillerborg, A., Modeer, M., Peterson, P.E.: Analysis of crack formation and crack growth in concrete by means fracture mechanics and finite elements. Cement Concr. Res. **6**, 773–782 (1976)
25. Bazant, Z.P., Oh, B.H.: Crack band theory for fracture of concrete. RILEM Mater. Struct. **16**, 155–177 (1983)
26. Cedolin, L., Poli, S., Iori, I.: Tensile behavior of concrete. J. Eng. Mechan. Div. ASCE **113**, 431–449 (1978)

Experimental Investigation on the Behaviour of Modified Connections at the Overhang Roof Under Uplift Pressure

T. L. Tan, T. A. Majid, and S. S. Zaini

Abstract The northern area of Peninsular Malaysia discovered that most of rural houses with the overhang roof was built without structural considerations, the overhang roof is the most vulnerable part of the roof system during strong wind events. Due to a lack of information about the behaviour of the modification connection to the overhang roof, this paper studies the behaviours of the prototype using modified connections with three variance parameters applied for the corrugated cladding and trapezoidal cladding with the same gauge. The three variance parameters are the type of reinforced material, number of reinforced materials, and the length of reinforced material. The PVC pipes and metal straps were added to the connection at the overhang roof part, and applied surface loading test was to simulate the uniformly uplift wind pressure. Results of modified connection using metal straps showed that the increment ratio on the total applied force for corrugated cladding in between 1.05 and 1.26, and the increment ratio for trapezoidal cladding in between 1.01 and 1.05 only. Moreover, results for modified connection using PVC pipe for corrugated cladding have an increment ratio for the total applied force of 1.47–1.79, while trapezoidal cladding only has a ratio of the increment of 1.20–1.60. In conclusion, the modified connection using PVC pipes was found to be the most efficient modified connection compared to metal straps, and metal straps are shown more suitable to be used for corrugated roof cladding.

Keywords Overhang roof cladding · Modified connections · Surface load test

1 Introduction

A rural house is a house located in a rural area or outside the city. In Malaysia, most of the rural house was constructed without appropriate structural considerations are classified as non-engineered structures. From the previous investigation data in the

T. L. Tan · T. A. Majid (✉) · S. S. Zaini
Universiti Sains Malaysia, Engineering Campus, 14300 Nibong Tebal, Penang, Malaysia
e-mail: taksiah@usm.my

N. Sabtu (ed.), *Proceedings of AWAM International Conference on Civil Engineering 2022—Volume 2*, Lecture Notes in Civil Engineering 385,
https://doi.org/10.1007/978-981-99-6018-7_30

Northern area of Peninsular Malaysia, it was discovered that most of the rural houses in Malaysia were built with the overhang roofs [1]. The overhang roofs included part of purlin spacing and the overhang of cladding as shown in Fig. 1 [2]. The overhang cladding refers to the portion of cladding extend from the last of the purlin at roof trusses. And then, the purlin spacing refers to the exterior of the surface wall to the last of the purlin. The function of the overhang roof is to blocking the sunlight to provide shade and prevent the rain away from the exterior wall.

When a house is subjected to such heavy winds, the roof is normally the most vulnerable component [3]. Wind loads on a low-rise building's roof would be significantly affected by the relationship between wind flow and building surfaces, as stated by Mahendran [4]. From the previous study regarding Malaysia's climate, it was founded that there are two main monsoons influence the wind flow pattern of the wind system, namely southeast monsoon and northwest monsoon which is from the middle of April until September and early November until March. Many thunderstorms happen during the transition between two main monsoons [5]. After the thunderstorms happen, a lot of roof failure cases were recorded in the northern area in peninsular Malaysia. The wind disaster events increased from the year 2007–2015 [6, 7]. Previous authors [6, 8–10] recorded that the failures modes on the roofing system, such as all the cladding were lost, partially of cladding was damaged, the cladding ripped off near to the edge of the roof structure and the serious damage when the roof structure is lost.

Furthermore, a previous study found that extreme wind flow can destroy a rural house in a variety of ways, including blown-off roofs, wall damage, and complete collapse [9]. The roof failure or roof was torn off was investigated by a previous researcher, and the overhang roof acts as a cantilever member that triggered the initial damages to the roof during strong wind event. This is because the overhang roof did not had adequate rafter or purlin to resist the uplift pressure acting on the surface of the roofing sheet [5]. The fastener connection of the roof can be fail in pull-through or pull-out, when fasteners under uplift/suction of wind pressure [11]. Figure 2a shows that the pull-through failure of the connection are the roofing sheeting punch holes

Fig. 1 Overhang roof for rural house [2]

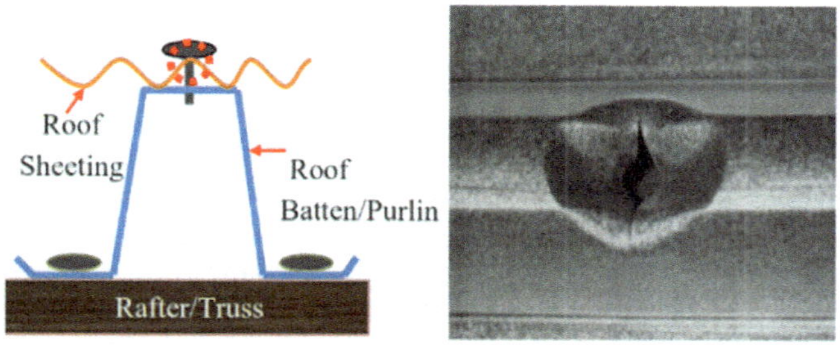

(a)

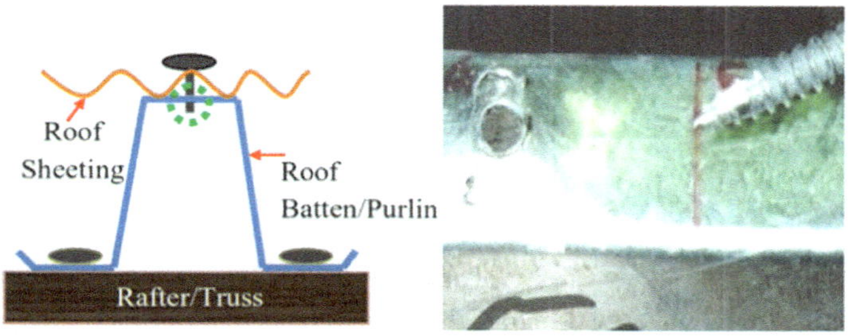

(b)

Fig. 2 Roof sheeting to batten/purlin connection failures. **a** Pull-through failures. **b** Pull-out failures [12]

through and screw fastener head pulled through. Figure 2b shows that the pull-out failure of the connection is the screw fastener pulled out from the roof purlin [12].

2 Methodology

This study was divided into two stages in experimental work, namely built prototypes and the surface loading test. The first stage is about built prototypes, this stage was included to propose the prototypes with modified connections and prepare the prototypes. The next stage is about surface loading test, this testing is carried out with the aim to determine the pull-through force, maximum applied force, and failure modes for the prototypes. The surface loading test was conducted using the under a set of loading frames with a load applicator set.

Table 1 List of materials for the prototypes

No.	Item	Type	Remarks
1	Timber purlin	Jelutong timber (SG6)	Dressed timber with dimension 45 mm × 45 mm
2	Roof cladding	Corrugated (G32)	Thickness: 0.23 mm
		Trapezoidal (G32)	Thickness: 0.23 mm
3	Fasteners	Self-tapping hexagon washer head screw with bonded washer	Washer diameter: 15.8 mm Diameter of screw: 5.6 mm Length: 63.5 mm
4	Reinforced material	Metal strap	17 mm width
		PVC pipe	External diameter: 21.50 mm Class D

2.1 Materials of Prototype

Table 1 shows all the materials for the prototype of the overhang roof. The type of timber is Jelutong timber was used as a timber purlin. The Jelutong type of timber is cheap and commonly used in construction. The cross-section of the dressed timber with dimension approximately 45 mm × 45 mm. In Malaysia Standard MS 544: Part 2:2001, this timber is classified in light hardwoods, and the strength group is SG6 [13]. There are two types of roof cladding with the same gauge of 32, which is corrugated and trapezoidal cladding were selected. There are two types of fastener connection commonly used in the rural house roofing system which are nails and screws [6]. The self-tapping roof screw is used in the experiment to connect the roof cladding and purlin, because this type of roof screw commonly used for timber material. Besides that, the roof screw can prevent fasteners failure during preparation, and the failure mode occurs at the connection instead of cladding during testing. The PVC pipes and metal straps were used in laboratory work as reinforced material, because there are simple and affordable, locally available, durable, and easy for installation work. Figure 3 shows the schematic diagram of modified connection using PVC pipes and metal straps at the overhang cladding part, and the schematic diagram was used for the prototype with corrugated cladding and trapezoidal cladding.

2.2 Surface Loading Test Set up

The surface loading test is referred to by previous researchers Siron et al. [14], and the researcher was referred to the concept of the mid-span load test [15, 16]. The fundamental of the mid-span load test is the prototype of the overhang roof can be referred to as the simple beam theory but not the shell theory. The surface loading was developed uniform distribution load transmission along the soffit of cladding

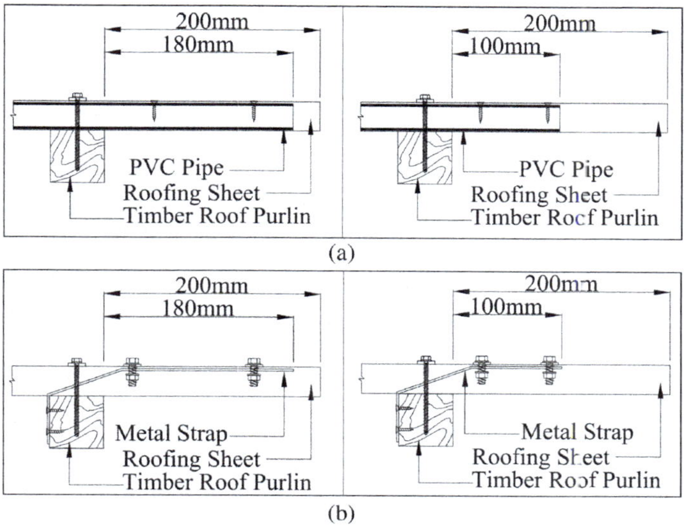

Fig. 3 Schematic diagram of the modified connection. **a** PVC pipes. **b** Metal straps

Fig. 4 Surface loading test setup

to simulate the wind uplift pressure. Therefore, the prototype was set up in upside down to apply the load by using a hydraulic jack as shown in Fig. 4.

3 Results and Discussions

This section presents the results of the laboratory tests was conducted in Heavy Structure Laboratory at the School of Civil Engineering. All the prototypes with modified connections were tested under the surface loading test. The results of the connection failure on the cladding, fastener pull-through force, and total applied

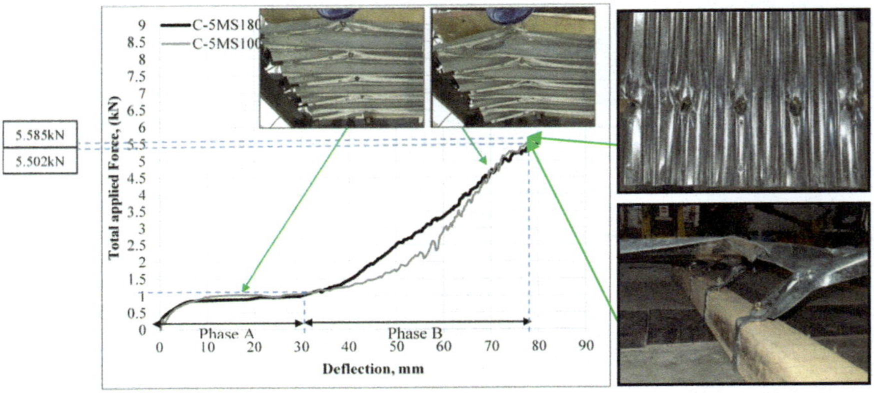

Fig. 5 Total applied force for prototype C-5MS100 and C-5MS180

force for the prototype, comparison between the types of modified connections are present in this section.

3.1 *Corrugated Roof Cladding with Modified Connection Using Metal Strap*

The results of prototype C-5MS180 are slightly similar to the prototype C-5MS100. The difference between both prototypes is in Phase B. When the pull-through force is in the range 0.13–0.85 kN, the line graph shows the prototype behave as linear elastic, and it was less deflection compared with a short metal strap. On the other hand, the applied force for prototype C-5MS100 higher than C-5MS180 is around 1.5% as shown in Fig. 5. Although the prototype C-5MS100 has a higher applied force, but the prototype C-5MS180 has more stiffness and ductile. For the failure mode of the prototype is similar to C-5MS100, the corrugated cladding experiences pull-through with a transverse crack around screw fasteners and cladding does not disengage with the purlin directly due to the support by a metal strap (Fig. 5). The maximum pull-through force is 0.937 kN, and the total applied force is 5.502 kN.

3.2 *Corrugated Roof Cladding with Modified Connection Using PVC Pipes*

The behaviour of prototypes C-5PVC100 and C-5PVC180 is divided into three phases which are Phase A, Phase B, and Phase C. In Phase A, the screwed crest behaves in linear elastic with a large gradient. The loading increases until the applied

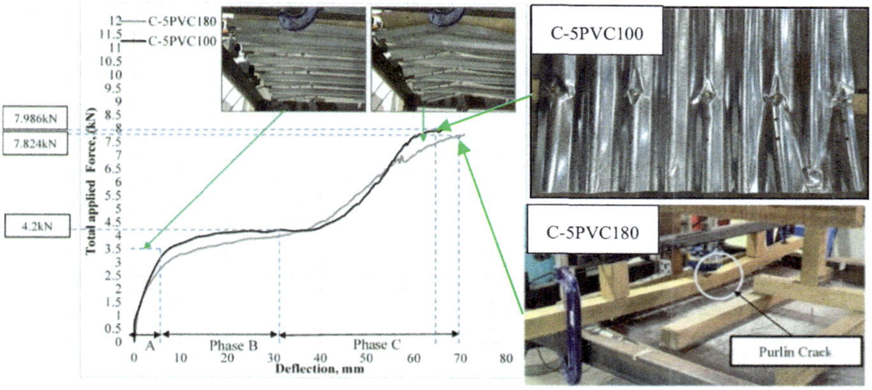

Fig. 6 Total applied force for prototype C-5PVC100 and C-5PVC180

force is around 3.5 kN and the deflection is around 7.4 mm. The loading increases continuously in phase B, there have some changes around the fastener of the prototypes is due to the material of the cladding starting to yield, and some dimpling formed around the fasteners. The behaviour in Phase B is slightly increasing until the total applied force is 4.2 kN, and the deflection is 32 mm. In Phase C, the load increases in parabolic until the pull-through failure occurs. Then, the applied force increases up to 7.986 kN, and the cladding experienced severe distortion and global plastic deformation at the screwed crest as shown in Fig. 6. The ultimate failure for the prototype C-5PVC100 occurred pull-through failure with transfer split when pull-through force is 1.116 kN. The graph for prototype C-5PVC180 is similar when compared with the short PVC pipe. The differences are the total applied force, pull-through force, and failure mode. When the applied force reaches 7.824 kN, the purlin crack appears. Therefore, the testing was terminated to prevent the devices and loading system damage. In comparison with the prototype C-5PVC100, the total applied force for C-5PVC180 is less than 2% only, but it allows more deformation as shown in Fig. 6.

3.3 Comparison of the Corrugated Cladding with Modified Connection

Based on Table 2, the increment ratio between the prototypes with 5 units of PVC pipe with length 100 mm over the control prototype with 5 units of screws 1.79 is the highest compare with the other prototype with 5 units of screws fastener. From the increment ratio, the modified connection by using long PVC is the most important compared with the others. Next, the increment ratio between the 3 units of PVC with length 100 mm over the control prototype with 3 units of screws is 1.64. Besides

Table 2 Comparison of all prototypes for the corrugated cladding with the control average values

Type of connection	Number of reinforced material	Pull-through force, (kN)	Total applied force, (kN)	Increment ratio
100 mm (PVC pipe)	2 nos	1.168	4.829	1.47
	3 nos	1.224	5.393	1.64
	5 nos	1.116	7.986	1.79
180 mm (PVC pipe)	2 nos	1.346	5.411	1.64
	3 nos	1.182	5.198	1.58
	5 nos	0.973	7.824	1.76
100 mm (Metal strap)	2 nos	1.163	3.562	1.08
	3 nos	0.899	3.580	1.09
	5 nos	1.044	5.585	1.26
180 mm (Metal strap)	2 nos	1.203	3.921	1.19
	3 nos	0.919	3.454	1.05
	5 nos	0.937	5.502	1.24
3 units of screws	(control)	0.820	3.292	
5 units of screws	(control)	0.818	4.450	

that, the prototype with 2 units of PVC pipes with a length of 180 mm also has the same increment ratio of 1.64. Therefore, the prototype C-3PVC100 and C-2PVC180 are equally important for the prototype with 3 screws fastener. It can be noticed that modified connection by using PVC pipe can be improved in terms of the uplift pressure and reduce the deflection at the overhang cladding part. The increment ratio for the prototype with 2 units and 3 units of the metal strap is in between 1.05 and 1.09, respectively. Alternatively, the increment for prototype installed 5 units of metal strap which is 1.26. From the increment ratio, the modified connection using metal strap also has improvement but less significant, due to the cost of material being cheaper than the PVC pipe. In summary, the modified connection for the corrugated cladding can be resisted more uplift pressure and reduced the deflection at the overhang cladding part, especially when there is a PVC pipe.

3.4 *Trapezoidal Cladding with Modified Connection Using Metal Strap*

The prototype T-5MS100 underwent two phases as shown in Fig. 7. During Phase A, the applied force maintains at around 1.3 kN. In the situation of Phases A, the crest flattens, and the valley of the cladding begins to deform into an arc of a circle. The pull-through force and total applied force increase in a vertical parabolic graph as the loading step progress to the next Phase B and the maximum applied force in

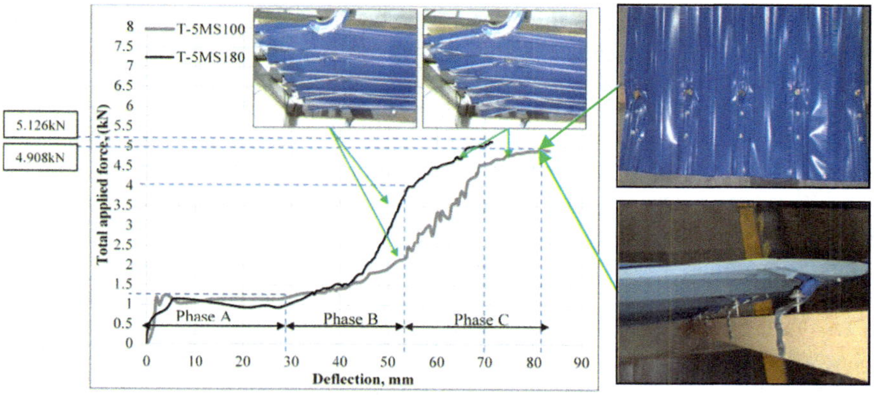

Fig. 7 Total applied force for prototype T-5MS100 and T-5MS180

phase B up to 4 kN. The crests and valleys of the cross-section were significantly damaged, and local plastic deformations (LPD) occurred around the screwed crest. In Phase C, the loading behaves linear elastic with low gradient and start. The cladding underwent severe distortion and global plastic deformation at the screwed crest when the total applied load was increased to 4.908 kN, the changing of the prototype in Phase C as shown in Fig. 7. Due to the support by the metal strap, the trapezoidal cladding experiences pull-through with transverse cracks around screw fasteners, and the cladding does not disengage with the purlin directly as shown Fig. 7.

Furthermore, both of the prototypes T-5MS100 and T-5MS180 have the similar behaviours and the pattern of the graph as shown in Fig. 7. The long metal strap gets results with less deflection when the same loading applies to the prototype. The maximum deflection of T-5MS180 is 71.4 mm only, but the deflection of short metal strap is around 83.2 mm, and the difference is around 16.5%. Moreover, the applied force for prototype T-5MS180 is large than T-5MS100 around 4.4%. The failure mode of similar to the short metal strap, and the ultimate failure occurred when the total applied force is 5.126 kN and the pull-through force is 0.866 kN.

3.5 *Trapezoidal Cladding with Modified Connection Using PVC Pipes*

Figure 8 depicts the behaviour of T-5PVC180 and T-5PVC100. Both prototypes are divided into two phases: Phase A and Phase B. The screwed crest of T-5PVC100 behaves linear elastic with a significant gradient in Phase A. As shown in Fig. 8, the loading increases until the fastening force is around 0.46 kN and the deflection is around 5.4 mm only. In phase B, with the applied force increases, there are some changes around the fasteners of the prototypes due to the cladding material starting to yield, with some dimpling forming around the fasteners. Further increase the applied load, the cladding crest flattens, and the valley of the cladding begins to deform into an arc of a circle. The pull-through force and total applied force increase in linear elastic with a low gradient in Phase B. In the observation of Phase B, the crests and valleys of the cross-sectional were significantly damaged, and the buckling at the overhang cladding was subjected to geometric nonlinearity, when the total applied force is 6.692 kN. The loading test is terminated before pull-through failure due to the loading system directly touching the support of the load cell shown in Fig. 9a. Thus, the reading of fastener force drops directly and cannot be accepted.

When compared to short PVC, the prototype T-5PVC180 is slightly similar trends. The total applied force, pull-through force, and failure mode have differences. In phase A, the applied force increases up to 3.8 kN in the linear elastic region, similar to the shorter PVC. In phase B, the applied force for prototype T-5PVC180 has increased by 17%, and the deflection also allows more than 11% compared with the

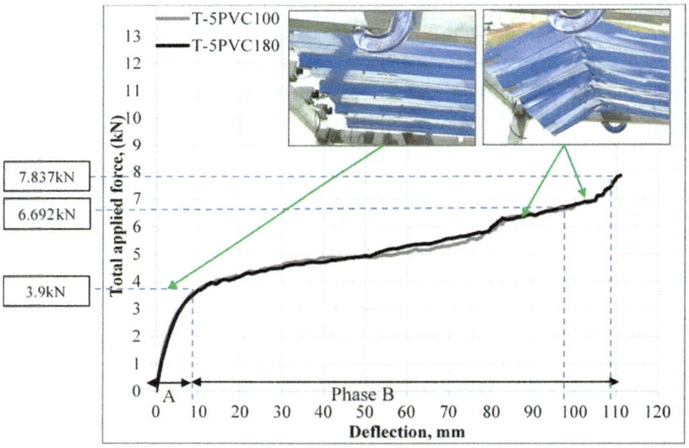

Fig. 8 Total applied force for prototype T-5PVC100 and T-5PVC180

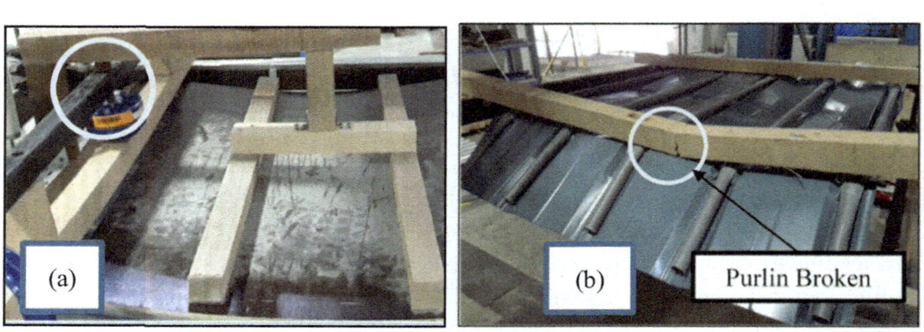

Fig. 9 Termination of test. **a** T-5PVC100. **b** T-5PVC180

prototype T-5PVC100. The prototype is more stiff and ductile as shown in Fig. 8. Furthermore, the purlin is broken when the applied force reaches 7.837 kN and the pull-through force is 0.905 kN, as shown in Fig. 9b. Since the cladding does not happen the pull-through failure at the screwed crest of cladding, the testing also terminates because the purlin cannot support the applied load anymore. Therefore, the trapezoidal cladding can summary that the longer the length of PVC pipes which is 180 mm, the highest the applied force.

3.6 *Comparison of the Trapezoidal Cladding with Modified Connection*

Refer Table 3 shown that the comparison to the other prototype with 5 units of screws fastener, the ratio of increment between the prototypes with 5 units of PVC with length 180 mm and the control prototype with 5 units of screws 1.60 is the highest. When comparing the increment ratios, the modified connection made with long PVC is the most significant. There is also the 1.43 ratio of increment between the three units of 180 mm PVC and the control prototype with 3 units of screws. As a result, the prototype T-3PVC180 is critical for the prototype with three screws. Using PVC pipe to modify the connection improves the uplift pressure and reduces the deflection at the overhang cladding.

Table 3 Comparison of all prototypes for the trapezoidal cladding with the control average values

Type of connection	Number of reinforced material	Pull-through force (kN)	Total applied force, (kN)	Increment ratio
100 mm (PVC pipe)	2 nos	0.944	4.344	1.20
	3 nos	1.226	5.041	1.40
	5 nos	0.871	6.692	1.37
180 mm (PVC pipe)	2 nos	1.037	4.658	1.29
	3 nos	1.123	5.180	1.43
	5 nos	0.905	7.837	1.60
100 mm (metal strap)	2 nos	0.885	3.693	1.02
	3 nos	1.098	3.711	1.03
	5 nos	0.912	4.908	1.01
180 mm (metal strap)	2 nos	1.063	3.750	1.04
	3 nos	0.997	3.703	1.02
	5 nos	0.886	5.126	1.05
3 units of screws	(control)	0.985	3.613	
5 units of screws	(control)	0.861	4.883	

The prototype inserted 2 and 3 units of the metal strap has only produced an increment ratio of 1.02–1.04. Alternatively, the prototype increment installed 5 units of the metal strap, around 1.01–1.05. In terms of the ratio of increment, the adjusted connection using metal strap improves, but only minimally, because the material is less expensive than PVC pipe. Moreover, the distance between the reinforced materials also can affect the performance of the prototype. In conclusion, the trapezoidal cladding can summarize that the longer the length of reinforced material, which is 180 mm, the higher the applied force or the uplift pressure to resist and reduce deflection at the overhang cladding the modified connection using PVC pipes.

4 Conclusion

From the results shown, the strength of the behaviour of overhang roof without reinforced material, the corrugated cladding, and trapezoidal cladding with 5 screws fastener resisted higher applied force compared with the 3 screws fastener. The difference in terms of the total applied force was calculated to be 35.2%. The prototype with 3 screws fastener showed larger deflection compared with the prototype with 5 screws fastener. Moreover, the total applied for trapezoidal cladding was higher than the corrugated cladding, and the difference was calculated to be 9.7%. This finding can be associated to the geometric profile where the actual dimension of trapezoidal cladding was wider compared to corrugated cladding. The prototype with trapezoidal cladding showed larger deflection compared with the prototype with corrugated cladding. The prototype with the metal strap length 180 mm showed some improvement on the deflection of cladding and the total applied, but the 100 mm length metal strap tended to increase the deflection compared with the control prototype failure mode. On the other hand, the PVC pipe with a length of 180 and 100 mm showed improvement in cladding deflection and total applied force compared to the control prototype.

The results for modified connection using PVC pipe for corrugated cladding showed increment ratio of total applied force of 1.47–1.79, while trapezoidal cladding showed increment ratio of 1.20–1.60. Therefore, the modified connection using PVC pipes was found to be the most efficient modified connection compared to the metal strap. In addition, results for modified connection using metal straps showed that the increment ratio on the total applied force for corrugated cladding is in between 1.05 and 1.26, and the increment ratio for trapezoidal cladding is in between 1.01 and 1.05 only. The modified connection using 5 units of the metal strap improved the resistance up to 26 and 5% for corrugated roof claddings and the trapezoidal cladding, respectively. Thus, the modified connection using metal straps is more suitable to be used for corrugated roof cladding.

Acknowledgements The authors would like to express their gratitude to Universiti Sains Malaysia for the financial support obtained from the RUI Grant (1001.PAWAM.8014023).

References

1. Majid, T.A., Zaini, S.S., Ismail, M.A., Deraman, N.C., Poon, J.L.: Numerical investigation on the effect of overhang roof around rural house. In: 2016 World Congress on Advances in Civil, Environmental, and Materials Research (2016)
2. Majid, T.A., Zaini, S.S., Muhammad, M.K.A., Nik Ahmad, N.A.S.: Experimental investigation on the pull through failure of roof cladding to purlin connection considering overhang roof. In: IOP Conference Series: Materials Science and Engineering, vol. 920, no. 1 (2020). https://doi.org/10.1088/1757-899X/920/1/012011
3. Yusof, W.M., Noratikah, S., Deraman, C.: Development of Windstorm Database System for Wind Damages in Malaysia (2014). https://doi.org/10.5923/c.jce.201402.36
4. Mahendran, M.: Behaviour and design of crest-fixed profiled steel roof claddings under wind uplift. Eng. Struct. **16**(5), 368–376 (1994). https://doi.org/10.1016/0141-0296(94)90030-2
5. Majid, T.A., Zakaria, S.A.S., Wan Chik, F.A., Deraman, S.N.C., Muhammad, M.K.A.: Past windstorm occurrence trend, damage, and losses in Penang, Malaysia. J. Eng. Sci. Technol. **11**(3), 397–406 (2016)
6. Muhammad, M.K.A., Majid, T.A., Ramli, N.I., Deraman, S.N.C., Wan Chik, F.A.: An overview of non-engineered buildings roofing system failure under wind loads. Appl. Mech. Mater. **802**(September), 89–94 (2015). https://doi.org/10.4028/www.scientific.net/amm.802.89
7. Majid, T.A., Ramli, N.I., Ali, M.I., Saad, M.S.H.: Malaysia Country Report 2012: Wind Related Disaster Risk Reduction and Wind Environmental Issues (2012)
8. Chik, F.A.W., Majid, T.A., Zakaria, S.A.S.: Case Study: Post Disaster Investigation on Wind Damage for Rural Housing Roofing System in Northern Peninsular Malaysia (2009)
9. Zaini, S.S., Majid, T.A., Che Deraman, S.N., Wan Chik, F.A., Azuan Muhammad, M.K.: Post windstorm evaluation of critical aspects causing damage to rural houses in the northern region of Peninsular Malaysia. In: MATEC Web Conference, vol. 103 (2017). https://doi.org/10.1051/matecconf/201710303016
10. Zaini, S.S., Humayunkabir, S.H., Majid, T.A., Muhammad, M.K.A.: Experimental investigation on the pull-out force of modified purlin to rafter connections. In: IOP Conference Series: Earth and Environmental Science, vol. 244, no. 1 (2019). https://doi.org/10.1088/1755-1315/244/1/012017
11. Mahaarachchi, D.: Behaviour and Design of Profiled Steel Cladding Systems Subject to Pull-Through Failures. Queensland University (2003)
12. Sivapathasundaram, M., Mahendran, M.: Development of suitable test methods for the screw connections in cold-formed steel roof battens. J. Struct. Eng. (2016)
13. Hong, L.T., et al.: Malaysian Standard Code of Practice for Structural Use of Timber (2001)
14. Siron, N., Majid, T.A., Zaini, S.S.: Experimental investigation on the effects of overhang length variations of steel roof cladding under uplift pressure. Structures **32**(April), 2032–2041 (2021). https://doi.org/10.1016/j.istruc.2021.04.007
15. Xu, Y.L., Reardon, G.F.: Test of screw fastened profiled roofing sheets subject to simulated wind uplift. Eng. Struct. **15**(6), 423–430 (1993). https://doi.org/10.1016/0141-0296(93)90060-H
16. Xu, Y.L., Reardon, G.F.: Behaviour of Different Profiled Roofing Sheets Subject to Wind Uplift.pdf (1994)

Development of Floating Platform for Water Quality Monitoring

Nazirul Mubin Zahari, Puventheran Sankar, Mohd Hafiz Zawawi, Nursyadzatul Tasnim Roslin, Nurhanani A. Aziz, Farah Nurhikmah, and Muhammad Husni Faiz Khamarudin

Abstract The fast development of the population and high-tech industries is driving water quality control unusually more critical. In addition, recent climate change is also a major factor in this issue. Regular inspections of water and collection of samples could also contribute greatly to water distribution management. Even though the water quality parameters are measured by the senor station automatically in the water reservoir, only a small number of the station is placed in the reservoir, and the "resolution" is not sufficient for modern applications. Thus, the major goal of this study is to design and develop a surface platform prototype for water quality monitoring. Hence, this paper proposed and developed a floating device which is well known as an unmanned surface vehicle to measure water quality parameters with several advanced sensors.

Keywords Floating · Prototype · Water quality

N. M. Zahari (✉) · P. Sankar · M. H. Zawawi · N. T. Roslin · N. A. Aziz · F. Nurhikmah
Department of Civil Engineering, College of Engineering Universiti Tenaga Nasional (UNITEN), 43000 Kajang, Selangor, Malaysia
e-mail: mubinzahari@gmail.com

P. Sankar
e-mail: puvenpo@gmail.com

M. H. Zawawi
e-mail: MHafiz@uniten.edu.my

N. T. Roslin
e-mail: tsyadzatul@uniten.edu.my

N. A. Aziz
e-mail: nurhanani_abdaziz@yahoo.com

F. Nurhikmah
e-mail: nurhikmah.ghazali@uniten.edu.my

M. H. F. Khamarudin
BIP Studio, 43650 Bandar Baru Bangi, Selangor, Malaysia
e-mail: sales@bipstudio.com

N. Sabtu (ed.), *Proceedings of AWAM International Conference on Civil Engineering 2022—Volume 2*, Lecture Notes in Civil Engineering 385,
https://doi.org/10.1007/978-981-99-6018-7_31

1 Introduction

Contamination of water is a major issue across the world. Water covers more than two-thirds of the Earth's surface, and as the world's population grows, so does the contamination of water resources. Water is an important necessity for survival and should be monitored regularly [1]. Every year, millions of people, most of them are children, die from diseases that are generally preventable due to water pollution, a lack of access to clean water and basic sanitation. The major sources of pollution include irresponsible human acts, industry waste dumped into rivers, unregulated garbage disposal, and hazardous waste dumped into rivers. According to [2], rivers provide 98% of total water, which must be managed in light of present river conditions.

As a result, environmental monitoring is a crucial procedure for monitoring environmental water quality, such as air and water quality [3]. Environmental monitoring is also used in the preparation of environmental impact assessments, as well as in a variety of other circumstances where human activities may disturb the natural environment. Therefore, numerous old water quality measurement techniques are still in use today, such as atomic absorption analysis, electrochemical processes, and gravimetric analysis. All of these methods are laboratory-based, requiring the delivery and evaluation of the sample. As a result, there is no way to do real-time analysis.

Our river, as we all know, is our primary source of water, and if the problem is not rectified, it will have an influence on our daily life. As a result, the Department of Environment (DOE) has been established by Malaysia to oversee environmental protection in which was founded 25 years ago. The objectives of the establishment are to prevent, eliminate, and regulate pollution, as well as to enhance the environment, in compliance with the objectives of the 1974 Environmental Quality Act [4]. 54 rivers were contaminated during 2010 according to the Department of the Environment [5]. As a result, this project initiated by integrating and enhancing individual sensors into a floating system powered by solar energy and transmitting live GSM data to a cloud processing base that can monitor and analyze water quality parameters such as temperature, TDS, and turbidity in real time.

Therefore, the main purpose of the study is to develop and construct a surface platform prototype for water quality monitoring.

2 Literature Review

Numerous research organizations worldwide have investigated on quality of water from a variety of angles for example traditional water sample collection and in-situ electronic monitoring enabled by sensor and communication networks advancements. In [6, 7], fixed and floating sensors were utilized to measure water quality. Li et al. [8] investigated a water quality monitoring system in real time with a cleaning mechanism that can identify unanticipated events. The past study predicted the quality of recreational beach water by focusing on a few characteristics such as turbidity, conductivity, turbidity, temperature, rainfall, and time since the last rainfall are all variables to consider [9]. The researchers discovered that wire-free sensor networks (WSNs) are still exceedingly appealing and effective for surface and groundwater quality monitoring applications by data collecting in high-resolution real time as compared to traditional manual sampling methods [10]. The authors of [11] examined a water quality monitoring system in real time in which monitors the quality using an independent network of sensors. In addition, their project's purpose is to track a variety of parameters in a river basin, such as conductivity, pH, dissolved oxygen, depth, turbidity, and temperature.

Using remote sensing equipment, a previous study [12] examined numerous meteorological parameters, including solar radiation, temperature, relative humidity, precipitation, wind speed, and direction, as well as hydrological parameters, including water temperature, water level, dissolved oxygen [DO], salinity, pH, and turbidity. From the study [13], a Lagrangian sampling system was built and attached to recreational ships such as kayaks, river rafts, and other watercraft to collect data continuously to monitor the temperature, pH, and oxidation–reduction potential at many locations as they floated along the river. Meanwhile, the past study [14] proposes a network of robotic Lagrangian floating sensors for real-time monitoring of water flows, environmental factors, and bathymetry in a shallow water situation. According to the authors of [12], low DO and high salinity levels have an influence on aquatic life. A lot of significant fish fatalities have occurred as a result of poor water quality.

3 Methodology

3.1 *Design of Floater Setup*

This floating platform is well known as an unmanned surface vehicle (USV) where it can be operated on the water's surface without crew. The floating platform was built up using readily available materials such as PVC pipes, solar panels, aluminum square frames, an electronic box, pipe holders, short and long stands, and pipe caps as shown in Table 1. Besides that, this prototype is capable of replacing human work in high-risk study locations such as areas contaminated with radioactive materials, as well as improving the human capability to gather data in difficult-to-reach places.

The conceptual design of the floating platform is shown in Figs. 1, 2, 3, and 4. The descriptions of components (1–8) shown in Fig. 1 are listed in Table 1. Since the monitoring software is based on open-source technology, it produces the most economical and affordable systems. These systems can be classified into essential parts such as the USV platform, water quality monitoring system, and application software.

Table 1 Component properties

Components	Descriptions	Quantity	Weight per unit (g)	Materials
1	Solar panel	1	2220.05	Aluminum (1060 alloys)
2	Square frame	2	3712.50	Aluminum (1060 alloys)
3	Box	1	5450.00	High-density polyethylene
4	Pipe holder	4	1197.45	Aluminum (1060 alloys)
5	Short stand	2	168.64	Aluminum (1060 alloys)
6	Long stand	2	253.01	Aluminum (1060 alloys)
7	Pipe	2	1638.23	PVC plasticized
8	Pipe cap	4	189.87	PVC rigid

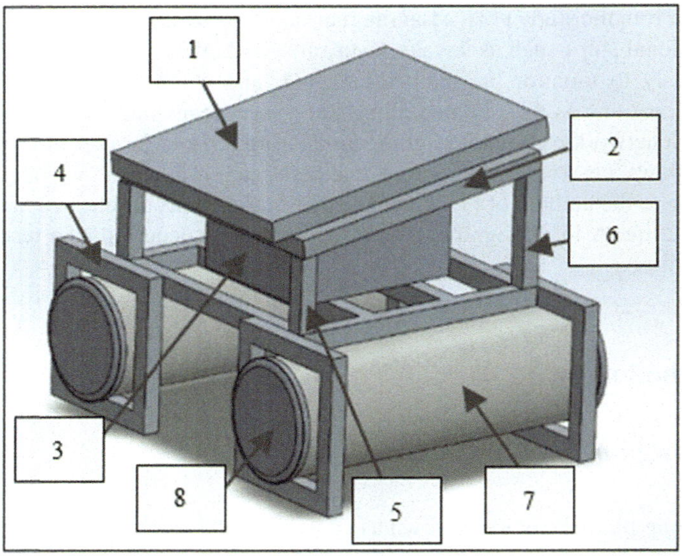

Fig. 1 Isometric view of floater

3.2 Operational System of Floater

Several phases of this project were accomplished in order to meet the goals. Its goal is to guarantee that the project's development goes as smoothly as possible. Firstly, the load distribution from top to bottom of the prototype was calculated at the beginning of this project. In the calculation of load distribution, equations of equilibrium such as summation of (Fx, Fy, M) were used to calculate axial forces exerted on each element. Secondly, the superposition method was used to analyze the shear and bending moment diagram since all the element's self-weights were included in the calculation. This caused all the elements subjected to more than two

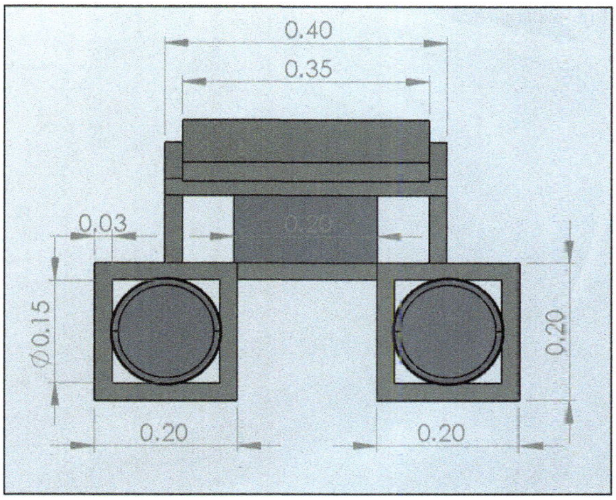

Fig. 2 Front view

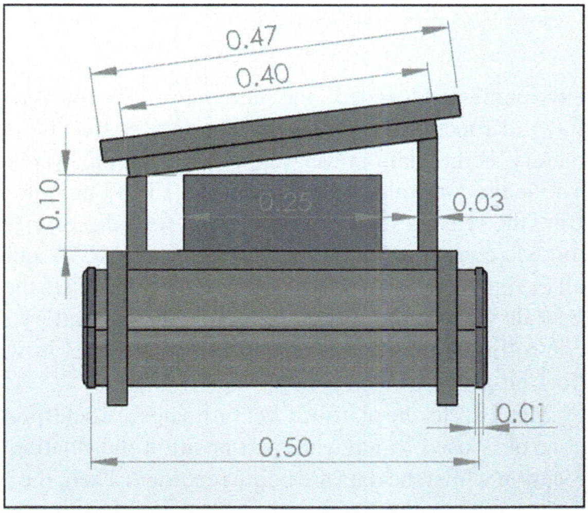

Fig. 3 Side view

loadings separately. Therefore, the superposition principle was a convenient method to use to calculate maximum shear and bending moment together with diagrams. After that, software was used to run the simulation and the maximum deflection, and stress and strain of the floating platform were obtained by inputting all the calculated loads. The results of the simulation will be discussed in the result section.

An electrical box was installed in the middle of the second frame for the prototype's design. The electrical components were encased in the electrical box to prevent the components from being exposed to the water as shown in Fig. 5. Figure 6 illustrates the electronic component flowchart of the floating platform to monitor the

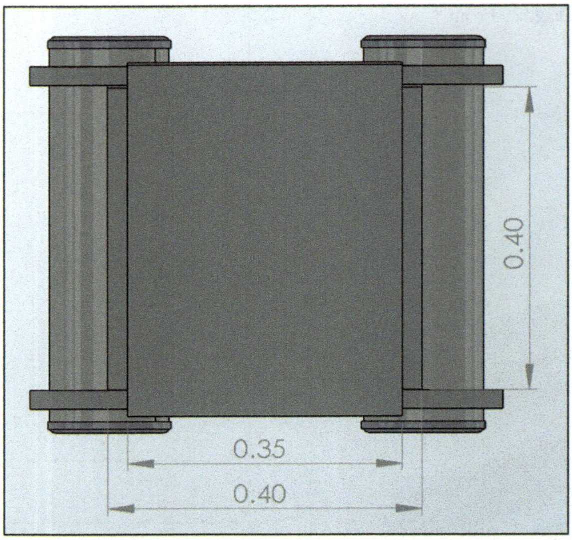

Fig. 4 Top view

environment. Moreover, the hardware in the electrical box allows the floating platform to function on the water's surface. After the components are connected, the battery as the main power supply turns on the system and detects the environment. The main controller in this design is TTGO module in which recorded all the data from the sensors such as temperature, total dissolved solid (TDS), and turbidity. The module consists of a microcontroller unit (MCU) and GSM module, which controls all the input sensors and transmits live GSM data to the cloud processing base. TTGO module is operated using a sealed lead acid battery and charged by solar charging controlling unit. The power is from solar panel in which location at the top of the floating platform.

Meanwhile, the platform's coordinates are determined by a GPS module. A remote control is used to navigate and position the floating platform to any point on site location whilst the data are being recorded. Then, the real-time data will be displayed and viewed on the Web site.

3.3 *Floating Stability Testing*

The floating device is subjected to initial testing before collecting data at the site location to determine whether the floating device can be floated. As shown in Fig. 7, the device floats on water, and the sensors are also put to the test to see if the data can be displayed on the Web site. Meanwhile, Fig. 8 indicates that the floating device has been tested and can be floated on the water at the site location. The sensors and electrical components are performing proficiently in terms of data collection.

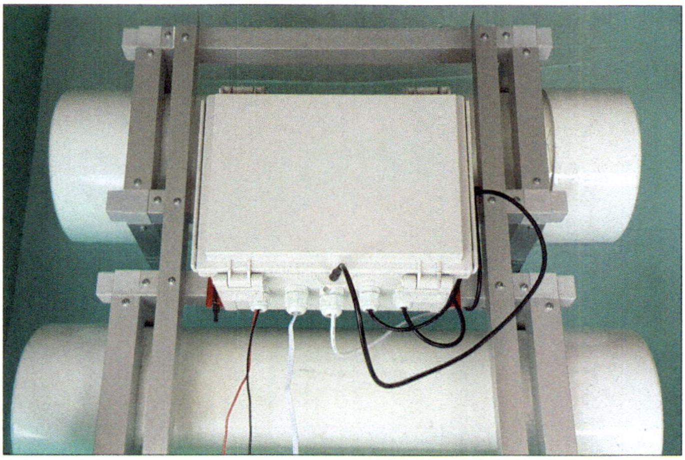

Fig. 5 Electronic components enclosed in an electrical box

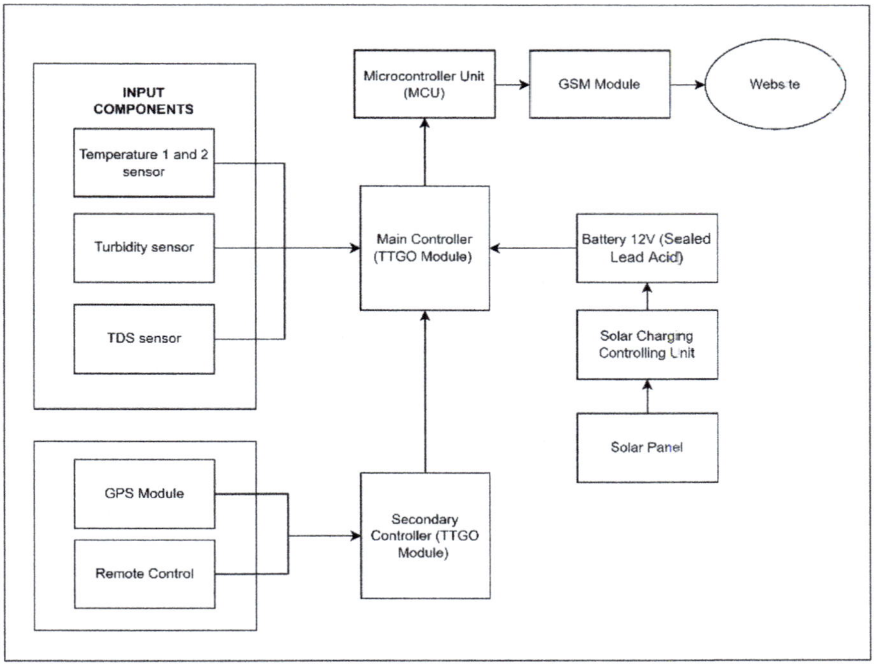

Fig. 6 Electronic components flowchart diagram

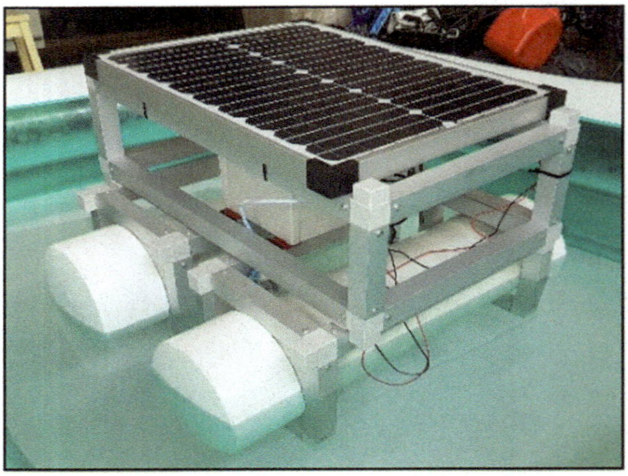

Fig. 7 Initial testing of floating device

Fig. 8 Deployment of floating device at the site location

4 Result and Discussion

4.1 Results

The pictures shown below in Figs. 9, 10, and 11 are the results of structural analysis parameters such as stress, displacement, and strain exerted on the developed prototype obtained from CAD software, respectively. The specified loads were set for each element on the top surface and simulated in the CAD software with applied materials. It produced comprehensive results of maximum stress, displacement, and strain. The maximum stress, displacement, and strain produced by the simulation were 0.3766 MPa, 0.004584 mm, and $2.010 \times 10 - 5$.

Summary of Load Distribution. The applied loads for each element that was calculated initially were shown in Table 2. During the initial calculation of load

Fig. 9 Stress result

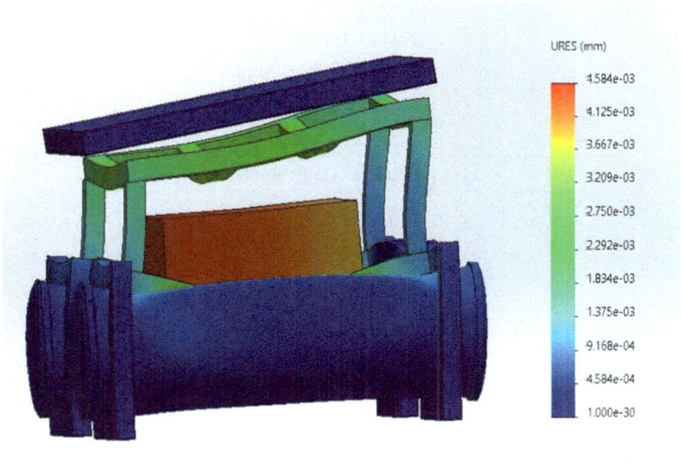

Fig. 10 Displacement result

distribution, loads exerted on each element were calculated accordingly. The self-weight of each element also has been included in the pre-calculations. Initially, the self-weight exerted as uniformly distributed on each element was changed to point load by multiplying with span length. This was due to the point load exerted in the mid-span and the element classified as symmetrical.

Fig. 11 Strain result

Table 2 Summary of load distribution

Level	Components	Elements	Total loads (N)
3	1	Solar panel	33.17
	2	Square frame	
2	5	Stand A (Short Stand)	9.95
	6	Stand B (long stand)	10.77
	5	Stand C (short stand)	9.95
	6	Stand D (long stand)	10.77
	3	Box	89.85
1	4	Bracket A	26.64
		Bracket B	26.64
		Bracket C	26.64
		Bracket D	26.64

5 Conclusion

As a conclusion, the floating platform can measure some parameters including temperature, humidity, water pH value, dissolved oxygen, and turbidity. The design and development of a water quality monitoring platform demonstrate its ability to produce more efficient and dependable results. The device is primarily designed to analyze the water quality in real time with 99% accuracy. Several phases that were planned were completed successfully and were able to operate according to the plan. The device is dependable and simple to operate, and it may also be used to monitor water contamination. Furthermore, as a result of the data and analysis that has been obtained, authorities will be able to undertake environmental evaluations more swiftly and efficiently with the help of this platform. Apart from that, the platform was built to meet Malaysia's environmental monitoring requirements.

Acknowledgements The authors would like to acknowledge Universiti Tenaga Nasional for providing the facilities and financial assistance under BOLD Research Grant 2021.

References

1. Meghana, M., Kumar, K., Verma, R., Kiran, D.: Design and development of real-time water quality monitoring system. In: 2019 Global Conference for Advancement in Technology GCAT GCAT 2019 (2019). https://doi.org/10.1109/GCAT47503.2019.8978414
2. Huang, Y.F., Ang, S.Y., Lee, K.M., Lee, T.S.: Quality of water resources in Malaysia. Res Pract Water Qual (2015). https://doi.org/10.5772/58969
3. Kafli, N., Othman, M.Z., Isa, K.: Development of a floating platform for measuring air and water quality. In: USYS 2016—2016 IEEE 6th International Conference on Underwater System Technology: Theory and Applications, pp. 177–182 (2017). https://doi.org/10.1109/USYS.2016.7893921
4. Department of Environment: Official Portal of Department of Environment (2015). http://www.doe.gov.my/portalv1/en/
5. Ngah, M.A., Ahmad, A., Abdul, A.: JurnalTeknologi. Full paper A Low Cost Method to Analyse Concentration of Carbon Monoxide (CO), vol. 3, pp. 43–47 (2015)
6. Ahmad, Z., Asad, E.U., Muhammad, A., Ahmad, W., Anwar, A.: Development of a low-power smart water meter for discharges in Indus basin irrigation networks. In: Communications in Computer and Information Science (2013)
7. Ahmad, Z., Muhammad, A.: Low power hydrometry for open channel flows. In: IECON Proceedings (Industrial Electronics Conference) (2014)
8. Li, T., Winnel, M., Lin, H., Panther, J., Liu, C., O'Halloran, R., Wang, K., An, T., Wong, P.K., Zhang, S., Zhao, H.: A reliable sewage quality abnormal event monitoring system. Water Res. (2017). https://doi.org/10.1016/j.watres.2017.05.040
9. He, L.M. (Lee), He, Z.L.: Water quality prediction of marine recreational beaches receiving watershed baseflow and stormwater runoff in southern California, USA. Water Res. (2008). https://doi.org/10.1016/j.watres.2008.01.002
10. Zia, H., Harris, N.R., Merrett, G.V., Rivers, M., Coles, N.: The impact of agricultural activities on water quality: a case for collaborative catchment-scale management using integrated wireless sensor networks. Comput. Electron. Agric. (2013)
11. O'Flynn, B., Regan, F., Lawlor, A., Wallace, J., Torres, J., O'Mathuna, C.: Experiences and recommendations in deploying a real-time, water quality monitoring system. Meas. Sci. Technol. (2010). https://doi.org/10.1088/0957-0233/21/12/124004
12. Glasgow, H.B., Burkholder, J.A.M., Reed, R.E., Lewitus, A.J., Kleinman, J.E.: Real-time remote monitoring of water quality: a review of current applications, and advancements in sensor, telemetry, and computing technologies. J. Exp. Mar. Bio. Ecol. (2004). https://doi.org/10.1016/j.jembe.2004.02.022
13. Schneider, J., Schultz, L.E., Mancha, S., Hicks, E., Smith, R.N.: Development of a portable water quality sensor for river monitoring from small rafts. In: OCEANS 2016 MTS/IEEE Monterey, OCE 2016 (2016)
14. Oroza, C., Tinka, A., Wright, P.K., Bayen, A.M.: Design of a network of robotic Lagrangian sensors for shallow water environments with case studies for multiple applications. Proc Inst. Mech. Eng. Part C J Mech. Eng. Sci. (2013). https://doi.org/10.1177/0954406213475947

Grillage Analysis of a Simply Supported Bridge Under Varying Skewing Angles and Span Lengths

Nur Ezzaryn Asnawi Subki, Hazrina Mansor, Nurul Qhairun Nisaq Kamarul Zaman, and Yazmin Sahol Hamid

Abstract Modern highways are designed to be as straight as possible to accommodate greater speed and safety in today's traffic. However, it is impossible to arrange for a bridge to span straight to the feature it crosses, especially when a relatively straight roadway alignment is required. Here, a 'skew' bridge is needed, but it will have a greater span length and create an angle at the support, the implications of which have received little attention in the current literature. The linear static behavior of a simply supported bridge with skewing geometry was investigated in this paper, and the effect of skew angles and their influence on the internal forces of the U-beam bridge at three different span lengths were observed. A three-dimensional (3D) grillage model of the proposed U-beam bridge was created in the STAAD.Pro software. Three different span lengths (15, 20, and 25 m) were investigated at three different skew angles (0°, 15°, and 40°), totaling nine grillage models studied. The BS 5400 and BD37/01 were used to design and analyze the bridge. The variation of bending moment and shear force was investigated using various skew angles and span lengths. The analysis has revealed that a non-skewed bridge behaves similarly to a one-way simply supported slab, in which load was transferred directly to the support. The analysis also showed that a skewed bridge developed a high shear force that was concentrated at the obtuse corners of the bridge deck. The load distribution of a skewed bridge was based on the shortest distance between the supports, which was located in between the obtuse corners. The grillage analysis was extended to a parametric study to further investigate the behavior of bridges under varying span lengths and skew angles. The study found that skew angle variation had a greater impact on bending moment demand and shear force demands than span length variation. This study also revealed that a bridge with a relatively low skew angle, specifically 15° or less, can be treated as a non-skewed bridge because the bending moment and shear force demands were found to be comparable.

N. E. A. Subki · H. Mansor (✉) · N. Q. N. Kamarul Zaman · Y. Sahol Hamid
School of Civil Engineering, College of Engineering, Universiti Teknologi MARA (UiTM) Shah Alam, Shah Alam, Malaysia
e-mail: hazrina4476@uitm.edu.my

N. Sabtu (ed.), *Proceedings of AWAM International Conference on Civil Engineering 2022—Volume 2*, Lecture Notes in Civil Engineering 385,
https://doi.org/10.1007/978-981-99-6018-7_32

Keywords Bridge engineering · Skew angle · Grillage analysis · Load path · Reinforced concrete bridge

1 Introduction

Skewed bridges were commonly adopted to address issues pertaining to geometry impediments, sudden change of road alignments, cost considerations, and environmental concerns. A skewed bridge is identified by its skewing angle [1]. The angle is defined as the angle between the edge of the bridge deck and the line normal to bridge support (i.e., abutment), as depicted in Fig. 1. The skewed bridge behavior is highly complex and bears no resemblance to the behavior of a straight bridge (i.e., a non-skewed bridge), which complicates the design of a skewed bridge. The load distribution for the non-skewed bridge differs from that of the skewed bridge. According to Sindhu et al. [2], the load path of a skewed bridge was often distributed to the nearest accessible support, which was noticeably at the deck corner with an obtuse angle (see Fig. 2).

A 'skew' bridge is commonly adopted when the geometry cannot align to form a straight bridge [3]. For a skew deck, the strip area that links the obtuse-angled corners is distributed by the load, and the slab bends along the joining line of the obtuse-angled corners [4]. The primary width of the bending strip is determined by the skew angle, skew span, and deck width ratio, resulting in a large twisting moment on the skewed bridge that must not be overlooked. As a result, the direction of the primary moment varies, as seen in Fig. 2.

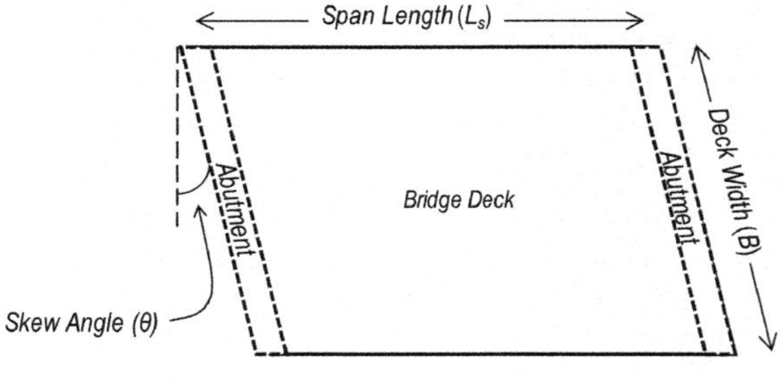

Fig. 1 Nomenclature of a skewed deck slab

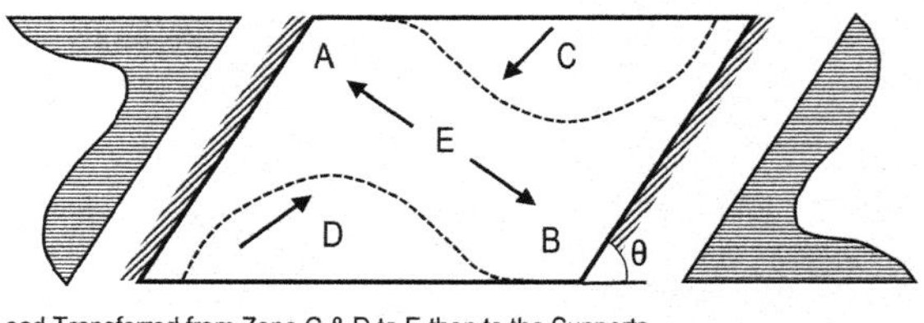

Fig. 2 Skew deck force flow (Reproduced image from [2])

The effect of skew angle on the response of completed structures has been well documented. As reported in [5, 6], a skew deck's moment distribution, support reaction, and shear force profiles with a skew angle less than 20° performed similarly to an orthogonal deck. However, the reactions and distributions of shear and moment on the abutment with a skew angle of 20° were notably different from orthogonal slabs, necessitating more advanced analysis approaches [7–10]. Research in [11] studied a theoretical analysis of a real ribbed bridge with two lanes under traffic loads based on Egyptian standards and practices recommendations. The study focused on a skew angle that ranged from 0° to 45°. Their research revealed that the bending moment response of a skewed bridge decreases with the increase of skew angle [11]. Similar findings were also reported in [12], which compared the bending moment between the skewed angle and the straight bridge decreases as skew increases for all spans. The influence of concrete grade on the overall behavior of bridge decks was also investigated in [11] using skewed angles of 0° and 30°. From the study conducted in [11], the results have revealed the skew angle and concrete grade had a significant influence on the overall behavior of the slab. Meanwhile, a study published in [2] looked at the effect of skew angle on a single-span reinforced concrete bridge and found that it resulted in a considerable reduction in bending moment and an increase in shear force.

In 2019, Xie et al. [13] highlighted that most existing codes (e.g., AASHTO LFRD-6 [14] or CSA S6-14 [15]) and design guidelines for skew slabs adopt the approach of first assuming the slab is orthogonal and then applying empirical factors to account for the influence of skew angle. However, the code [14, 15] lacked an amplification factor for shear force and provided no guidance. The majority of existing theoretical work on skews bridge decks uses finite element analysis (FEA) [5, 6, 12] or the grillage analogy method [16–19]. Although both the FEA and grillage methods produce accurate results, they are complex and time-consuming. They do not provide mathematical expressions that can be easily generalized for guidelines or codes of practice.

Previous literature revealed that the majority of researchers examine the bending behavior of different types of bridges under various skewing angles. However, previous studies did not investigate into great detail about the fundamentals of a skewed bridge's load distribution. In this work, the authors were interested investigating the load-resisting mechanisms of a simply supported U-beam skewed bridge designed according to the code of practice used in Malaysia (i.e., BS5400-4:1990 [20] and BD37/01 [21]). After understanding the fundamental behavior of skewed bridge, the study was then extended to a parametric study to investigating the influence of skewing angle and span length on the structural behavior of a simply supported U-beam skewed bridge. This paper adopted the grillage method executed in the STAAD.Pro software to analyze the linear static behavior of the simply supported U-beam skewed bridge. The grillage method is explained further in the following section.

2 Investigation Approach: Grillage Method

A simply supported slab with the variation of skew angles and span lengths were considered in this paper. Figure 3 illustrated the plan view of the proposed bridge deck (straight bridge, $\theta = 0°$) with its cross-sectional view. The bridge deck was comprised of three 2930 mm wide lanes and two 450 mm wide parapets, which leads to the total deck width of 9.7 m (Refer to Fig. 3). The bridge deck was supported by four numbers of U8 prestressed beams. For the interest of the parametric study, the bridge span was varied to 15, 20, and 25 m, while skew angle was 0°, 20°, and 40°. The straight bridge ($\theta = 0°$) was included in this study as a control specimen for comparison purposes. A total of nine bridge models were included in the parametric study, as given in Table 1.

In this study, the grillage analysis using the STAAD.Pro software was executed on each bridge model in Table 1. As depicted in Fig. 4, the grillage model was constructed by laying down the longitudinal beam element on top of the crosshead, while nine

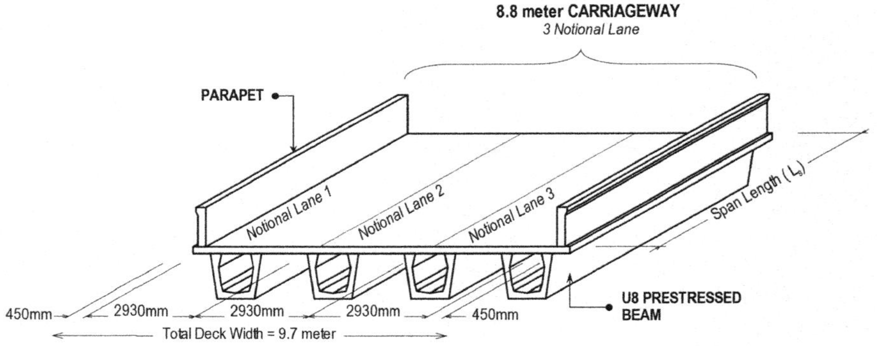

Fig. 3 Structural drawing of the proposed bridge deck

Table 1 Structural properties of the bridge deck used in the parametric study

No	Span length, L_s (m)	Skew angle, θ (°)
1	15	0
2		15
3		40
4	20	0
5		15
6		40
7	25	0
8		15
9		40

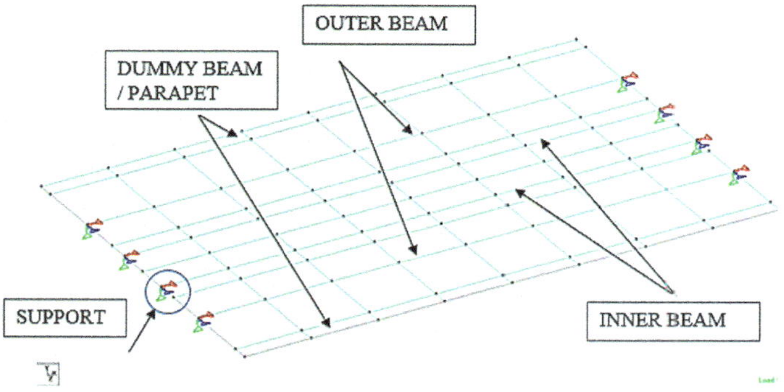

Fig. 4 The grillage model of a straight bridge ($\theta = 0°$)

beam elements were allocated in the transverse direction with equal spacing. Dummy beams were added at the outer side of the grillage model (see Fig. 4) to represent the parapets. All beam elements in the grillage model were rigidly connected at the intersecting nodes. The pinned boundary conditions were assigned at the short edge of the bridge deck (See Fig. 4). The grillage models analyzed the bending moment and shear force at each prestressed beam that consists of two outer and inner beams shown in Fig. 4. The grillage modelling techniques were similar for all types of bridge in Table 1 (Fig. 5).

The self-weight of the bridge deck, road surfacing, parapet, and traffic loads (i.e., live load) were all considered in the grillage loading. The self-weight load was applicable to composite sections of the prestressed beam and end diaphragm. The bridge deck self-weight load was applied to the whole grillage model as depicted in Fig. 6a. The road surfacing consisted of a 50 mm thick asphalt wearing course with a regulating course of 50 mm thickness. The density of asphalt was 22.0 kN/m^3, and the total asphalt thickness was 100 mm. The estimated road surface load was 2.2 kN/m^2, and it was assigned as uniformly distributed load (UDL) on the grillage model as depicted in Fig. 6b. The parapet load was estimated based on the product of the concrete density (25.0 kN/m^3) and the cross-sectional area of the parapet (4250 cm^2). The parapet load (22.2 kN/m^2) was also assigned as a UDL load, but on the dummy beams (see Fig. 6c). The traffic loads that comprised of HA loads and HB load were determined using BD37/01 [21]. According to BD37/01 [21], HA loading represents a normal traffic load, while HB loading is an abnormal vehicles unit loading. HA loading consisted of two types; a uniformly distributed load (UDL) and knife end load (KEL). HA UDL was determined following clause 6.2 of BD37/01 [21], which was later applied as a patch load in the grillage model. The HA KEL load was designed according to clause 6.2.2 of BD37/01 [21] that applied as patch load per notional lane width. In this study, the KEL per notional lane was taken as 120 kN and the HA KEL was calculated to be 40 kN/m. The distribution of the UDL and KEL shall be taken to occupy one notional lane, uniformly distributed over the full width of the lane (see Fig. 6d).

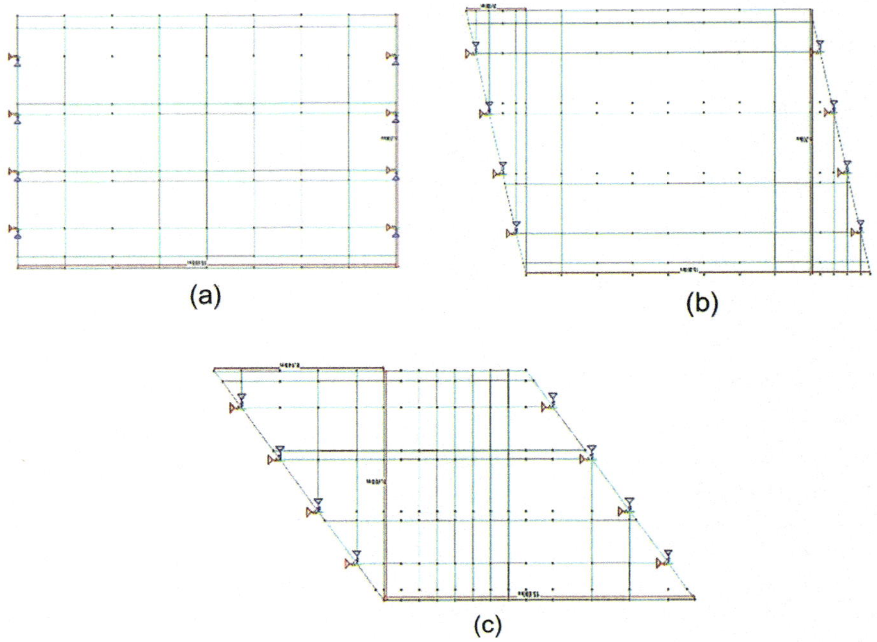

Fig. 5 Grillage model of the bridge deck with the skewing angle of **a** 0°, **b** 15°, and **c** 40°

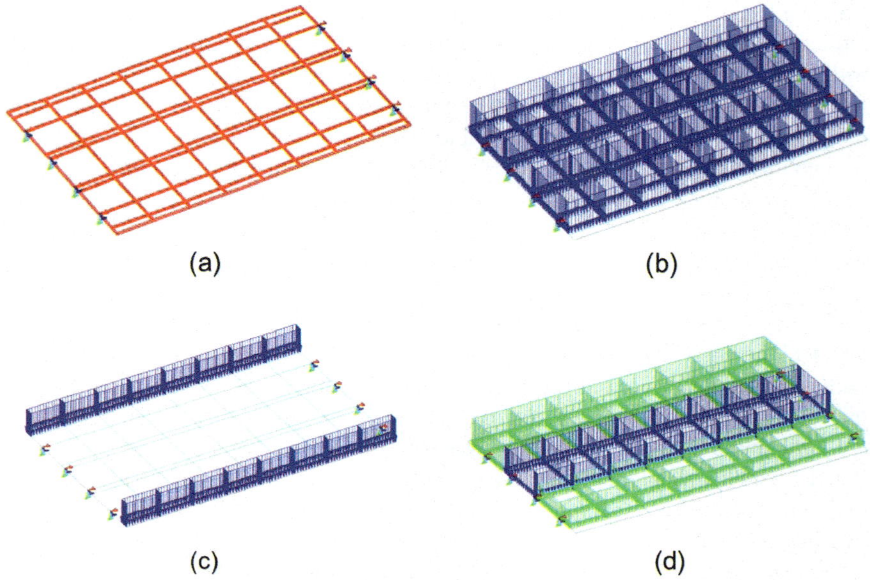

Fig. 6 Loading scheme for the: **a** bridge deck self-weight, **b** road surfacing, **c** parapet and **d** traffic loads

3 Load-Resisting Mechanism of Non-skewed and Skewed Bridges

The results of 25 m length non-skewed and skewed bridges were selected to compare and discuss the load-resisting mechanisms of a bridge under linear static condition. The load-resisting mechanisms were presented as bending moment diagram (i.e., moment about the transverse axis) and shear force diagram in Figs. 7 and 8, respectively for the non-skewed bridge ($\theta = 0°$), and Figs. 9 and 10, respectively, for the skewed bridge ($\theta = 40°$). It should be highlighted that the dummy beams (see Fig. 4) were not plotted in the aforementioned figures.

The data show that a non-skewed bridge has a uniform distribution of bending moment and shear force throughout the width of the bridge deck (see Figs. 7a and

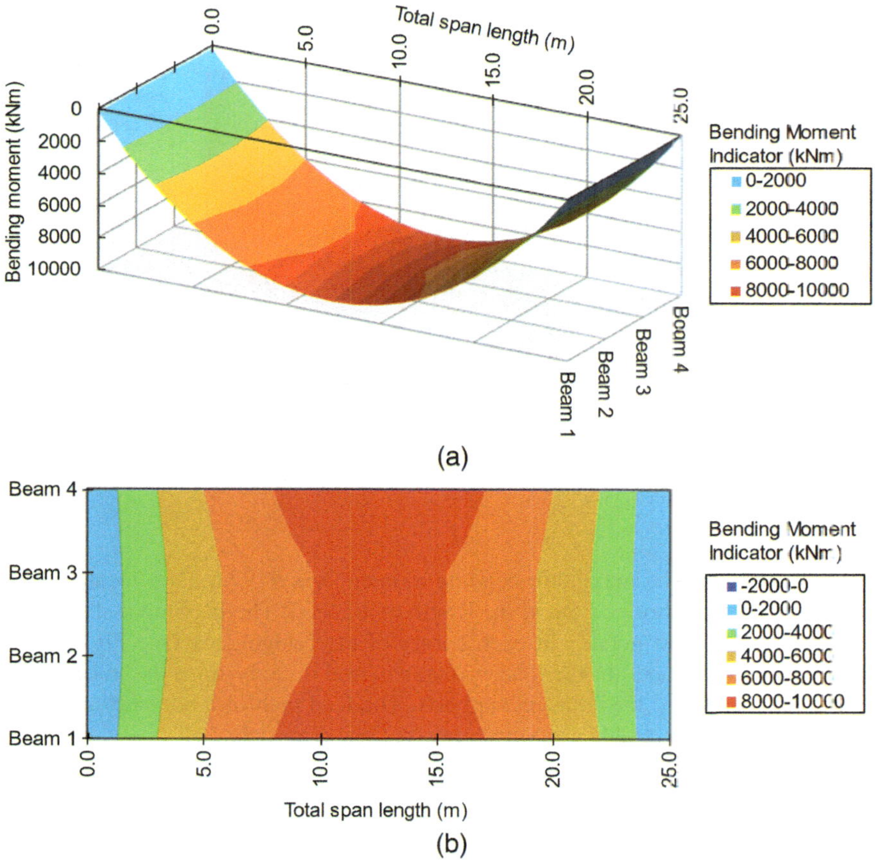

Fig. 7 Bending moment contour plot of a 25 m non-skewed bridge (skew angle, $\theta = 0°$) in **a** three-dimensional view and **b** top view

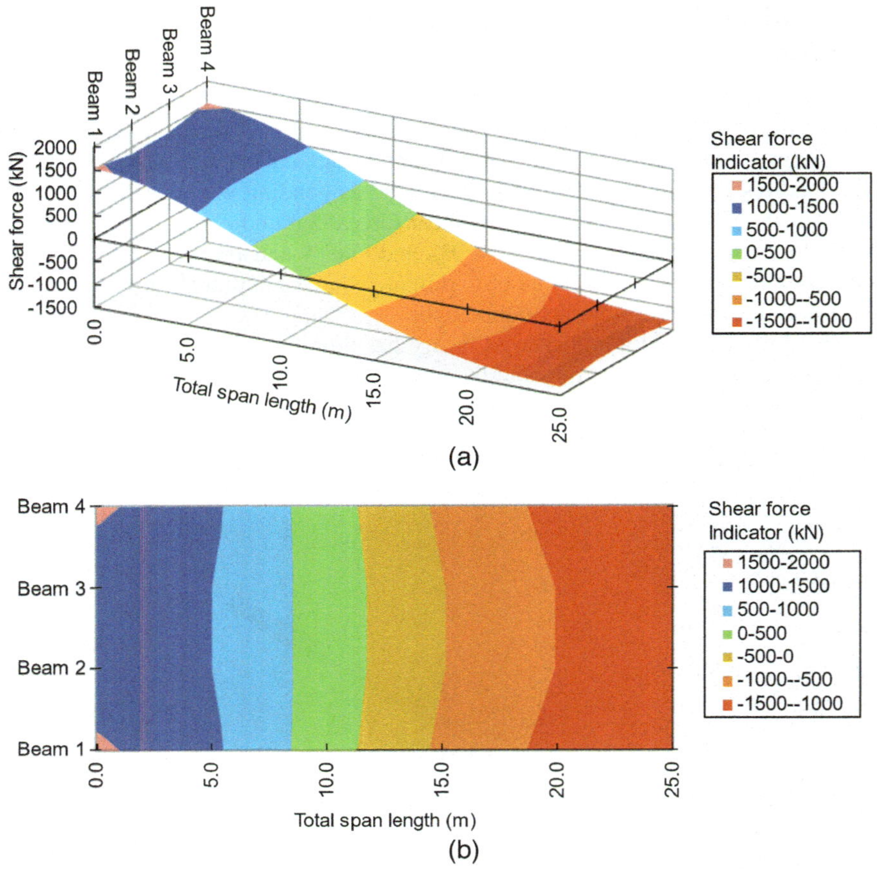

Fig. 8 Shear force contour plot of a 25 m non-skewed bridge (skew angle, $\theta = 0°$) in **a** three-dimensional view and **b** top view

8a, respectively). The maximum bending moment was 9197.6 kNm located at the middle span of the bridge (i.e., at the length of 12.5 m). The maximum shear force was 1619.1kN, most notably at each corner of the bridge deck (see Fig. 8b). As shown in Figs. 7b and 8b, the load was transferred directly from the middle span of the bridge deck to the adjacent supports. The load path of a non-skewed bridge essentially would resemble a simply supported one-way slab.

The skewed bridge, however, demonstrates distinctive bending moment and shear force distribution across the width of the bridge deck, as shown in Figs. 9a and 10a, respectively. The analysis revealed that the maximum bending moment was 993.4 kNm located at the middle span of the outer beams (i.e., Beam 1 and Beam 4) as shown in Fig. 9b. The recorded maximum shear force was 5876.5 kN and located at the obtuse corners of the bridge deck as shown in Fig. 10b. As illustrated

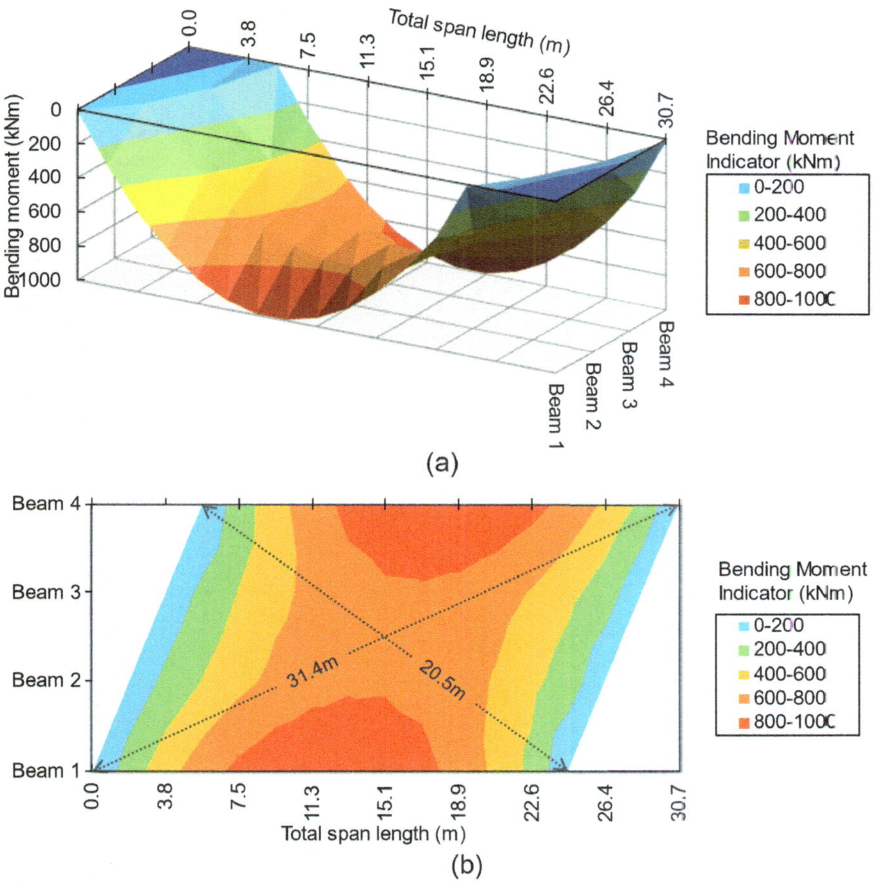

Fig. 9 Bending moment contour plot of a 25 m non-skewed bridge (skew angle, $\theta = 40°$) in **a** three-dimensional view and **b** top view

in Fig. 9b, it can be observed that the load initially flowed from the middle span of outer beams (i.e., Beam 1 and Beam 4) into the inner beams (Beam 2 and Beam 3) via bending action. The load was then redistributed to the adjacent supports, with the load concentrating at the obtuse corners (see and Fig. 10b). It is worth noting that the shear force (i.e., reaction force) at the acute corner of the bridge deck was approximately 80% lower than the shear force at the obtuse corner. Previous researchers [1, 2, 22] claimed that the shortest load path on a skewed bridge deck was between the inner beams and the obtuse corner, which resulted in the concentration of shear force at the obtuse corner. Their claim can be supported based on the data in Fig. 9b, whereby the shortest path for the 25 m skewed bridge ($\theta = 40°$) was located in between the obtuse corners with the total length of 20.5 m. The load path presented in this analysis, as depicted in Figs. 9b and 10b, exhibits similarity with the previous findings in [4, 23].

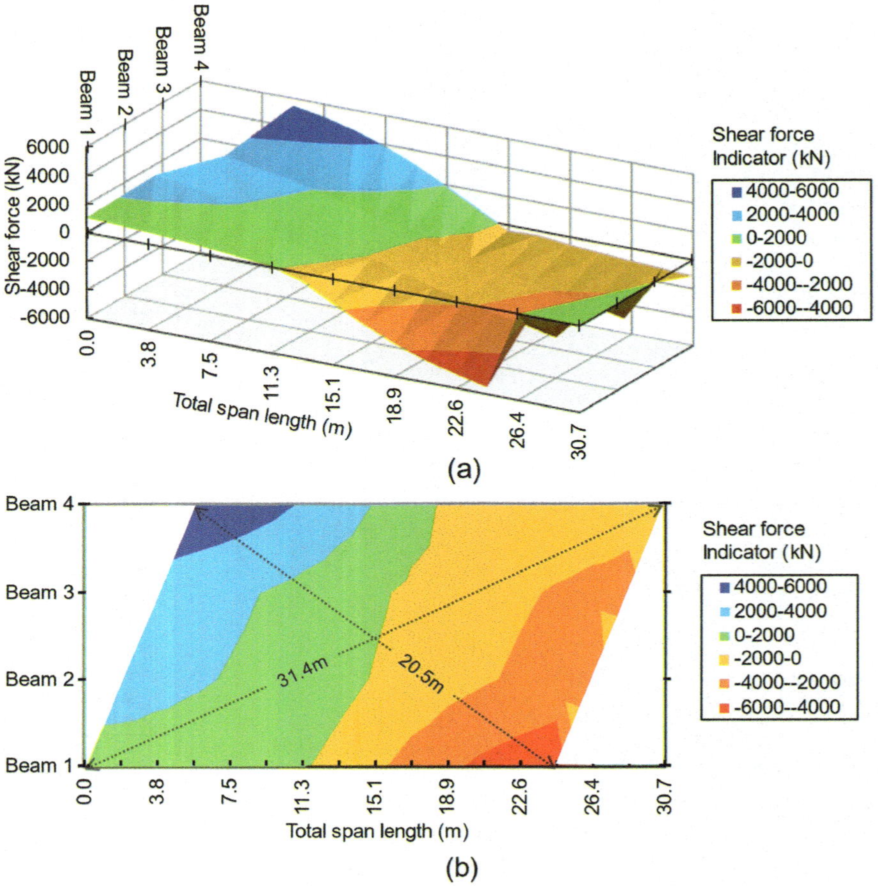

Fig. 10 Shear force contour plot of a 25 m non-skewed bridge (skew angle, $\theta = 40°$) in **a** three-dimensional view and **b** top view

The results presented in Figs. 7, 8, 9 and 10 depict the behavioral differences between non-skewed and skewed bridges. A non-skewed bridge ($\theta = 0°$) was observed to have a high bending resistance, but with a reduced shear force development (see Figs. 7, 8). The load-resisting mechanism of a skewed bridge, on the other hand, was dominated by the shear force development, with bending moment having a minor influence (Figs. 9, 10). The data have shown that the maximum shear force of a skewed bridge ($\theta = 40°$) was 3.6 times higher than a non-skewed bridge. The maximum bending moment was significantly reduced by 89% for the skew bridge. The force demands on the skewed bridge change accordingly because the shortest distance between the supports (i.e., load path) was reduced from 25.0 to 20.5 m, which minimize the bending behavior in the principal axis and accumulates shear force at the obtuse corners.

4 Parametric Study

The grillage analysis was executed on various bridge models to investigate the effect of span length and skew angle on the linear static behavior of the bridge. The two structural responses that will be discussed in this paper are bending moment and shear force. It is important to note that the bending moment and shear force diagrams presented in this section represent the highest magnitude of force along the stretch of the bridge, regardless of the load combination (i.e., HA or HA + HB) or location (i.e., Beam 1, Beam 2, Beam 3, or Beam 4) from which it was extracted. For instance, as illustrated in Fig. 11, the highest magnitude of force was extracted from the bending moment and shear force diagrams of 25 m skewed bridge (skew angle, $\theta = 40°$). Essentially, the force diagrams depict the largest force demand exerted along the bridge's length.

4.1 *Effect of Span Length*

The bending moment and shear force diagrams under varying span lengths are given in Figs. 12 and 13, respectively. The plots in Fig. 12 show that the moment demand along the stretch of the bridge increased notably as longer span length was utilized. It was inferred that a long span bridge constituted a lower overall bending stiffness, which prompt higher moment demand across the bridge to accommodate the loads. The highest moment demand (i.e., maximum bending moment) was most prominent at the middle span of the bridge deck ($L/L_s = 0.5$) for all models. Based on Fig. 14a, irrespective of the skewing angle, it is evident that the maximum bending moment increased as the bridge span length increased. However, the variation is more pronounced when a lower skew angle was used. As recorded in Table 2, the maximum

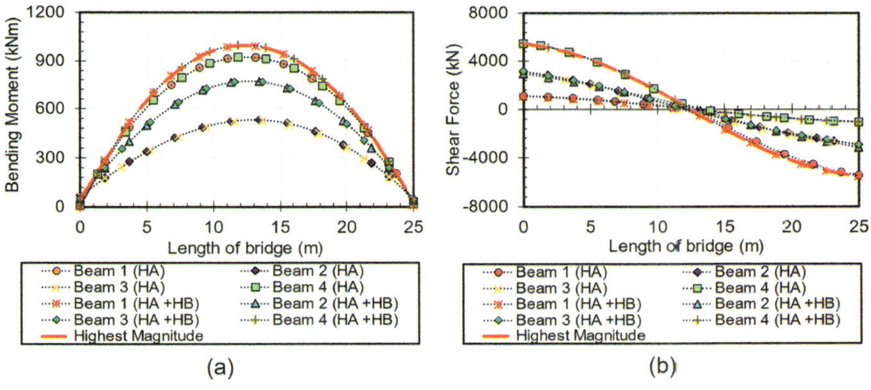

Fig. 11 Extraction of largest force demands along the 25 m skewed bridge (skew angle, $\theta = 40°$) from the **a** bending moment diagram and **b** shear force diagram

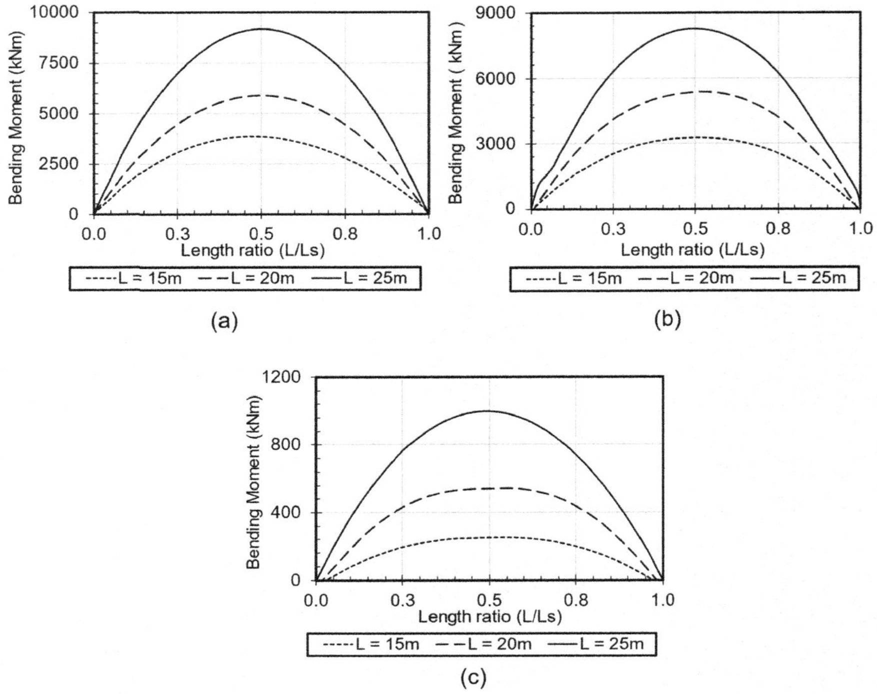

Fig. 12 Comparison of bending moment diagrams for bridges with varied span lengths at skewed angle of: **a** $\theta = 0°$, **b** $\theta = 15°$, and **c** $\theta = 40°$

bending moment resistance could drop by 64% on average when the bridge span was reduced by 60% from 25 m length.

As the span length changed, the shear force demand (see Fig. 13) varied in a similar manner to the bending moment demand (see Fig. 12), with the longer span contributing to higher shear force demand. The data in Fig. 13 shows that the highest shear force demand (i.e., max shear force) for all models was located at the supports. Regardless of the skew angle, the maximum shear force would consistently increase with span length, as depicted in Fig. 14b. However, the maximum shear force variation is more evident for the $\theta = 40°$ skewed bridge. Based on the data in Table 3, the maximum shear force can increase by 70% as the bridge span was extended by 60% from 15 m length.

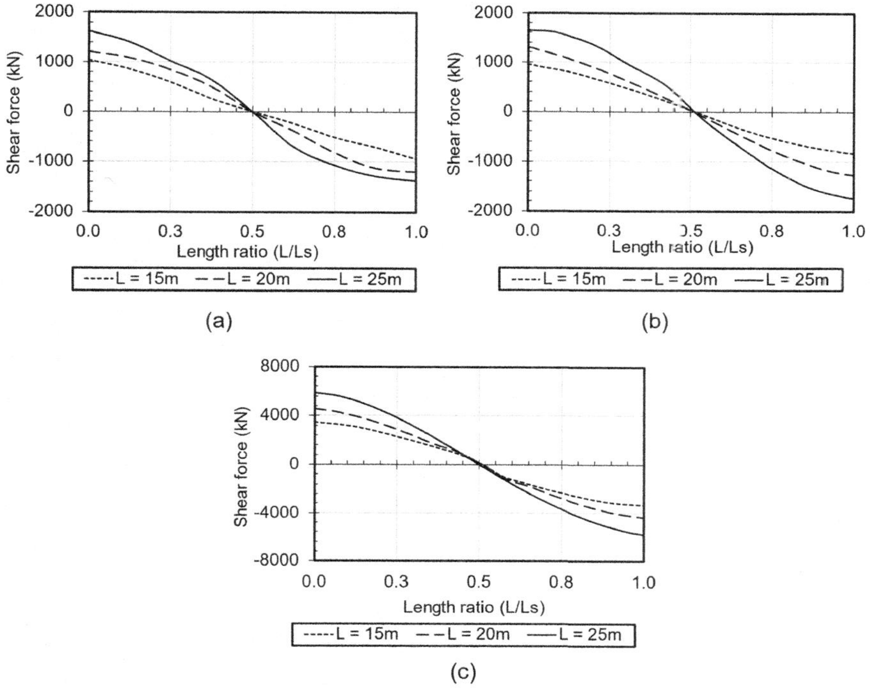

Fig. 13 Comparison of shear force diagrams for bridges with varied span lengths at skew angle of: **a** $\theta = 0°$, **b** $\theta = 15°$, and **c** $\theta = 40°$

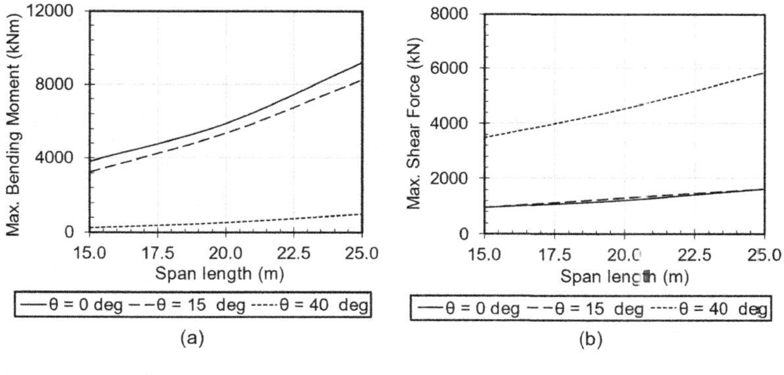

Fig. 14 Comparison of maximum resistance for bridges with varying span lengths: **a** maximum bending moment, and **b** maximum shear force

Table 2 Maximum bending moment of all bridge models

Maximum bending moment (kNm)			
	Skew angle (°)		
Span length (m)	0°	15°	40°
15	3825.2	3465.1	254.2
20	5898.4	5350.1	540.5
25	9197.6	8233.0	993.4

Table 3 Maximum value of shear force in the outer beam

Maximum shear force (kN)			
	Skew angle (°)		
Span length (m)	0°	15°	40°
15	945.2	956.4	3486.0
20	1209.0	1311.7	4552.0
25	1619.1	1643.2	5876.5

4.2 Effect of Skew Angle

As presented in Figs. 15 and 16, the bending moment and shear force demands for varying skew angles have different responses than the variation of span length. The bending moment demand, as depicted in Fig. 15, was substantially amplified when the skew angle changed from $\theta = 40°$ to $\theta = 15°$, regardless of the bridge span length. Contrastingly, the shear force demand was reduced significantly when a lower skew angle was adopted, as shown in Fig. 16. These variations occurred because the shortest distance between supports (i.e., load path) was lengthened for a skewed bridge with a lower skew angle. For example, as the skew angle changed from $\theta = 40°$ to $\theta = 15°$, the shortest distance between the supports for the 25 m skewed bridge increased from 20.5 m to 24.5 m, which is a 16.3% increment. Essentially, the skew bridge will rely on bending resistance to accommodate the loads.

However, regardless of the span length, the bending moment and shear force demands demonstrated minimal changes as the skew angle decreased from $\theta = 15°$ to $\theta = 0°$ (refer to Figs. 15 and 16). By comparing the results between the $\theta = 15°$ and $\theta = 0°$ bridges, the biggest percentage difference in bending moment and shear force demands was 10.5% and 7.8%, respectively (Refer to Tables 2 and 3). It is inferred that the shortest distance between supports for the $\theta = 15°$ and $\theta = 0°$ bridges was both comparable, which does not significantly affect the load path on the bridge deck. For instance, the shortest distance between the supports for the 25 m skewed bridge ($\theta = 15°$) was 24.2 m, which is 3.2% shorter than the supports distance for the non-skewed bridge. This proves that a bridge with a lower skew angle, specifically below than $\theta = 15°$ can be analyzed and designed as a non-skewed bridge, which further support the suggestion made by Abozaid et al. [11].

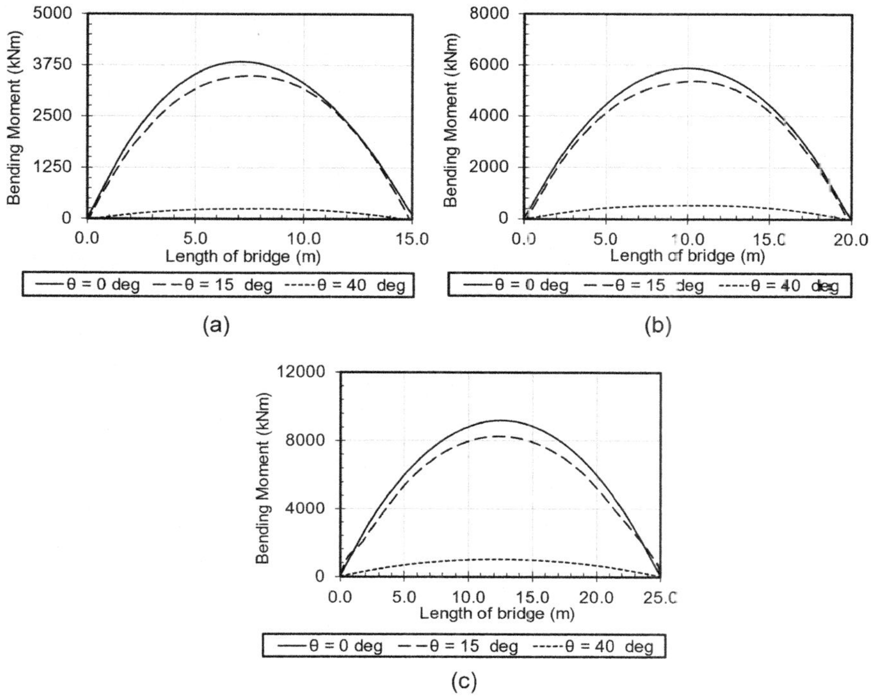

Fig. 15 Comparison of bending moment diagrams for bridges with varied skew angles at span length of: **a** $L = 15$ m, **b** $L = 20$ m, and **c** $L = 25$ m

The variation of maximum bending moment and maximum shear force against the skew angle was recorded in Fig. 17a, b, respectively. Generally, both plots in Fig. 17 demonstrated a similar data trend regardless of the bridge span length. A previous study [4] also produced similar results as in Fig. 17. By comparing the data in Figs. 14 and 17, it can be deduced that the skew angle has more significant effects on the maximum bending moment and maximum shear force as compared to span length variation.

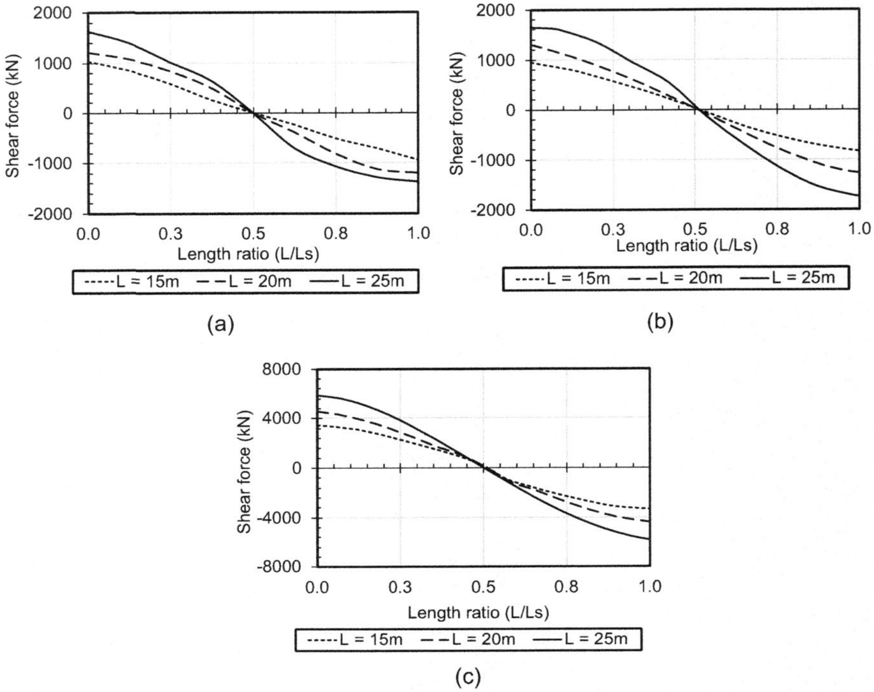

Fig. 16 Comparison of shear force diagrams for bridges with varied skew angles at span length of: **a** $L = 15$ m, **b** $L = 20$ m, and **c** $L = 25$ m

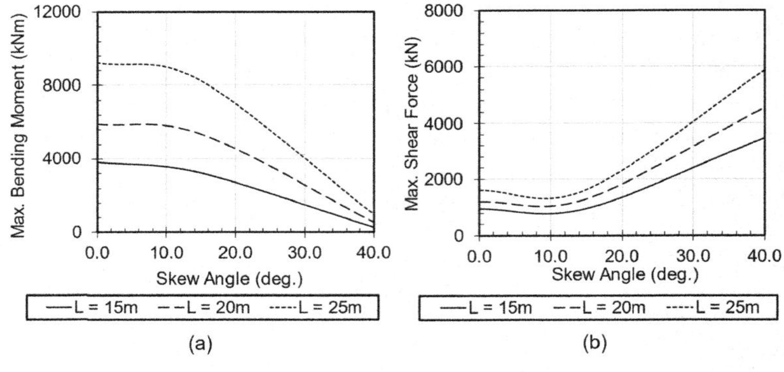

Fig. 17 Comparison of maximum resistance for bridges with varied skew angles: **a** maximum bending moment, and **b** maximum shear force

5 Conclusion

The linear static behavior of simply supported non-skewed and skewed bridges was investigated in this paper. Each bridge was analyzed using the grillage method in the STAAD.Pro software. The analysis revealed that the non-skewed bridge developed a high bending moment with comparatively low shear force to accommodate the loads. The load path of a non-skewed bridge was comparable to a one-way simply supported slab, whereby the load was directly transferred to the adjacent supports. The skewed bridge, on the other hand, accommodated the load by developing substantial shear force at the obtuse corners with noticeably low bending moment. The skewed bridge distributed the loads selecting the shortest distance between the adjacent supports, which is located in between the obtuse corners.

A parametric study also has been executed to investigate the effects of varying skew angles and span length on the linear static behavior of a simply supported bridge. The results showed that longer span length will contribute to the increase in demand for bending moment and shear force. The results also revealed that a bridge with a large skew angle (i.e., $\theta = 40°$ in this case) had a high shear force demand, but a low bending moment demand. The behavior of a bridge with a skew angle of $\theta = 15°$ and below was similar to a non-skewed bridge, whereby bending moment and shear force demands were comparable. In comparison with span length variation, this study found that skew angle had a greater impact on the bending moment and shear force demands on the bridge.

Acknowledgements The authors would like to show gratitude to UiTM Research Management Centre (RMC) for providing financial support through *Geran Inisiatif Penyeliaan*, GIP (File no: 600-RMC/GIP 5/3 (089/2021)).

References

1. Manoharan, Joseph, G.: Analysis of skew bridge with varying skew angles. Int. J. Sci Eng. Res. **7**(10) (2016)
2. Sindhu, V., Ashwin, K.N., Dattatreya, J.K., Dinesh, S.V.: Effect of skew angle on static behaviour of reinforced concrete slab bridge decks. Int. J. Res. Eng. Technol. (IJRET) 50–58 (2013)
3. Sravani, Venkateswarlu, D.: Effect of skew angle on RCC T-Girder bridge using STAAD PRO. Int. J. Innovat. Technol. Res. (IJITR) **5**(6), 7525–7532 (2017)
4. Kar, V.K., Maiti, P.R., Singh, P.K.: Study on effect of skew angle in skew bridges. Int. J. Eng. Res. Dev. **2**(12), 13–18 (2012)
5. Menassa, M.M., Tarhini, K., Frederick, G.: Influence of skew angle on reinforced concrete slab bridges. J. Bridge Eng. **12**(2) (2007)
6. Dhar, M.M., Chowdhury, M., Karmakar, S.: Effect of skew angle on longitudinal girder (support shear, moment, torsion) and deck slab of an IRC skew bridge. Ind. Concrete J. (2013)
7. Gupta, T., Misra, A.: Effect on support reactions of T-beam skew bridge decks. ARPN J. Eng. Appl. Sci. (2007)
8. Nouri, G., Ahmadi, Z.: Influence of skew angle on continuous composite girder bridge. J. Bridge Eng. **17**(4) (2012)

9. Wei, X., Bruneau, M.: Analytical investigation of buckling restrained braces' applications in bidirectional ductile end diaphragms for seismic performance of slab-on-girder bridge. Eng. Struct. **141**, 634–650 (2017)
10. Shahu, D., Joshi, S.V., Pachpor, P.D.: Analysis and behaviour of skew bridges with different skew angle. Int. J. Current Eng. Sci. Res. (IJCESR) **3**(10), 27–32 (2016)
11. Abozaid, L.A., Hassan, A., Abouelezz, A.Y., Abdel Hafez, L.M.: Nonlinear behaviour of a skew slab bridge under traffic loads. World Appl. Sci. J. **30**(11), 1479–1493 (2014)
12. Harba, S.I.: Effect of skew angle on behavior of simply supported R. C. T-beam bridge decks. J. Eng. Appl. Sci. **6**(8) (2011)
13. Xie, T., Mohamed Ali, M.S., Visintin, P.: Behaviour and analysis of ultra high performance fibre reinforced concrete (UHPFRC) skew slabs. Eng. Struct. **199**, 109588 (2019)
14. AASHTO, AASHTO LRFD Bridge Design Specifications, 8th ed., Washington (2017)
15. Canadian Standards Association (CSA), CSA S6:14 - Canadian Highway Bridge Design Code, 11th ed., Toronto (2014)
16. Théoret, P., Massicotte, B., Conciatori, D.: Analysis and design of straight and skewed. J. Bridge Eng. **17**(2) (2012)
17. Maleki, S.: Deck modeling for seismic analysis of skewed slab-girder bridges. Eng. Struct. **24**(10), 1315–1326 (2002)
18. Bakht: Analysis of some skew bridges as right bridges. J. Struct. Eng. **114**(10) (1988)
19. Pandey, K., Maru, S.: Modelling of skew bridge deck slab by grillage. Int. J. Eng. Res. General Sci. **3**(4) (2015)
20. British Standard Institution, Steel, concrete and composite bridges—Part 4: Code of practice for design of concrete bridges (1990)
21. The Highway Agency, Design Manual for Roads and Bridges - Volume 1 Section 3 Part 14 (BD 37/01) Highway structures: approval procedures and general design. General design. Loads for highway bridges, London (2001)
22. Raj, K., Phani, R.G.: Design and study on behaviour of skew slab bridges with various skew angles. Res. Rev. J. Appl. Sci. Innovat. **1**(2), 37–49 (2017)
23. Gupta, T., Kumar, M.: Flexural response of skew-curved concrete box-girder bridges. Eng. Struct. **163**, 358–372 (2018)

Investigation of Waste Clay Brick Powder as Supplementary Cementitious Material in Self-compacting Concrete

Murat Tuyan and Mehdi Maghfouri

Abstract The disposal of construction waste has emerged as one of the most critical environmental concerns in recent years. The demolition of clay bricks in building structures generates a significant volume of waste. This waste clay brick material can be effectively utilized in powdered form as a supplementary cementitious material (SCM), due to its inherent pozzolanic activity, which allows it to interact with the hydration products of cement. In essence, the waste clay brick can be recycled and reused in the construction industry, especially in the concrete production. The present communication aims to investigate the effect of using waste clay brick as a supplementary cementitious material in the production of self-compacting concrete (SCC). In this study, SCC mixtures were produced with 10, 20, 30, 40, and 50% waste clay brick powder and class F fly ash as a replacement of cement, based on weight. Consequently, a series of tests were conducted in the fresh state, including slump flow and V-funnel tests, as well as assessments of compressive strength and freeze-thaw resistance. From the obtained results, it was concluded that the fresh state performance of waste clay brick-based SCC mixtures was lower than control SCC mixture and fly ash-based SCC mixtures. Additionally, the compressive strength of the waste brick powder-based SCC mixtures was found to be lower compared to the fly ash-based SCC mixtures. In addition, freeze–thaw resistance of the mixtures increased by increasing the substitution level of waste clay brick powder in SCC mixtures.

Keywords Waste clay brick · Self-compacting concrete · Fresh state properties · Compressive strength · Freeze–thaw resistance

M. Tuyan · M. Maghfouri (✉)
Department of Civil Engineering, University of Ottawa, 161 Louis-Pasteur Private, Ottawa, ON K1N 6N5, Canada
e-mail: m.maghfouri@gmail.com

M. Maghfouri
Hume Concrete Sdn Bhd, Wisma Hume, 46100 Petaling Jaya, Malaysia

N. Sabtu (ed.), *Proceedings of AWAM International Conference on Civil Engineering 2022—Volume 2*, Lecture Notes in Civil Engineering 385,
https://doi.org/10.1007/978-981-99-6018-7_33

1 Introduction

Self-compacting concrete, also known as self-consolidating concrete (SCC) is a special type of concrete developed in Japan during the early 1980s. Its primary advantage over traditional concrete lies in its exceptional flow properties, which enable it to effortlessly fill moulds [1]. Moreover, the superior performance of SCC in its fresh state positively affects the hardened concrete properties by providing a more void-free concrete [2]. This remarkable flow ability is primarily achieved through a high binder material content and the incorporation of a new generation superplasticizer admixture [3]. With the development of concrete chemical admixture production technology day by day, the costs of chemical admixtures decrease over time. As a result, these admixtures, which improve various properties of concrete, have started to be used more in the concrete sector [4]. In addition, due to complexity of the concrete structures, the use of SCC within the construction industry has been increasing over the years.

Although SCC has many technical advantages over conventional concrete, the higher cement content increases the cost of these concretes [5] compared to conventional concrete. In order to reduce the costs, the performance of such concretes is determined by replacing cement with various types of mineral additives and supplementary cementing materials (SCM) from by-product and waste materials [6–8]. As a result of additives and SCMs substitution for cement, binder systems with a more complex hydration structure emerge. However, the features of those materials such as grain size, shape, chemical, and mineralogical composition and intergranular space structure affect the properties of the binder structure [9].

In the realm of building construction, the main constituents of demolition waste are concrete and brick generated from the construction of new buildings, demolition and renovation of old buildings, and from natural disasters [10–12]. Brick, following concrete, is one of the most widely used building materials in the construction industry. During brick production and the demolition of existing structures, significant brick fragmentation occurs, resulting in substantial waste generation. Therefore, disposal of this waste material has become an important matter for the environment. In response, many studies have been carried out in recent years on reusing of the brick shards in the construction industry. In the studies, waste brick shards were used as replacement of cement. It has been determined that the mechanical properties of the concrete decrease significantly after the amount of waste brick dust reaches to a certain level. It has been highlighted that the use of waste brick dust/powder as replacement of cement by more than 20% (by weight) results in a reduction of mechanical properties of concrete [13].

A study conducted by Heikal et al. [14] investigated the compressive strength and rheological properties of SCC mixtures incorporating 12.5, 25 and 37.5% waste brick powder. As the content of waste brick dust increased, the compressive strength of the mixtures decreased gradually. In addition, it has been reported that the shear stress of fresh concrete mixes gradually increased as a result of the increase in the content of waste brick dust.

In today's world, the disposal of waste materials ranks among the foremost global environmental concerns. The utilization of waste clay brick in powdered form as cement replacement in concrete represents a subject that warrants further investigation. Furthermore, the scarcity of studies on the use of waste clay brick powder in self-compacting concrete within the international literature, especially those examining the fresh state and freeze-thaw resistance properties of such concrete, underscores the importance of this study in addressing this knowledge gap .

2 Materials and Methods

2.1 Materials

In this study, up to 50% by weight of waste clay brick powder (WCBP) is used in self-compacting concrete as replacement of cement. Previous researches have indicated that the optimal replacement level for WCBP is below 50% [15–17]. Since the original dimensions of broken waste clay brick are larger than cement, they can be reduced to cement-like fineness and suitable aggregate gradation through various crushing and grinding processes.

In addition, the use of class F fly ash was compared with WCBP in the study. The oxide composition of WCBP and fly ash is given in Table 1. It can be said that the highest oxide of both materials is silicon oxide. The silicon oxide content in clay brick is high as it prevents cracking, shrinking, and wrapping of raw bricks. Furthermore, silica imparts a uniform shape to the bricks; however, excess of silica makes the bricks brittle.It is noteworthy that the predominant oxide in both materials is silicon oxide. The presence of a high silicon oxide content in clay bricks serves to mitigate cracking, shrinking, and warping of raw bricks while also imparting a uniform shape. However, an excessive amount of silica can render the bricks brittle.

The morphological structure of waste clay brick powder and fly ash is shown in Fig. 1. In general, it was determined that the waste brick dust had angular particles and fly ash had spherical particles. This also affects the fresh state and hardened properties of concrete.

Table 1 Oxide composition of WCBP and fly ash

Oxide composition (%)	WCBP [18]	Fly ash [19]
SiO_2	62.8	58.6
Al_2O_3	10.4	28.2
Fe_2O_3	16.3	3.4
CaO	1.7	2.2
MgO	2.2	0.3

(a) WCBP [17] (b) Fly ash [18]

Fig. 1 SEM images of WCBP and class F fly ash

Table 2 Typical range of SCC mix composition [20]

Constituent	Typical range by mass (kg/m^3)	Typical range by volume (liters/m^3)
Powder	300–600	–
Paste	–	300–380
Water	150–210	150–210
Coarse aggregate	750–1000	270–360
Fine aggregate (sand)	Content balance the volume of the other constituents, typically 48–55% of total aggregate weight	
Water/powder ratio by Vol		0.85–1.10

2.2 *Mix Proportions*

The mix proportions of self-compacting concrete mixtures to be produced in the study were designed according to the EFNARC [20] guideline. Table 2 presents the typical range of the SCC mix composition. It can be seen from Table 1 that there should be much more fine material content in SCC than in conventional concrete.

In this study, the mix design was calculated using the absolute volume method in accordance with the SCC criteria specified in EFNARC [20] Guideline in Table 3. The water to binder ratio of 0.40 (by weight) was considered for the mix design.

2.3 *Test Methods*

Slump flow test and V-funnel test were performed on fresh SCC mixtures according to the EFNARC [20] Guideline. Determination of compressive strength on hardened

Table 3 Mix proportions of SCC mixtures

Materials	kg/m^3
Binder (Cement + SCM)	490
Water	196
Fine aggregate (filler: 0–125 μ)	92
Fine aggregate (125μ-4 mm)	700
Coarse aggregate	801
Super plasticizer	8–10
Total	2289

concrete was carried out on samples with 150 × 300 mm cylindrical specimens according to ASTM C39 [21] Standard. Additionally, freeze–thaw test was carried out on 100 mm cubic concrete specimens according to ASTM C666 [22] Standard, by subjecting them to a total of 300 freeze–thaw cycles.

3 Results and Discussion

In this study, the fresh state properties of SCC mixtures were investigated. The slump flow values of all mixtures were produced to be 650 mm ± 20 mm. To achieve the desired slump spread values (as depicted in Fig. 2), varying quantities of superplasticizer admixtures were employed for each concrete mixture. It is noteworthy that in order to reach the target slump flow value, a higher quantity of admixtures was necessary for WCBP-based SCC mixtures compared to the control SCC mixture. In essence, the workability of the mixtures decreased as the use of WCBP in SCC mixtures increased. It is thought that one of the important results is that the void structure of WCBP is higher than that of cement. In the results of the slump flow test of fly ash-based SCC mixtures, less superplasticizer admixtures were needed than the control SCC mixture. Figure 1 shows that fly ash particles generally have a spherical structure. Therefore, it can be said that the use of spherical fly ash improves the workability of the concrete mixture. Moreover, as can be seen in Fig. 1, WCBP has a more angular shape than fly ash, and it is thought that an important factor reducing the workability of WCBP-based mixtures is due to this morphological structure.

The results of the V-funnel flow time test of SCC mixtures are shown in Fig. 3. The V-funnel flow time test allows us to get an idea about the viscosity and flow properties of SCC mixtures [20]. Mixtures with the same slump flow value may have different viscosities. As a matter of fact, the experimental results confirm this situation. A significant increase in the V-funnel flow time was observed as a result of the use of WCBP in SCC mixtures. Furthermore, the viscosity of the concrete mixtures increased by increasing of the superplasticizer's dosage. Although a higher V-funnel flow time value was obtained for the fly ash-based mixtures, these values remained at a very low level compared to WCBP-based SCC mixtures. It can be

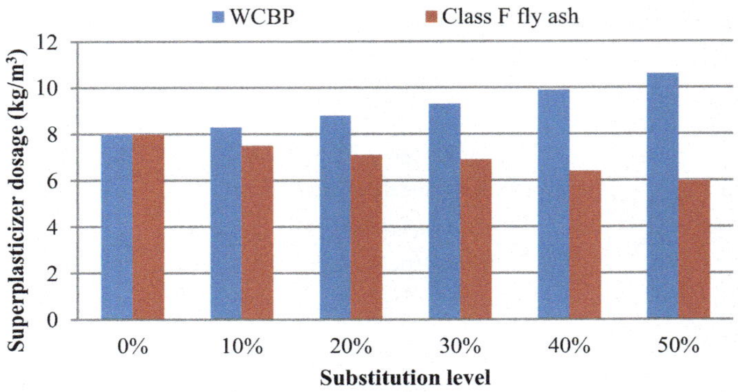

Fig. 2 Superplasticizer dosage with the substitution level of SCM's

noted that the V-funnel flow time for the fly ash-based mixtures was recorded longer than the control mixtures, although less superplasticizer was used. This shows that the cohesiveness of the SCC mixtures increases as a result of using fly ash in the concrete mixtures.

The 28-day compressive strength results of SCC mixtures are presented in Fig. 4. In general, a progressive reduction in compressive strength was observed as the use of WCBP in SCC mixtures increased. Prior research on the use of mineral admixtures as cement replacement has attributed this gradual decline in compressive strength to the dilution effect on the binder product (CSH) as the amount of mineral admixture increases [15]. The potential of WCBP to show pozzolanic activity due to its chemical structure (high SiO_2 ratio) is also known from the literature [16]. However, it has been determined that the substitution of more than 30% WCBP instead of cement significantly reduces the compressive strength of the SCC mixtures. It is known that the contribution of pozzolanic materials to the strength of concrete at later ages is

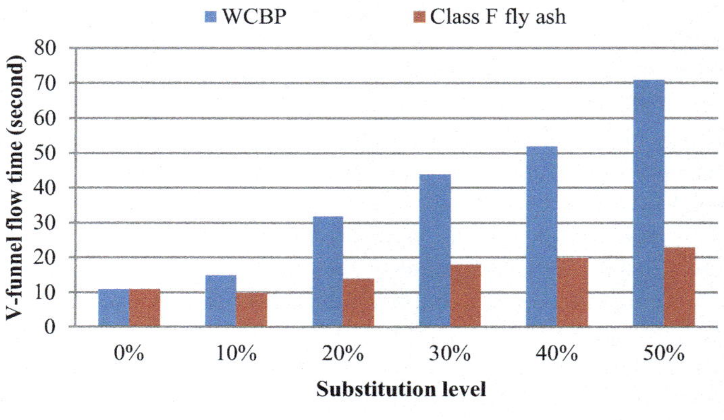

Fig. 3 V-funnel flow time test results of SCC mixtures incorporating SCM

higher than that of early age strength [23, 24]. It is predicted that this effect will occur as a result of using WCBP instead of cement in SCC mixtures. As a result of the use of fly ash in SCC mixtures, there was a decrease in strength compared to the control SCC mixture. According to the test results, the compressive strength of the fly ash-based SCC mixtures was higher than the WCBP-based SCC mixtures. It can be attributed to the higher chemical reaction between reactive silica and alumina present in the fly ash particles and portlandite.

The freeze–thaw resistance test results of SCC mixtures are shown in Fig. 5. It was determined that the weight loss resulting from the freeze–thaw effect of both WCBP and fly ash-based SCC mixtures was less than that of the control SCC mixture. When the mixtures incorporating WCBP and fly ash are compared, it has been observed that the weight loss of the mixtures with WCBP is less than the mixtures with fly ash mineral admixture. It is considered that the void structure of WCBP may slightly act as an air-entraining admixture.

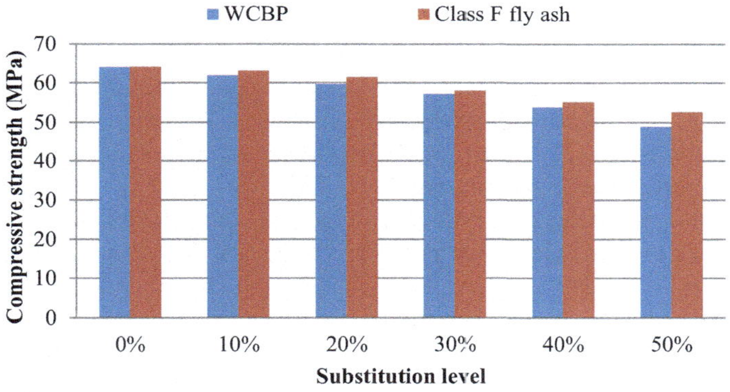

Fig. 4 Compressive strength of SCC mixtures incorporating SCM

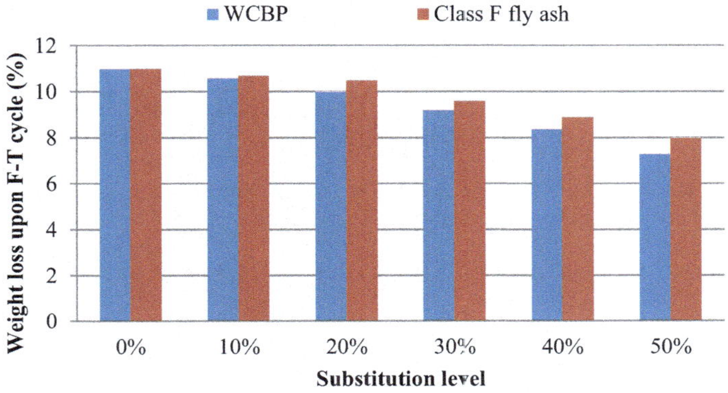

Fig. 5 Freeze–thaw resistance of SCC mixtures incorporating SCM

4 Conclusions

In this study, the effect of waste clay brick as supplementary cementitious material on the fresh state properties, compressive strength, and freeze–thaw resistance of self-compacting concrete mixtures were studied and the following conclusions were drawn:

1. In order to obtain a constant slump flow value in SCC mixtures, different amounts of superplasticizer admixtures were used for each SCC series. Notably, a higher dosage of superplasticizer was required in SCC mixtures containing waste clay brick powder in comparison to the control SCC mixture and fly ash-based SCC mixtures.
2. According to the V-funnel flow time test results, the viscosity of the SCC mixtures increased as the use of waste clay brick increased in the mixtures containing waste clay brick.
3. The fresh state performance of waste clay brick-based SCC mixtures was found to be inferior to both the control SCC mixture and fly ash-based SCC mixtures.
4. Compressive strength tests revealed that SCC mixtures containing waste brick powder exhibited lower compressive strength values when compared to the fly ash-based SCC mixtures. This difference can be attributed to the increased chemical reactivity between the reactive silica and alumina present in fly ash particles and the portlandite in the mixture.
5. It was determined that the freeze–thaw resistance of the mixtures improved as the usage rate of waste clay brick powder as a mineral admixture in SCC mixtures increased.

References

1. Okamura, H., Ozawa, K.: Self-compacting high performance concrete. Struct. Eng. Int. **6**(4), 269–270 (1996)
2. Domone, P.L.: A review of the hardened mechanical properties of self-compacting concrete. Cem. Concr. Res. **29**(1), 1–12 (2007)
3. Şahmaran, M., Christianto, H.A., Yaman, İÖ.: The effect of chemical admixtures and mineral additives on the properties of self-compacting mortars. Cement Concr. Compos. **28**(5), 432–440 (2006)
4. Yamada, K., Takahashi, T., Hanehara, S., Matsuhisa, M.: Effects of the chemical structure on the properties of polycarboxylate-type superplasticizer. Cem. Concr. Res. **30**(2), 197–207 (2008)
5. Domone, P.L.: Self-compacting concrete: an analysis of 11 years of case studies. Cement Concr. Compos. **28**(2), 197–208 (2006)
6. Uysal, M., Yilmaz, K.: Effect of mineral admixtures on properties of self-compacting concrete.Cement Concrete Compos. **33**(7), 771–776 (2011)
7. Le, H.T., Müller, M., Siewert, K., Ludwig, H.-M.: The mix design for self-compacting high-performance concrete containing various mineral admixtures. Mater. Des. **72**, 51–62 (2015)

8. Maghfouri, M., Alimohammadi, V., Azarsa, P., Asadi, I., Doroudi, Y., Balakrishnan, B.: Impact of fly ash on time-dependent properties of agro-waste lightweight aggregate concrete. J. Compos. Sci. **5**(6), 156 (2021)
9. Rizwan, S.A., Bier, T.A.: Self-compacting mortars using various secondary raw materials. ACI Mater. J. **106**(1), 25–32 (2009)
10. Zengfeng, Z.H.A.O., Courard, L., Groslambert, S., Jehin, T., Leonard, A., Jianzhuang, X.I.A.O.: Use of recycled concrete aggregates from precast block for the production of new building blocks: an industrial scale study. Resour. Conserv. Recycl. **157**, 104786 (2020)
11. Zhao, Z., Remond, S., Damidot, D., Courard, L., Michel, F.: Improving the properties of recycled concrete aggregates by accelerated carbonation. Proc. Inst. Civ. Eng.-Construct. Mater. **171**(3), 126–132 (2018)
12. Liu, Q., Tong, T., Liu, S., Yang, D., Yu, Q.: Investigation of using hybrid recycled powder from demolished concrete solids and clay bricks as a pozzolanic supplement for cement. Constr. Build. Mater. **73**, 754–763 (2014)
13. Wong, C.L., Mo, K.H., Yap, S.P., Alengaram, U.J., Ling, T.-C.: Potential use of brick waste as alternate concrete-making materials: a review. J. Clean. Prod. **195**, 226–239 (2018)
14. Heikal, M., Zohdy, K.M., Abdelkreem, M.: Mechanical, microstructure and rheological characteristics of high performance self-compacting cement pastes and concrete containing ground clay bricks. Constr. Build. Mater. **38**, 101–109 (2013)
15. Naceri, A., Hamina, M.C.: Use of waste brick as a partial replacement of cement in mortar. Waste Manage. **29**(8), 2378–2384 (2009)
16. Lin, K.-L., Chen, B.-Y., Chiou, C.-S., Cheng, A.: Waste brick's potential for use as a pozzolan in blended Portland cement. Waste Manage. Res. **28**(7), 647–652 (2010)
17. Olofinnade, O.M., Ede, A.N., Ndambuki, J.M., Bamigboye, G.: Structural properties of concrete containing ground waste clay brick powder as partial substitute for cement. Mater. Sci. Forum **866**, 63–67 (2016)
18. Zhao, Z., Grellier, A., El Karim Bouarroudj, M., Michel, F., Bulteel, D., Courard, L.: Substitution of limestone filler by waste brick powder in self-compacting mortars: properties and durability. J. Build. Eng. **43**, 102898 (2021)
19. Siddique, R.: Properties of self-compacting concrete containing class F fly ash. Mater. Des. **32**(3), 1501–1507 (2011)
20. European SCC Guidelines.: Specifications and guidelines for self-compacting concrete. EFNARC, Association House, 99 West Street, Farnham, UK (2005).
21. ASTM C39.: standard test method for compressive strength of cylindrical concrete specimens. Annual Book of ASTM Standard, USA (2021)
22. ASTM C666.: Standard test method for resistance of concrete to rapid freezing and thawing. Annual Book of ASTM Standard, USA (2016)
23. Mehta, P.K., Monteiro, P.J.M.: Concrete: Microstructure, Properties, and Materials. U.S.A., The McGraw-Hill Companies, New York (2006)
24. Tuyan, M., Andiç-Çakir, Ö., Ramyar, K.: Effect of alkali activator concentration and curing condition on strength and microstructure of waste clay brick powder-based geopolymer. Compos. B Eng. **135**, 242–252 (2018)
25. Gökçe, H.S., Tuyan, M., Ramyar, K., Nehdi, M.L.: Development of eco-efficient fly ash–based alkali-activated and geopolymer composites with reduced alkaline activator dosage. J. Mater. Civ. Eng. **32**(2), 04019350 (2020)
26. Khalaf, F.M., DeVenny, A.S.: Properties of new and recycled clay brick aggregates for use in concrete. J. Mater. Civ. Eng. **17**(4), 456–464 (2005)

Self-compacting Concrete Using Eggshell Ash and Rice Husk Ash as Partial Cement Replacement

N. M. Husain, Y. Yajid, S. N. C. Deraman, S. A. Masjuki, S. A. Saad, S. L. Ibrahim, N. K. Basri, and M. Taib

Abstract The usage of self-compacting concrete (SCC) is highly recommended in heavy construction as it helps to ease concreting process during construction and improves productivity, energy consumption, as well as the working environment. This research study is to show the efficiency of self-compacting concrete made of waste products as cement replacement. Concrete structure is becoming a vital element in the construction industry, hence the increase in cement demand. As proven in various studies, cement production leads to environmental pollution. Thus, a new material made up of waste material is proposed as a partial cement replacement in concrete. The percentage of cement content in SCC will be reduced by replacing it with eggshell ash (ESA) and rice husk ash (RHA) in the mix design mainly to improve the concrete strength. Thus, the optimum percentage of RHA and ESA will be determined throughout this research. Several tests were conducted to test

N. M. Husain (✉) · Y. Yajid · S. N. C. Deraman · S. A. Masjuki · S. A. Saad · S. L. Ibrahim · N. K. Basri
Cataclysmic Management and Sustainable Development (CAMSDE) Research Group, Department of Civil Engineering, Kulliyyah of Engineering, International Islamic University Malaysia, 53100 Kuala Lumpur, Malaysia
e-mail: drnadiah@iium.edu.my

S. N. C. Deraman
e-mail: snoratikah@iium.edu.my

S. A. Masjuki
e-mail: aliyyah@iium.edu.my

S. A. Saad
e-mail: asmahanisaad@iium.edu.my

S. L. Ibrahim
e-mail: saerahany@iium.edu.my

N. K. Basri
e-mail: nurkhairiyahbasri@iium.edu.my

M. Taib
School of Housing, Building and Planning, Universiti Sains, Penang, Malaysia
e-mail: mariati@usm.my

N. Sabtu (ed.), *Proceedings of AWAM International Conference on Civil Engineering 2022—Volume 2*, Lecture Notes in Civil Engineering 385,
https://doi.org/10.1007/978-981-99-6018-7_34

the mechanical and physical properties of the specimen and comparing with the SCC control. It was proven that SCC with 3% ESA 8% RHA produced maximum strength compared to the control and other design mixes with 0%, 6%, and 9% ESA, respectively. The increment of 14.31% compressive strength on the 28th day compared to the control mix has created the optimum silica and calcium hydroxide reaction for the hydration process.

Keywords Self-compacting concrete · ESA · RHA · Mineral admixture

1 Introduction

Self-compacting concrete (SCC) or also known as self-consolidation concrete is a type of concrete that can be placed easily and consolidated by its gravity in a formwork. Using SCC gives benefits on fast construction, noise reduction, energy effectiveness, and good formability. Thus, it is an example of technological innovation to produce new types of structural elements with more complex shapes, which is probably not possible to be produced traditional concrete. Based on this theme, this research will elaborate more on the new material innovation from SCC by using the waste products which are eggshell ash (ESA) and rice husk ash (RHA). Recent studies proved that the combination of these materials may produce more eco-friendly concrete, low cost, and has high early strength. RHA which is rich in silica and eggshells contain identical amount of calcium oxide in cement resulting in the formation of calcium silicates which are responsible for achieving higher strength [1].

Self-compacting concrete is one of the advanced cementitious composites that archive the durability problem of concrete in construction work. However, this type of concrete costs much higher than regular concrete as SCC requires higher powder and chemical admixtures content. Besides, as the construction industry grows vigorously, the cement demand also increases by days. Meanwhile, the production of cement creates pollution to the environment such as cement dust, air pollution, waste pollution, water pollution, ground vibration, and resource depletion due to raw material extraction and noise pollution. The production of cement needs a huge quantity of energy and emits a large amount of carbon dioxide gas (CO_2) due to the process of calcination. This process releases 0.53 tons of CO_2 into the atmosphere to process 1 ton of Portland cement [2]. Moreover, according to [3], the production of the world's cement in 2007 was estimated at 2.77 billion tons. Therefore, in this research, RHA and ESA which are local powders could be new innovative materials that can reduce the cost and cement usage in SCC production and at the same time reduce the greenhouse gas effect caused by cement production. Replacing cement with RHA and ESA which are environmentally friendly and efficient, by-product materials not only ensure greener construction practices but also improve SCC performance and at the same time is considered an effective utilization and management of these wastes and ensures economic benefits.

As stated by [4], eggshell consists of calcium compounds which are very similar to cement, where 93.7% is of calcium carbonate content. Hence, eggshells can become an alternative for cement replacement in a concrete mix due to the similar calcium carbonate content in eggshells and cement. Calcium carbonate provides a good accelerator as it contains extra calcium oxide. Calcium oxide also enhanced the setting time. Short setting time duration is very desirable in construction as unpredictable weather can interrupt the construction process on-site. According to [5], a combination of eggshell powder and RHA was confirmed to have equal strength with conventional concrete. Hence, RHA which acts as an admixture will be added to the concrete mix. Besides, [4] stated that RHA and ESA were utilized as partial cement replacements, and it was shown that when mixed with cement, the duo had strength that is comparable to conventional concrete in mixtures. The pozzolanic reactivity of cement can be enhanced with the help of RHA [6]. In terms of chemical reaction, the silica in RHA will react with the excess calcium hydroxide created by the cement-eggshell ash hydration process to form extra calcium silicate hydrates, which is critical for strength growth. The suggested equation for the chemistry between silica and calcium hydroxide, the primary product of cement and eggshell ash hydration, is as follows:

$$3ca(OH)_2 + 3SiO_2 \rightarrow 3CaO.SiO_2.H_2O$$

The main purpose of this research study is to show the efficiency of SCC with low cement content with the usage of eco-friendly material. The specific objectives of this research are to determine the optimum percentage of RHA and ESA content as partial cement replacement and to evaluate the physical and mechanical properties of self-compacting concrete by comparing the performance of this concrete using these two materials with conventional concrete. In order to measure its compressive strength, an SCC mixture encompassing eggshell powder was developed and evaluated.

2 Methodology

2.1 *Materials Selection*

This research study is conducted hands-on in producing self-compacting concrete with additional mineral materials as cement replacement. The laboratory experiment is carried out using the main ingredient of concrete mix design which are fine aggregate, coarse aggregate, ordinary Portland cement, water, superplasticizer (Sp), and viscosity-modifying admixture. The additional mineral materials are ESA and RHA which are used as a partial substitute for cement. Several tests were conducted such as a sieve analysis test for the coarse and fine aggregate, fresh SCC tests; slump flow

Table 1 Mixture proportion of concrete with glass powder as cement replacement

Concrete mix	w/c	Eggshell Ash (kg)	RHA (kg)	Cement (kg)	Water (kg)	Fine agg. (kg)	Coarse agg. (kg)	Sp (kg)	VMA (kg)
Normal SCC (control)	0.4	–	–	8.20	3.30	21.90	14.30	0.15	0.0165
SCC 1 (3% eggshell 8% RHA)	0.4	0.25	0.70	7.30	3.30	21.90	14.30	0.15	0.0165
SCC 2 (6% eggshell 8% RHA)	0.4	0.50	0.70	7.10	3.30	21.90	14.30	0.15	0.0165
SCC 3 (9% eggshell 8% RHA)	0.4	0.75	0.70	6.80	3.30	21.90	14.30	0.15	0.0165
Total mix		1.50	2.10	29.40	13.20	87.60	57.20	0.60	0.066

test and T500 flow test. While for the hardened concrete, tests were the compressive strength test, ultrasonic pulse velocity (UPV) test, and the rebound hammer test. The eggshells are collected and incinerated into ash at a temperature of about 100 °C, ground and sieved through a 75 μm sieve. Meanwhile, the RHA is obtained from a supplier which has been burnt through the combustion process. A normal self-compacting concrete mix is placed in 150 × 150 × 150 mm cube mold and observed the strength at 3-, 7-, and 28-days curing. Meanwhile, mixes of SCC with 8% of RHA and different percentages of ESA content as partial cement replacement are prepared using the same mold size and same curing days. Then, the properties of self-compacting concrete with different replacement quantities are evaluated and compared to conventional concrete in terms of physical and mechanical properties. Finally, the optimum quantity of RHA and ESA is suggested and discussed according to the results obtained.

For producing the SCC, the mixtures were designed according to American Concrete Institute. The characteristic compressive strength of the SCC designed was 45 N/mm^2 on the 28th day. The mix design is presented in Table 1.

2.2 *Fresh Concrete Test*

The slump test was performed to determine the consistency of the fresh concrete. It was done in accordance with the ASTM C 1611/C1611M, 2009. The fresh concrete was tested for slump flow test to measure the workability in terms of the filling ability of the fresh concrete and consistency of the fresh concrete mix produced before it is set. The T500 test on the other hand is the same as the slump flow test. The difference is only at taking the time taken for the concrete to spread until the diameter reaches 500 mm using a stopwatch.

2.3 Hardened Concrete Test

Two non-destructive tests (NDT) were done before the actual compressive strength test. Using UPV test can assess the strength of the concrete and detect any presence of damage mechanisms such as cracks. It was done in accordance with BS 1881-203:1986. The method is based on the propagation of a high-frequency sound wave that passes through the concrete. The speed of the wave is then measured to know any discontinuity in cross-sections such as cracks and cover concrete delamination. The rebound hammer test or also called Schmidt hammer is a non-destructive concrete testing procedure that provides a convenient and simple indicator of the concrete's compressive strength from the value of the rebound number obtained. It was done in accordance with BS 12504-2:2012. After conducting the NDT test, these 24 cubes, 150 × 150 × 150 mm hardened concrete samples, will be tested for compressive strength test after 3rd, 7th, and 28th days of curing. The compressive strength of any material could be defined as the resistance to failure under any action of compressive forces such as crack or deflection. The result obtained indicating the compressive strength of concrete is an important parameter to determine the performance of the material during service conditions. It was done in accordance with BS 1881-116:1983.

3 Results and Discussion

3.1 Sieve Analysis

Based on the result obtained, a grading curve of coarse aggregate can be plotted as shown in Fig. 1. As the coefficient of gradation, Cc obtained is 1.22 which is in between 0.5 and 2.0, this coarse aggregate sample is indicated as a well-graded soil. A well-graded aggregate has a particle size gradation that spans the size range from the smallest to the largest. The arrangement of the particles becomes more compact and creates only a small number of voids. The lower number of voids creates denser particles that can resist high force as this concrete mixture has high strength against normal force.

From the particle size distribution curve for fine aggregate in Fig. 2, the value of Cu obtained is 4.62 which is less than 6, and Cc, 0.98 which is less than 1; hence, this fine aggregate or usually called sand can be categorized as SP group that containing

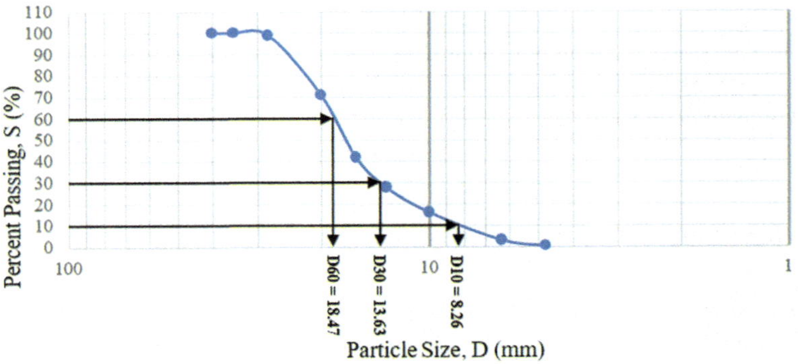

Fig. 1 Particle size distribution curve for coarse aggregate

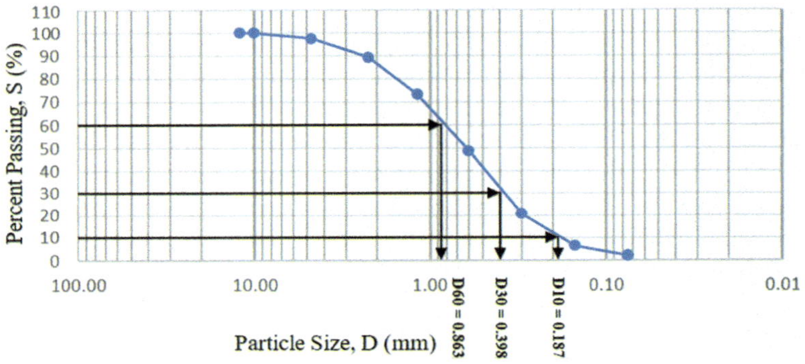

Fig. 2 Particle size distribution curve for fine aggregate

less than 5% fines which means poorly graded sands, gravelly sands, little or no fines according to Unified Soil Classification System. SCC can be used with any regular concreting sand. Crush and rounded sands are also acceptable options [7]. Hence, this type of fine aggregate is good to be used for SCC purposes.

3.2 Slump Values

A slump flow test has been carried out in this project to study the effect of RHA and ESA used in SCC on the filling and passing ability and also segregation resistance of the SCC. The result of the bar chart in Fig. 3 illustrates that the combination of 8% of RHA and 9% of ESA (SCC 3) has the same flowability and filing ability as the 100% cement content of SCC. Besides, increasing the ESA dosage has increased the flowability of the SCC. This is due to the fact that higher ESA content has slowed down the initial setting time; thus, the SCC mixture can retain in a fresh mix state

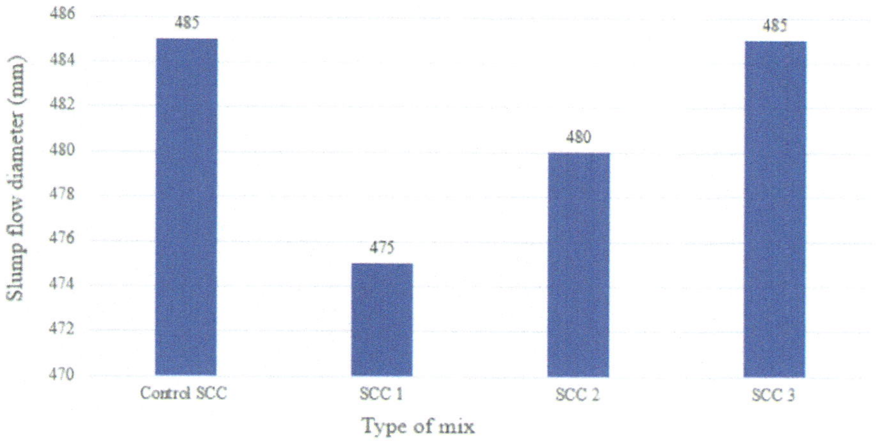

Fig. 3 Slump flow diameter of SCC mixes with different percentages of eggshell ash

and has better liquidity [8]. Higher dosages of ESA also reduce the friction resistance between all powders and coarse aggregates [9].

3.3 T500 Slump Flow

A test called T500 mm slump flow has also been carried out to evaluate the flowability, passing ability, or viscosity and also filling ability of the SCC mixture produced. Time taken for the SCC mixes to spread to the diameter of 500 mm has been plotted in Fig. 4. SCC 3 with 8% RHA and 9% ESA produced the lowest time of flow to reach 500 mm diameter. The trend also illustrates that the higher the percentage of ESA used in SCC, the shorter the slump flow time. The viscosity of the fresh mix is dependent on the time T500. The shorter the time T500, the higher the flow rate of SCC during placement [10]. Hence, this result indicates that 9% ESA and 8% RHA used to replace cement produced better flowability and filling ability than normal SCC which contained 100% cement.

3.4 Ultrasonic Pulse Velocity

The results of time taken for an ultrasonic pulse passing through the SCC from three different sides were summed and recorded on average according to their curing ages and plotted as in Fig. 5. The higher the ESA percentage used in SCC, the lower the UPV obtained on day 28 of concrete age. On all three ages of cubes, it can be seen clearly that SCC 1 with 3% of ESA and 8% of RHA achieved the highest ultrasonic

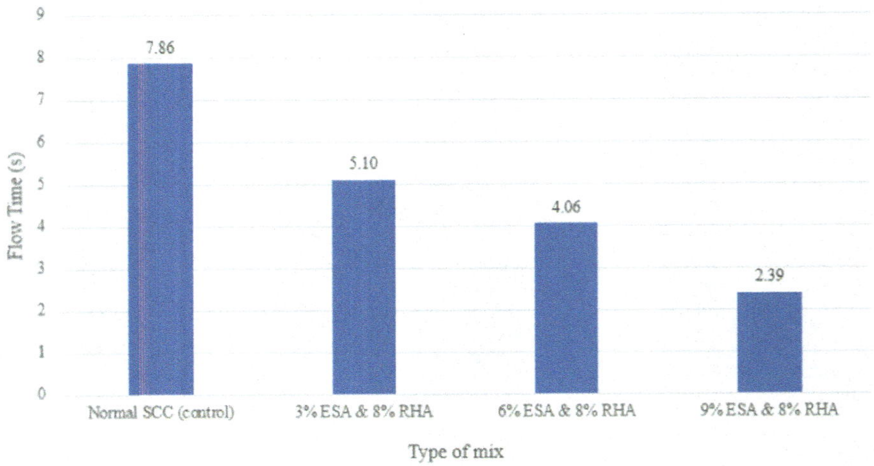

Fig. 4 Flow time of SCC mixes with different percentages of eggshell ash

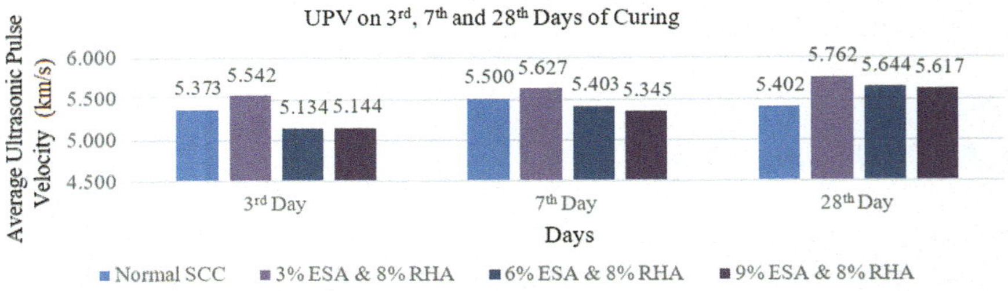

Fig. 5 Average ultrasonic pulse velocity with different percentages of eggshell ash

pulse velocities which are 5.54, 5.63, and 5.76 km/s than others. According to Gehlot et al. [11], the quality of concrete can be classified into four ranges based on the pulse velocity which are excellent, good, medium, and doubtful. From the data collected throughout this experiment, all the SCC with different proportions of ESA and fixed RHA belongs to the excellent concrete quality since all pulse velocity is above 4.5 km/s. This shows that all SCC mixtures produced have high internal structure packing densities and low voids content.

3.5 Rebound Hammer

The rebound number indicates the concrete's compressive strength. The average rebound number and estimated compressive strength obtained from the digital rebound hammer on the 3rd, 7th, and 28th day of curing are illustrated in Fig. 6.

Based on the result obtained, all types of SCC mixtures have rebound numbers in the range of 18–60 in all three days of SCC ages. According to Mishra and Siddiqui [12], concrete quality can be classified into five types based on the average rebound number obtained. From the result of the average rebound number obtained on day 28, it can be interpreted that all types of SCC mixtures have a very good hard layer as the rebound numbers were more than 40. However, among all types of mixtures, SCC 1 with 3% of ESA and 8% of RHA has the highest average rebound numbers.

From the chart shown in Fig. 7, the estimated compressive strength obtained from the digital hammer shows that all of the mixture except for SCC 2 with 6% of ESA have achieved the specified strength needed for the mix which is 45 MPa. The highest compressive strength on the 28th day is 58.7 MPa, followed by 57.1 MPa, 49.2 MPa, and 44.3 MPa on the SCC that were incorporated with 3%, 0%, 9%, and 6% of eggshell ash, respectively.

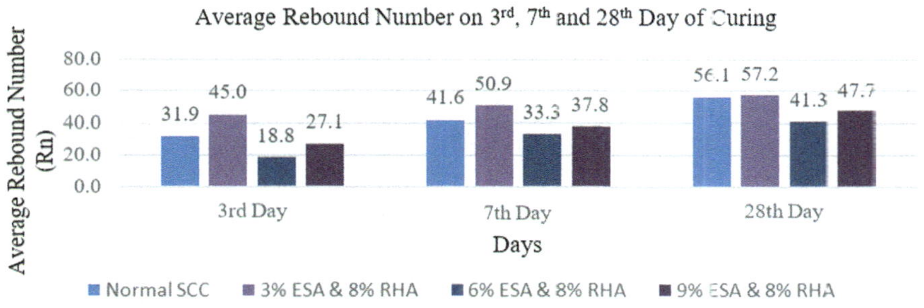

Fig. 6 Average rebound number

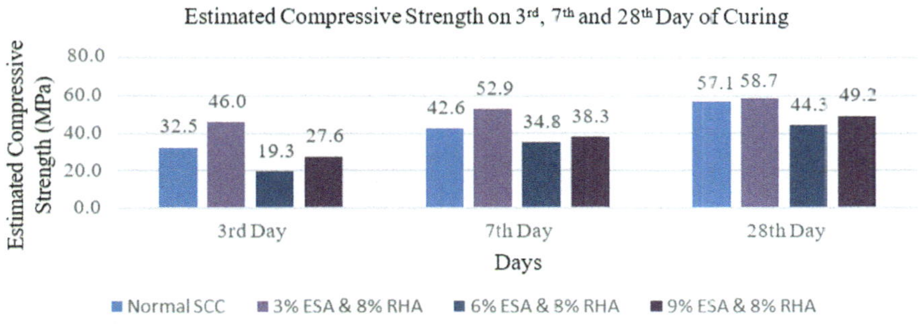

Fig. 7 Estimated compressive strength

3.6 *Compressive Strength*

In this project, two SCC cubes for each type of mixture have been prepared to evaluate the compressive strength for the 3rd, 7th, and 28th days of curing. The average values of every SCC mix have been calculated and simplified according to the test days as in Fig. 8. Due to RHA's pozzolanic nature and lack of calcium, an attempt to utilize eggshell ash in addition with RHA as cement replacement in concrete was done [1].

From the result obtained, using 8% RHA with increasing percentages of eggshell ash, reducing the SCC compressive strength on the 28th day or in other words, the higher the percentage of eggshell ash used, the lower the SCC strength on the 28th day. From the SCCs that were incorporated with the RHA and eggshell ash, SCC 1 with 8% RHA with 3% eggshell replacement has the best result and any further replacement has yielded a lower compressive strength. Besides, this mixed batch created better strength development than the control SCC which is without any mineral admixture replacement with a 14.31% improvement [13].

The presence of calcium carbonate ($CaCo_3$) in ESA, when mixed with cement, acts as an excellent accelerator in the mix probably the major reason for the rise in strength development. An increasing amount of ESA has not increased the concrete strength which might be due to an unsuitable amount of high calcium in the eggshell to react effectively with cement and the fixed RHA. From the results achieved, it is proven that it is possible to use both materials together and substitute some amount of cement in producing green concrete. To summarize all the data that has been presented for the compressive strength test, SCC 1 with 8% of RHA and 3% ESA yield the best compressive strength.

Correlation Between Rebound Hammer Test and Compressive Strength Test.

A relationship has been established and demonstrated using data acquired from samples used for concrete strength and rebound hammer testing on the 3rd, 7th, and

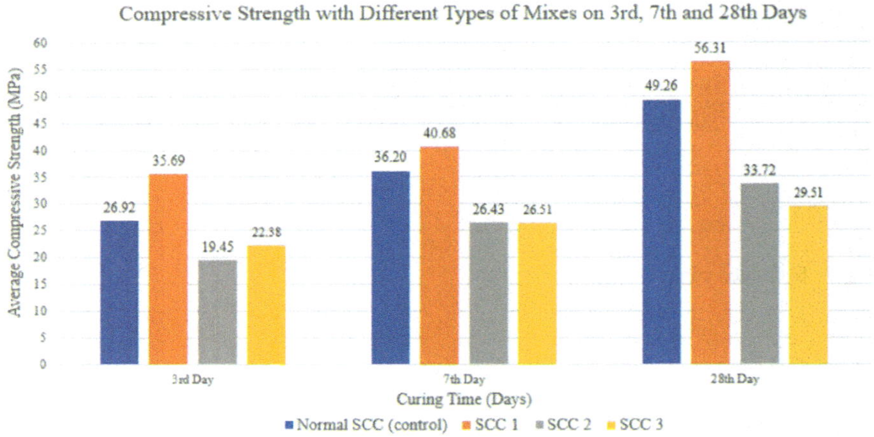

Fig. 8 Compressive strength development

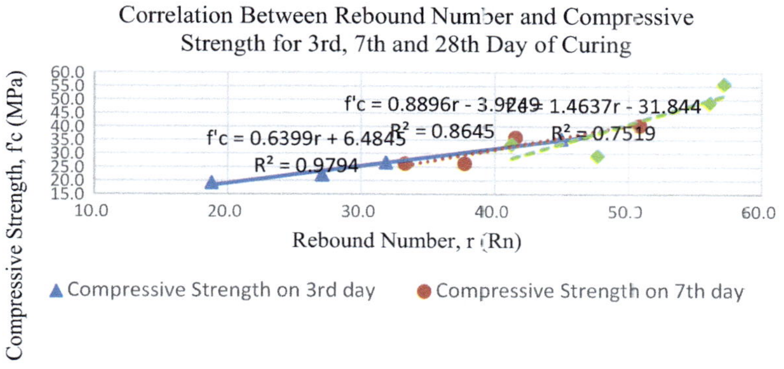

Fig. 9 Rebound hammer and compressive strength correlation

28th days of curing. From the linear plot, as shown in Fig. 9, the R2 value for samples on the 3rd day of curing is 0.9794, using the equation f′c = 0.6399r + 6.4845 (where f′c = compressive strength and r = rebound number). Meanwhile, on the 7th day, the equation obtained is f′c = 0.8896r - 3.9249 with an R2 value of 0.8645. On the 28th day of curing, the R2 value of the samples was found to be 0.7519, with the equation f′c = 1.4637r – 31.844. All three equations further confirmed that the link between the compressive strength test and the rebound hammer test is acceptable since all of the R2 (correlation coefficient) values were near one (1).

3.7 *Correlation Between Compressive Strength and UPV*

The linear regression was applied to establish the relationship between compressive strength and UPV. Figure 10 shows the compressive strength versus UPV relationship at the age of 3rd, 7th, and 28th days. The correlation equations for the graph of the 3rd day is f′c = 35.119 V – 159.96, while for the 7th day is f′c = 55.828 V – 272.87, and for the 28th day is f′c = 4.6365 V + 16.204 where f′c represents the compressive strength and V represents ultrasonic pulse velocity. Their R2 coefficients of determination are 0.9472, 0.9282, and 0.003, respectively, demonstrating a strong correlation between data points and regression curves on SCCs age 3rd and 7th days while poor relationship on the age of 28th day. The relationship between the compressive strength test and UPV test on day 3 and 7 of SCC ages is acceptable since both R2 (correlation coefficient) is close to 1.

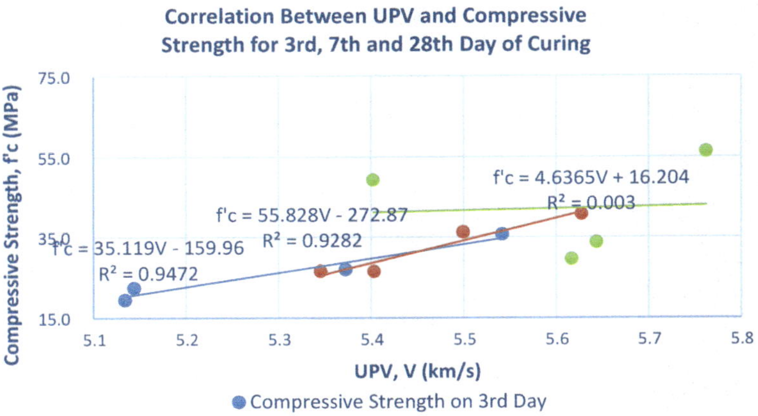

Fig. 10 Compressive strength and UPV correlation

4 Conclusion

The main objective which is to determine the optimum percentage of RHA and ESA content as partial cement replacement has been accomplished along with all other objectives. SCC 3 with 9% of ESA and 8% of RHA showed the same flowability as the normal SCC with 0% cement replacement. In terms of concrete strength, SCC 1 with 3% of ESA with 8% RHA substitution showed the maximum compressive strength. Furthermore, the satisfactory strength of SCC-type replacement outperformed the control sample strength. SCC 1 containing 3% of ESA with 8% of RHA was found to produce the maximum strength of self-compacting concrete, producing a strength gain of 14.31% at 28 days of curing. This shows that the combination of 8% RHA and 3% ESA created optimum silica and calcium hydroxide reaction for the hydration process which was crucial for strength development. Furthermore, from the non-destructive tests conducted, the ultrasonic pulse velocity and rebound number showed the same outcome, where the SCC 1 with 3% of ESA with 8% RHA produced the best performance in terms of high internal structure packing densities, low voids content, and highest concrete hardness. The correlations between compressive strength and UPV and between compressive strength and rebound number have been established in which the correlation coefficients obtained are close to 1. Overall, the results collected from this project suggested that a combination of RHA and ESA gives a good alternative for cement in SCC production.

References

1. Jhatial, A.A., Goh, W.I., Mo, K.H., Sohu, S., Bhatti, I.A.: Green and sustainable concrete—the potential utilization of rice husk ash and egg shells. Civil Eng. J. **5**(1) (2019). www.CivileJournal.org
2. Zainudeen, N., Jeyamathan, J.: Cement and its effect on the environment: a case study in Sri Lanka. In: Proceedings from International Conference on Building Education and Research (BEAR), pp. 1408–1416 (2008). http://www.irbnet.de/daten/iconda/CIB11336.pdf
3. Sakai, E., Yamada, K., Ohta, A.: Molecular structure and dispersion adsorption mechanisms of comb-type superplasticizers used in Japan. J. Adv. Concr. Technol. **1**(1), 16–25 (2003)
4. Okonkwo, U.N., Odiong, I.C., Akpabio, E.E.: The effects of eggshell ash on strength properties of cement-stabilized lateritic. Int. J. Sustainable Construction Eng. Technol. **3**(1), 18–25. ISSN: 2180-3242
5. Afizah Asman, N.S., Dullah, S., Lynn Ayog, J., Amaludin, A., Amaludin, H., Lim, C.H., Baharum, A.: Mechanical properties of concrete using eggshell ash and rice husk ash as partial replacement of cement. MATEC Web Conf. **103**, 4–9 (2017). https://doi.org/10.1051/matecconf/201710301002
6. Saad, S.A., Nuruddin, M.F., Shafiq, N., Ali, M.: Pozzolanic reaction mechanism of rice husk ash in concrete—a review. Appl. Mech. Mater. **773–774**, 1143–1147 (2015). https://doi.org/10.4028/www.scientific.net/amm.773-774.1143
7. EFNARC. (2002). Specification and Guidelines for Self-Compacting Concrete. Report from EFNARC,44(February),32. http://scholar.google.com/scholar?hl=en&btnG=Search&q=intitle:Specification+and+Guidelines+for+Self-Compacting+Concrete#0
8. Neville, A.M.: Properties of Concrete (4th ed.): Pearson Education Ltd. (2005)
9. Pan, S., Chen, D., Chen, X., Ge, G., Su, D., Liu, C.: Experimental study on the workability and stability of steel slag self-compacting concrete. Appl. Sci. (Switzerland), **10**(4). (2020). https://doi.org/10.3390/app10041291
10. Schutter, G.D., Bartos, P.J.M., Domone, P., & Gibbs, J.: Self-compacting concrete (2008)
11. Gehlot, T., Sankhla, D.S.S., Gehlot, D.S.S., Gupta, A.: Study of concrete quality assessment of structural elements using ultrasonic pulse velocity test. IOSR J. Mech. Civ. Eng. **13**(05), 15–22 (2016). https://doi.org/10.9790/1684-1305071522
12. Mishra, S., & Siddiqui, N.A.: A Review On Environmental and health impacts of cement manufacturing emissions. Int. J. Geol. Agric. Environ. Sci. **2**(3 June 2014), 2 (2014). www.woarjournals.org/IJGAES
13. Ujin, F., Ali, K.S., Harith, Z.Y.H.: Influence of addition eggshells ash as partial replacement cement on the durability of concrete. J. Eng. Appl. Sci. **13**(4), 809–812 (2018). https://doi.org/10.3923/jeasci.2018.809.812

Effect of Pressure Coefficient on a Rural House Due to the Neighboring Building's Arrangement Using Low-Speed Wind Tunnel Test

S. N. C. Deraman, M. A. A. Mohamad Zaki, N. M. Husain, S. A. Saad, S. A. Masjuki, N. K. Basri, and S. L. Ibrahim

Abstract Most of the rural houses in Malaysia especially in the Northern region of Peninsula Malaysia experienced severe damage during windstorm events due to the lack of engineering consideration. Moreover, these types of houses are also mostly made from timber or a combination of concrete and wood. Therefore, the study on the effects of wind flow surrounding the rural houses as the low-rise building becomes the focus of the present study. Generally, rural houses will be surrounded by other structures such as houses and trees. Thus, the main objective of this study is to analyze the wind tunnel simulation of wind pressure on a rural house due to the effect of the surrounding building's arrangements. The rural house model that has been scaled down has been tested in the wind tunnel to simulate the wind flow toward the rural house, resulting in pressure distribution and flow visualization surrounding the house. By comparing the value of pressure coefficient (C_P) and the flow visualization, this study comes out with the most suitable location and arrangement of neighboring buildings for the rural house which is Case 2 with $C_P = 2.179$ for the highest pressure

---

S. N. C. Deraman (✉) · M. A. A. Mohamad Zaki · N. M. Husain · S. A. Saad · S. A. Masjuki · N. K. Basri · S. L. Ibrahim
Cataclysmic Management and Sustainable Development (CAMSDE) Research Group, Department of Civil Engineering, Kulliyyah of Engineering, International Islamic University Malaysia, Jalan Gombak, 53100 Gombak, Selangor, Malaysia
e-mail: snoratikah@iium.edu.my

N. M. Husain
e-mail: drnadiah@iium.edu.my

S. A. Saad
e-mail: asmahanisaad@iium.edu.my

S. A. Masjuki
e-mail: aliyyah@iium.edu.my

N. K. Basri
e-mail: nurkhairiyahbasri@iium.edu.my

S. L. Ibrahim
e-mail: saerahany@iium.edu.my

N. Sabtu (ed.), *Proceedings of AWAM International Conference on Civil Engineering 2022—Volume 2*, Lecture Notes in Civil Engineering 385,
https://doi.org/10.1007/978-981-99-6018-7_35

and C_P = -1.997 for the highest suction that occurred at the roof house. It can be concluded that the arrangement for Case 2 has the balance pressure and suction among the three cases since the pressure and suction that occurred at the roof house are not too high compared to other arrangements.

Keywords Rural house · Wind tunnel test · Arrangement · Pressure coefficient

1 Introduction

Windstorm is one of the natural disasters that always occurred in Malaysia. The majority of damage incurred during severe windstorms appears to affect non-engineered rural houses that are made from timber or a combination of concrete and wood. In the newspapers, almost every year some districts in Malaysia have been reported to suffer roof damage and truss failure due to windstorms event, especially in rural houses. Rural houses in Malaysia are listed as non-engineered buildings and are prone to damage during windstorms [1]. Based on statistics by Wan Chik et al. [2], from 2007 to 2012, most cases occurred in Peninsula Malaysia in Kedah. Figure 1 shows the recent windstorm that hit a rural house in Parit Buntar, Perak in 2021, causing the uplift of the roof.

There is very little priority on infrastructure design, such as roofing and cladding, to minimize wind-induced damage to buildings [4]. The most current problems of wind load on houses are the roof structure which affected significantly the durability of the house. Wan Chik et al. [2] stated that most low-rise rural Malaysian buildings are constructed of timber and that during a wind event they are exposed to great

Fig. 1 Roof uplift due to strong wind in Parit Buntar [3]

damage. The non-engineered structures were constructed without referring to any design code or guidelines of structure design as long as it can be a shelter to the owner. This type of structure is likely to be blown off first during a post-windstorm. Then, the unique and traditional design of rural houses in Malaysia that consists of overhang to prevent the weariness of house façade during rainy days and additional area for kitchen house may affect the cause of roof blown off issues. The wind pressures on the building can be affected by several factors such as overhang, roof geometries, roof pitches, house geometries, wind direction [5], neighboring buildings, and terrain categories. Wang et al. [5] conducted a study on the roof overhang of low-rise buildings using a wind tunnel test and full-scale measurement by comparing with the results stipulated in ASCE 7-16 and found that the suction for both methods is lower than those generated from the standard. Moreover, the maximum magnitudes of RMS and minimum pressure coefficients increased with the turbulence intensity.

Generally, low-rise buildings are surrounded by numerous obstructions in urban or rural areas. In rural areas, buildings are surrounded by trees, while in urban areas, buildings are surrounded by adjacent houses and other structures. Chang and Meroney [6] investigated the effect of pressure on different separation distances of building using wind tunnel tests and full-scale measurements. The models were arranged with various symmetric configurations with different separation distances in the wind tunnel. From this study, the central building surrounded by the buildings was found to be significant. Moreover, Kim et al. [7] studied the characteristics of the wind pressures on the building due to the large group of surrounding buildings. The result shows that with the presence of surrounding buildings, the wind pressure decreases as the target model goes to the downstream side, and the local top wind speed also decreases. Surry and Lin [8] proposed that roof suction is reduced by the existence of such obstructions. The most significant element among them is the arrangement of the building structure, which can affect the flow directly [9]. Neighbouring structures may either increase or decrease the flow-induced forces on building, depending mainly on the geometry and arrangement of these structures, their orientation with respect to the direction of the flow, and terrain conditions. The main parameters affecting the interaction between adjacent buildings are the types of terrain, size and shape of the building, the incident wind direction, and last but not least the building arrangement and spacing [10]. This research addressed Sustainable Development Goals (SDG) which can improve the social effect and economic prosperity of societies and produce sustainable cities and communities. It can be shown in the previous studies conducted by Irtaza et al. [11] where the model with a roof pitch of 30° among the various models was found to have maximum wind pressure. This finding is important to the communities in Malaysia since Malaysia is a country with heavy rain during the monsoon season. Because of that, this research focuses on three objectives which are to study the wind flow visualization on the rural house due to the arrangements of a neighboring building, to generate the pressure coefficient around the rural house by using a wind tunnel test, and to determine the optimum arrangement of neighboring building around the rural house in preventing severe damage during the windstorm.

2 Methodology

2.1 Data Collection

Data collection is very important in order to build and develop the rural house model. The data was obtained from previous site surveys by other researchers. Some useful data was adopted from the post-windstorm site survey conducted by Deraman [12] to develop the model. This study simulated three (3) rural house models which are labeled as Case 1, Case 2, and Case 3 with different arrangements of a neighboring building. Figure 2a, b show the schematic diagram of the rural house model with the wind profile perpendicular to the house.

2.2 Experimental Study

The experiments were carried out in a low-speed, closed-circuit wind tunnel at International Islamic University Malaysia (IIUM). The wind tunnel has a working section of 2.3 m wide, 1.5 m high. and 6.0 m long as shown in Fig. 3a. Models are constructed to a geometric scale of 1:25. The scale was selected to avoid the excessive distortion of the flow due to the blockage effect. The free stream velocity of 26.4 m/s was blown. This value of wind velocity is according to the Beaufort scale, where the selection of the value is based on the effects on the surrounding during the windstorm event. Wind pressures on the roofs of these three (3) buildings were measured for wind directions of 0° to facilitate measurements over the entire roofs and sidewalls. Figure 3b shows the location of the vortex generator or as known as spires used as an obstacle to creating a wind profile and the position of the concrete block as a neighboring building.

The dimension of the rescaled model is given in Table 1. The aspect ratio (wall length to wall width) of the building models is chosen as 3:2. The blockage ratio for the models in the wind tunnel is 2.783% which is less than 3.0%. This ratio is the most preferred and used by the previous researchers to get fewer blockage effects, and no blockage correction of the measured wind pressures was carried out.

The dimension of the cube concrete block that acts as a neighboring building is 4.5 m^3 for the real model while the prototype model is 15 cm^3 as illustrated in Fig. 4. The neighboring building can represent structures such as a gazebo, a small stall, and toilet.

The model has a total number of 22 nodes of pressure tap with the gap set at 2.5 cm on the top of the model's roof to obtain the pressure value as shown in Fig. 5. To measure pressure distributions on the models, pressure taps of 10 mm long pieces with 3.81 mm external diameter and 2.39 mm internal diameter stainless-steel tubing are inserted into the holes drilled in the Perspex.

The 2D diagrams of the arrangement of the neighboring building are shown in Fig. 6a–c for Case 1, Case 2, and Case 3, respectively. The direction of wind 0° was

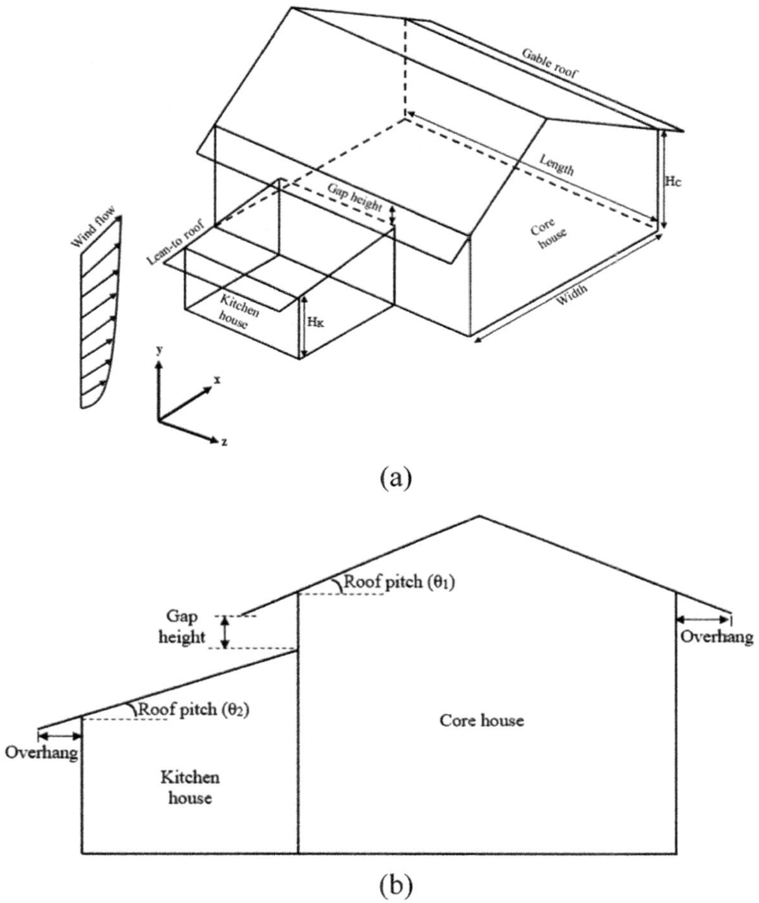

Fig. 2 Schematic diagram of **a** 3D rural house model **b** cross-section of the rural house with additional kitchen house [12]

chosen as the natural direction of the wind, which influenced the house from the front view of the house model.

For Case 1, the neighboring building was placed beside the main building with a distance of 150 mm, while for Case 2, the neighboring building was placed in front of the rural house model with a distance of 150 mm. The neighboring building was placed at the corner of the main building with a 150 mm distance at the front and at the side of the main building for Case 3.

(a) (b)

Fig. 3 **a** Test section of IIUM wind tunnel. **b** Arrangement of the neighboring building

Table 1 Dimension of rural house model

Parameters	Real model		Prototype model	
	Core house	Kitchen house	Core house	Kitchen house
House dimension	12 m (L) × 8 m (W) × 4 m (H_C)	6 m (L) × 4 m (W) × 2 m (H_K)	48 cm (L) × 32 cm (W) × 16 cm (H_C)	24 cm (L) × 16 cm (W) × 8 cm (H_K)
Roof pitch	27°	Varies with gap height	27°	Varies with gap height
Overhang	1.0 m	1.0 m	4.0 cm	4.0 cm
Gap height	1.0 m	–	4.0 cm	
Position of kitchen house		Center of the core house		Center of the core house

Fig. 4 3D schematic diagram of neighboring building

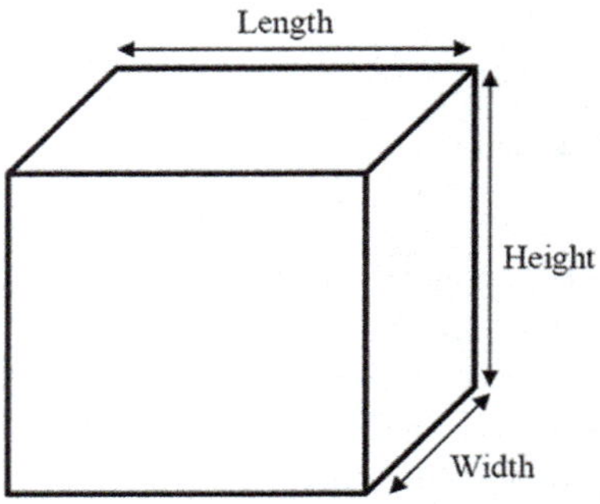

2.3 Determination of Pressure Coefficient

The wind pressure coefficient is defined as a dimensionless function of the wind pressure difference and dynamic pressure measured at the average roof height [13].

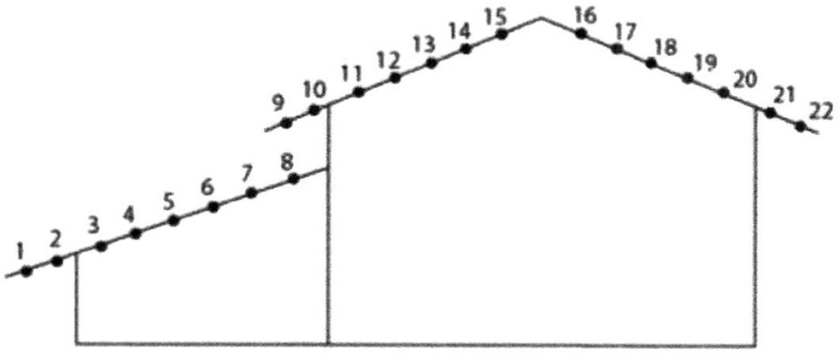

Fig. 5 Schematic diagram for the location of pressure nodes [12]

The equation of wind pressure coefficient (C_P) is written in Eq. 1 based on the Bernoulli theorem in Kim et al. [13], Liu [14], Xu et al. [15], and Holmes [16] similarly.

$$C_P = \frac{P_s - P_{ref}}{0.5\rho U_{He}^2} \tag{1}$$

where

C_P pressure coefficient at the wall surfaces,
P_s static pressure at the middle of wall surfaces (Pa),
P_{ref} reference pressure located at the reference height of the building (Pa),
ρ density of air (kg/m^3),
U_{He} velocity at the building height, H_e (m/s).

The pressure coefficient (C_P) is a non-dimensional parameter that describes the loading effect on buildings [16]. The pressure node along the core house upwind slope to leeward overhang experienced the negative value of C_P known as suction. Figure 7 shows the illustration of the pressure and suction that occurred to the pressure coefficient of the roof. The positive value of the pressure coefficient indicated the pressure occur, while the negative value indicates the suction occur on the roof.

3 Results and Discussion

The experimental investigation was carried out on the turbulent flow fields around the rural house building models with gabled roofs having different arrangements of neighboring buildings immersed in the atmospheric boundary layer. Information on the flow structure was obtained by flow visualization and measurements of surface pressure distribution prediction.

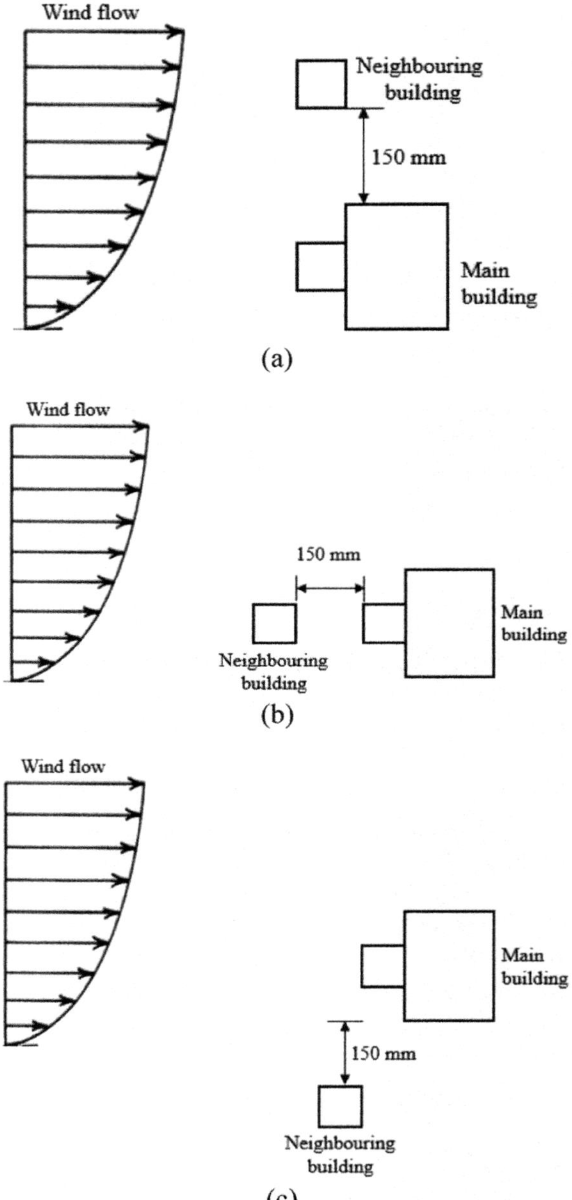

Fig. 6 Arrangement of neighboring building for **a** Case 1 **b** Case 2 **c** Case 3

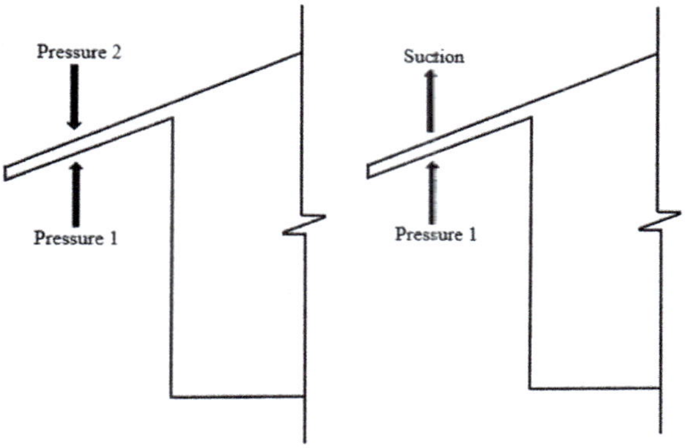

Fig. 7 Pressure and suction action on the roof [1]

3.1 *Pressure Coefficient Distribution*

Figure 8 compares the spatial distributions of the static pressure as expressed by the pressure coefficient (C_P) distribution profile of the house with four different roof pitches, divided into six (6) zones namely Zone A (overhang-kitchen), B (upwind-kitchen), C (overhang-core), D (upwind-core), E (downwind-core), and F (overhang-core). The incoming flow at Zone A and B produces a similar trend and magnitude of pressure coefficient with positive C_P at the upwind roof of the kitchen house.

It shows that positive pressure coefficient C_P occurs in Zone A, B, and C with similar patterns and magnitude of C_P. The highest suction ($-C_P$) occurs at node 13 with $C_P = -2.259$ for Case 3 where the neighboring building was placed at the corner of the main building with a 150 mm distance at the front and at the side of

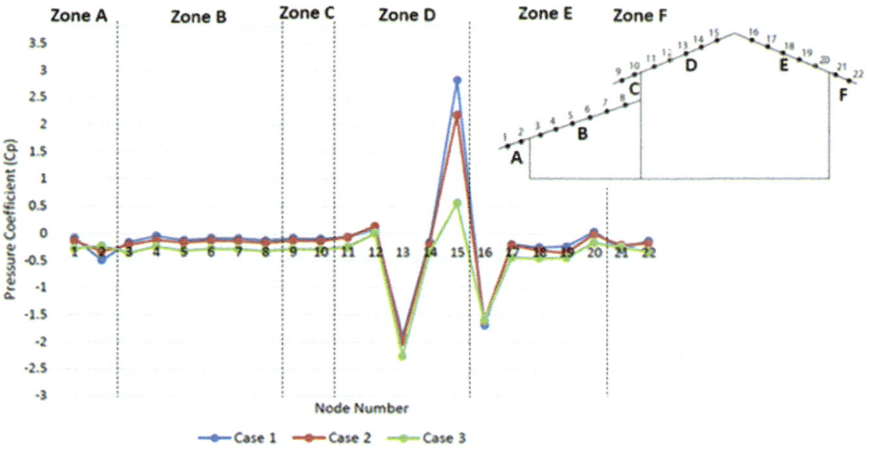

Fig. 8 Graph of C_P value along the roof with different neighboring building arrangement

the main building followed by Case 2 where the neighboring building was placed in front of the rural house model.

It shows that positive pressure coefficient C_P occurs in Zone A, B, and C with similar patterns and magnitude of C_P. The highest suction ($-C_P$) occurs at node 13 with $C_P = -2.259$ for Case 3 where the neighboring building was placed at the corner of the main building with a 150 mm distance at the front and at the side of the main building followed by Case 2 where the neighbouring building was placed in front of the rural house model with the distance of 150 mm. Case 3 has the highest suction occurred at the upwind slope of the house model due to the wake region of the neighboring building. The high value of suction that occurred will be contributed to severe damage such as roof uplift. The C_P is extremely increased at node 15in Zone *D* with $C_P = 2.830$. The C_P value decreases to -1.688 at node 16 in Zone E near the ridge of the roof. This is because the upstream flow impinged the separation point in Zone C causing the separation bubble in Zone D. When the flow approached Zone E, the pressure drastically dropped from node 15 to node 16 due to the roof peak. As proved by Surry and Lin [17], the surroundings lead to the mean C_P of 50% to 65% near the corner. C_P values at node 17 to node 22 produce a similar pattern and magnitude for all cases.

From the three arrangements of the neighboring buildings which are at the side, at the center and front, and at the corner of the rural house model with 150 mm distance, the optimum location of the neighboring building is at the center and at the front with 150 mm of the rural house model because it has the balance pressure and suction behavior that occurred at the roof of the house model with $C_P = 2.179$ for the highest pressure and $C_P = -1.997$ for the highest suction. As can be seen from the graph, the value of $-C_P$ for case 3 is higher than for Case 1 and Case 2. This happens maybe because the size of the neighboring is not too big or equal in size to the rural house model. So, the neighboring building is not giving that much effect on the house model. As stated by [9], the wind force on a building can either be reduced or increased by nearby buildings.

3.2 *Flow Visualization*

The pattern of the flow visualization along the side view of the rural house model for three different arrangements of the neighboring building is shown in Fig. 9. The flow visualizations are not clearly captured due to some technical errors during the experiment. The direction of wind flow had changed after hitting the neighboring building and will interfere with the wind flow at the main building. As stated by [9], the turbulent flow generated in the downstream neighboring building may interfere with the well-organized vortex shedding formation in the main building. The reverse flow was generated at the wake region leading to the formation of swirling [18]. Based on Fig. 9, the separation zone occurred at the ridge of the roof and cause turbulence in the wake region. Based on the explanation by [18], the shielding effect depends on the distance between the main building and the neighboring buildings and the number of buildings located in front of the main building from the airflow. Figure 9b, c shows the main building which is in the wake of the neighboring building. The value of C_P on the windward of the kitchen house and the core house changes from positive to negative due to the shielding effect of the adjacent building. Even the negative force was generated when the adjacent building was quite close to the main building [9].

4 Conclusion

In conclusion, the study was carried out to investigate the wind tunnel simulation of wind pressure on a rural house due to the effect of the surrounding building's arrangements. The wind tunnel test has been used in this study because many past researchers used the computational fluid dynamics (CFD) method in their study. The practical testing using WTT was performed to simulate the wind flow similar to a windstorm event toward a rural house, resulting in a pressure distribution on the roof house profile. The turbulent flow at the wake region generated in the downstream neighboring building interfered with the well-organized vortex shedding formation in the main building. It can either reduce or increase the wind force of the main building. The closer the neighboring building and the measured building, the higher the shielding effect and even negative force can be generated. Most of the C_P values generated are negative, and only two of the C_P values are positive. It means that more suction occurred at the roof house than pressure behavior. Among the three arrangements of neighboring buildings, Case 1 has the highest pressure with $C_P = 2.830$ and the lowest suction with $C_P = -1.892$ occurred at the roof. Case 3 has the lowest pressure with $C_P = 0.572$ and the highest suction with C_P = -2.259 occurred on the house roof. So, the optimum arrangement of the neighboring building should be Case 2 which has the balance pressure with $C_P = 2.179$ and suction with $C_P = -1.997$ occurred around the rural house roof model. Hence, it is important for the rural house to experience the least damage during windstorms due to the arrangement of the neighboring's buildings.

Fig. 9 Flow visualization of airflow hit the windward wall of the main building and neighboring building for **a** Case 1 **b** Case 2 **c** Case 3

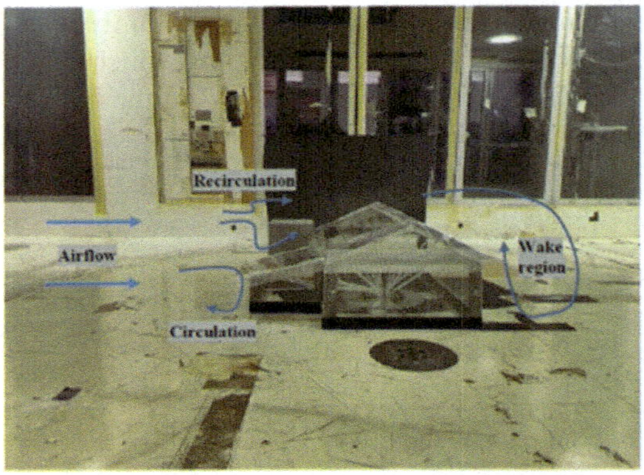

(a)

(b)

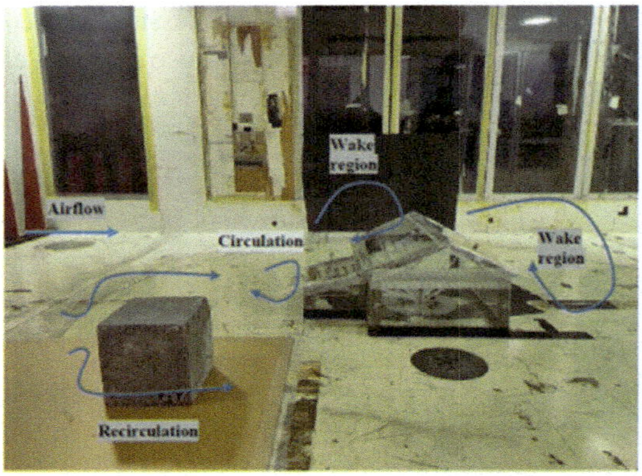

(c)

Acknowledgements The authors would like to express their deep gratitude to the staff from the Department of Civil Engineering and Department of Mechanical Engineering for their help in finishing this study, International Islamic University Malaysia for providing the facilities, the financial support of the Research Management Centre (RMC) Grant 2020 (RMCG20-071-0071) from International Islamic University Malaysia.

References

1. Zaini, S.S., Majid, T.A., Deraman, S.N., Wan Chik, F.A., Muhammad, M.K.A.: Post windstorm evaluation of critical aspects causing damage to rural houses in the northern region of peninsular Malaysia. MATEC Web of Conf. **103**, 03016 (2017)
2. Wan Chik,F.A., Che Deraman, S.N.A., Ramli, N.I., Muhammad, M.K.A., Majid, T.A., Zulkarnain, N.Z.: Development of windstorm database system for wind damages in Malaysia, **4**:214–217 (2014)
3. NST Homepage, https://www.nst.com.my/news/nation/2021/03/676616/storm-damages-houses-kuala-kurau. Last accessed 21 Mar 2021
4. Majid, T.A., Ramli, N.I., Ali, M.I., Saad, M.S.H., Hashim, M., Zakaria, I.: Malaysia country report 2010
5. Wang, X., Li, Q., Li, J.: Field monitoring and wind tunnel study of wind effects on roof overhang of a low-rise building. Struct Control Health Monit (2019)
6. Chang, C.-H. Meroney, R.N.: The effect of surroundings with different separation distances on surface pressures on low-rise buildings. J. Wind Eng. Ind. Aerodyn. **91**, 1039–1050 (2003)
7. Kim, Y.C., Yoshida, A., Tamura, Y.: Characteristics of surface wind pressures on low-rise building located among large group of surrounding buildings. Eng. Struct. **35**, 18–28 (2012)
8. Surry, D., Lin, J.X.: J. Wind Eng. Ind. Aerodyn **58**, 113–138 (1995)
9. Cho, K., Hong, S., Hwang, K.-S.: Effects of neighboring building on wind loads effects of neighboring building on wind loads. Council on Tall Buildings and Urban Habitat, (2004)
10. Pundhir, R., Barde, N.: Study of wind pressure on tall building due to change in relative position of interfering building. Int. J. Civ. Eng. Res. **7**(2), 105–115 (2016)
11. Irtaza, H., Javed, M.A., Jameel, A.: Effect on wind pressures by variation of roof pitch of low-rise hip-roof building. IJREAT Int. J. Res. Eng. Adv. Tech. **3**(1) (2015)
12. Deraman, S.N.C.: Distribution of pressure coefficient along the external profile of rural house exposed to windstorm in Malaysia using COMPUTATIONAL FLUID Dynamics (CFD). Ph.D. thesis, Universiti Sains Malaysia (2019)
13. Kim, R.W., Kim, D.W., Kwon, K.S., Ha, T., Lee, I.B.: Evaluation of wind pressure coefficients of greenhouses using wind tunnel test and numerical model. Trans. ASAE **47**, 1–124 (2015)
14. Liu, H.:Wind Engineering A Handbook for Structural Engineers. Prentice-Hall, Inc. A Pearson Education Company Upper Saddle River, New Jersey (1991)
15. Xu, W., Peng, H., Ming, G.: Field measurement of wind loads on low-rise building with adjustable roof pitch. In: 7th International Colloquium on Bluff Aerodynamics and Applications (BBAA7), pp. 1197–1208, Shanghai, China (2012)
16. Holmes, J.D.: Wind Loadings of Structures. Spon Press, London, UK (2001)
17. Surry, D., Lin, J.X.: The effect of surroundings and roof corner geometric modifications on roof pressures on low-rise buildings. J. Wind Eng. Ind. Aerodyn. **58**, 113–138 (1995)
18. Majid, T.A., Zaini, S.S., Ismail, M.A., Deraman, S.N.C., Poon, J.L.: Numerical investigation on the effect of overhang roof around rural house. Adv Civ Environ Mater Res (ACEM16), 2–8 (2016)

Compressive Strength Development and Homogeneity of Concrete Incorporating Fly Ash and Used Engine Oil as Chemical Admixture

S. A. Saad, A. N. Jamaluddin, S. A. Masjuki, N. M. Husain, W. N. F. W. Hassan, W. N. A. W. Azahar, N. Kasim, and S. Beddu

Abstract In this paper, the hardened properties of concrete with the utilization of waste materials of pulverized fly ash (PFA) and used engine oil (UEO) as chemical admixtures are conducted in order to improve concrete properties in terms of its compressive strength development and its homogeneity. Ordinary Portland cement (OPC) is replaced by PFA at the percentage of 20% and 40%, respectively. Meanwhile, UEO is added by the dosage of 0.15 and 0.3%, subsequently. Slump test was done for determination of rheological properties of fresh concrete. Compression test and ultrasonic pulse velocity (UPV) test are conducted to analyze the mechanical properties of hardened concrete and its homogeneity. Based on the non-destructive test, UPV test recorded the highest value with concrete sample of 20% PFA and 0.3% UEO of 4.19 km/s at 28 days, which demonstrate a high integrity of a concrete. For

S. A. Saad (✉) · A. N. Jamaluddin · S. A. Masjuki · N. M. Husain · W. N. F. W. Hassan · W. N. A. W. Azahar · N. Kasim
Department of Civil Engineering, Kulliyyah of Engineering, International Islamic University Malaysia, Jalan Gombak, 53100 Gombak, Selangor, Malaysia
e-mail: asmahanisaad@iium.edu.my

S. A. Masjuki
e-mail: aliyyah@iium.edu.my

N. M. Husain
e-mail: drnadiah@iium.edu.my

W. N. F. W. Hassan
e-mail: wanfirdaus@iium.edu.my

W. N. A. W. Azahar
e-mail: aifa@iium.edu.my

N. Kasim
e-mail: ayukasim@iium.edu.my

S. Beddu
Department of Civil Engineering, College of Engineering, Universiti Tenaga Nasional, Jalan Ikram-Uniten, 43000 Kajang, Malaysia
e-mail: salmia@uniten.edu.my

N. Sabtu (ed.), *Proceedings of AWAM International Conference on Civil Engineering 2022—Volume 2*, Lecture Notes in Civil Engineering 385,
https://doi.org/10.1007/978-981-99-6018-7_36

a destructive test, it also follows the same trend having concrete sample of 20% PFA and 0.3% UEO recorded the highest compressive strength which was at 42.74 MPa at 90 days. It can be concluded that, utilization of these waste materials improves properties of fresh concrete without much adverse effect on properties of hardened concrete.

Keywords Pulverized fuel ash · Used engine oil · Compressive strength · Concrete homogeneity · UPV test

1 Introduction

High demand of energy for domestic and industrial uses produces million tonnes of fly ash from the combustion of pulverized coal in electric power generation plants. In 2013, it was recorded that Malaysia had produced more than 54 million metric tonnes of pulverized fuel ash (PFA) and 57.2% of the totally produced PFA was disposed in landfills which leads to soil and groundwater contamination [1]. Therefore, the optimum utilization of the PFA in concrete production will contribute in solving this problem. Production of Portland cement contributes to the emission of carbon dioxide (CO_2). Cement production consumes huge energy and causes about 7% of the total green-house gas emission in the world. Hence, utilization of self-cementitious materials such as PFA as a cement replacement material (CRM) is being researched extensively over the last few decades to enhance durability and sustainability [2]. As for the automotive industry, approximately 45% of used engine oil (UEO) is being collected worldwide, while the remaining 55% is thrown by the end user to the environment directly [3]. The improper disposal of UEO cause water and soil contamination and hence affect human, and plant life. Therefore, incorporation of UEO in concrete can contribute towards solving the issue of improper UEO disposal. Meanwhile, in the production of concrete, admixtures are added to give special properties and improve certain characteristics of concrete. Utilization of superplasticizer (SP) as a water reducing admixture for example is costly and therefore causes escalation in concrete cost production. UEO has similar properties when compared with SP and researchers are looking forward towards this opportunity as a means of solution for the problem [4]. Moreover, UEO has the functions of reducing the water content used and entraining microscopic air bubbles in concrete. However, when the UEO dosage surpasses its optimal level, the hydrophobicity of the UEO significantly influences its dispersion in fresh concrete mixtures [5]. In this regard, the effect of UEO incorporation in concrete as water reducing admixture needs to be further explored. Most of the research studied on the concrete production are incorporated or associated either with the utilization of PFA or UEO solely. Hence, this research will highlight on incorporating both of the materials mentioned which expected to induce the improvement of the concrete strength in hardened state. Therefore, the primary objective of this study is to determine the effect of pulverized fuel ash (PFA) and used engine oil (UEO) incorporated concrete towards the homogeneity as well as compressive strength development behaviour.

2 Methodology

2.1 Material Selection and Mix Proportion

Mix proportion in a concrete design plays an important role in order to ensure precise consistency of fresh mix concrete produced. The mix proportion applied in this experimental works is given in Table 1.

Cement from Tasek Corporation Berhad conforming to Malaysian Standard, MS 522: Part 1: 2003, type I ordinary Portland cement (OPC) was used for this study. Crushed type coarse aggregate and fine aggregate from natural sand used in this research works were having a maximum size of 20 mm and 5 mm, subsequently. Both coarse and fine aggregates were obtained from Civil Engineering Lab, IIUM which conforms to BS EN 12620:2002 + A1:2008 [6]. As for the admixture incorporated in this study, pulverized fuel ash (PFA) and used engine oil (UEO) were selected. The PFA was acquired from Sultan Salahuddin Abdul Aziz Shah Power Station in Kapar, Selangor. Meanwhile, the UEO was collected from local motorcycle workshops located around Gombak, Selangor. In this experimental work, three groups of different concrete samples (100% OPC, 80% OPC + 20% PFA and 60% OPC + 40% PFA) were prepared. Two sets of concrete samples were incorporated

Table 1 Mix proportions for samples preparation

Mix type	OPC (kg/m^3)	PFA (kg/m^3)	CA (kg/m^3)	FA (kg/m^3)	W (kg/m^3)	UEO (%)
Control (100% OPC)	436	0	1169	584	205	0
20PFA/ 0.15UEO	348.8	87.2	1169	584	205	0.15
20PFA/ 0.3UEO	348.8	87.2	1169	584	205	0.30
40PFA/ 0.15UEO	261.1	174.4	1169	584	205	0.15
40PFA/ 0.3UEO	261.1	174.4	1169	584	205	0.30

[a] *OPC* Ordinary Portland cement
PFA Pulverized fuel ash
CA Coarse aggregate
FA Fine aggregate
W Water
UEO Used engine oil

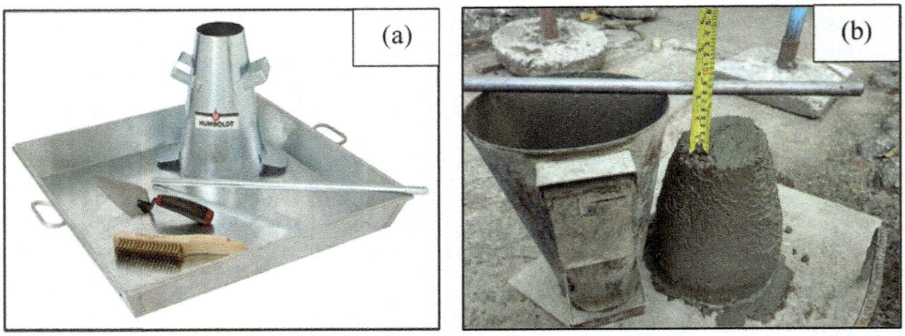

Fig. 1 Workability testing set-up. **a** Slump apparatus, **b** measurement of slump

with 0.15% and 0.3% dosage of UEO, subsequently. 100% OPC was set as control. Selection of determined percentages used was based on earlier research [7].

2.2 *Slump Test*

Workability of fresh concrete was measured by conducting slump test in accordance with BS EN 12350-2:2019 [8]. Frustum cone mould of 305 mm high with smaller opening at the top as shown in Fig. 1a was used. The cone was filled by three layers of fresh concrete with each layer being compacted with 25 times tamping using standard steel rod of 16 mm diameter and 600 mm long. Top surface of the final layer was levelled using a trowel. The cone was slowly lifted. The level difference between the height of the mould and the highest point of the subsided concrete was directly measured. The difference in height in mm was recorded as slump of concrete as shown in Fig. 1b.

2.3 *Ultrasonic Pulse Velocity (UPV) Test*

The homogeneity and relative quality of concrete is assessed using Portable Ultrasonic Non-Destructive Digital Indicating Tester (PUNDIT) in accordance with BS 12504-4:2021, providing information on the presence of cavities, void, cracks and defect [9].

2.4 Compression Test

Compressive strength test was conducted to measure the strength development of hardened concrete. Concrete cubes of size 150 mm × 150 mm × 150 mm were cast using standard steel moulds for evaluation of the compressive strength of concrete. The test was conducted using digital compressive testing machine available at Civil Engineering Lab, IIUM in accordance with BS EN 12390-3:2019 at 7, 28 and 90 days, subsequently [10].

3 Results and Discussion

3.1 Effect of Pulverized Fuel Ash (PFA) and Used Engine Oil (UEO) on Fresh Concrete Workability

Based on the workability test of PFA and UEO incorporated concrete as shown in Fig. 2, generally, it can be said that, with the constant percentages of cement replacement of 20% and 40% PFA, the slump values show a positive increment from the addition of UEO from 0.15 to 0.3%. Combination of 20% PFA and 0.3% UEO mix almost double the slump values of the control mix with the highest value recorded at 59 mm. This trend is supported by researches conducted which justify that indeed with the addition of UEO will enhance the slump values of concrete.

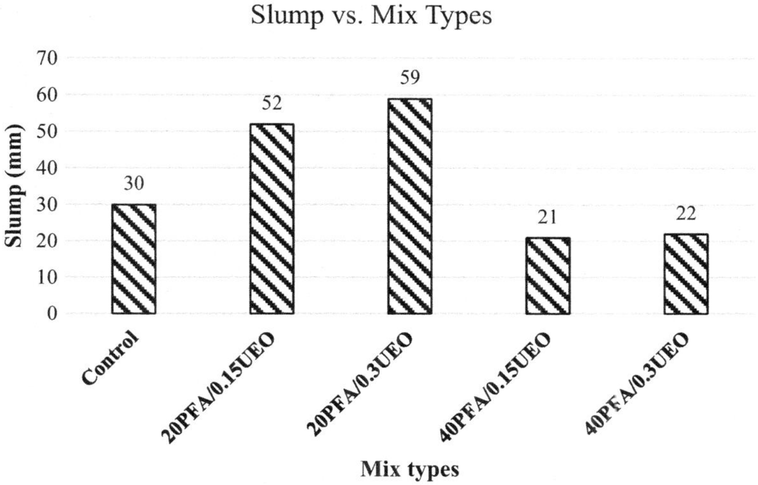

Fig. 2 Effect of PFA and UEO on workability of fresh concrete

Table 2 Slump classification and its application [12]

Slump (mm)	Class	Application suitability
0–50	436	Massive section, little reinforcement
50–130	348.8	Columns, beams, retaining walls
> 130	348.8	Heavily reinforced section, flowable concrete

Beddu et al. [4] concluded that up to 0.5% use of UEO will impart the characteristics of superplasticizers [4]. In addition, Shafiq et al. [4] also reported that with the addition of UEO will result in increment of slump of all concrete mixes in relation with the control mix [7]. The enhancement of slump values is contributed to the sulphur trioxide (SO_3) content in UEO that is found out to be relatively high. Therefore, with the addition of UEO can possibly act as water reducer and lubricant that improves concrete workability. Furthermore, in other research, it is reported that no segregation, bleeding or sedimentation were observed in any of the UEO-added concrete mixes during the concrete manufacturing process [11]. The designed concrete can be associated to the slump classification and its application suitability in various areas as tabulated in Table 2 [12]. Overall slump values are within the range of 0–50 mm which falls under the class of stiffed slump.

3.2 *Effect of Pulverized Fuel Ash (PFA) and Used Engine Oil (UEO) on Non-destructive Test of Hardened Concrete*

For non-destructive test, the analysis of hardened concrete is being conducted by using ultrasonic pulse velocity (UPV). Based on the results, it can be observed that UPV readings are increasing over the days and this phenomenon is due to the reaction of PFA with $Ca(OH)_2$ which acts as micro filler which fills the microscopic voids between cement particles that are present in the concretes. This then helps to reduce porosity of the concretes and leads to higher UPV results. Nevertheless, it can be noticed that all readings of incorporated concretes are lower than the control with an exceptional to the case of 20FA/0.3UEO for 28 and 90 days. This is because the pulse velocity will be affected by the introduction of a pozzolanic material into the concrete. According to Panzera et al. [13], a unit volume of cement has more contribution towards the pulse velocity through the concrete when compared with an equivalent volume of the ash [13].

Overall, 40FA/0.15UEO incorporated concrete yields the best value of pulse velocity. This is depicted in Table 3 where the results are in accordance with different percentages of incorporation of FA and UEO with respect to number of days of curing follow up by the representation of the graph variation of the UPV test results in Fig. 3.

Table 3 UPV readings with respect to different specimens over various curing duration

Specimens/days	Velocity (km/s)		
	7	28	90
Control	3.80	3.95	3.92
20PFA/0.15UEO	3.65	3.78	3.82
20PFA/0.3UEO	3.68	4.19	4.17
40PFA/0.15UEO	3.55	3.66	3.76
40PFA/0.3UEO	3.59	3.68	3.82

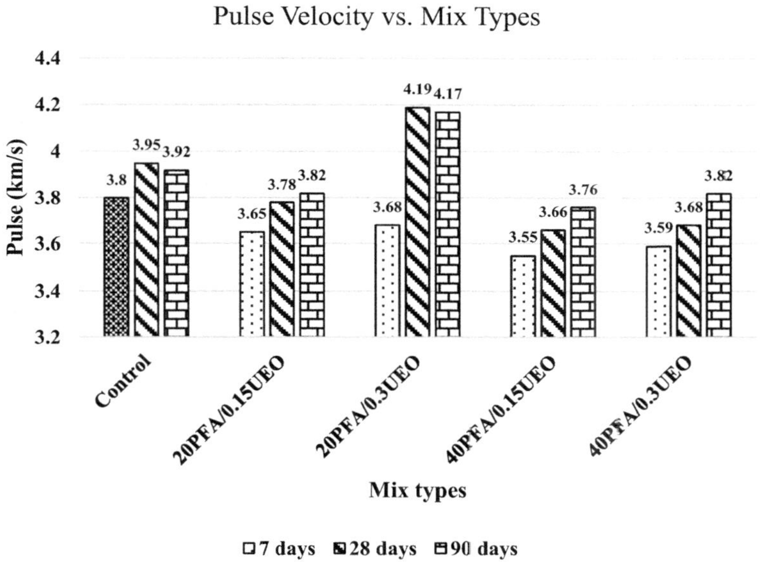

Fig. 3 Ultrasonic pulse velocity (UPV) test results with various fly ash and used engine oil percentage incorporation

3.3 Effect of Pulverized Fuel Ash (PFA) and Used Engine Oil (UEO) on Destructive Test of Hardened Concrete

The compressive strength of concrete is an important characteristic in investigating the mechanical properties of concrete. Therefore, scrutinizing the effect of PFA and UEO towards the compressive strength of concrete is an essential element included in this project. In analyzing the concrete destructive test, compression test was being conducted for all concrete samples by compression test machine at the age of 7, 28 and 90 days. The results are presented in Fig. 4.

Generally, for all mix types, the result shows a positive increment of compressive strength through the process of curing by ponding method. As for the concrete with the incorporation of PFA and UEO, it caused a variation ranging from 3 up to 54%

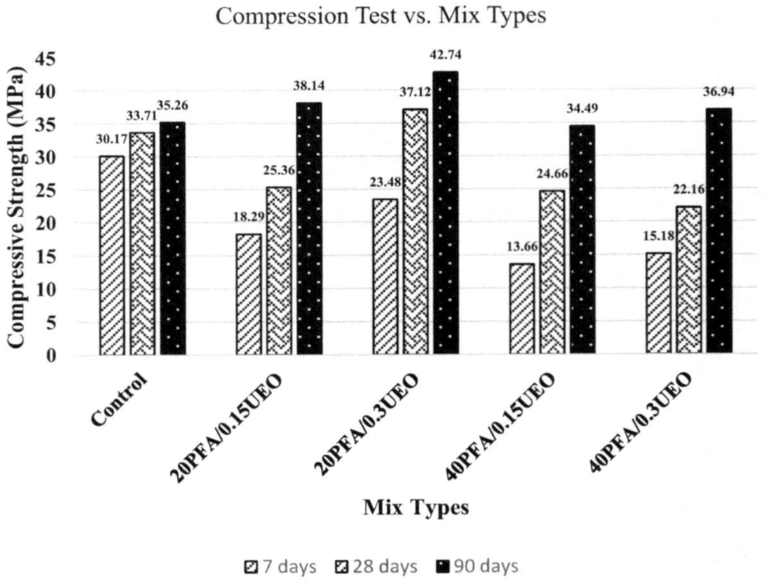

Fig. 4 Effect of PFA and UEO on compressive strength of concrete of different mix types at different ages of curing period

when compared with the compressive strength of concrete control mix at later age, i.e. 90 days. The trend of the compressive strength of concrete incorporated with PFA and UEO shows a significant reduction both at the age of 7 and 28 days with an average reduction of 32% when compared with the control concrete mixes. Concrete with the combination of 40% PFA and 0.15% UEO recorded the lowest value of 13.66 MPa at 7 days of curing, which indicated the biggest reduction of 54% from the control samples. This is due to the Pozzolanic reaction of PFA that occurred at later age of concrete, i.e. after 28 days. In addition, this phenomenon may be influenced by various factors such as type, chemical composition of the PFA, etc. According to Ayub et al. [14], the reason behind the slower development of compressive strength gain of concrete is due to the texture of PFA having lesser surface area and slower reactivity [14]. Moreover, it has also been found out by Siddique [15] that the higher the cement replacement level with PFA, the bigger the reduction in compressive strength [15].

On the other hand, looking at the usage of UEO with a cement replacement-based concrete, it is also contributed towards the adverse effect of compressive strength development. Even though it shows a positive trend of increment in compressive strength with the increment of its dosage from 0.15 to 0.3%, except for the mix having 40% PFA with a reduction of 10% at 28 days, overall, it resulted in lower compressive strength when compared with the control mixes. This could possibly be due to the chemical composition of UEO having high amount of Zinc (Zn). This resulted in reduction of concrete setting time and hence lowering the compressive strength. In addition, the slower rate of strength development is also due to slow rate of hardening with high dosage of phosphorus pentoxide (P_2O_5), affecting the concrete hydration process at the first place [4].

However, even though the results indicated a lower value in compressive strength test than that of the control mixes, it is noticeable that, with the case of the cubes sample of 20% PFA and 0.3% UEO, an increment of 3% from the control sample was recorded at the highest value of compressive strength of 37.12 MPa at the age of 28 days. This outcome was likely to occur due to the lower percentages of PFA used (20%). In this study, despite the fact that the PFA incorporated concrete sample having low compressive strength, Shafiq et al. [5] have reported vice versa [5]. Based on the study, it is shown that better concrete strength developed after initiation of pozzolanic reactivity which usually occurs at later concrete age, i.e. after 28 days. This can be justified by having all samples resulting in a compressive strength higher than control at the age of 90 days except for the case of 40% PFA with 0.15% UEO which nearly reaches to the control value. Furthermore, with the combination of 0.3% UEO which yields the best result, it is parallel to the research by Assaad [16] that once has concluded in his paper that 0.3% dosage of UEO is proven to be the optimum amount of dosage for which detrimental effects on compressive strength can be encountered [16]. These two factors could possibly be the backup reason that underlies the positive outcome of the respective cubes sample.

4 Conclusion

Considering PFA as a cementitious replacement and UEO as an alternative to chemical admixtures in concrete production, it is a positive remark towards a greener concrete technology in construction industry. Therefore, utilization of PFA and UEO as constituent in concrete will minimize the abundance of by-product and wasted material over the world. While examining the hardened properties of the concrete, based on the non-destructive test, UPV test recorded the highest value with concrete sample of 20% PFA and 0.3% UEO of 4.19 km/s at 28 days which demonstrates high integrity of a concrete. For a destructive test, it also follows the same trend having concrete sample of 20% PFA and 0.3% UEO recorded the highest compressive strength which was at 42.74 MPa at 90 days. It is then proven that concrete incorporated with UEO has better strength development at later age, i.e. after 28 days.

Acknowledgements The author would like to extend her gratitude to the Ministry of Higher Education, Malaysia (MoHE), IIUM Fundamental Research Grant Scheme with MyGRANTS reference code FRGS/1/2019/TK06/UIAM/03/1 for financial support and staffs of the Civil Engineering Department, International Islamic University Malaysia for facilitating with adequate facilities and support for the accomplishment of this study.

References

1. Fauzi, A., Nuruddin, M.F., Malkawi, A.B., Abdullah, M.M.A.B.: Study of fly ash characterization as a cementitious material. Procedia Eng. **148**, 487–493 (2016). https://doi.org/10.1016/j.proeng.2016.06.535
2. Nath, P., Sarker, P.: Effect of fly ash on the durability properties of high strength concrete. Procedia Eng. **14**, 1149–1156 (2011). https://doi.org/10.1016/j.proeng.2011.07.144
3. Abdelaziz, G.E.: Utilization of used-engine oil in concrete (2015)
4. Beddu, S., Shafiq, N., Nuruddin, M., Kamal, N., Sadon, S.: Effects of used engine oil as an admixture in concrete durability. Br. J. Appl. Sci. Technol. **15**(6), 1–10 (2016). https://doi.org/10.9734/BJAST/2016/20738
5. Chen, H., Chow, C.L., Lau, D.: Developing green and sustainable concrete in integrating with different urban wastes. J. Clean. Prod. **368**, 133057 (2022). https://doi.org/10.1016/j.jclepro.2022.133057
6. European Standards, BS EN 12620:2002+A1:2008 Aggregates for concrete (2008)
7. Shafiq, N., Choo, C.S., Isa, M.H.: Effects of used engine oil on slump, compressive strength and oxygen permeability of normal and blended cement concrete. Constr. Build. Mater. **187**, 178–184 (2018). https://doi.org/10.1016/j.conbuildmat.2018.07.195
8. European Standards, BS EN 12350-2:2019 Testing fresh concrete Slump test (2019)
9. European Standards, BS EN 12504-4:2021 Testing concrete in structures Determination of ultrasonic pulse velocity (2021)
10. European Standards, BS EN 12390-3:2019 Testing hardened concrete Compressive strength of test specimens (2019)
11. Chen, H., Qin, R., Lau, D.: Recycling used engine oil in concrete design mix: an ecofriendly and feasible solution. J. Clean. Prod. **329**, 129555 (2021). https://doi.org/10.1016/j.jclepro.2021.129555
12. American Society for Testing and Materials, ASTM C143/C143M-12 Standard test method for slump of hydraulic-cement concrete (2012)
13. Panzera, T.H., Christoforo, A.L., Cota, F.P., Borges, P.H.R., Bowen, C.R.: Ultrasonic Pulse velocity evaluation of cementitious materials. In: Těšinova, P. (ed.) Advances in Composite Materials. IntechOpen (2011) https://doi.org/10.5772/17167
14. Ayub, T., Khan, S.U., Memon, F.A.: Mechanical characteristics of hardened concrete with different mineral admixtures: a review. Sci. World J. **2014**, 1–15 (2014). https://doi.org/10.1155/2014/875082
15. Siddique, R.: Performance characteristics of high-volume Class F fly ash concrete. Cem. Concr. Res. **34**, 487–493 (2004). https://doi.org/10.1016/j.cemconres.2003.09.002
16. Assaad, J.J.: Disposing used engine oils in concrete—Optimum dosage and compatibility with water reducers. Constr. Build. Mater. **44**, 734–742 (2013). https://doi.org/10.1016/j.conbuildmat.2013.03.078

Effect of Incorporating Supplementary Cementitious Materials on Properties of High-Performance Fine-Grained Cement Mortar

Salem Giuma Ibrahim Sagr, M. A. Megat Johari, and M. J. A. Mijarsh

Abstract Using supplementary cementing materials (SCMs) as partial cement replacement is a common practice in the construction industry to achieve more durable mortar and less CO_2 emissions associated with producing mortar materials. This study investigated the properties of high-performance cement mortar (HPM) containing SCMs, which include fly ash (FA), silica fume (SF), palm oil fuel ash (POFA), and ground-granulated blast-furnace slag (GGBS). The examined properties include flowability, compressive strength, ultrasonic pulse velocity, porosity, and absorption for four replacement levels (5, 10, 15, and 20%). At replacement levels up to 20%, mortar containing SF and GGBS exhibits significantly better and enhanced properties than the control mix without any cement replacement material, while the incorporation of POFA and FA led to a decrease in all mortar properties. However, both of POFA and FA could be used at lower replacement ratios < 10%. The cementitious and pozzolanic properties of SF and GGBS permit higher cement replacement levels (> 10%) which can achieve equivalent or superior properties and durability performance than the conventional cement mortar. Based on the overall results, HPM containing SCMs can exhibit a similar effect and therefore reducing global warming potential of the cement-based materials.

Keywords Supplementary cement materials · High performance cement mortar · Fly ash · Silica fume · Palm oil fuel ash · Ground-granulated blast-furnace slag

S. G. I. Sagr · M. A. Megat Johari (✉)
School of Civil Engineering, Engineering Campus, Universiti Sains Malaysia, Nibong Tebal, Penang, Malaysia
e-mail: cemamj@usm.my

S. G. I. Sagr
Higher Institute of Sciences and Technology, SoqAlkamisMsehel, Tripoli, Libya

M. J. A. Mijarsh
Civil Engineering Department, Faculty of Engineering, Al-Merghab University, Al-Khums, Libya
e-mail: mjalmijarsh@elmergib.edu.ly

N. Sabtu (ed.), *Proceedings of AWAM International Conference on Civil Engineering 2022—Volume 2*, Lecture Notes in Civil Engineering 385,
https://doi.org/10.1007/978-981-99-6018-7_37

1 Introduction

Over the past decade, extensive research has been conducted to explore alternative binder materials to reduce the heavy reliance on Portland cement [1]. Presently, the industrial and agricultural sectors are overwhelmed with huge quantities of wastes, presenting disposal challenges and various health issues [2]. However, the chemical and physical analyses of some of these residues have revealed characteristics such as filler effect and pozzolanic reactivity that make them useful as the potential supplementary cementitious materials (SCM) [3]. Pozzolans, as defined by the American Society for Testing and Materials (ASTM), encompass siliceous or siliceous and aluminous materials, which in themselves possess little or no cementitious properties. However, in the presence of water, it reacts with calcium hydroxide (portlandite) that produced through cement hydration, which produce more cementitious material (calcium silicate hydrates or C–S–H) [4]. Examples of pozzolans include fly ash (FA), silica fume (SF), ground-granulated blast-furnace slag (GGBS), and palm oil fuel ash (POFA) [5].

Different approaches have been used in the recent years for the development of high strength cementitious materials, including mechanical densification to produce very homogeneous minimal porosity structures [6]. One of the most important materials for application in repairing, retrofitting, and strengthening of damaged concrete structures is high-strength or high-performance mortar (HPM) [7]. HPM can be produced by using ultrafine particles and superplasticizing admixtures which enable reduction in water–cement (w/c) ratio to less than 0.3 [8]. HPM requires a huge amount of cement on its production reaching up to 1000 kg/m^3 [9]. As a consequence, the cost of this material will greatly increase in comparison with conventional mortar. This problem can be solved by utilizing alternative by-product based binders, combined with the latest curing technology to obtain a low-cost HPM without the loss of quality in term of the strength and performance of the material [6].

Chu and Kwan [10] studied the use of industrial SF as a replacement to cement in HPM, and reported significant increase in term of the strength of the HPM. In different study, Zhang et al. [11] observed a significant decrease in the porosity of HPM containing 4% SF as a replacement to cement. Kim and Lee [12] recommended both GGBS and FA to be used in HPM. Danish and Mosaberpanah [13] mentioned that replacing cement by 20% of FA in HPM enhanced the fluidity by 14.3%, but reduced the compressive strength by 10%. In other study, Supit et al. [14] assessed the workability and compressive strength of mortar containing 40% of FA. They observed that, at 40% replacement ratio, flow diameter improved by 2% and the compressive strength increased by 2.8%. Abdelli et al. [15] observed an increase in compressive strength of HPM by around 10% when cement has been replaced by at 20% GGBS. Wang et al. [16] stated that, in HPM, when the GGBS replacement ratio increased from 0 to 20%, the flow and compressive strength increased considerably. Shaladi et al. [17] replaced cement by 30% of ultra-fine treated POFA to produce HPM with compressive strength of 81.70 MPa at 28 days. Rukzon and Chindaprasirt

[18] reported that carbonation depth of HSM containing 20% POFA was less than reference samples.

The aim of this study is to assess the properties of HPM using industrial and agricultural by-product based pozzolans as partial replacement for Portland cement, with constant w/b ratio. To achieve this, flow table, compressive strength, ultrasonic pulse velocity, absorption, and porosity of HPM were investigated.

2 Methodology

2.1 Materials

Portland cement (PANDA MS EN 197-1 CEM I 52.5N) was used in the present study. This type of cement was used to ensure the highest quality of mortar, as it is the highest grade available in the market in terms of compressive strength grade (52.5 N) with specific gravity of 3.16. Natural siliceous sand was used as fine aggregates with a maximum size of 300 μm. The natural sand was sieved using a sieve shaker to obtain the highly fine sand. The water absorption and specific gravity were 0.06 and 2.70, respectively. The surface moisture content and evaporable moisture content were 0.20 and 0.25, respectively. High-performance superplasticiser Sika® ViscoCrete®-2044 was used to ensure excellent flowability obtained from its powerful water reducing capability.

Figure 1 shows all the SCMs used in preparing the HPM mixtures. Table 1 provides the chemical compositions for the cement and SCMs used in this study. The FA used in this study was supplied by Lafarge Malaysia Berhad (Rawang Plant) and it complied with ASTM C618–19. The GGBFS was supplied by YTL Cement Marketing Sdn. Bhd. which conformed to the specification of BS EN 15167-1:2006. POFA was collected from a palm oil mill located in Nibong Tebal, Penang, Malaysia. The POFA was subjected to multi-stages of thermal and mechanical treatments to obtain the ultrafine POFA as previously reported in past study [19].

(a) (b) (c) (d)

Fig. 1 **a** SF; **b** FA; **c** GGBS; **d** POFA

Table 1 Chemical compositions of cement and SCMs

Comp	Cement	SF	FA	GGBS	POFA
	Mass (%)	Wt. (%)	Wt. (%)	Wt. (%)	Wt. (%)
CaO	64.89	0.45	4.84	38.42	8.19
SiO_2	21.01	93.48	49.45	37.14	65.01
Al_2O_3	4.68	1.32	25.21	14.32	5.72
SO_3	3.66	0.34	0.32	4.18	0.33
Fe_2O_3	3.20	0.65	13.87	0.37	4.41
K_2O	1.17	1.34	1.15	0.39	6.48
MgO	0.81	0.82	1.04	3.72	4.58
TiO_2	0.22	–	1.36	0.38	0.25
Na_2O	0.086	0.42	0.27	0.062	0.07
P_2O_5	0.079	–	0.87	0.03	4.69
$CaO + SiO_2 + Al_2O_3$	90.58	95.25	79.5	89.88	78.92
LOI	0.48	1.42	2.13	0.601	2.53

SF is silica fume, *FA* is fly ash, *GGBS* is ground-granulated blast-furnace slag, and *POFA* is palm oil fuel ash

2.2 Mix Proportions and Samples Preparation

Seventeen mixes were prepared and characterized to achieve the aim of this study including a reference mix (Table 2) with no SCMs, in addition to four mixes with SCMs replacement levels to cement of 5, 10, 15, and 20% by volume for each of the SCMs namely SF, FA, POFA, and GGBS. The control mix was labelled as CM, whereas the other mixes by the name of the SCM material + (replacement level) such as SF15 refers to HPM containing 15% SF replacement to cement. The CM was designed to achieve about 90 N/mm^2 compressive strength at 28 days with flowability index of fresh mortar between 1.9 and 2. The corresponding water/cement ratio (w/c) was 27%. The superplasticizer (Sp) was added to all mortar mixtures by mixing it with water in order to maintain comparable range of workability. The SP to binder ratio (Sp/binder) was 2.2% by weight. The sand to binder ratio s/b was set to be 1.5. For all mixtures, the fine aggregates and SCM were accurately weighed and added in a mortar mixer. The components were mixed for two minutes in dry condition and then mixed with half of the water for three more minutes. Finally, the Sp was mixed to the second half of the remaining quantity of water and then added to the mix. The mixing was continued for three more minutes to reach the required consistency. After casting, each specimen of the mortar was cured in moist cabinets for 24 h with 20°C before demoulding. After that, all the mortar specimens were placed in a curing basin, where the temperature was maintained at room temperature.

Table 2 Mixture proportions (kg/m^3)

Mix	OPC kg	POFA kg	GGBS kg	FA kg	SF kg	Sand kg	Water kg	SP kg
OPC	867.68	–	–	–	–	1301.53	234.27	19.09
POFA5	824.0.30	35.7	–	–	–	1301.53	234.27	18.92
POFA10	780.92	71.39	–	–	–	1301.53	234.27	18.75
POFA15	737.53	107.09	–	–	–	1301.53	234.27	18.58
POFA20	694.15	142.78	–	–	–	1301.53	234.27	18.41
GGBS5	824.0.30	–	39.68	–	–	1301.53	234.27	19.01
GGBS10	780.92	–	79.35	–	–	1301.53	234.27	18.93
GGBS15	737.53	–	119.03	–	–	1301.53	234.27	18.84
GGBS20	694.15	–	158.71	–	–	1301.53	234.27	18.76
FA5	824.0.30	–	–	33.22	–	1301.53	234.27	18.87
FA10	780.92	–	–	66.45	–	1301.53	234.27	18.64
FA15	737.53	–	–	99.67	–	1301.53	234.27	18.42
FA20	694.15	–	–	132.9	–	1301.53	234.27	18.2
SF5	824.0.30	–	–	–	31.99	1301.53	234.27	18.84
SF10	780.92	–	–	–	63.98	1301.53	234.27	18.59
SF15	737.53	–	–	–	95.97	1301.53	234.27	18.34
SF20	694.15	–	–	–	127.53	1301.53	234.27	18.09

2.3 Experimental Procedures

The flow table test was used for evaluating the consistency and flow of the Portland cement mortars containing different types and proportions of SCMs. This test was applied for fresh mixes at all SCM substitution levels which was performed following the procedures stipulated in ASTM C1437 and ASTM C230. The compressive strength test was performed according to ASTM C109. The compressive strength was determined at 7 and 28 days of samples age. For each mix, three cubes of 50 × 50 × 50 mm (length, width, and height, respectively) were tested for every curing age. The average compressive strength of the three samples was considered as the reported value.

Non-destructive analyses were conducted with the aid of the ultrasonic pulse velocity (UPV) test on the HPM specimens with a direct transmission method of measurement. This test was applied on mortar cubes with dimensions of 50 mm for two curing ages (7 and 28 days). The average of UPV values obtained by testing three cubes was recorded and presented in this study. The test was performed according to ASTM C597. The absorption of water by immersion of a hardened mortar under vacuum is defined as the difference between the wet mass (saturated with water in a vacuum medium) and the dry mass of a given sample of hardened mortar, expressed in terms of the dry sample volume. The procedures stipulated in ASTM C642 were adopted to determine both the water absorption and porosity of mortar samples via the vacuum saturation method on 50 mm mortar cubes. The absorption and porosity were tested at 7 and 28 days of samples curing age in water.

3 Results and Discussion

3.1 Influence of SCMs on Flowability of HPM

Figure 2 shows the measured workability determined by the usage of flow table test. Flowability of mortar slightly decreased as the SF content increased. At 20% replacement ratio of SF, the lowest flow diameter of around 150 mm was recorded. This was due to the higher specific surface of SF, which needed more water to achieve similar workability [20]. Both FA and GGBS improved the workability, which concurred with the findings of previous studies [21, 22]. Replacing cement by 15% of FA and GGBS increased the flow diameter by 7% and 4%, respectively. In the case of the HPM containing POFA, the flow reduced with the inclusion of POFA, and the effect was more noticeable at the higher POFA content. This could be attributed to the greater surface area of the POFA and variation in its physical properties, where the number of angular particles increased as more POFA was included to the mix. As a result, increasing the level of POFA from 0 to 20% reduced the flow diameter from 175 to 161 mm. These observations agreed with the ones reported in earlier studies [23, 24].

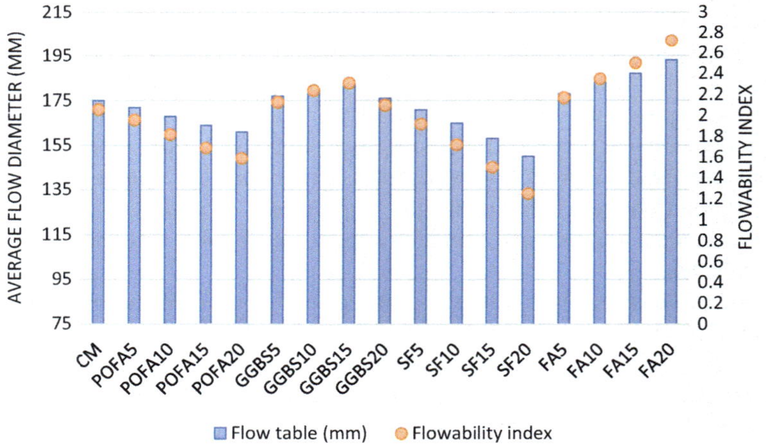

Fig. 2 Flow characteristics of fresh HPM; cement (CM), palm oil fuel ash (POFA), ground-granulated blast-furnace slag (GGBS), silica fume (SF), fly ash (FA)

3.2 Influence of SCMs on Compressive Strength of HPM

The results obtained from the compressive strength tests for all specimens at 7 and 28 days are shown in Fig. 3. The strength of the OPC specimens at 7 and 28 days was 76.6 and 93.1 MPa, respectively. The 28-day strength of the HPM with 10% SF increased by 22.8% higher than the control mix, and at higher SF content than 10% the strength decreased. However, all HPM mixtures containing SF had a higher strength than the CM. Similar trends were observed in previous study [20]. Three mechanisms had been linked as responsible for the strength enhancement by SF: (i) strength enhancement by pore size refinement and matrix densification, (ii) strength enhancement by reduction in the content of $Ca(OH)_2$ (CH), and (iii) strength enhancement by cement paste-aggregate interfacial zone refinement [20]. In the case of the effect of FA, the CM showed higher strength than all mortar mixtures containing FA. Using 20% FA as partial replacement of cement led to reduction in compressive strength by 12%, a trend that agreed with earlier study [13]. The reduction in compressive strength due to the partial replacement of cement with FA could be attributed to the dilution effect as part of the cement was replaced with FA and the slower rate of pozzolanic reaction of FA.

The mortar containing 5–20% of GGBS had comparable or superior strength result than the reference mortar at 7 days, but higher strength was recorded at 28 days for all GGBS mixtures. At early age, the GGBF did not instantly react with portlandite produced by the hydration of C_3S and C_2S of cement, it required a few days to have the ideal conditions to allow this activity. The chemical effect or pozzolanic of GGBFS that occurred during this time was not significant. HPM containing POFA

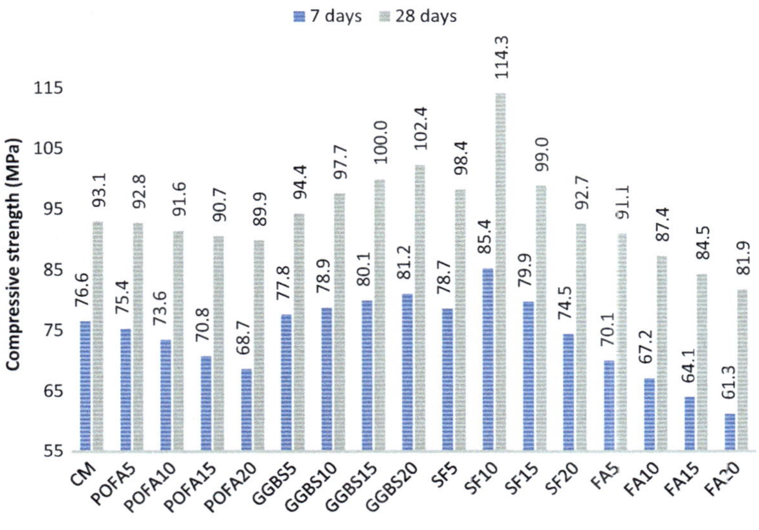

Fig. 3 Compressive strength of HPM mixes at 7 and 28 days

had lower compressive strengths than the control OPC mortar at the same curing age, and the strength reduction was greater with mortars with higher content of POFA. At 20% replacement level, the compressive strength reduced by 3.3% at 28 days of HPM age with strength of 89.1 MPa. According to Zeyad et al. [19], low content of Al_2O_3 could be the reason behind the reduction in strength.

3.3 Influence of SCMs on Ultrasonic Pulse Velocity of HPM

The level of packing of the paste and aggregates can be determined using the UPV test. Figure 4 shows the UPV of specimens at 7 and 28 days of curing age. The 28-day ultrasonic pulse velocity (UPV) values, which determines the quality of concrete, were in the range of 4500–5300 m/s. Higher velocity indicates good quality of concrete and vice versa. The lowest value in this study belongs to the FA group of specimens and the highest ones to the SF group of specimens. Since SF particles were much finer than OPC, they filled the micropores in cement paste, and hence, the mechanical properties and durability of mortar were improved, by reducing permeability and porosity. The UPV of GGBFS mixtures increased with age and content. The UPV of POFA20 mix, which contains 20% POFA replacement to cement, decreased by 1.2% at 28 days curing age. This finding is in great agreement with other researchers' findings [25, 26]. Furthermore, it has been mentioned in previous studies that cement mortar typically has a similar interfacial transition zone (ITZ) and uniformly distributed voids within the mortar containing POFA [25].

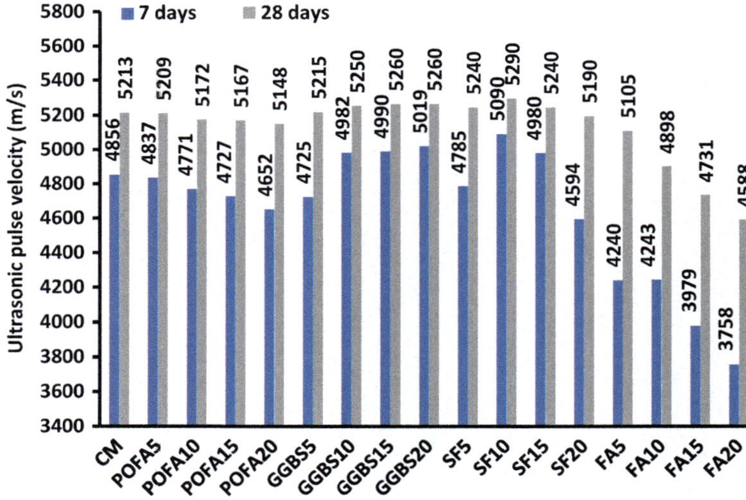

Fig. 4 UPV for various HPM mixes at 7 and 28 days

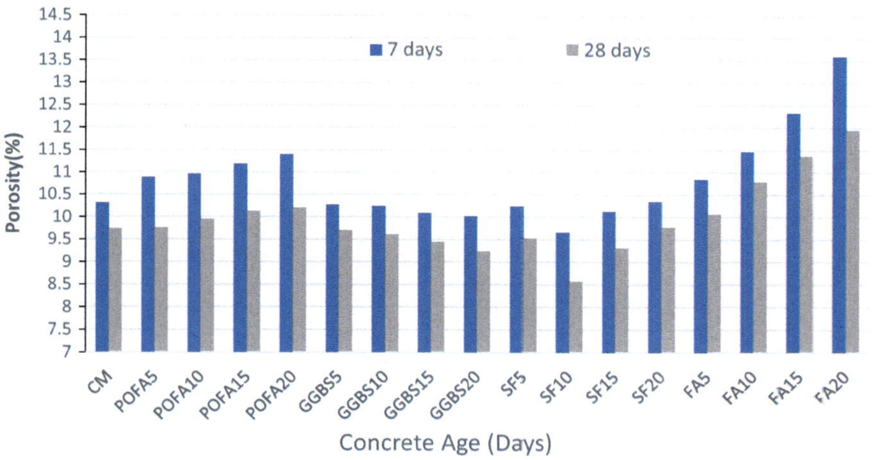

Fig. 5 Porosity of HPM mixes at different ages: (7 and 28 days)

3.4 Influence of SCMs on Porosity of HPM

Figure 5 shows the effect of the variation of SCMs replacement levels on the porosity of the HPM. The porosity of mortars containing SF and GGBS was found to be lower than that of the CM as well as those of mortar mixtures containing FA and POFA. The mortar containing 10% SF recorded the lowest porosity at both 7 and 28 days followed by mortar containing 20% GGBS. The observed effect of SF and GGBS concurs with previous findings on the use of similar SCMs [27]. Also, at 20% replacement ratio of GGBS, the lowest porosity was observed in this group with 5.1% lower than the reference mortar. The FA had an adverse effect on the porosity of the HPM with a recorded increase of 22.8% over the CM after 28 days of curing. As the POFA content increased from 5 to 20%, the porosity values increased by 0.2%–4.8% in comparison with the CM at the age of 28 days. The same observation on the proportionality between porosity and POFA content was indicated in previous study [24]. Also, Kroehong et al. [28] confirmed that adding POFA particles having the same size as the cement resulted in a paste with higher total porosity than the one containing OPC.

3.5 Influence of SCMs on Absorption of HPM

The effect of replacing the SCMs to the OPC on the water absorption rate of hardened specimens is shown in Fig. 6. Using SF and GGBS led to a reduction in the absorption at all ages. At 10% replacement ratio and after 28 days of curing, the absorption declined by 3% and 4.3% for GGBS and SF, respectively. Adding 15% and 20% FA

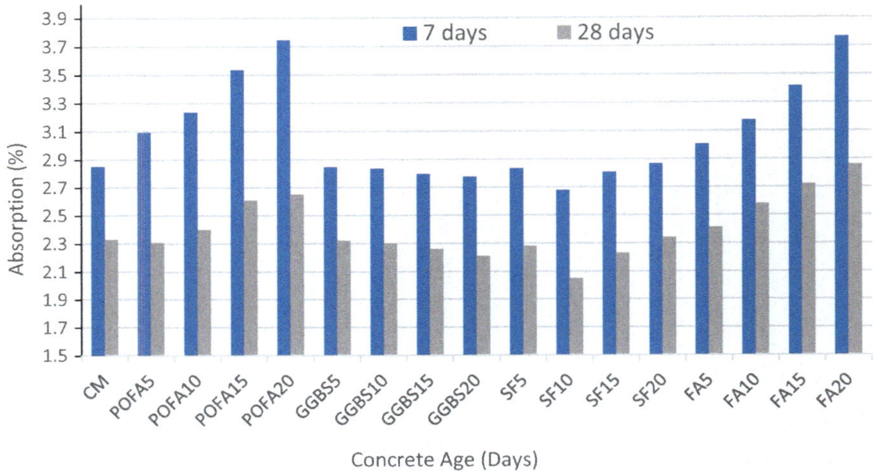

Fig. 6 Absorption for various mixes of HPM at different ages (7 and 28 days)

as replacement to cement increased the water absorption by 16.7% and 22.7% higher than the CM after 28 days.

At 28 days, the water absorption of HPM containing 5% POFA decreased by 1%. However, using 10% of POFA increased the absorption by 3% to reach 2.4% at 28 days. While the level of fineness of cement and POFA particles marginally varied at different intervals, it mostly did not influence the water intake properties. The results of water absorption agreed with earlier study [25].

4 Conclusions

Based on the results presented earlier, it can be concluded that the workability of HPM experienced a reduction when replacing cement with SF and POFA, and the effect magnified as the SF content increased. However, at 20% replacement level, the HPM achieved 1.25 and 1.59 flowability index for POFA and SF, respectively. On the other hand, both GGBS and FA increased the flow diameter of the HPM. The compressive strength at all ages for the HPM was increased by the SF and GGBS incorporation. However, replacing cement by 20% SF recorded a slight reduction in strength at 28 days. In contrast, the compressive strength of FA and POFA was adversely affected by the POFA incorporation. However, the mortar mixtures containing 20% POFA and FA recorded compressive strength of 89.91 MPa and 81.89 MPa, respectively, at 28 days. UPV increased as content of SF and GGBS increased, indicating an improvement in mortar quality to reach up to 5170 m/s and 5250 m/s, respectively, at 10% replacement level. UPV decreased as the content of FA and POFA increased, indicating a reduction in mortar quality. The UPV went down from 5213 to 5068 m/s, reflecting the negative effect of POFA on the quality of HPM. Both SF and GGBS reduced the porosity of the HPM mixtures. However, using > 15% SF in HPM led to reduce the porosity lower than CM. For GGBS, at 20% replacement level, the porosity of HPM was 9.24% with a 5.1% drop lower than the reference mix. Moreover, using FA and POFA led to increase the porosity at all ages. The absorption decreased as GGBS and SF level increased in the mortar mixes. At 15% replacement level, the absorption of HPCM containing GGBS and SF was 3% and 4.29%, which was lower than the CM. The lowest water absorption was 2.05% recorded by the mixture containing 5% SF at 28 days. The POFA and FA adversely affected the absorption of the HPM. All in all, SCMs can be used in the production of HPM at high replacement rates reaching 20%. This approach reduces the adverse effect of cement on the environment. Also, this reduces CO_2 emissions due to cement manufacturing. Besides, this provides a green cement mortar that reduces the vast amounts of wastes from the agriculture and industrial sectors that require landfilling, when they are utilized as pozzolanic materials in the production of mortar or concrete.

Acknowledgements The authors gratefully acknowledge the Ministry of Higher Education, Malaysia, and Universiti Sains Malaysia for providing the financial support through the Fundamental Research Grant Scheme (203/PAWAM/6071365) and University Bridging Grant Scheme (304/PAWAM/6316313), respectively, for the undertaking of the research work. Special thanks are due to United Palm Oil Industries for providing the palm oil fuel ash.

References

1. Panesar, D.K., Zhang, R.: Constr. Build. Mater. **251**, 118866 (2020)
2. Yang, H., Wang, Y., Liu, Z., et al.: Bioresources and Bioprocessing **4**, 1 (2017)
3. Adjei, S., Elkatatny, S.: J. Petrol. Sci. Eng. **195**, 107911 (2020)
4. Skibsted, J., Snellings, R.: Cem. Concr. Res. **124**, 105799 (2019)
5. Aprianti, E., Shafigh, P., Bahri, S., Farahani, J.N.: Constr. Build. Mater. **74**, 176 (2015)
6. Edwin, R., Kimsan, M., Pramono, B., Masud, F.: IOP conference series: earth and environmental science. IOP Publishing (2021)
7. Ramli, M., Dawood, E.T.: Concrete Res. Lett **2**, 249 (2011)
8. Husain, H.M., Raouf, Z.A., Khalil, W.I.: Eng. Tech. J. **26**, 1036 (2008)
9. Li, Z., Afshinnia, K., Rangaraju, P.R.: Constr. Build. Mater. **102**, 631 (2016)
10. Chu, S., Kwan, A.: Constr. Build. Mater. **197**, 716 (2019)
11. Zhang, B., Tan, H., Shen, W., Xu, G., Ma, B., Ji, X.: Cement Concr. Compos. **92**, 7 (2018)
12. Kim, H.-K., Lee, H.-K.: J. Mater. Civ. Eng. **25**, 662 (2013)
13. Danish, A., Mosaberpanah, M.A.: Constr. Build. Mater. **279**, 122458 (2021)
14. Supit, S.W., Shaikh, F.U., Sarker, P.K.: Constr. Build. Mater. **51**, 278 (2014)
15. Abdelli, K., Tahlaiti, M., Belarbi, R., Oudjit, M.N.: Energy Procedia **139**, 224 (2017)
16. Wang, H.-Y., Wang, W.-C., Wang, J.C., Chen, Y.-W.: Constr. Build. Mater. **280**, 122498 (2021)
17. Shaladi, R.J., Johari, M.A.M, Ahmad, Z.A., Mijarsh, M.: AWAM international conference on civil engineering. Springer (2019)
18. Rukzon, S., Chindaprasirt, P.: Indoor Built Environ. **18**, 313 (2009)
19. Zeyad, A.M., Johari, M.A.M., Alharbi, Y.R., et al.: J. Build. Eng. **38**, 102204 (2021)
20. Rao, G.A.: Cem. Concr. Res. **33**, 1765 (2003)
21. Ezziane, K., Ngo, T.-T., Kaci, A.: Constr. Build. Mater. **70**, 549 (2014)

22. Wang, H.-Y., Kuo, W.-T., Lin, C.-C., Po-Yo, C.: Constr. Build. Mater. **41**, 532 (2013)
23. Salami, B.A., Johari, M.A.M., Ahmad, Z.A., Maslehuddin, M., Adewumi, A.A.: Constr. Build. Mater. **190**, 65 (2018)
24. Huseien, G.F., Tahir, M.M., Mirza, J., Ismail, M., Shah, K.W., Asaad, M.A.: Constr. Build. Mater. **175**, 174 (2018)
25. Kanadasan, J., Abdul Razak, H.: Materials **8**, 8817 (2015)
26. Usman, J., Sam, A.R.M., Hussin, M.W.: J. Mater. Civ. Eng. **29**, 04016201 (2017)
27. Igarashi, S.-i., Watanabe, A., Kawamura, M.: Cement Concr. Res. **35**, 513 (2005)
28. Kroehong, W., Sinsiri, T., Jaturapitakkul, C., Chindaprasirt, P.: Constr. Build. Mater. **25**, 4095 (2011)

Feasibility Evaluation on Potential High Volume of FA-Based Binder as Early Strength at Ambient Cured

N. Archanaah, M. N. Noor Azline, A. B. Nabilah, S. Nor Azizi, and Mohd Saleh Jaafar

Abstract In general, strength development of fly ash-based geopolymer composite is still in focus of some researchers due to its low strength gain under ambient condition, particularly at early age of strength. An alternative way in formulating mix design for fly ash-based geopolymer with the aim to enhance the geopolymerization reactivity of fly ash in alkaline system is still in the depth of research. Many efforts have been conducted in combining various aluminosilicate materials in fly ash-based geopolymer with variation of content levels such as ground granulate blast furnace slag (GGBS), pulverzied oil palm ash (POFA), metakaolin (MK) and silica fume (SF). Despite the selection of aluminosilicate and its content level, alkaline activator concentration also gives major influence in the development of engineering properties of geopolymer composites. Therefore, it is crucial to determine the potential formulation in achieving high early strength without compromising fresh properties. Thus, this research is focusing on the evaluation of the potential binder combination of fly ash-based geopolymer paste with fly ash as the primary binder. Also, this study intended to explore the potential of POFA in enhancing the early engineering properties of fly ash-based geopolymer paste. In relation with this, the engineering properties have been evaluated which includes setting time, flowability, and compressive strength. Also, study evaluated the compatibility between the molarity and geopolymer binder combinations in achieving the promising engineering properties at the early engineering properties development. The results show that 12 NaOH concentration plays a viable role in the performance of geopolymer paste. The most pre-eminent combination is 60F40G and 60F30G10MK where both performed well in early and matured age of compressive strength. Besides that, it is also found that

N. Archanaah · A. B. Nabilah
Department of Civil Engineering, Faculty of Engineering, Universiti Putra Malaysia, UPM Serdang, 43400 Darul Ehsan, Selangor, Malaysia
e-mail: gs51044@student.upm.edu.my

M. N. Noor Azline (✉) · S. Nor Azizi · M. S. Jaafar
Housing Research Center, Department of Civil Engineering, Faculty of Engineering, Universitu Putra Malaysia, UPM Serdang, 43400 Darul Ehsan, Selangor, Malaysia
e-mail: nazline@upm.edu.my

N. Sabtu (ed.), *Proceedings of AWAM International Conference on Civil Engineering 2022—Volume 2*, Lecture Notes in Civil Engineering 385,
https://doi.org/10.1007/978-981-99-6018-7_38

POFA has potential in replacement up to 30%, however combination of 60F40P with 12 M performed well at 28 days.

Keywords FA-based geopolymer paste · Early age strength · Geopolymer paste · Molarity · Setting time · Flowability · FA/GGBS geopolymer paste · FA/POFA geopolymer paste

1 Introduction

Development of geopolymer cement composite has been in great interest from researchers all over the globe and its application has been expanded in many areas. Geopolymer technology is categorized as one of the research filed where it utilized the solid waste and industrial by-product by producing low-cost geopolymer materials and on the same hand became a preventive measure to environmental problems where the hazardous materials have been treated. It is well-understood that the geopolymer concept consists of two major components, which are aluminosilicate sources and alkaline activators. Thus, many research has been conducted using industrial by-product and waste material such as fly ash (FA), ground granulated blast furnace (GGBS), metakaolin (MK), palm oil fuel ash (POFA) and other silica aluminosilicate materials. Despite the mentioned aluminosilicate sources, fly ash (FA) and ground granulated blast furnace slag (GGBS) are the two main industrial by-products that have triggered interest for most of the researchers.

Nevertheless, the usage of fly ash in geopolymer system reacts incredibly slow in ambient temperature, but better performances under heat curing which is not practical curing method for construction application. In order to broaden its application and be more practical for construction industry, countless researchers have attempted to develop fly ash-based geopolymers at ambient curing temperature without external source of heating [1]. Many studies have been conducted in formulating mix design for fly ash-based geopolymer with the aim to enhance the geopolymerization reactivity of fly ash in alkaline system by incorporating more additional aluminosilicate material and by incorporating other additional aluminosilicate materials in the geopolymer system. These additional aluminosilicate materials could be either secondary or third component in the geopolymer system. The common additional aluminosilicate materials are GGBS, MK, POFA, silica fume, etc. [2]. On the other hand, literature also indicates that by including a source of calcium signification influences the engineering properties of geopolymer composites, particularly for fly ash due to the variation of calcium content [1, 3].

A lot of research has been conducted on the inclusion of GGBS blending with fly ash in order to produce geopolymer binder and the combination of aluminosilicate materials more than two also widely explored. Palm oil fuel ash (POFA) is one of potential aluminosilicate source that can be used as an alkali-activated cement. POFA is considerable potential aluminosilicate source due to its similarity properties with fly ash. However, the availability of POFA is limited due to its original sources, which is a key agricultural industries is hugely concentrated in Asia regions such as Malaysia, Thailand and Indonesia [2]. Many studies are using POFA as a cementitious material in alkali activation system since this agricultural waste is rich in silica and provides opportunity for producing sustainable building materials. Furthermore, some work on POFA in geopolymer system with blend of FA and GGBS has been explored [2, 4].

Few researchers found that POFA can be classified as a Class N pozzolanic material according to ASTM C618 (2004a) since it has total summation of the SiO_2, Al_2O_3, and Fe_2O_3 around 70% and the SO_3 and loss of ignition which is not higher than 4 and 10% [2, 4, 5]. In another study, POFA was categorized under Class F pozzolanic supplementary cementing material [6, 7]. Besides that, it has also been reported that ultrafine POFA has superior pozzolanic properties and can be used in high volume which is up to 60% in the production of high-strength geopolymer concrete [8]. Literature reported that compressive strength development for 100% POFA geopolymer mortar generates the lowest compressive strength which is less than 20 Mpa at 28 days. However, inclusion of GGBS resulted gain in compressive strength with addition of POFA, but then further increment of GGBS more than 70% shows reduction in strength [4, 9].

Despite the aluminosilicate materials combinations, alkaline activator concentration also is widely investigated in the geopolymer composite mixtures [3, 10]. The issue with alkaline activator is more complex in conjunction with variation of binder combination. Within the variation of binder combination, the major binder also plays an essential role in the geopolymerazation reactivity due to its chemical compositions. As result, the properties of geopolymer composites are influenced by (1) selection of aluminosilicate material, (2) percentage of combination of aluminosilicate material and (3) alkaline activator concentration.

Thus, this research is focusing on the evaluation of the potential binder combination of fly ash-based geopolymer paste with fly ash as the primary binder. Also, this study intended to explore the potential of POFA in enhancing the early engineering properties of fly ash-based geopolymer paste. Investigating the paste is crucial as this is the essential component that is binding the aggregates in forming the final product for construction applications. In relation with this, the engineering properties have been evaluated which includes setting time, flowability and compressive strength. Also, study evaluated the compatibility between the molarity and geopolymer binder combinations in achieving the promising engineering properties at the early engineering properties development.

2 Experimental

2.1 Material and Specimen Preparation

The major component of the geopolymer mix combinations was dominant by FA which collected from Kapar Energy Venture Sdn Bhd and the secondary binder were either by GGBS or POFA which was collected from local supplier in Malaysia. The third component of binder applied was metakaolin (MK) which was available commercially in Malaysia. Table 1 presents the physical and chemical properties of geopolymer binders, whereas Table 2 describes the geopolymer paste parameter and the details of binder combinations. A total of 24 mixes with variation in the binder components with 3 molarity of sodium hydroxide, 10, 12 and 14, were investigated. The paste specimens were prepared with binder to liquid ratio (B/L) of 0.43 and sodium hydroxide to sodium silicate ratio (SH/SS) of 2.5. For mix procedure of paste, the binders were dry-mixed for 3–4 min, then subsequently the combined alkaline solution was added, then mixing of the geopolymer paste was carried out for another 5–6 min until the mixture became homogenous. Flow table and setting time test were conducted on the freshly mixed geopolymer paste, then the remaining paste mixture has been cast into 50 * 50 * 50 mm steel molds in two layers with equal heights. In order to achieve sound and compact specimens, each layer has been vibrated by using vibration table for 25 s. After 24 h, the specimen was demolded and stored at ambient temperature for age at 1 and 28 days.

Table 1 Physical and chemical compositions of aluminosilicate binders

	FA	GGBS	POFA	MK
Physical properties				
Type	Class F	–	–	–
Color	Gray	White	Dark	White
Specific gravity (g/cc)	2.24	2.67	2.18	2.6
Specific surface area (m_2/g)	4.54	5	3.95	19–20
Chemical composition (% by mass)				
SiO_2	38.32	16.03	32.11	60.44
Al_2O_3	16.48	–	–	28.11
Fe_2O_3	22.48	1.05	32.5	3.55
CaO	12.39	76.25	19.85	1.10
MgO	0.28	0.75	0.661	–
K_2O	2.79	0.59	3.03	1.05
TiO_2	3.78	1.37	4.12	3.48
P_2O_5	–	–	5.109	–
SO_3	1.06	3.21	1.53	1.19

Table 2 Details of geopolymer parameter and mix combinations

Geopolymer paste parameter					
Alkaline activator parameter					
Sodium hydroxide concentration (molarity)	10 M, 12 M and 14 M				
Sodium hydroxide to sodium silicate ratio (SH/SS)	2.5				
Binder to liquid ratio (B/L)	0.43				
Curing method	Ambient				
Geopolymer paste mix combinations					
	OPC (%)	FA (%)	GGBS (%)	POFA (%)	MK (%)
Control	100	–	–	–	–
Binary system	–	60	40	–	–
	–	40	60	–	–
	–	60	–	40	–
	–	40	–	60	–
Ternary system	–	60	30	–	10
	–	40	50	–	10
	–	60	–	30	10
	–	40	–	50	10

2.2 Test Procedures

The initial setting time of the geopolymer paste was measured according to the standard test method for setting time of hydraulic cement using Vicat Needle ASTM C191-21. The flowability of geopolymer paste was measured by using flow table test and was carried out according to the standard test method for flow of hydraulic cement mortar ASTM C230-21. The compressive strength test was carried out in accordance with ASTM C109-21 at 1 and 28 days. The compressive machine with a loading capacity of 1500 kN and loading rate of 0.05 kN/s was used for the test. The average compressive strength value was reported for three specimen per batch of testing.

3 Result and Discussion

3.1 *Setting Time*

3.1.1 Fly Ash/GGBS-Based Binders

The initial setting time was obtained in accordance with specification of ASTM C191-21, where the initial setting time is specified at 10–30 min for CEM Type-1. The influences of molarity on the initial setting time show that all the mixtures obtained lower initial setting time at high molarity except for 100FA where the obtained results were 3000 min, equally same for three types of molar, which is beyond the specification of ASTM C191-21, as can be seen in Fig. 1. This is due to the fly ash being characterized by very high fineness and low presence of calcium; as a consequence of this, initial setting time of paste for 100FA shows longer initial setting time as it contains low calcium oxide (CaO) content which is only 12.39%, nevertheless, it is also found that with the increasing percentage level of FA, the water demand is decreasing. Thus, high molarity contributes to reduction in initial setting time [11]. Contrarily, addition of GGBS in FA-based geopolymer paste shows remarkable reduction in initial setting time regardless of molarity. The presences of CaO, remarkably has low initial setting time compared with those with higher content of FA. The fast precipitation of C–A–S–H gels at the hardening stage, which corresponded to high CaO content and lowest initial setting time were obtained for high amount of GGBS content, regardless to molarity. CaO which introduced by GGBS, increases the heat development during the activation and polymerization, this process helps remarkably in reducing the prolonged setting time [3]. Apart from GGBS contribution, it is also observed that inclusion of 10% MK decreases the initial setting time, where its due to MK has the highest specific surface area where leads to high reactivity and consumption of liquid in very early stage.

3.1.2 Fly Ash/POFA-Based Binders

A reduction in initial setting time was observed as the molarity increased from 10 to 14 M, except for 100FA, as can be seen in Fig. 2. This is due to the excess NaOH amount in the geopolymer paste system, causing a rapid polymerization reaction as similar findings were reported by Huseien et al. [12]. The initial setting time of FA/POFA-blended geopolymer paste with various proportions of FA and POFA experience long initial setting time in between 380 and 38 min; however, there is slight decrease with an increase of sodium hydroxide molarity from 10 to 14 M. This is because of slow chemical reaction under an ambient temperature by FA and POFA, this shows the mixture requires longer time to initial set. Using POFA as the main component prolongs the setting process as can be seen in proportion of 80 and 60% POFA content. However, a slight decline is observed as the molarity of NaOH increased to 14 M. Dramatic retardation in initial setting time was observed with

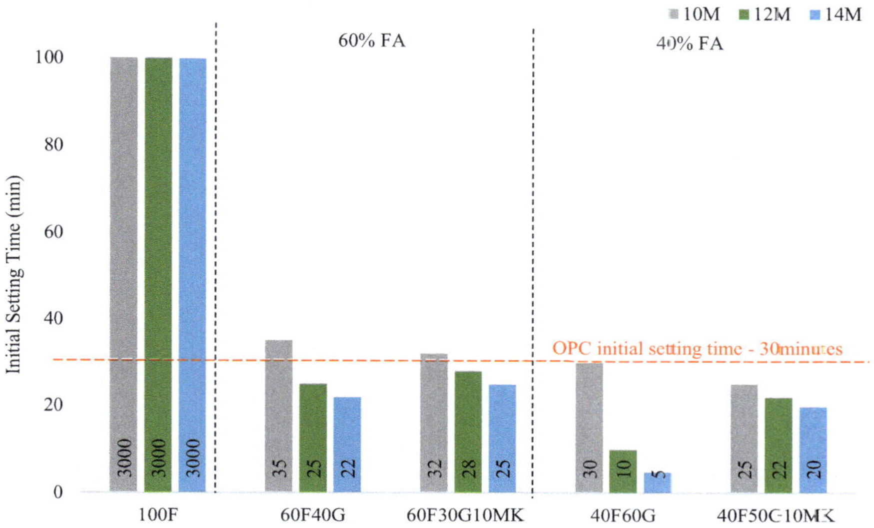

Fig. 1 Initial setting time for FA/GGBS-based geopolymer paste including 10% MK with 10 M, 12 M and 14 M. Note: FA-fly ash; G-GGBS and MK-metakaolin

the addition of 10% MK in these geopolymer paste, where this caused by MK has the highest specific area which leads to high reactivity and consumption of liquid at very early stage. Based on the obtained results, the initial setting time for all the mixtures was not in accordance with the standard range as specified in ASTM C191-21. Therefore, high content of POFA is not suggested as prime material, as it will delay the setting reaction compared with FA/GGBS-based geopolymer paste.

3.2 *Flowability*

The flow percentage of all the geopolymer mixture was obtained in accordance with specification of ASTM C230-21, standard flow percentage were specifying as 90–100% for CEM Type-1. Figure 3 shows the flowability behavior of geopolymer paste for all combination at 10, 12 and 14 M molarity. The flowability investigation revealed that 10 M developed higher flow percentage compared with 12 and 14 M. Therefore, the findings suggested that higher molarity contributes to lower flowability of the geopolymer paste mixture. From obtained results, it can be seen that 100 FA has the highest flow percentage compared with other binder combination regardless to molarity, where this is concluding that higher FA content shows a positive impact toward the workability. Inclusion of GGBS has diminished the flowability, with an increase in GGBS content and a reduction of FA content, and flowability trended downward. Increasing GGBS content contributes large numbers of angular shape

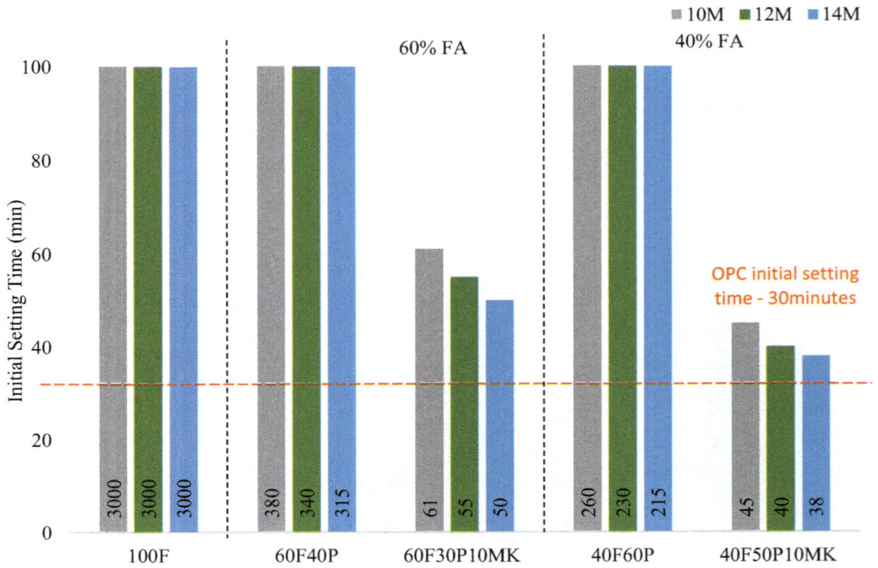

Fig. 2 Initial setting time for FA/POFA-based geopolymer paste including MK with 10 M, 12 M and 14 M. Note: FA-fly ash; P-POFA and MK-metakaolin

particles increased and spherical particles of FA reduces which results in dramatic reduction in flow percentage. On the other hand, POFA inclusion also shows reduction in flow percentage. This is attributed due to porous and spongy nature of POFA where it absorbs the liquid at very early stage and results in low-flow percentage [12]. However, the addition of 10% MK enhances the flow percentage where it's noticeably seen on 60F30G10MK mixture, this scenario is likely due to the differences in the physical properties and chemical reaction of the mixture. It is also observed higher water demand is required for high percentage of GGBS and POFA content, this is due to production of higher amount of amorphous phase in the combination which increases the chemical reaction, thereby demands high consumption of water as the reaction rate became higher and finally resulting lower flowability, similar findings have been reported by Caroline et al. From the results, 100F, 60F40G and 60F30G10MK geopolymer paste had reached the standard flowability percentage with > 100%, as specified in accordance with ASTM C230. Saying as above, GGBS and POFA as a prime material gives less favorable results toward the flowability of the geopolymer paste, and with FA as primer material the desired flowability can be achieved.

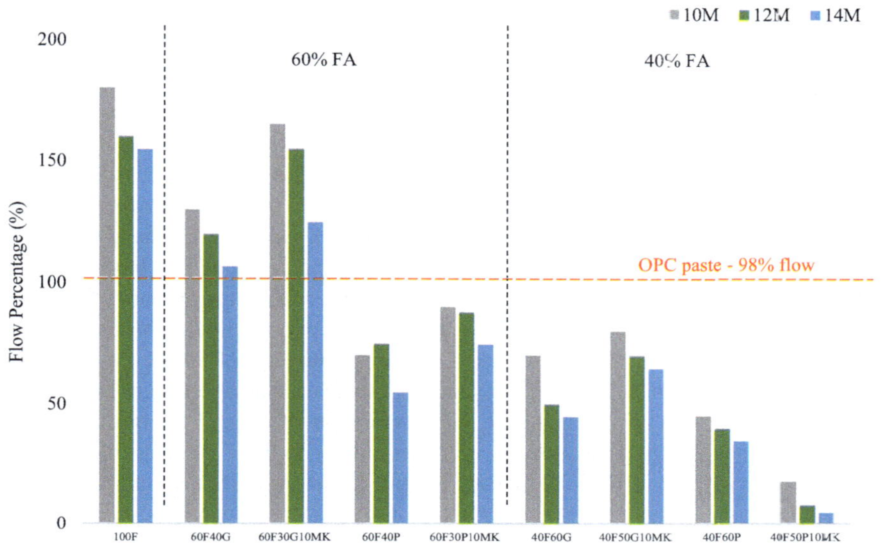

Fig. 3 Flowability of geopolymer paste including 10%MK with 10, 12 and 14 M. *Note* FA-fly ash; G-GGBS; P-POFA and MK-metakaolin

3.3 Compressive Strength

The compressive strength was conducted in accordance with ASTM C109-21, where CEM Type-1 cement paste compressive strength resulted 10.4 MPa at 1 day and 39.5 MPa at 28 days, all the results obtained for geopolymer paste mixtures where compared with CEM Type-1 cement paste compressive strength result. The effect of molarity on the compressive strength for the age of 1 day is presented in Fig. 4, on the other hand, the effect of molarity on the compressive strength at 28 days is presented in Fig. 5. According to the obtained results, it shows that 12 M shows the highest strength gain at day 1 and a similar trend is observed at age 28 days, the compressive strength is observed higher at 12 M for all the geopolymer paste mixtures except for 40F60P and 40F50P10MK. Impressive compressive strength in both early and mature age of the geopolymer paste has been observed at 12 M, saying such it is concluded that 12 M is an appropriate molarity for the compressive strength development of FA-based geopolymer paste. 60F30G10MK at 12 M shows extremely great compressive gain from the early age to mature age with the development percentage of 160%. On the other hand, for 60F30P 10MK shows 35% strength development from day 1 to day 28. From the obtained results, inclusion of GGBS drastically improves and shows a significant increment in compressive strength in both early and matured age of geopolymer paste with up to 40% of GGBS content, where this is due to GGBS that helped forming C–S–H gel and accelerated the reaction which results in increase of strength (Pathi et al., 2022). However, as the GGBS content gets higher more than

40% of decrease of compressive strength is observed, this scenario can be explained as the addition of GGBS content increased it contributes to increment to higher rate of Ca_{2+} ions in the geopolymer system, and with alkaline activator the calcium ions diluted; similar finding has been reported by Nath et al. [13]. Even though, CaO is an important role in contributing to compressive strength and soundness of hardened geopolymer paste, excess CaO content causes expansion and disintegration, where due to the speedy setting of paste resulting in poor initial matrix framework hindered the strength development (Chindaprasirt et al. 2018). On the other hand, inclusion of POFA in geopolymer paste more than 30% shows low strength at both early and mature age compared with other paste mixtures, this is due to the amount of CaO and Al_2O_3 that has a tremendous effect on the compressive strength. The character of POFA affected the compressive strength due to its low ratio of SiO_2 compared with FA. Based on the results, POFA is not suggested to be used as a secondary competent. However, the usage of the material, which consists of Si and Al in conjunction with POFA has shown a favorable result, particularly to the FA/GGBS blended. As well as inclusion of MK as third supplementary material shows enhancement in strength in both 1 day and 28 days for all geopolymer paste mixes, however highest strength development was observed for mix 60F30G10MK at 12 M for both 1 and 28 days. Inclusion of 10% MK improved the compressive strength tremendous as MK has finer grain size where it enhances the rate of alkaline activation and the mechanism of gel formation improves which results in improvement of mechanical properties. From Table 1, MK has approximately 88.5% combined with SiO_2 and Al_2O_3 where higher concentration of SiO_2 and Al_2O_3 in MK increase the alkaline activation reaction, resulting in formation of C–S–H, C–A–S–H and N–A–S–H gels, where this comparably enhance the density and its mechanical properties.

4 Conclusion

The following conclusion is made from the study:

- Initial setting time is affected by increasing the content of GGBS or POFA and high NaOH concentration (molarity). The inclusion of 10% metakaolin as third material significantly enhances the initial setting time by decreasing the setting process.
- Higher molarity contributes to low flowability, but with the presence of high content fly ash mitigates the low flow percentage and significantly increases the flowability of geopolymer paste.
- The flowability of geopolymer paste decreased with high levels of GGBS within the fly ash-based geopolymer system. Where else, POFA content more than 30% either as binary or ternary component affected the flowability of the fly ash-based geopolymer system.

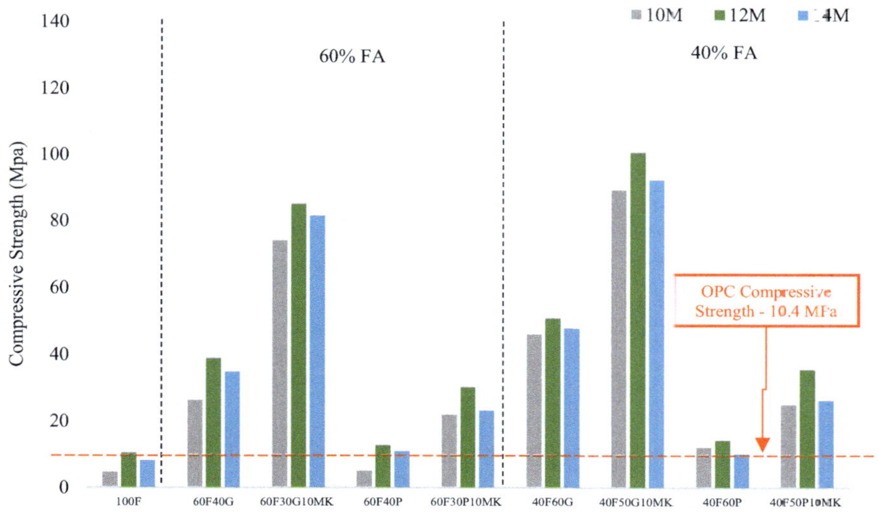

Fig. 4 Compressive strength geopolymer paste at age 1 day. *Note* FA-fly ash; G-GGBS, P-POFA and MK-metakaolin

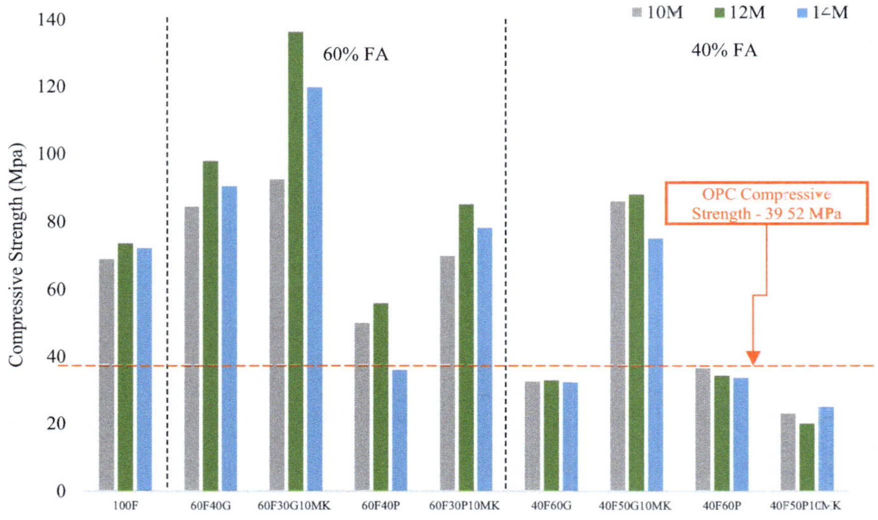

Fig. 5 Compressive strength of geopolymer paste at age 28 days. *Note* FA-fly ash; G-GGBS; P-POFA and MK-metakaolin

- The inclusion of 10% metakaolin as third material significantly enhances flowability of fly ash-based geopolymer paste regardless of molarity level except for combination of 40F50P10MK.
- 12 M shows the most favorable compressive strength development at both 1 and 28 days of the geopolymer paste. The highest compressive strength at 1 day age was observed for 40F50G10MK (12 M), however strength reduction over time was observed.
- The second highest compressive strength was observed for 60F30G10MK (12 M), however the compressive strength over time improved to be the highest among all the other mixes at 28 days.
- 60F40G and 60F30G10MK with 12 M shows an excellent early age strength as well as strength at 28 days and has potential to extend for mortar study.
- 60F30P10MK shows good strength gain at age of 28 days regardless of molarity and has potential in extending to mortar study.
- Inclusion of MK as third supplementary material shows enhancement in strength in both 1 day and 28 days for all the ternary geopolymer mixes; however, highest strength development was observed for mix 60F30G10MK with 12 M at 28 days.
- A combination of FA/POFA at high FA usage (60%) and POFA at 30% had a potential strength development with the presence of MK as third material.

Acknowledgements The authors would like to acknowledge Fundamental Research Grant Scheme given by the Ministry of Education (Ref: FRGS/1/2019/TK01/UPM/02/2) for the financial support given.

Declaration of Conflict of Interests The author(s) declare(s) that there is no conflict of interest. They have no known competing financial interests or personal relationships that could have appeared to influence the work reported in this paper.

References

1. Ge, X., Xiang, H., Shi, C.: Impact of micro characteristics on the formation of high-strength class F fly ash-based geopolymer cured at ambient conditions Constr. Build. Mater. **352**, 129074 (2022)
2. Salih, M.A., Farzadnia, N., Demirboga, R., Abang Ali, A.: Effect of elevated temperatures on mechanical and microstructural properties of alkali-activated mortar made up of POFA and GGBS. Construct. Build. Mater. **328** 127040 (2022)
3. Alexander, A.E., Shahikala, A.P.: Studies on the microstructural and durability characteristics of ambient cured FA-GGBS based geopolymer mortar. Construc. Build. Mater. **347**, 128538 (2022)
4. Isa, M.N., Awang, H.: Characteristic of palm oil fuel ash geopolymer mortar activated with wood ash lye cured at ambient temperature. J. Build. Eng. **66**, 105851 (2023)
5. Tangcirapat, W., Jaturapitakkul, C., Chinaprasirt, P.: Use of palm oil fuel ash as a supplementary cementitious material for producing high strength concrete. Const. Build. Mater. **23**(7), 2641–2646 (2009)

6. Abdullah, K., Hussin, M., Zakaria, F., Muhamad, R., Abdul Hamid, Z., POFA: A potential partial cement replacement material in aerated concrete. In: Proceedings of the 6th Asia-Pacific Structural Engineering and Construction Kuala Lumpur, Malaysia (2006)
7. Ranjbar, N., Mehrali, M., Johnson Alengram, U., Metselaar, H.S.C., Jumaat, M.Z.: Compressive strength and microstructural analysis of fly ash/palm oil fuel ash based geopolymer mortar under elevated temperatures. Construct. Build. Mater. **65**, 114–121 (2014)
8. Megat Johari, M.A., Zeyad, A.M, Muhamad Bunnori, N., Ariffin, K.S. Engineering and transport properties of high strength green concrete containing high volume of ultrafine palm oil fuel ash. Construct. Build. Mater. **30**(0), 281–288 (2012)
9. Islam, A., Alengaram, U.J., Jumaat, M.Z., Bashar, I.I.: The development of compressive strength of ground granulated blast furnace slag-palm oil fuel ash-fly ash based geopolymer mortar. Mater. Des. **56**, 833–841 (2014)
10. Sadangi, S.C., Pradhan, B.: Effect of admixed chloride and molarity of NaOH solution on early-age strength and reinforcement corrosion in FA/GGBS geopolymer concrete. Mater. Today: Proc. **65**, 1528–1533 (2022)
11. Wang, W., Fan, C., Wang, B., Zhang, X., Liu, Z.: Workability, rheology and geopolymerization of fly ash geopolymer: role of Alkali content, modulus, and water-binder ratio Constr. Build. Mater. **367**, 130357 (2023)
12. Huseien, G.F., Tahir, M.M., Mirza, J., Ismail, M., Wei Shah, K., Asaad, M.A., Effects of POFA replaced with FA on durability properties of GBFS included alkali activated mortars. Const. Build. Mater. **175**,174–186 (2018)
13. Nath, P., Sarker, P.K.: Effect of GGBFS on setting time, workability and early strength properties of fly ash geopolymer concrete cured in ambient condition. Constr. Build. Mater. **66**, 163–171 (2014)
14. Carolina Co, A., Trindade, S., Ribeiro de Avillez, R., Letichevsky, S., de Andrade Silva, F.: Influence of precursor material on the fresh state and thermos-chemo-mechanical properties of sodium-based geopolymers. Ceramic Int. **48**, 19806–19817 (2023)
15. Hosan, F.U.A., Shaikh: Influence of nano silica on compressive strength, durability, and microstructure of high-volume slag and high-volume slag-fly ash blended concretes. Struct. Concr. **22**(S1) (2021). https://doi.org/10.1002/suco.202000251
16. ASTM C191-21: Standard test methods for time of setting of hydraulic cement by vicat needle (2021)
17. ASTM C109-21: Standard test method for compressive strength of Hydraulic Cement Mortars (Using 2-in. or 50mm Cube Specimens)
18. ASTM C230-21: Standard specification for flow table for use in test of hydraulic cement (2021)
19. Bakhrey, T.: Durability of geopolymer materials in sodium and magnesium sulfate solutions. Cem. Concr. Res. **35**(6), 1233–1246 (2005)
20. Bakhrev, T.: Thermal behavior of geopolymers prepared using class F fly ash and elevated temperature curing. Cement and Concrete Res. **36**(6), 1134–1147
21. BS EN 1015-3. Methods of test for mortar for masonry determination of consistence of fresh mortar (by flow table) (1999)
22. BS EN 12390-3: Testing hardened concrete—compressive strength of test specimens (2019)
23. Nagiah, C., Azmi, R.: A review of smallholder oil palm production: challenges and opportunities for enchancing sustainability- a Malaysian prespective. J. Oil Palm Environ. **3**, 144–120 (2012)
24. Chi, M., Huang, R.: Binding mechanism and properties of alkali-activated fly ash/slag mortars. Construct. Build. Mater. **40**, 291–298. https://doi.org/10.1016/j.conbuildmat.2012.11.003 (2013)
25. Davidovits, J.: 30 Years of successes and failures in geopolymer applications. Market trends and potential breakthroughs. In: Geopolymer Conference October 28–29 Melbourne Australia (2002)
26. Davidovits, J.: Geopolymeric reactions in archaeological cements and in modern blended cements. Geopolymer **1**, 93–106 (1998)

27. Ded, P.S., Nath, P., Sarker, P.K.: The effect of ground granulated blast furnace slag blending with fly ash and activator content on the workability and strength properties of geopolymer concrete cured at ambient temperature. Mater Des. **62**, 32–39 (2014). https://doi.org/10.1016/j.matdes.2014.05.001
28. Xiaz, E.I., Allouche, E.N., Eklund, S.: Factors affecting the suitability of fly ash as source material for geopolymers. Fuel **89**(5), 992–996 (2010). https://doi.org/10.1016/j.fuel.2009.09.012
29. Fahim, H., Mirza, G., Ismail, M., Ghoshal, S.K., Abdulameer Hussein, A.: Geopolymer mortars as sustainable repair material: a comprehensive review. Renew. Sustain. Energy Rev. **80**, 54–74 (2017). https://doi.org/10.1016/j.rser.2017.05.076
30. Hardjito, D., Cheak, C.C., Ing, C.H.L.: Strength and setting time of low calcium fly-ash based geopolymer mortar. Modern Appl. Sci. **2**(4), P3 (2009)
31. Poloju, K.K., Srinivasu, K.: Influences of GGBS and concentration of sodium hydroxide on strength behavior of geopolymer mortar. Mater. Today: Proc. **65**, 702–706 (2022)
32. Khater, H.: Effect of calcium on geopolymerization of Aluminosilicate Waste. J. Mater. Civ. Eng. **24**(1), 92–101 (2011)
33. Khale & Chaudhary: Mechanism of geopolymerization and factors influencing its development: a review. J. Mater Sci. **42**, 729–746 (2007)
34. Kumar, S., Kumar, R., Mehrotra, S.: Influence of granulate blast furnace slag on the reaction, structure and properties of fly ash based geopolymer. J. Mater. Sci. **45**(3), 607–615 (2010)
35. Lim, S.K., Tan, C.S., Lim, O.Y., Lee, Y.L.: Fresh and hardened properties of lightweight foamed concrete with palm oil fuel ash as filler. Constr. Build. Mater. **46**, 39–47 (2013)
36. Mustafa Al Bakri, A.M., Kamarudin, H., Abdulkareem, O.A.K.A., Ruzaidi, C.M., Rafiza, A.R., Norazian, M.N.: Optimization of alkaline activator/fly ash ratio on the compressive strength of manufacturing fy-ash based geopolymer (2011)
37. Ranjbar, N., Mehrali, M., Behnia, A., Johnson Alengaram, U., Zamin Jumaat, M.: Compressive strength and microstructural analysis of fly ash/palm oil fuel ash based geopolymer mortar. Mater. Des. **59**, 532–539 (2014)
38. Nazri, A.: Compressive strength of geopolymers produced by ordinary Portland cement: application of genetic programming for design. Mater. Des. **43**, 356–366 (2013)
39. Munjal, P., Hau, K.H., Arthur, C.C.H.: Effect of GGBS and curing conditions on strength and microstructure properties of oil well cement industry. J. Build. Eng. **40**, 102331 (2021)
40. Phoo-ngemkham, T., Chindaprasit, P., Sata, V., Pangdaeng, S., Sinsiri, T.: Properties of high calcium fly ash geopolymer paste with Portland cement as an additive. Int. J. Miner. Metall. Mater. **20**(2), 214–220 (2013)
41. Puertas, F., Fernandez-Jimenez, A.: Mineralogical and microstructural; characteristic of alkali-activated fly ash/slag pastes. Cement Concr. Compos. **25**(3), 287–292 (2003)
42. Rashad, A.M.: Alkali activated metakaolin: a short guide for civil engineer-an overview. Constr. Build. Mater. **41**, 751–765 (2013)
43. Rovnaník, P.: Effect of curing temperature on the development of hard structure of metakaolin-based geopolymer. Const. Build. Mater. **24**(7), 1176–1186 (2010)
44. Sata, V., Jaturapitakkul, C., Kiattikomol, K.: Utilization of palm oil fuel ash in high strength concrete. J. Mater. Civ. Eng. **16**(6), 623–628 (2004)
45. Sata, V., Jaturapitakkul, C. And Kiattikomol, K., Influences of pozzolan from various by-product materials on mechanical prpoerties of high strength concrete. Const. Build. Mater. **21**(7), 1589–1598 (2007)
46. Temuujin, J., Van Riessen, A., Williams, R.: Effect of mechanical activation of fly ash on the properties of geopolymer cured at ambient temperature. J. Mater. Process. Technol. **209**(12), 5276–5280 (2009)
47. Temuujin, J., Williams, R., Van Riessen, A.: Influence of calcium compounds on the mechanical properties of fly ash geopolymer paste. J. Hazard. Mater. **167**(1), 82–88 (2009)

48. Tunerner, L., Collins, F.: Cardon dioxide equivalent emissions: a comparison between geopolymer and OPC cement concrete. Construct. Build. Mater. **43**, 125–130 (2013)
49. Zarina, L., Yang Cai, D., Xu, Y., Wei Fan, C., Qin Peng, X.: Novel method for preparation of calcined kaolin intercalation compound-based geopolymer. Appl Clay Sci. **101**, 637–642 (2016)

Mechanical and Leaching Properties of Solidified Clinical Waste Incinerator Fly Ash Mortar

Ezliana Ghazali, Megat Azmi Megat Johari, Mohd Azrizal Fauzi, and Noorsuhada Md Nor

Abstract With the rise of COVID-19 pandemic cases, the volume of clinical waste delivered to incinerator plants from hospitals, clinics, laboratories, temporary quarantine centres, and research institutes worldwide has increased significantly. These incinerator plants will produce clinical waste incinerator fly ash (CWIFA), which will overburden the landfill if all is placed there, but it has the potential to be utilized as a supplementary cementitious material in cement mortar or concrete. The goal of this cement-based mixed CWIFA (0, 2.5, 5, 10, and 15% as a partial cement replacement) research is to study the strength, the leachability of heavy metals, and the resistance to chloride ion penetration of the mortars with CWIFA inclusion. This will enable this CWIFA to be used in construction, while being disposed of safely in landfills that only need minimal environmental protection. The solid mortar leaching behaviour and mechanical properties were investigated with a ratio W/C of 0.485. Results show that adding 2.5% CWIFA as cement replacement in mortar mixture achieved the highest compressive and flexural strength of 41.04 MPa and 8.00 MPa, respectively. Further increase in CWIFA content in the mortars reduced compressive and flexural strengths. The rapid chloride permeability (RCP) values also increased with the inclusion of CWIFA in the mortar mixtures. A toxicity characteristic leaching procedure (TCLP) test performed on all solidified mortar mixtures portrays that the

E. Ghazali (✉) · M. A. Megat Johari
School of Civil Engineering, Universiti Sains Malaysia, Engineering Campus, 14300 Nibong Tebal, Pulau Pinang, Malaysia
e-mail: ezlianaghazali85@student.usm.my

M. A. Megat Johari
e-mail: cemamj@usm.my

E. Ghazali
Engineering Services Division, Ministry of Health, 62590 Putrajaya, Malaysia

M. A. Fauzi · N. Md Nor
School of Civil Engineering, College of Engineering, Universiti Teknologi MARA, Cawangan Pulau Pinang, 13500 Permatang Pauh, Pulau Pinang, Malaysia
e-mail: azrizal@uitm.edu.my

N. Md Nor
e-mail: ida_nsn@uitm.edu.my

N. Sabtu (ed.), *Proceedings of AWAM International Conference on Civil Engineering 2022—Volume 2*, Lecture Notes in Civil Engineering 385,
https://doi.org/10.1007/978-981-99-6018-7_39

heavy metals investigated (Cd, Cu, Ni, Pb, and Zn) are strongly immobilized by the cementitious matrix. Therefore, 2.5% CWIFA as cement replacement in mortar is better to be utilized in mortar due to good strength, RCP, and leaching results.

Keywords Clinical waste · Leaching · Strength · Rapid chloride permeability

1 Introduction

Municipal and clinical solid wastes are burned to reduce their volume due to the difficulty of procuring regulated and uncontrolled landfill sites for trash disposal. The incineration process reduces the volume of solid waste by 65–70% [1, 2] and produces clinical waste incinerator bottom ash and clinical waste incinerator fly ash (CWIFA). 80–85% of the stiff solid mass of actual waste is reduced using incinerators [3]. In India, the incineration procedure entails reducing collected clinical wastes to ash before transporting it to various landfills for disposal [4]. Therefore, toxic metals and inorganic compounds are released into the environment when the ash is deposited in landfills [4].

CWIFA is the flue gas that typically enters the baghouse filter and is collected at the hopper. CWIFA is an adsorbent for the toxic heavy metals emitted when clinical waste (CW) is burned. In addition, CWIFA must be appropriately disposed of in sanitary landfills to prevent the seepage of soluble and toxic elements into the subsoil because of its tiny particle size and high concentration of dangerous heavy metals. The high maintenance expenses of such facilities increase the total cost of the burning process. In addition, thermal treatment methods like incinerators-activated carbon and sodium bicarbonate will treat CWIFA by neutralizing acid and absorbing heavy metals and dioxins [5]. According to Ababneh [6], CWIFA can be used as a pozzolanic ingredient to enhance the strength of cementitious products.

Therefore, CWIFA can be employed as a supplemental cementitious material in the construction industry. Nonetheless, further study is required on this substance and its impact on cementitious materials and mortar characteristics. However, the use of CWIFA in the construction sector contributes to decreasing landfill space necessary for disposal and preserving non-renewable natural resources required for cement manufacture. This work is considered crucial in the twenty-first century, focusing on sustainable development. Therefore, substantial research [6] is being performed to find the safest and most practical technique of inertizing this material before disposing of it in landfills with less strict standards than untreated CWIFA and examining its possible usage as a building material. Currently, the cement-based solidification/stabilization (S/S) method is one of the better solutions [7, 8] for safely disposing of heavy metal-containing wastes.

There have been initiatives [6] for solidifying matrices based on CWIFA combinations using cement or mixed cement mixes. However, there has been a rise in awareness and concern over the possibility that the CWIFA might harm human health and the environment due to the probable leakage of toxic compounds into the environment

[9–12]. However, the long-term stability of these precipitates may be influenced by dangerous circumstances containing elements such as soil acids and acid rain. In the outer zones of a solidified matrix, the acidic environment may transform insoluble metal hydroxides into more soluble carbonates or even hydrocarbonates. Dangerous heavy metals in cementitious matrices must meet physical and chemical tests to assure their stability. The physical characteristic is the impermeability of the hardened product. At the same time, the chemical factor is the high alkalinity of the pore solution within the cementitious matrix, which allows heavy metals to be converted into insoluble compounds. As a result, the risk of contamination due to short-term leaching has been reduced.

CWIFA is categorized as a pozzolan, one of the primary components used to create the mortar. SiO_2 nanoparticles can function as nano-fillers to fill the spaces between C–S–H gel particles [6]. As a partial alternative for cement, CWIFA decreases the amount of Portland to facilitate a more progressive hydration process. Due to the presence of calcium and silica, as well as the filling effect of tiny particles, pozzolanic characteristics have been proven to boost compressive strength. The compressive strength values satisfy the criteria for solidified waste material, and the solidified waste material satisfies the 0.414 MPa requirement of the landfill [13]. At low SiO_2, FeO_3, and Al_2O concentrations, CWIFA lacks pozzolanic participation [6]. Utilizing CWIFA as a partial replacement for cement in the mortar will provide inexpensive goods. Most published research focuses on fly ash from municipal solid garbage (MSW). CWIFA from CW, which often includes higher sulphates and chlorides than MSW, is poorly understood. CWIFA S/cement-based S's strategy appears to be the most effective method for immobilizing hazardous chemicals in CWIFA wastes This research presents the outcomes of a solidification/stabilization (S/S) procedure for CWIFA applied to cement in various quantities to determine the optimal conditions for inertization. Mechanical and leaching experiments have been conducted to evaluate the S/S product's features. It also examines the potential future uses of CWIFA and the obstacles and possibilities in this field of study.

2 Materials

2.1 Sampling of CWIFA

The CWIFA used in this study was created at the RADICARE (M), Sdn Bhd Plant in Teluk Panglima Garang, Selangor, which serves as one of Malaysia's largest CW incinerators. The material was collected from the baghouse filter throughout a three-day sampling period to allow for potential modifications in incinerator feed composition. The sodium bicarbonate and activated carbon are injected to neutralize acid and absorb heavy metals and dioxins before flue gas enters the baghouse filter. This flue gas is called CWIFA and is collected at the hopper. Figure 1 shows the proses of the incineration plant. During sampling, the operating conditions were kept as consistent as feasible. The material was quartered to generate homogeneous samples for characterization and prepare specimens for mechanical and leaching testing.

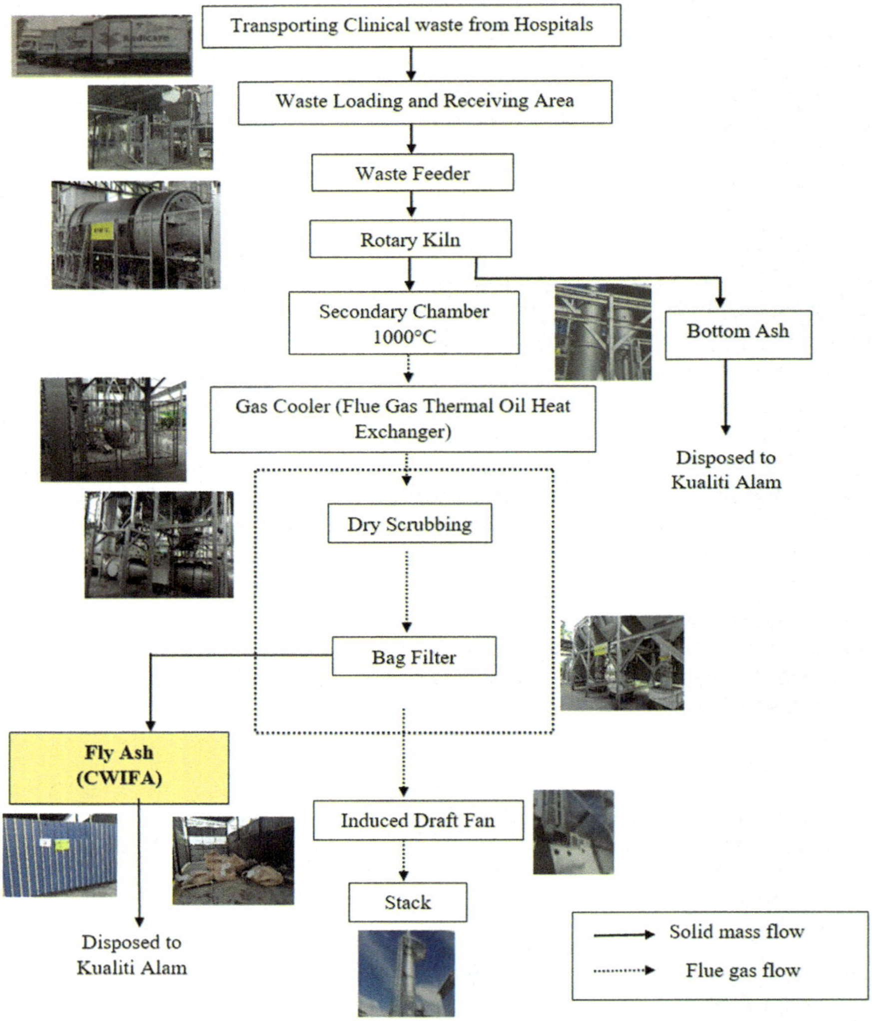

Fig. 1 Proses of incineration CW at Radicare Sdn Bhd incinerator plant, Telok Panglima Garang, Malaysia

2.2 Characterizations

Table 1 provides the elemental compositions and particle size distribution of the CWIFA. The predominant heavy metal in the CWIFA is chromium (121.407 mg/kg), followed by nickel (2.441 mg/kg). Suitability for landfills and utilization in construction material are determined by assessing the leachability of these heavy metals and other metals. Tables 2 and 3 show the compositions and physical characteristics of cement and sand. The particle size distribution of the CWIFA sample shown in Fig. 2a indicates a median particle size of 36.99 μm. The thermal analysis curves (differential thermal analysis (DTA) and thermogravimetry analysis (TGA)) results are shown in Fig. 2b. The results in Fig. 2b show the TGA results that include three steps of weight loss caused by dehydration (30–150 °C), crystallization (400–500 °C), and burning of organic matter (600–800 °C). For the first peak, the mass loss can be attributed to the movement of the moisture (free H_2O) contained in the CWIFA, which is closer to that measured by oven drying at 30–150 °C shows 3.17% moisture removal. The TGA's corresponding weight loss in the second peak region, mainly due to the dehydration of $Ca(OH)_2$ contained in the CWIFA at 400–500 °C, shows 1.24%. In the third range, the weight loss may belong to the decomposition of $CaCO_3$, which could be the presence of large quantities of chlorides and sulphates in CWIFA and TGA analysis at 600–800 °C is 5.895%. The DTA curve shows an endothermic break due the oxidation and structural breakdown. The images of CWIFA collected at the baghouse filter, CWIFA sample, and SEM image of CWIFA are shown in Fig. 3. The CWIFA particle displayed as spherical granules with numerous microparticles adsorbed on the surface, various forms, and quite a few CWIFA are the same amorphous micrograph as silica fume or calcined clay as shown in Fig. 3c.

Table 1 Physical characteristics and chemical compositions of CWIFA obtained from the clinical waste incinerator

Element (%)	Ca	Ti	Si		S	P
	53.230	1.824	3.303		0.792	0.550
	K	Mg	Fe		Mn	Al
	0.886	5.018	0.539		0.032	0.346
Element (mg/kg)	As	Cd	Co		Cr	Cu
	UD	UD	UD		121.407	0.134
	Mo	Ni	Pb		Se	Zn
	3.051	2.441	UD		UD	0.104
Moisture content (%)	Specific surface area (m^2/g)		Loss on ignition (%)	Colour	pH	Density (kg/m^3)
2.63	0.61		29.14	Dark grey	9.93	1834

UD Under Detection

Table 2 Physical and chemical characteristics of Portland cement used in the mortar mixes

Oxide	CaO	SiO_2	Fe_2O_3	Al_2O_3	SO_3	MgO	K_2O	Na_2O
%	64.56	20.8	3.79	4.77	2.34	2.52	0.4	0.02
Density (kg/m^3)		Specific surface area (m^2/g)		Moisture content (%)		Loss on ignition (LOI)		pH
3060		0.216		1.16		10.02		12.4

Table 3 Physical characteristics of sand used in the mortar mixes

Property	Mass passing 75-μm sieve	Bulk density (kg/m^3)	Specific gravity	Water absorption (%)	Moisture content (%)
	1.0%	1485	2.65	1.1	2.16

3 Methodology

The CWIFA was immobilized in mortar mixes made with cement and sand—whose chemical and physical characteristics are given in Tables 1, 2, and 3, respectively as well as water. The percentage of the CWIFA used ranges from 0, 2.5, 5, 10, and 15% as partial cement replacement, while the highest content used was 15%. In this study, five mixes of mortar mixture were prepared according to ASTM C109M [14]. All specimens were cast with sand-to-cementitious material (cement and CWIFA) ratio of 2.75:1, and a water-to-cementitious ratio was kept at 0.485. The whole sets of samples are given in Table 4.

3.1 Compressive and Flexural Strength

The 50 × 50 × 50 mm cube samples were cast to measure the compressive strength of the mortar mixes following ASTM C109M [14]. For flexural strength testing, 40 × 40 × 160 mm prism samples were cast by ASTM C348 [15]. After 24 h, the samples were demoulded and cured in water at a temperature of 2.25 °C and an appropriate depth. The compressive and flexural strengths of the samples were evaluated at 3, 7, 28, 91, and 180 days after curing.

3.2 Rapid Chloride Permeability

At 3, 7, 28, 91, and 180 days, chloride permeability was determined following ASTM C1202 [16]. The 50 mm thick specimens used for the test were obtained by cutting concrete cylinders 100 mm in diameter and 200 mm in height with a diamond-impregnated concrete cutter. The water-saturated samples were exposed to a 60-V

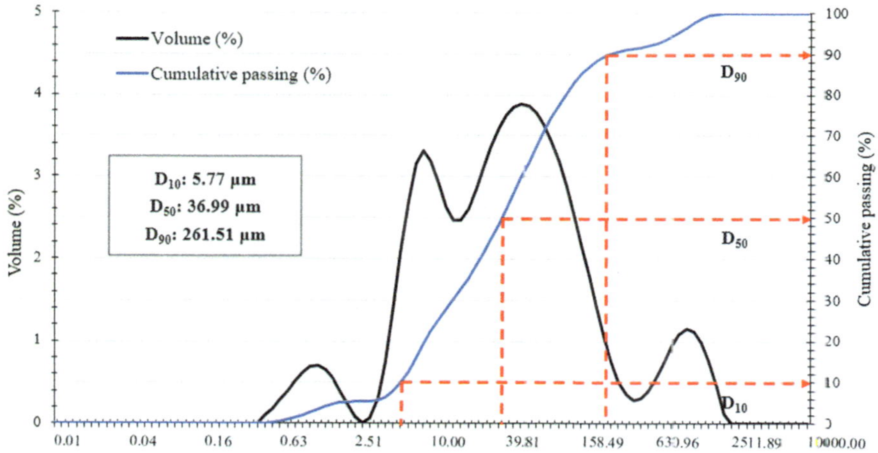

(a) Particle size distribution of the CWIFA sample

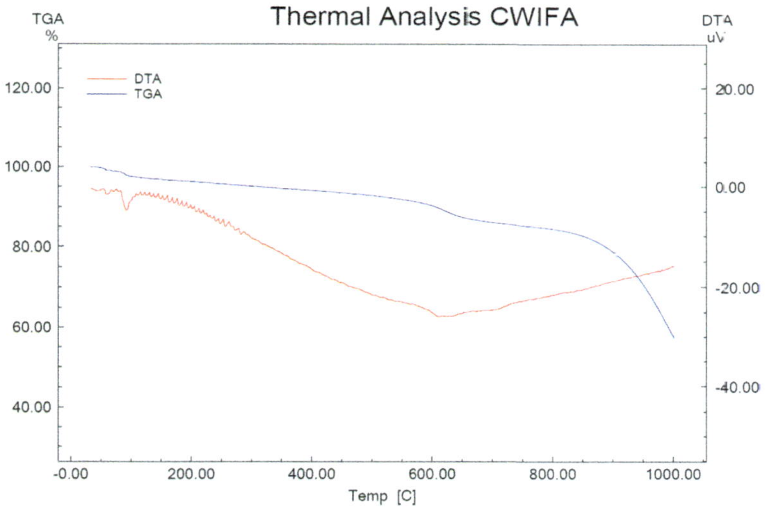

(b) Thermogravimetry results for CWIFA sample

Fig. 2 **a** Particle size distribution and **b** Thermogravimetry of CWIFA sample

electric potential for six hours. The specimens' chloride penetration was represented as the total charge passed (TCP) in Coulombs throughout the test time. The samples were immobilized and sealed adequately between the two-part chloride permeability cells. One side of the cell was filled with a 3% sodium chloride solution (which will be linked to the cathode terminal of the power supply), while the other was filled with 0.3 N sodium hydroxide solution and connected to the anode terminal. The interpretation is that the bigger the TCP or charge transmitted during the test, the

(a) CWIFA collected from bag house filter

(b) CWIFA

(c) CWIFA particles using Scanning Electron Microscopes (SEM)

Fig. 3 Sources collected (**a**), Image (**b**) and SEM image (**c**) of CWIFA

Table 4 Mortar mix proportions

Sample	OPC (kg/m^3)	CWIFA (kg/m^3)	Sand (kg/m^3)	Water (kg/m^3)	W/C ratio	CWIFA (%)
CWIFA-0 (Control)	540	0	1485	261.90	0.485	0
CWIFA-2.5	527	14	1485	261.90	0.485	2.5
CWIFA-5	513	27	1485	261.90	0.485	5
CWIFA-10	486	54	1485	261.90	0.485	10
CWIFA-15	459	81	1485	261.90	0.485	15

greater the permeability of the material—the more porous the mortar, the greater the charge transfer, and vice versa. The chloride permeability rating according to ASTM C1202 [16] is displayed in Table 5.

Table 5 Rating of chloride permeability (as per ASTM C1202-12)

Total charge passed (Coulombs)	Chloride ion penetrability
> 4000	High
2000–4000	Moderate
1000–2000	Low
100–1000	Very low
< 100	Negligible

3.3 Leaching Test

The specimens were tested for their leachability of toxic metals using the U.S. Environmental Protection Agency (US EPA) of solid waste toxicity characteristic leaching procedure (TCLP) (SW 846) by method 1311. These tests are used to evaluate the behaviour of the S/S mortar products during utilization in the construction industry and either disposed of in landfills where they are subjected to acid rains and CO_2 aggressive waters, which may dissolve the immobilized hazardous metals causing seepage of harmful substances into the subsoil. First, TCLP reduced the sample size to minus 9.5 mm particle size. Then, 100 grams of minus 9.5 mm samples were placed into an extraction bottle, and an extraction fluid was added to the extraction bottle at a weight equal to 20 times the weight of the sample particles. The extraction fluid consisted of 5.7 mL glacial CH_3CHOOH, 64.3 mL of NaOH, and 930 mL of reagent water used extraction fluid with the pH 4.93 ± 0.05. The extraction bottle was rotated at 30 ± 2 revolutions per minute for 18 ± 2 hours at a temperature of 23 ± 2 °C. The contents of the extraction bottle were filtered through a membrane filter with 0.45 μm pore size after the extraction period and finally analyzed to ascertain the heavy metal concentration.

4 Results and Discussion

4.1 Compressive and Flexural Strength

Figures 4 and 5, respectively, show the results of the compressive and flexural strength tests performed on mortar samples after 3, 7, 28, 91, and 180 days of curing. In all cases, the compressive strength of the specimens containing 2.5% CWIFA is higher than the control specimens (Portland cement mix) at all ages, from 3 to 180 days. At CWIFA replacement levels greater than 2.5%, the development of compressive strength of mortar containing various levels of replacement of CWIFA reduced, particularly the higher CWIFA contents. After 91 days of age, the strength increase is very small and consistent until 180 days in all the mixtures. In this study, compressive strength values at 180 days for CWIFA-0 (control sample), CWIFA-2.5, CWIFA-5,

CWIFA-10, and CWIFA-15 were 36.49 MPa, 41.04 MPa, 34.02 MPa, 25.46 MPa, and 21.24 MPa, respectively.

The strength of the CWIFA-5 was less than that of the reference mortar specimen because the water to cementitious and sand was the same in all the mixtures. However, CWIFA-5 mortar achieved over 75% of the control mortar's strength at testing ages 3, 7, 28, 91, and 180 days. This suggests that CWIFA-2.5 and CWIFA-5 values could be used in the mixes, provided that increased CWIFA does not inhibit cement hydration. Besides, the compressive strength values of all mortar mixtures meet the standard stipulated for solidified waste material, which after 28 days of solidification, is 0.414 MPa [13].

The flexural strength test results are shown in Fig. 5. The values were in the range of 3.9–8.0 MPa. The results show that adding 2.5% CWIFA to mortar increased

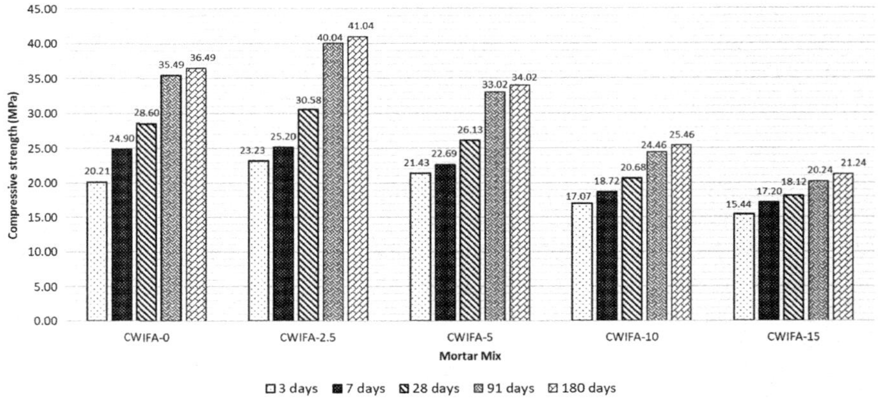

Fig. 4 Development of compressive strength of mortar containing different percentages of CWIFA

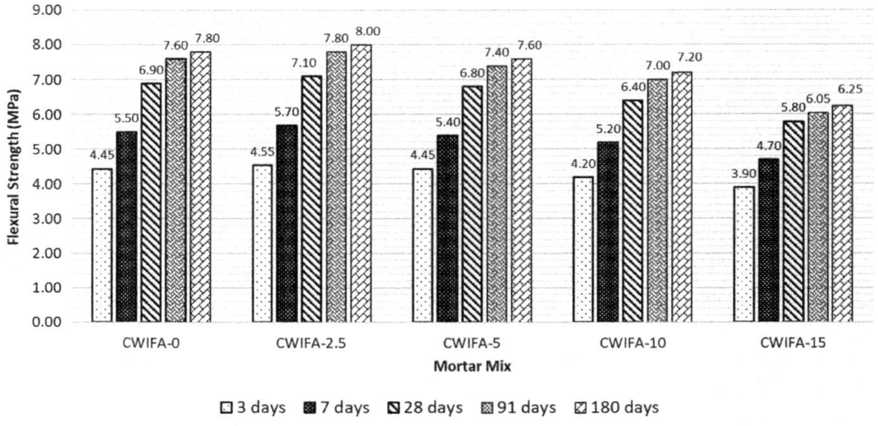

Fig. 5 Development of flexural strength of mortar containing different percentages of CWIFA

flexural strength. The increase in compressive and flexural strength at 2.5% CWIFA content is due to the increased pozzolanic reaction and the packing ability of the CWIFA particles. Nonetheless, adding 5–15% of CWIFA greatly reduces the strength but exerts no inhibiting effect on the cement hydration process. The CWIFA was observed in terms of accelerating the hydration process except that the CWIFA acted more like a filler than as a binder. Thus, increasing the ash content led to an increase in the surface area of the mortar filler to be bonded by the same amount of cement as that of the control [17].

4.2 *Rapid Chloride Permeability*

Figure 6 indicates a significant increase in chloride ion permeability for the mortar specimens containing CWIFA, indicating lower resistance to chloride ion penetration of the mortar samples containing CWIFA, particularly those containing higher CWIFA content, where the total charge passed (TCP) greatly increased with higher CWIFA content in the mortar mixtures. Nonetheless, the TCP significantly reduced at the longest curing period of 180 days for all mortar mixtures, where TCP values of 934, 1243, 1473, 1644, and 2317 coulombs were recorded for mixtures CWIFA-0, CWIFA-2.5, CWIFA-5, CWIFA-10, and CWIFA-15, respectively. The reduction in TCP with curing time could be associated with a reduction in porosity and permeability of the mortars from the prolonged cement hydration for the control mixture and continuous cement hydration, as well as the CWIFA pozzolanic reaction for the mixtures containing CWIFA. Based on the 180 days TCP values, the chloride ion penetrability of the mortar mixtures could be classified as very low for the CWIFA-0 mixture, low for the CWIFA-2.5, CWIFA-5, and CWIFA-10 mixtures, as well as moderate for the CWIFA-15 mixture. The CWIFA-2.5 in the mortar mixtures could be suitable for many applications. However, the inclusion of CWIFA from 5 to 15% in the mortar mixtures could still be suitable for many applications, particularly those not related to chloride exposure. This is due to a higher TCP result that will be a side effect of the corrosiveness.

The increase in TCP at all CWIFA replacement levels is inconsistent with the compressive and flexural strength results, which exhibited enhancement in strength at 2.5% CWIFA content. The heavy metal constituents could have contributed to the increase in the TCP in the mortars containing CWIFA, especially those with high CWIFA content, which could have promoted current flow passing the samples during the RCP test, leading to higher TCP.

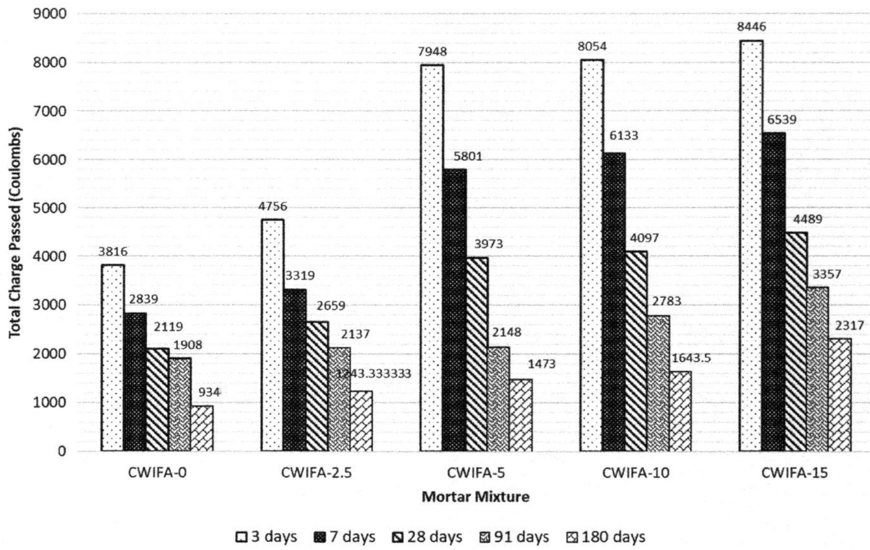

Fig. 6 Development of RCP at different percentages of CWIFA in mortar

4.3 Leaching Behaviour of the Mortar

Toxicity Characteristic Leaching Procedure (TCLP) Test

The results of the TCLP leaching test are presented in Table 6, along with the limit values established by the US EPA standard limit and the metal concentrations in the TCLP leaching solutions. The final pH of leaching solutions at the end of TCLP tests is also presented in Table 6. The results obtained on the raw CWIFA show concentrations of chromium (6.07037 mg/l), zinc (0.0052 mg/l), and nickel (0.122062 mg/l). Chromium exceeds the raw limits, so it is necessary to dispose of the CWIFA in a landfill with a high level of environmental protection. However, the results of the test on the mortar specimens, prepared with different percentages of CWIFA, indicate that the concentrations of all the heavy metals are well below the limits established by US EPA legislation. This proves that the stabilization process in mortar is effective, even after a relatively short curing time (3 days). These results show that no metals were leached over the USEPA limits. Curing the material improves its ability of the encapsulation of heavy metals. The excess acid could attack the matrix where heavy metals were encapsulated and induce metal release. However, the metal concentrations measured in all cases were much lower than the solubility, thus indicating that there is chemical entrapment of heavy metals in the hydrated cement paste phases where the heavy metals are strongly immobilized.

Table 6 Results of TCLP leaching test on CWIFA, Portland cement, sand, and products; curing time 3, 7, 28, 91, and 180 days

Element (mg/l)	As	Cd	Cr	Ni	Pb	Zn
Raw CWIFA	UD	UD	6.07037	0.122062	UD	0.0052
Sand	UD	UD	0.010363	UD	0.05325	0.27397
Portland cement	0.143394	UD	0.204675	UD	0.087617	UD
US EPA limit (mg/l)	5.0	1.0	5.0	–	5.0	–
3 days						
CWIFA-0	0.061434	UD	0.221441	0.081877	0.040741	UD
CWIFA-2.5	0.144258	UD	0.335525	0.099826	0.059980	UD
CWIFA-5	0.148040	UD	0.348412	0.112055	0.065658	UD
CWIFA-10	0.186238	UD	0.364804	0.117892	0.078451	UD
CWIFA-15	0.192967	UD	0.468077	0.120214	0.080303	UD
7 days						
CWIFA-0	UD	UD	0.072421	0.034706	UD	UD
CWIFA-2.5	0.067284	UD	0.110513	0.039979	UD	UD
CWIFA-5	0.105782	UD	0.175011	0.089855	UD	UD
CWIFA-10	0.149824	UD	0.236391	0.106690	0.027781	UD
CWIFA-15	0.189177	UD	0.332559	0.119907	0.034809	UD
28 days						
CWIFA-0	UD	UD	0.054796	UD	UD	UD
CWIFA-2.5	0.066494	UD	0.080475	UD	UD	UD
CWIFA-5	0.073030	UD	0.202227	0.024583	UD	UD
CWIFA-10	0.120002	UD	0.260064	0.050991	UD	UD
CWIFA-15	0.173201	UD	0.312379	0.110046	UD	UD
91 days						
CWIFA-0	UD	UD	0.048055	UD	UD	UD
CWIFA-2.5	UD	UD	0.035222	UD	UD	UD
CWIFA-5	UD	UD	0.066784	0.018046	UD	UD
CWIFA-10	UD	UD	0.115512	0.025714	UD	UD
CWIFA-15	UD	UD	0.168168	0.046900	UD	UD
180 days						
CWIFA-0	UD	UD	0.021199	UD	UD	UD
CWIFA-2.5	UD	UD	0.029944	UD	UD	UD
CWIFA-5	UD	UD	0.052842	UD	UD	UD
CWIFA-10	UD	UD	0.058138	0.015072	UD	UD
CWIFA-15	UD	UD	0.075621	0.030440	UD	UD

UD Undetected

5 Conclusions

The compressive and flexural strength of the mortar mixtures increases by adding 2.5% of CWIFA as cement replacement in the mortar mixture due to the pozzolanic activity, and CWIFA will act as a micro filler in the mortar matrix. On the other hand, the RCP values decrease with increased curing duration for all the mixtures due to increased hydration. The cement replacement by CWIFA was carried out in this study on a mass-for-mass basis; the modified mortars are expected to have lower porosities at early ages because of the replacement materials than the cement it replaces. However, at later ages, the pozzolanic reaction of the replacement materials results in pore structure densification. Besides, the chloride contained in CWIFA affected the increase of RCP values.

The TCLP test was performed on materials at 3, 7, 28, 91, and 180 days to investigate the influence of the curing duration on the retention of S/S of mortar pollutants. It is observed that concentrations of heavy metals in S/S mortar materials are reduced compared with the concentrations of elements leached from the raw waste. The process of S/S applied to waste has decreased the polluting character and the release of heavy metals into the environment considerably. The excess acid could attack the matrix where heavy metals were encapsulated and induce metal release. With the increase in strength, the heavy metals are better encapsulated, inhibiting or reducing leaching. Therefore, the mortar could be used in the construction where high chloride resistance is not required. As evidenced by the results of the TCLP leaching test performed on the S/S mortar, the heavy metals investigated (Cd, Cu, Ni, Pb, and Zn) are strongly immobilized by the cementitious matrix. The S/ S products can thus be disposed of in safety landfills with a lower level of environmental protection than the untreated waste needs. Besides, the product could be utilized in the construction industry, where CWIFA has the potential as a replacement for cement in mortar products.

Acknowledgements The authors would like to thank the Ministry of Health (MOH, Civil Engineering Division) for their study's scholarship, encouragement, and support. Besides, thanks to Clinwaste Sdn Bhd and Radicare Sdn Bhd for the supply of CWIFA and information regarding this study.

References

1. Ali Jawaid, S.M., Kaushik, J.: Solidification/stabilisation of hospital solid waste incinerator ash for utilization in geotechnical construction, GeoCongress 2012 © ASCE 2012 pp. 3806–3815 (2012)
2. Xie, Y., Zhu, J.: The detoxification of medical waste incineration fly ash using self-propagating reaction, procedia. Environ. Sci. **16**, 222–228 (2012). https://doi.org/10.1016/j.proenv.2012.10.031
3. Rahman, H.A.: Incinerator in Malaysia: really needs? Int. J. Chem. Environ. Biol. Sci. **1**, 678–681 (2013)

4. Kaur, H., Siddique, R., Rajor, A.: Influence of incinerated biomedical waste ash on the properties of concrete. Constr. Build. Mater. **226**, 428–441 (2019). https://doi.org/10.1016/j.conbuildmat.2019.07.239
5. Tri Ecoedge Sdn Bhd, Detailed environmental impact assessment (EIA) Report for proposed clinical waste incinerator replacement project at radicare (M) Sdn Bhd Plant in Lot 5, Jalan Waja 16, Kawasan Perindustrian, 42500, Teluk Panglima Garang, Selangor (2018)
6. Ababneh, A., Al-Rousan, R., Gharaibeh, W., Abu-Dalo, M.: Recycling of pre-treated medical waste fly ash in mortar mixtures. J. Mater. Cycles Waste Manag. **22**, 207–220 (2020). https://doi.org/10.1007/s10163-019-00928-z
7. Malviya, R., Chaudhary, R.: Factors affecting hazardous waste solidification/stabilization: a review. J. Hazard. Mater. **137**, 267–276 (2006). https://doi.org/10.1016/j.jhazmat.2006.01.065
8. Akyıldız, A., Köse, E.T., Yıldız, A.: Compressive strength and heavy metal leaching of concrete containing medical waste incineration ash. Constr. Build. Mater. **138**, 326–332 (2017). https://doi.org/10.1016/j.conbuildmat.2017.02.017
9. Rajor, A., Xaxa, M., Mehta, R.: Kunal, an overview on characterization, utilization and leachate analysis of biomedical waste incinerator ash. J. Environ. Manage. **108**, 36–41 (2012). https://doi.org/10.1016/j.jenvman.2012.04.031
10. Rozumová, L., Motyka, O., Čabanová, K., Seidlerová, J.: Stabilisation of waste bottom ash generated from hazardous waste incinerators. J. Environ. Chem. Eng. **3**, 1–9 (2015). https://doi.org/10.1016/j.jece.2014.11.006
11. Sobiecka, E., Obraniak, A., Antizar-Ladislao, B.: Influence of mixture ratio and pH to solidification/stabilisation process of hospital solid waste incineration ash in Portland cement. Chemosphere **111**, 18–23 (2014). https://doi.org/10.1016/j.chemosphere.2014.03.057
12. Rajan, D.T., Robin, M.: Vandanarani, biomedical waste management in Ayurveda hospitals—current practices and future prospectives. J. Ayurveda Integr. Med. **10**, 214–221 (2019). https://doi.org/10.1016/j.jaim.2017.07.011
13. Anastasiadou, K., Christopoulos, K., Mousios, E., Gidarakos, E.: Solidification/stabilisation of fly and bottom ash from medical waste incineration facility. J. Hazard. Mater. **207–208**, 165–170 (2012). https://doi.org/10.1016/j.jhazmat.2011.05.027
14. ASTM C109/C109M(b), Standard Test Method for Compressive Strength of Hydraulic Cement Mortars (Using 2-in. or [50 mm] Cube Specimens), ASTM Int. West Conshohocken, PA, 2020, www.Astm.Org (2020). https://doi.org/10.1520/C0109
15. ASTM C348, Flexural strength of hydraulic-cement mortars. Am. Soc. Test. Mater. 04 (2020) 1–6
16. ASTM C1202, Standard test method for electrical indication of concrete's ability to resist chloride ion penetration. Am. Soc. Test. Mater. 1–8 (2012). https://doi.org/10.1520/C1202-12.2
17. Udoeyo, F.F., Hilary, I., Young, D., Oparadu, E. Potential of wood waste ash as an additive in fibre reinforced concrete. Int. J. Eng. Res. V4 605–611 (2015). https://doi.org/10.17577/IJERTV4IS120443

GPR Assessment Method of Reinforced Concrete Structures: A Review

Ahmad Zaki

Abstract Reinforced concrete (RC) structural damage has become a global issue. As a result, there is an increasing need for actual health monitoring methods. The development of reliable monitoring tools for assessing the damage and condition of RC structures has been identified as a pressing requirement. The ground penetrating radar (GPR) method is a valuable non-destructive testing method that depicts excellent potential for being a part of an automated assessment component in RC structures. This paper explores the application of GPR for assessment in RC structures (i.e., localization and estimation of reinforcement bars (rebar), corrosion detection, and void detection). However, the success has been confined to simple scenarios, necessitating the usage of post-processing techniques, i.e., image processing, signal processing, and classification technique. Combining GPR with other NDT methods has long been recognized as having the ability to increase the value of test combinations. This study aims to review GPR as an assessment method of RC structures.

Keywords RC · Damage · Assessment · Post-processing · NDT · GPR

1 Introduction

Throughout the preceding century, reinforced concrete (RC) structures were used in numerous civil engineering constructions with exceptional durability [1]. However, degradation affects the RC structures' durability. Furthermore, degradation significantly impacts RC structures' service life and maintenance costs [2]. After a limited service life, the deteriorating structures must be repaired [3, 4]. Repairing RC structures is almost as crucial as building new ones. An appropriate evaluation technique is required early on because the deterioration has substantially harmed the structures'

A. Zaki (✉)
Department of Civil Engineering, Faculty of Engineering, Universitas Muhammadiyah Yogyakarta, Bantul 55183, Daerah Istimewa Yogyakarta, Indonesia
e-mail: ahmad.zaki@umy.ac.id

N. Sabtu (ed.), *Proceedings of AWAM International Conference on Civil Engineering 2022—Volume 2*, Lecture Notes in Civil Engineering 385,
https://doi.org/10.1007/978-981-99-6018-7_40

functioning. In general, the assessment method provides data on RC structures' current condition to estimate their future performance. Furthermore, a proper evaluation technique is essential for efficient and cost-effective rehabilitation of existing RC structures [5–7].

Non-destructive testing (NDT) methods are objective assessment methods. NDT methods should assess RC structures without causing damage and be applied to both new and aged structures. The primary use of NDT methods for new constructions is expected to be the quality control of material and building process. Furthermore, a proper evaluation technique is essential for the efficient and cost-effective rehabilitation of existing RC structures [8–11]. Some of the NDT methods include visual inspection, infrared thermography (IRT), half-cell potential (HCP), acoustic emission (AE), ultrasonic pulse velocity (UPV), and ground penetrating radar (GPR) [6, 12, 13].

Visual examination is a popular method of determining the extent of damage in RC constructions. Still, it is highly dependent on the operator's expertise and cannot detect unseen damage like corrosion of the steel reinforcement bar (rebar) [14]. Meanwhile, the HCP test has been the most extensively used for assessing the damage (i.e., rebar corrosion). The HCP values reflect the likelihood of steel reinforcement corrosion activity [15]. On the other hand, one of the NDT methods, GPR, has become a helpful tool for assessing RC structures in recent years [16–18]. The GPR method can detect the damage directly and early. However, most damage detections were performed in the laboratory, thus requiring advanced studies on GPR use in all cases [19]. This paper reviews the assessment of RC structures using the GPR method.

2 Ground Penetrating Radar

GPR, commonly known as surface-penetrating radar or EM reflection, is an electromagnetic (EM) investigative technology, as displayed in Fig. 1. When an antenna transmits a signal into the RC structure, it is employed in reflection mode. Dielectric permittivity affects both the velocity and the intensity of reflected waves. The dielectric permittivity of concrete is influenced by its dry state. Temperature, moisture, and chloride impact the permittivity, influenced by EM characteristics [17, 20, 21].

When EM waves pass through the interface of two media with various dielectric constants, a portion of the waves is reflected at the receiving antenna. The scattering quantity is determined by the dielectric characteristics contrast between the two media and the velocity of EM waves in concrete when they pass through the interface of two media with various dielectric constants. EM waves' velocity in concrete is given by Eq. 1 [22]. Table 1 displays some common dielectric constant values for a variety of materials.

$$v = \frac{c}{\sqrt{\varepsilon r}} \tag{1}$$

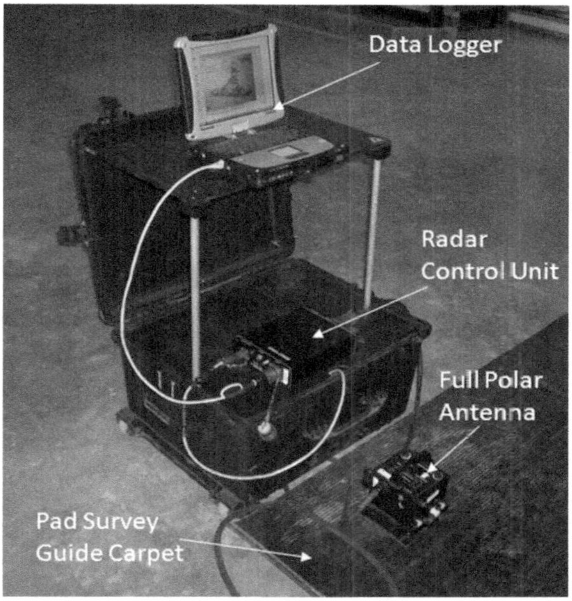

Fig. 1 GPR

With: v = the velocity of EM waves in material (m/s); c = the speed of EM waves in vacuum (3×10^8 m/s); and ε_r = the dielectric constant of material.

The dielectric characteristics of the material created scattered and reflected waves, which were recorded and studied [12, 23]. The received waves were then transformed into a voltage trace (*a-scan*), as demonstrated in Fig. 2a. By analyzing the material along the lines, the GPR system created fast successions of traces displayed as a radargram (*b-scan*). The others were exhibited as time slices by evaluating material and a grid pattern (*c-scan*) [24]. Figures 2b and c exhibit the radargram and time slice illustrations. Investigating the material in a grid pattern has created three-dimensional data records (3-D cube), as depicted in Fig. 2d.

Table 1 Dielectric constants of some common materials [20]

Medium	ε_r
Air	1
Asphalt	3
Concrete	4–11 (5)
Granite	4–7
Limestone	4–8
Basalt	8–9
Soils and sediments	4–30
Fresh and saltwater	81

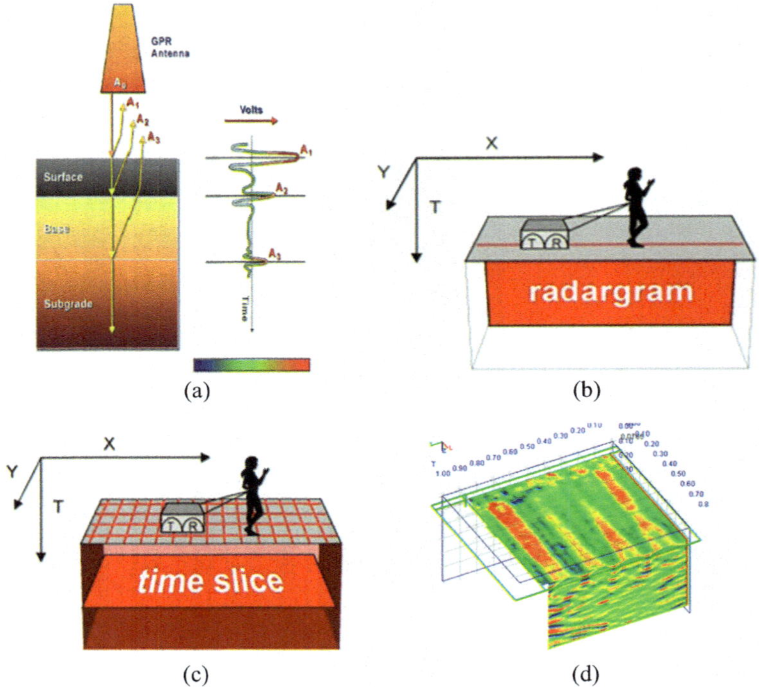

Fig. 2 **a** Single trace (*a*-scan) [25], **b** radargram (*b*-scan), **c** time slice (*c*-scan) [26], and **d** 3-D cube [24]

The GPR method has developed during the past 25 years. In the 1980s, this method was developed for civil engineering [27, 28]. Cantor [29] and Clemena [30] utilized this method for testing RC structures. It has been proposed to inspect and maintain RC structures regularly [31]. Furthermore, it has been employed for various applications, including rebar localization and estimate, rebar corrosion detection, void localization and evaluation, crack localization, and honeycombed localization [12, 16, 17, 32–37].

3 Assessment of RC Structures Using GPR

3.1 Localization and Estimation of Rebars

The concrete had been an anisotropic medium, but the rebar was considered an abnormal object in RC structures [38, 39]. Because of the significant anomaly between the two media, the GPR method would be shapely reflected at the rebar-concrete contact. A succession of signals returning to the receiving antenna when a GPR antenna was translated across the concrete surface might then be displayed as raw findings [19, 40]. One of the most notable uses of the GPR method in civil engineering has been the detection of rebars in RC constructions. Nonetheless, interpreting the findings is quite tricky. To provide reliable results, they can employ time-consuming post-processing procedures and subjective signature expertise [41]. There are many post-processing techniques for localization and estimations of rebars in RC structures, i.e., image processing [34, 42–47], signal processing [48], classification techniques, such as the neural network [45, 49–51], deep learning [52], machine learning [53], multilayer perceptron [41, 54], and other techniques such as reverse time migration [55], hyperbolic [56–60], and X/N amplitude ratios [61].

For image processing, Shihab and Al-Nuaimy [42] achieved it by GPR radargrams subjected to various image processing processes before being fitted with a hyperbola-specific curve-fitting technique. The study discovered that this method could accurately estimate the rebar's depth and radius. In addition, Chang et al. [43] achieved it by processing GPR radargrams through several digital image processing phases, followed by varying power reflectivity inside the energy zone during the motion of the GPR antenna moves across the RC surface. The findings suggest that this model might determine the rebar radius. On the other hand, a pattern recognition algorithm for the image processing technique was proposed by Pasolli et al. [46], Ma et al. [44]. The method begins by selecting a rebar hyperbola template from the GPR data that automatically changes throughout data processing to improve rebar detection and material classification.

Xu et al. [48] have developed a set of signal processing techniques to automate GPR data characterization and increase radar operating efficiency and accuracy, including data pretreatment for systematic noise and interference reduction, target area detection, and linear hyperbola fitting. Experiments with multiple test sets have validated the design. For classification techniques, Shaw et al. [41], Asadi et al. [54] revealed that automating the identification and position of embedded steel rebars from a GPR study utilizing a multilayered perceptron (MLP) for data classification technique was highly successful, as illustrated in Fig. 3. The dielectric properties of the concrete must be thoroughly understood to predict depth or cover with accuracy. A post-processing technique was created to lower false detection rates and mark the exact location of rebars.

On the other hand, the system applied a combination of image processing and neural network approaches proposed by Al-Nuaimy et al. [45], Dinh et al. [51]. As exhibited in Fig. 4, the methodologies used to generate a high-resolution view

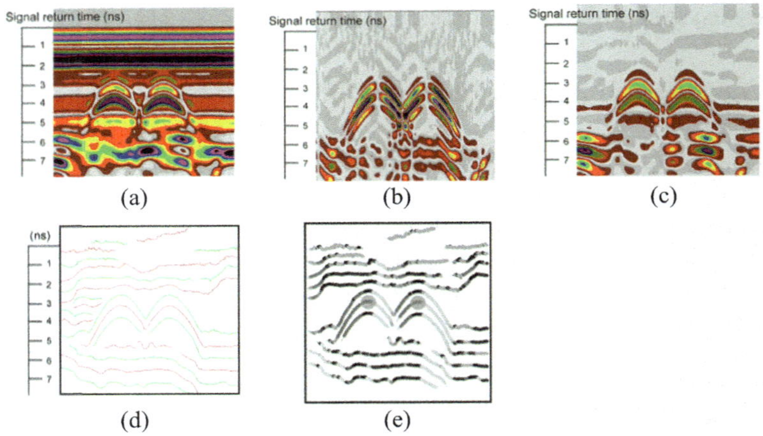

Fig. 3 MLP networks: **a** GPR image, **b** high pass filtering, **c** background subtraction, **d** reduction of the image, and **e** identification of the resulting hyperbolas [41]

of the subsurface in near real-time resulted in straightforward data interpretation and exact rebar depth and azimuth position information [45]. Dinh et al. [51] have demonstrated how to use deep convolutional neural networks (CNN) with standard image processing approaches. Image processing methods like migration, normalized cross-correlation, and thresholding are utilized in the initial step to discover pixels with possible rebar peaks. The second stage begins with the extraction of windowed pictures containing the accessible pixels from the raw GPR scans used in the previous phase. After that, a CNN expert classifies them. The suggested technique performed admirably when applied to the analysis of GPR data for twenty-six bridge decks.

Wiwatrojanagul et al. [56], Mechbal and Khamlichi [58], Ramya et al. [59] suggested a method based on the hyperbolic signature technique to determine the location of the rebar. Wiwatrojanagul et al. [56] created the approach by altering an

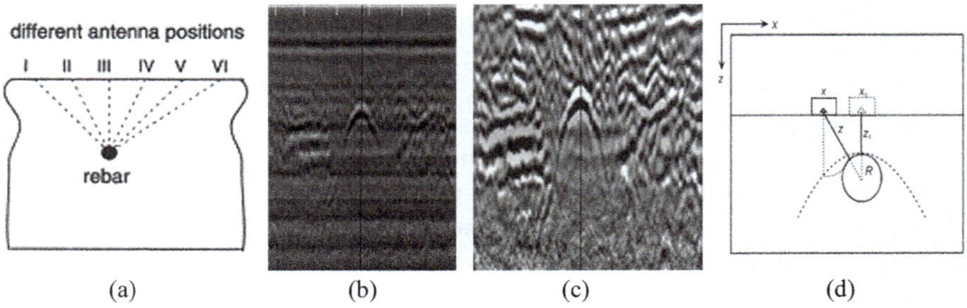

Fig. 4 **a** Rebar detection, **b** GPR data, **c** GPR data with image processing, and **d** estimation of rebar size [45]

existing hyperbolic signatures model to detect steel rebars and estimate cover thickness efficiently. In addition, Mechbal and Khamlichi [58] examined the sensitivity of the inverse problem solution connected to seeing rebars embedded in homogeneous concrete. The hyperbola trace obtained in a B-scan radargram was also analyzed using a closed-form parameterization. On the other hand, Ramya et al. [59] has presented a new method for extracting the high-intensity hyperbolic signal in the radargram, indicating the presence of the rebar and then determining its location and size. Experiments with a single rod and three rods on standard blocks were conducted to better understand the behavior of the GPR waves and to validate the suggested method.

3.2 Corrosion Detection

Many factors influence rebar corrosion in RC structures, such as concrete moisture and chloride content, concrete cracking, and spalling. The primary prerequisite for utilizing GPR to identify rebar corrosion is changing water and chloride content and delamination surrounding the rebar [17]. The radar signals reflected from the embedded rebar are attenuated by the water and chloride concentration [24, 62–65]. The above condition likewise reduces the average velocity of reflected signals. As a result, corrosion is indicated by reduced relative signal reflection and a longer travel time [24, 66, 67]. Sbartai et al. [68], Sossa et al. [69], Hubbard et al. [70], Zaki et al. [24], Senin et al. [71, 72] revealed that when the chloride content in corrosion of steel rebars increased, the amplitude of direct and reflected waves (i.e., a-scan and b-scan) decreased. The amplitude of rebar reflection in the sound section has increased the amplitude of the corrosion or degraded region due to free chloride ions, transporting electricity created by the GPR antenna's electrical field and increasing the degree of radar amplitude attenuation.

Narayanan et al. [73], Martino et al. [74] evaluated rebar reflection characteristics to obtain corrosion condition information to distinguish corroded rebar. A threshold level was set. In Franklin County, Missouri, GPR was utilized by Kim et al. [75] to assess Bridge A-2138. The deteriorated rebar inside concrete bridge decks was not directly resolved by GPR. It might, however, be employed to identify and map zones with a high dielectric constant (low EM velocity) and high attenuation.

Moreover, the effects of chloride contamination on GPR signal intensity in corroded RC structures were investigated by Lai et al. [76], Hong et al. [77, 78]. The journey time from the surface to the bar was delayed. The direct waves' peak-to-valley amplitudes were reduced [76]. The area with unchanged signal intensity is referred to as an uncorroded area, while the area with significantly decreased GPR signal intensity is termed a corroded area [77]. Moreover, Lai et al. [79] utilized the GPR method, a novel methodology for evaluating reinforcing corrosion in concrete was presented. The continuous corrosion process in concrete was monitored using pulse GPR. The bar reflections' amplitudes, trip durations, and short-time Fourier transform (STFT) spectrograms were continually monitored. The results provided

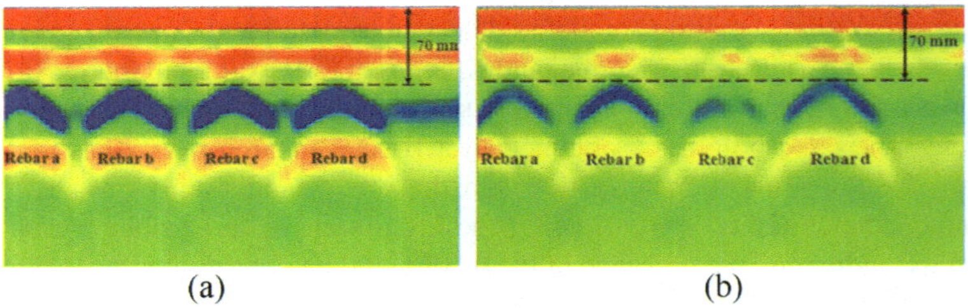

Fig. 5 B-scan of corroded RC slab **a** before corrosion and **b** after corrosion [24]

in this research imply that the parameters obtained by GPR can expose and justify the electrochemical corrosion process.

GPR's capacity to detect corrosion damage was tested by Arndt et al. [80], Ghani et al. [81], Hong et al. [78, 82]. The increased chloride concentration resulted in lower EM wave velocity and signal attenuation, as seen by the GPR. Corrosion was detected in the form of blurring of rebar image characteristics or a contraction of the trace by Arndt et al. [83], Mechbal and Khamlichi [84], Zaki et al. [24]. Furthermore, as exhibited in Fig. 5, the GPR raw signal, b-scan, and post-processed c-scans all demonstrate a clear indication of corrosion in the blurring of rebar image features [24, 85].

Liu et al. [86, 87] asserted that polarimetric GPR could identify early-stage rebar corrosion in RC structures. A polarization GPR system was built and calibrated during an accelerated corrosion experiment to capture polarimetric radar data. A-scans and B-scans were the two types of scans available. The signal variance associated with reinforcing concrete degradation was examined by comparing GPR data recorded over corroded and reference rebars. Finally, the polarimetric GPR data were analyzed, and the properties of corroded rebar were classified using H-Alpha polarization decomposition. The GPR method can detect steel rebar corrosion in RC constructions directly and early. Combining GPR with other methods such as impact echo (IE), half-cell potential (HCP), and electrical resistivity [88], infrared thermography (IRT) [89, 90], IE, HCP, and electrical resistivity [91], and others have been generally acknowledged as having the ability to improve the value of test combinations. According to the application, steel rebar corrosion and other problems with RC constructions could be detected.

3.3 Localization and Assessment of Voids

The GPR method has been unveiled as the most effective for identifying a void in various RC structure subsurface conditions [92–101], as illustrated in Fig. 6. There is a relation between the arrival time and the size of the vacancies. The time it takes for the top of the voids to arrive increases as the size of the voids becomes greater. There were inversions and substantial negative reflections in the GPR data where

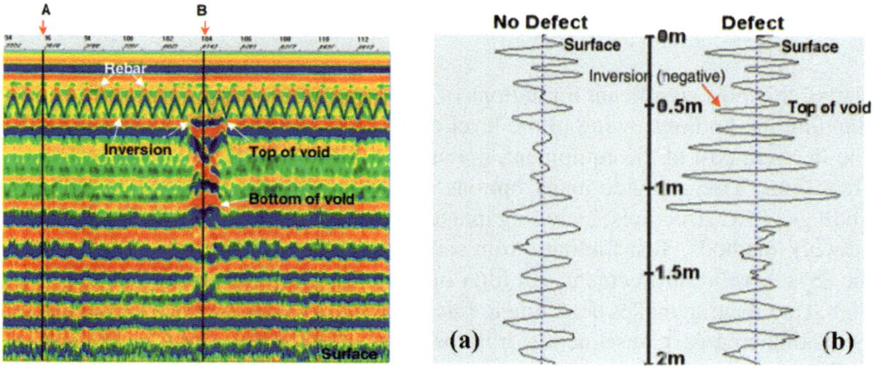

Fig. 6 GPR data of (**a**) voids detection dan (**b**) comparisons of the electromagnetic spectrums for locations with no defect (normal) and defect (voids)

the defect (air-filled void) was discovered [100]. GPR and impact echo (IE) were employed by Razak et al. [92] to identify void defects. The study aimed to examine the accuracy and limits of both methods when identifying the size and depth of air spaces. The sample was a grade 40 concrete slab with air gaps incorporated. Air voids were created by placing air-void plastic balls in the concrete slab. Following the findings, GPR could identify air voids greater than 20 mm in diameter, with an error range of − 8.9 to 30% of their real diameters.

A two-stage regression technique was developed by Yang et al. [96]. Spectral analysis and connection evaluation between echo amplitude and dimension characterization of small-size voids were applied to develop a GPR-based small-scale void-size detection method in RC constructions. The considerable reflection interference of the rebar echo was the most critical hurdle to discovering void flaws in RC structures. According to spectrum analysis, the frequency of the rebar echo was dispersed mainly in the high-frequency sector of the spectrum; on the other hand, the frequency of the middle part of the empty echo was primarily distributed in the low-frequency area. Garg and Misra [95] evaluated the applicability of the GPR method for detecting air voids in grouted ducts of RC members by combining it with ultrasonic pulse velocity (UPV) and IRT techniques. A GPR scan in plastic ducts could provide conclusive results on the existence of air spaces. The length and depth of the voids could be determined relatively precisely. When the number of strands was increased, the effectiveness of void detection improved. The efficacy of void detection decreased as the duct diameter shrank.

4 Limitations and Prospectives

One of the most significant limitations of employing GPR is the difficulty in understanding the findings owing to the level of expertise necessary to evaluate the data and the high cost of the equipment, notwithstanding recent advances in automation algorithms. They provide many options for automating the interpretation of GPR findings of RC structures, including image processing, signal processing, and neural network methods. Test findings from samples must be compared to the results of the tests. Steel reinforcement will filter out any characteristics not worth recording. Signal attenuation makes detecting low-level signals from tiny objects more difficult as depth increases. Consequently, it is costly and unsuitable for small-scale surveys.

Rebar factors such as size and placement are still refined to be precise and genuinely beneficial. GPR has emerged as a viable method for detecting rebar deterioration directly and early. Nonetheless, most of the experiments were carried out in the lab before being effectively adopted and coupled with traditional methods. The reflection amplitude threshold utilized to distinguish between excellent and corroded rebar varied from item to object. Many elements impacted it, including the rebar cover's water and chloride content and the concrete buried in the rebar. On the other hand, it has been confined to combining GPR with some methods for void identification, and ultrasonic testing was depicted to be more effective than impact echo analysis in detecting voids.

GPR can quickly scan huge surfaces for reinforcement, voids, and cracks. Identifying and detecting the reinforcement, voids, and cracks requires optimizing antenna orientation to take advantage of signal polarization. For example, combining GPR with IRT or impact echo (IE) has been recognized as improving the value of test combinations. Simultaneously, a variety of IE and radiography methods have been suggested. Combining the NDT methods is beneficial for two reasons. The first is to utilize a well-developed method to confirm the results of a less well-developed one, resulting in increased dependability or other advantages. The second purpose is to take advantage of technique synergies. Techniques cover the same area of a building, and the information acquired by one method can be utilized to interpret or calibrate the results of another technique.

References

1. Flatt, R.J., Roussel, N., Cheeseman, C.R.: Concrete: an eco material that needs to be improved. J. Eur. Ceram. Soc. **32**(11), 2787–2798 (2012)
2. Angst, U.M.: Challenges and opportunities in corrosion of steel in concrete. Mater. Struct. **51**(1), 4 (2018)
3. Elsener, B., Andrade, C., Gulikers, J., Polder, R., Raupach, M.: Hall-cell potential measurements—potential mapping on reinforced concrete structures. Mater. Struct. **36**(7), 461–471 (2003)
4. Alexander, M., Beushausen, H.: Durability, service life prediction, and modelling for reinforced concrete structures–review and critique. Cem. Concr. Res. **122**, 17–29 (2019)

5. Hamdia, K.M., Arafa, M., Alqedra, M.: Structural damage assessment criteria for reinforced concrete buildings by using a fuzzy analytic hierarchy process. Underground Space **3**(3), 243–249 (2018)
6. Zaki, A., Chai, H.K., Aggelis, D.G., Alver, N.: Non-destructive evaluation for corrosion monitoring in concrete: a review and capability of acoustic emission technique. Sensors (Basel, Switzerland) **15**(8), 19069–19101 (2015)
7. Kashif Ur Rehman, S., Ibrahim, Z., Memon, S.A., Jameel, M.: Non-destructive test methods for concrete bridges: A review. Constr. Build. Mater. **107**, 58–86 (2016)
8. Buyukozturk, O.: Imaging of concrete structures. NDT and E Int. **31**(4), 233–243 (1998)
9. Schabowicz, K.: Non-destructive testing of materials in civil engineering. Multidisciplinary Digital Publishing Institute, p. 3237 (2019)
10. Ramesh, G., Srinath, D., Ramya, D., Vamshi Krishna, B.: Repair, rehabilitation and retrofitting of reinforced concrete structures by using non-destructive testing methods. Mater. Today Proc. (2021)
11. El Masri, Y., Rakha, T.: A scoping review of non-destructive testing (NDT) techniques in building performance diagnostic inspections. Constr. Build. Mater. **265**, 120542 (2020)
12. Lai, W.W.-L., Derobert, X., Annan, P.: A review of ground penetrating radar application in civil engineering: a 30-year journey from locating and testing to imaging and diagnosis. NDT and E Int. **96**, 58–78 (2018)
13. Ahmed, H., La, H.M., Gucunski, N.: Review of non-destructive civil infrastructure evaluation for bridges: State-of-the-art robotic platforms, sensors and algorithms. Sensors (Switzerland) **20**(14), 1–38 (2020)
14. Omar, T., Nehdi, M.L.: Condition assessment of reinforced concrete bridges: current practice and research challenges. Infrastructures **3**(3), 36 (2018)
15. Verma, S.K., Bhadauria, S.S., Akhtar, S.: Monitoring corrosion of steel bars in reinforced concrete structures. Sci. World J. **2014**, 957904 (2014)
16. Dinh, K., Gucunski, N.: Factors affecting the detectability of concrete delamination in GPR images. Constr. Build. Mater. **274** (2021)
17. Tešić, K., Baričević, A., Serdar, M.: Non-destructive corrosion inspection of reinforced concrete using ground-penetrating radar: a review. Materials **14**(4), 1–20 (2021)
18. Jiao, L., Ye, Q., Cao, X., Huston, D., Xia, T.: Identifying concrete structure defects in GPR image. Measur. J. Int. Meas. Confederation **160** (2020)
19. He, X.-Q., Zhu, Z.-Q., Liu, Q.-Y., Lu, G.-Y.: Review of GPR rebar detection. In: Progress in Electromagnetics Research Symposium. PIERS proceedings, Beijing, China (2009)
20. Maierhofer, C.: Nondestructive evaluation of concrete infrastructure with ground penetrating radar. J. Mater. Civ. Eng. **15**(3), 287–297 (2003)
21. Klewe, T., Strangfeld, C., Kruschwitz, S.: Review of moisture measurements in civil engineering with ground penetrating radar—applied methods and signal features. Constr. Build. Mater. **278**, 122250 (2021)
22. Pedret Rodés, J., Martínez Reguero, A., Pérez-Gracia, V.: GPR spectra for monitoring asphalt pavements. Remote Sens. **12**(11), 1749 (2020)
23. Hugenschmidt, J., Mastrangelo, R.: GPR inspection of concrete bridges. Cement Concr. Compos. **28**(4), 384–392 (2006)
24. Zaki, A., Johari, M.A.M., Hussin, W.M.A.W., Jusman, Y.: Experimental assessment of rebar corrosion in concrete slab using ground penetrating radar (GPR). Int. J. Corros. **2018** (2018)
25. He, X., Zhu, Z., Lu, G., Lu, Q.: Bridge management with GPR. In: 2009 International Conference on Information Management, Innovation Management and Industrial Engineering, ICIII 2009. Xi'an (2009)
26. Taffe, A., Maierhofer, C.: Guidelines for NDT methods in civil engineering. In: The International Symposium on Non-Destructive Testing in Civil Engineering, NDT-CE 2003. Berlin, Germany (2003)
27. Ulricksen, C.P.F.: Application of impulse radar to civil engineering. In: Department of Engineering Geology. Lund University of Technology. Sweden, Lund (1982)

28. Forde, M.C.: Ground penetrating radar. In: Introduction to Nondestructive Evaluation Technologies for Bridges, Transportation Research Board Pre-conference Workshop (2004)
29. Cantor, T.R. (ed.): Review of penetrating radar as applied to non-destructive testing of concrete. In: Malhotra, V.M. (ed.) ACI SP-82, Insitu Nondestructive Testing of Concrete, American Concrete Institute. 20 (1984)
30. Clemena, G.G. (ed.): Short pulse radar methods. In: Malhotra, V.M., Carino, N.J. (ed) Handbook on Non-destructive Testing of Concrete, Chap. 11. Boston, CRC Press (1991)
31. Wiggenhauser, H., Reinhardt, H.W.: NDT in civil engineering: experience and results of the for 384 research group. In: Review of Progress in Quantitative Nondestructive Evaluation, vol. 29. AIP, Kingston (Rhode Island) (2010)
32. Barrile, V., Pucinotti, R.: Application of radar technology to reinforced concrete structures: a case study. NDT and E Int. **38**(7), 596–604 (2005)
33. Rathod, H., Debeck, S., Gupta, R., Chow, B.: Applicability of GPR and a rebar detector to obtain rebar information of existing concrete structures. Case Stud. Constr. Mater. **11**, e00240 (2019)
34. Zaki, A., Kabir, S.: Radar-based quantification of corrosion damage in concrete structures. In: Progress in Electromagnetics Research Symposium (2011)
35. McCabe, T., Erdogmus, E., Kodsy, A., Morcous, G.: Early detection of honeycombs in concrete pavement using GPR. J. Perform. Constr. Facil. **35**(1), 04020138 (2021)
36. Dawood, T., Zhu, Z., Zayed, T.: GPR-based deterioration mapping in subway networks. In: Construction Research Congress 2018, 351–361 (2018)
37. Rasol, M.A., Pérez-Gracia, V., Fernandes, F.M., Pais, J.C., Santos-Assunçao, S., Santos, C., Sossa, V.: GPR laboratory tests and numerical models to characterize cracks in cement concrete specimens, exemplifying damage in rigid pavement. Measur. J. Int. Measur. Confederation **158** (2020)
38. Millard, S.G., Shaw, M.R., Giannopoulos, A., Soutsos, M.N.: Modeling of subsurface pulsed radar for non-destructive. J. Mater. Civ. Eng. **10**(3) (1998)
39. Shaw, M.R., Molyneaux, T.C., Millard, S.G., Bungey, J.H., Taylor, M.J.: Automatic analysis of GPR scans on concrete structures. In: Proceedings of GPR (1998)
40. Zhou, D., Zhu, H.: Application of ground penetrating radar in detecting deeply embedded reinforcing bars in pile foundation. Adv. Civ. Eng. **2021**, 4813415 (2021)
41. Shaw, M., Millard, S., Molyneaux, T., Taylor, M., Bungey, J.: Location of steel reinforcement in concrete using ground penetrating radar and neural networks. NDT and E Int. **38**(3), 203–212 (2005)
42. Shihab, S., Al-Nuaimy, W.: Radius estimation for cylindrical objects detected by ground penetrating radar. Subsurf. Sens. Technol. Appl. **6**(2), 151–166 (2005)
43. Chang, C.W., Lin, C.H., Lien, H.S.: Measurement radius of reinforcing steel bar in concrete using digital image GPR. Constr. Build. Mater. **23**(2), 1057–1063 (2009)
44. Ma, X., Liu, H., Wang, M.L., Birken, R.: Automatic detection of steel rebar in bridge decks from ground penetrating radar data. J. Appl. Geophys. **158**, 93–102 (2018)
45. Al-Nuaimy, W., Huang, Y., Nakhkash, M., Fang, M., Nguyen, V., Eriksen, A.: Automatic detection of buried utilities and solid objects with GPR using neural networks and pattern recognition. J. Appl. Geophys. **43**(2–4), 157–165 (2000)
46. Pasolli, E., Melgani, F., Donelli, M.: Automatic analysis of GPR images: a pattern-recognition approach. IEEE Trans. Geosci. Remote Sens. **47**(7), 2206–2217 (2009)
47. Kabir, S., Zaki, A.: Detection and quantification of corrosion damage using ground penetrating radar (GPR). In: Progress in Electromagnetics Research Symposium (2011)
48. Xu, X., Xia, T., Venkatachalam, A., Huston, D.: Development of high-speed ultrawideband ground-penetrating radar for rebar detection. J. Eng. Mech. **139**(3), 272–285 (2013)
49. Xiang, Z., Rashidi, A., Ou, G.G.: An improved convolutional neural network system for automatically detecting rebar in GPR Data. In: Computing in Civil Engineering 2019: Data, Sensing, and Analytics-Selected Papers from the ASCE International Conference on Computing in Civil Engineering 2019 (2019)

50. Ahmed, H., La, H.M., Gucunski, N.: Rebar detection using ground penetrating radar with state-of-The-Art convolutional neural networks. In: 9th International Conference on Structural Health Monitoring of Intelligent Infrastructure: Transferring Research into Practice, SHMII 2019-Conference Proceedings (2019)
51. Dinh, K., Gucunski, N., Duong, T.H.: An algorithm for automatic localization and detection of rebars from GPR data of concrete bridge decks. Autom. Constr. **89**, 292–298 (2018)
52. Liu, H., Lin, C., Cui, J., Fan, L., Xie, X., Spencer, B.F.: Detection and localization of rebar in concrete by deep learning using ground penetrating radar. Autom. Constr. **118** (2020)
53. Asadi, P., Gindy, M., Alvarez, M.: A machine learning based approach for automatic rebar detection and quantification of deterioration in concrete bridge deck ground penetrating radar B-scan images. KSCE J. Civ. Eng. (2019)
54. Asadi, P., Gindy, M., Alvarez, M., Asadi, A.: A computer vision based rebar detection chain for automatic processing of concrete bridge deck GPR data. Autom. Constr. **112** (2020)
55. Shen, R., Zhao, Y., Hu, S., Li, B., Bi, W.: Reverse-time migration imaging of ground-penetrating radar in ndt of reinforced concrete structures. Remote Sens. **13**(10) (2021)
56. Wiwatrojanagul, P., Sahamitmongkol, R., Tangtermsirikul, S., Khamsemanan, N.: A new method to determine locations of rebars and estimate cover thickness of RC structures using GPR data. Constr. Build. Mater. **140**, 257–273 (2017)
57. Drobiec, Ł., Jasiński, R., Mazur, W.: Accuracy of eddy-current and radar methods used in reinforcement detection. Materials **12**(7) (2019)
58. Mechbal, Z., Khamlichi, A.: Sensitivity of the inverse problem solution related to detection of rebars buried in concrete by using GPR scanning. In: MATEC Web of Conferences (2018)
59. Ramya, M., Balasubramaniam, K., Shunmugam, M.S.: On a reliable assessment of the location and size of rebar in concrete structures from radargrams of ground-penetrating radar. Insight: Non-Destructive Test. Condition Monit. **58**(5), 264–270 (2016)
60. Xie, Z., Wei, X., Zhang, Y.: Automatic extraction of hyperbolic signatures in ground penetrating radar images. In: ASME 2012 Conference on Smart Materials, Adaptive Structures and Intelligent Systems, SMASIS 2012 (2012)
61. Utsi, V., Utsi, E.: Measurement of reinforcement bar depths and diameters in concrete. In: Proceedings of the Tenth International Conference on Grounds Penetrating Radar, 2004. GPR 2004. IEEE (2004)
62. Klysz, G., Balayssac, J.P., Laurens, S.: Spectral analysis of radar surface waves for non-destructive evaluation of cover concrete. NDT and E Int. **37**(3), 221–227 (2004)
63. Sbartai, Z.M., Laurens, S., Balayssac, J.P., Ballivy, G., Arliguie, G.: Effect of concrete moisture on radar signal amplitude. ACI Mater. J. **103**(6), 419–426 (2006)
64. Lai, W.L., Kou, S.C., Tsang, W.F., Poon, C.S.: Characterization of concrete properties from dielectric properties using ground penetrating radar. Cem. Concr. Res **39**(8), 687–695 (2009)
65. Dérobert, X., Villain, G.: Effect of water and chloride contents and carbonation on the electromagnetic characterization of concretes on the GPR frequency band through designs of experiment. NDT and E Int. **92**, 187–198 (2017)
66. Barnes, C.L., Trottier, J.-F., Forgeron, D.: Improved concrete bridge deck evaluation using GPR by accounting for signal depth-amplitude effects. NDT and E Int. **41**(6), 427–433 (2008)
67. Lai, W.L., Kind, T., Wiggenhauser, H.: Detection of accelerated reinforcement corrosion in concrete by ground penetrating radar. In: Proceedings of the 13th Internarional Conference on Ground Penetrating Radar, GPR 2010 (2010)
68. Sbartai, Z.M., Laurens, S., Balayssac, J.P., Arliguie, G., Ballivy, G.: Ability of the direct wave of radar ground-coupled antenna for NDT of concrete structures. NDT and E Int. **39**(5), 400–407 (2006)
69. Sossa, V., Pérez-Gracia, V., González-Drigo, R., Rasol, M.A.: Lab non destructive test to analyze the effect of corrosion on ground penetrating radar scans. Remote Sens. **11**(23), 2814 (2019)
70. Hubbard, S.S., Zhang, J., Monteiro, P.J.M., Peterson, J.E., Rubin, Y.: Experimental detection of reinforcing bar corrosion using nondestructive geophysical techniques. ACI Mater. J. **100**(6), 501–510 (2003)

71. Senin, S.F., Hamid, R., Ahmad, J., Rosli, M.I.F., Yusuff, A., Rohim, R., Abdul Ghani, K.D., Mohamed Noor, S.: Damage detection of artificial corroded rebars and quantification using non-destructive methods on reinforced concrete structure. J. Phys. Conf. Ser. (2019)
72. Senin, S.F., Hamid, R.: Ground penetrating radar wave attenuation models for estimation of moisture and chloride content in concrete slab. Constr. Build. Mater. **106**, 659–669 (2016)
73. Narayanan, R.M., Hudson, S.G., Kumke, C.J.: Detection of rebar corrosion in bridge decks using statical variance of radar reflected pulses. In: Seventh International Conference on Ground-penetrating Radar. University of Kansas, Lawrence, Kansas (1998)
74. Martino, N., Maser, K., Birken, R., Wang, M.: Determining ground penetrating radar amplitude thresholds for the corrosion state of reinforced concrete bridge decks. J. Environ. Eng. Geophys. **19**(3), 175–181 (2014)
75. Kim, W., Ismail, A., Anderson, N.L., Atekwana, E.A., Buccellato, A.: Non-destructive testing (NDT) for corrosion in bridge decks using GPR. In: 3rd International Conference on the Application of Geophysical Methodologies and NDT to Transportation Facilities and Infrastructure, Geophysics 2003. Orlando, Florida (2003)
76. Lai, W.L., Kind, T., Wiggenhauser, H.: Effects of high chloride contamination on direct wave of ground penetrating radar. www.ndt.net (2010)
77. Hong, S., Wiggenhauser, H., Helmerich, R., Dong, B., Dong, P., Xing, F.: Long-term monitoring of reinforcement corrosion in concrete using ground penetrating radar. Corros. Sci. **114**, 123–132 (2017)
78. Hong, S., Lai, W.L., Helmerich, R.: Experimental monitoring of chloride-induced reinforcement corrosion and chloride contamination in concrete with ground-penetrating radar. Struct. Infrastruct. Eng. **11**(1), 15–26 (2015)
79. Lai, W.L., Kind, T., Wiggenhauser, H.: Detection of accelerated reinforcement corrosion in concrete by ground penetrating radar. In: 13th International Conference on Ground Penetrating Radar (GPR) (2011)
80. Arndt, R.W., Jalinoos, F., Cui, J., Huston, D.: Periodic NDE for bridge maintenance. In: 13th International Conference on Structural Faults & Repair. Engineering Technics Press, Edinburgh, UK (2010)
81. Ghani, A., Fithry Senin, S., Hamid, R.: Attenuation of ground penetrating radar signal amplitude in monitoring reinforced steel corrosion. Jurnal Teknologi **65** (2013)
82. Hong, S., Lai, W.W.L., Wilsch, G., Helmerich, R., Helmerich, R., Günther, T., Wiggenhauser, H.: Periodic mapping of reinforcement corrosion in intrusive chloride contaminated concrete with GPR. Constr. Build. Mater. **66**, 671–684 (2014)
83. Arndt, R.W., Cui, J., Huston, D.R.: Monitoring of reinforced concrete corrosion and deterioration by periodic multi-sensor non-destructive evaluation. In: Proceedings of 37th Annual Review of Progress in Quantitative Nondestructive Evaluation, QNDE 2010 (2011)
84. Mechbal, Z., Khamlichi, A.: Identification of corrosion in reinforcement rebars by using the GPR. MATEC Web Conf. **16**, 07005 (2014)
85. Cui, J., Huston, D.R., Arndt, R.: Early detection of concrete bridge deck corrosion using ground penetrating radar, half-cell potential and anode ladder. In: 90th Annual Meeting Transportation Research Board. Transportation Research Board, Washington DC (2011)
86. Gucunski, N., Pailes, B., Kim, J., Azari, H., Dinh, K.: Capture and quantification of deterioration progression in concrete bridge decks through periodical NDE surveys. J. Infrastruct. Syst. **23**(1), B4016005 (2017)
87. Solla, M., Lagüela, S., Fernández, N., Garrido, I.: Assessing rebar corrosion through the combination of non-destructive GPR and IRT methodologies. Remote Sens. **1**(14) (2019)
88. Omar, T., Asce, S., Nehdi, M., Zayed, T., Asce, M.: Performance of NDT techniques in appraising condition of reinforced concrete bridge decks. J. Perform. Constr. Facil. **31** (2017)
89. Rashidi, M., Azari, H., Nehme, J.: Assessment of the overall condition of bridge decks using the Jensen-Shannon divergence of NDE data. NDT and E Int. **110**, 102204 (2019)
90. Razak, N.A., Senin, S.F., Hamid, R.: Detection of sizes and locations air voids in reinforced concrete slab using ground penetrating radar and Impact-Echo methods. Jurnal Teknologi **74**(3), 63–67 (2015)

91. Liu, L.B., Qian, R.Y., Li, J., Sun, M.G., Ge, S.C.: GPR detection of subsurface voids and its validation based on similarity principle. In: Proceedings of 2016 16th International Conference of Ground Penetrating Radar, GPR 2016 (2016)
92. Abbasghorbani, M., Bamdad, A., Tabatabaee, N.: Non-destructive detection of voids under airfield pavement. In: Bearing Capacity of Roads, Railways and Airfields-Proceedings of the 10th International Conference on the Bearing Capacity of Roads, Railways and Airfields, BCRRA 2017 (2017)
93. Garg, S., Misra, S.: Efficiency of NDT techniques to detect voids in grouted post-tensioned concrete ducts. Non-destructive Test. Eval. **36**(4), 366–387 (2021)
94. Yang, Y., Lu, J., Li, R., Zhao, W., Yan, D.: Small-scale void-size determination in reinforced concrete using GPR. Adv. Civ. Eng. **2020** (2020)
95. Stefan, M., Salvador, V., David, S.: Validation of artificial defects for Non-destructive testing measurements on a reference structure. MATEC Web Conf. (2018)
96. Lai, W.W.L., Chang, R.K.W., Sham, J.F.C.: A blind test of non-destructive underground void detection by ground penetrating radar (GPR). J. Appl. Geophys. **149**, 10–17 (2018)
97. Ge, Z., Wang, H., Yao, H., Zhang, Q., Zhang, L.: Air void detection using ground-penetrating radar and its relationship to moisture damage. J. Test. Eval. **44**(1), 635–646 (2015)
98. Chen, D.H., Tang, C., Xiao, H.B., Ying, P.: Utilizing electromagnetic spectrum for subsurface void detection—case studies. Arab. J. Geosci. **8**(9), 7705–7717 (2015)
99. Cassidy, N.J., Eddies, R., Dods, S.: Void detection beneath reinforced concrete sections: the practical application of ground-penetrating radar and ultrasonic techniques. J. Appl. Geophys. **74**(4), 263–276 (2011)
100. Liu, H., Zhong, J., Yang, Z., Meng, X., Ding, F.: Application of full-polarimetric GPR to rebar corrosion detection. In: 2021 IEEE International Geoscience and Remote Sensing Symposium IGARSS (2021)
101. Liu, H., Zhong, J., Ding, F., Meng, X., Liu, C., Cui, J.: Detection of early-stage rebar corrosion using a polarimetric ground penetrating radar system. Constr. Build. Mater. **317** (2022)

The manufacturer's authorised representative in the EU is Springer Nature Customer Service Centre GmbH, Europaplatz 3, 69115 Heidelberg, Germany. If you have any concerns regarding our products, please contact ProductSafety@springernature.com

Printed and bound by CPI Group (UK) Ltd, Croydon, CR0 4YY
01/10/2026
02225867-0017